BASICS OF ENDOVASCULAR INTERVENTION

Edited by

Immad Sadiq, MD

Associate Professor of Medicine
University of Connecticut School of Medicine
Director Vascular & Endovascular Medicine
Director Non-invasive Vascular Laboratory
Division of Cardiology
Hartford Hospital
Hartford, Connecticut

ISBN: 0-9887314-0-1
ISBN 13:978-0-9887314-0-0

To

My Parents

Without whose love , nurturing and guidance

I would not be where I am.

IS

CONTENTS

CONTRIBUTORS

Talhat Azemi, MD, RPVI
Interventional Cardiologist,
Assistant Professor of Medicine, University of Connecticut School of Medicine,
Hartford Hospital
Hartford, Connecticut

Feras A. Elbash, MD, RPVI
Interventional Cardiologist,
Reid Hospital,
Richmond, Indiana

Francis J. Kiernan, MD, FACC, FSCAI
Associate Professor of Medicine, University of Connecticut School of Medicine,
Director Cardiac Catherization Laboratory,
Director Interventional Cardiology Fellowship
Hartford Hospital
Hartford, Connecticut

Firas S. Almahasneh, MD, RPVI
Interventional Cardiologist,
United Hospital Center,
Bridgeport, West Virginia

Immad Sadiq, MD, FACC, FSCAI, FSVM, RPVI
Associate Professor of Medicine, University of Connecticut School of Medicine,
Director Vascular and Endovascular Medicine,
Director Non-invasive Vascular Laboratory,
Hartford Hospital
Hartford, Connecticut

PREFACE

Despite an explosive growth in the field of endovascular intervention there has been a paucity of written material to guide the beginning interventionalist. While many specialities are jumping into this field, the training programs have not formally incorporated the teachings of basic technique in their curricula. As a result, most practitioners often learn 'on-the-job'.

The purpose of this book is to serve as a 'how-to' guide for endovascular procedures and to equip the future interventionalist with fundamentals of well-rounded endovascular technique and skills.

Due to emerging evidence, the role of different endovascular procedures in a particular vascular bed may change. However, while tools for intervention and evidence behind their efficacy may improve with time, the basic concept of a procedure and the skill set required to perform it will remain more or less the same. We encourage the reader to familiarize themselves with current endovascular technology & literature and hope that they will practice evidence-based medicine. It is expected that with increasing experience the reader will not only acquire superior techniques but also refine the existing standards!

The first four chapters of this book provide an introduction to the angiography suite basics, patient setup, diagnostic and interventional equipment and vascular access. The rest of the textbook has been organized in chapters that pertain to interventions relevant to a particular vascular territory. A brief introduction is followed by relevant vascular anatomy and a simple, step-wise discussion of diagnostic and interventional equipment and technique. Towards the end of each chapter illustrative cases using the aforementioned technique are presented.

We hope that this book will serve to enhance the reader's understanding, technique and practice of endovascular procedures. It is not intended as a definitive text on endovascular medicine as there are many books available that fulfill this need. Rather, this work represents a collection of our individual experiences and should serve as a practical guide to those wishing to practice endovascular procedures. Hence, we have omitted specific literature refrences.

While we have attempted to cover several fundamental topics, we look forward to feeback from our readers via email at: BEI2012Book@gmail.com.

Immad Sadiq

ACKNOWLEDGEMENTS

Many people have contributed to this book in one way or another. I would like to thank them all.

I would like to thank Drs. Robert Schainfeld and Peter Soukas from the bottom of my heart for teaching me the art and science of vascular medicine and endovascular intervention.

Moreover, I would like to thank the physicians of Shaikh Zayed Hospital in Lahore, Pakistan, in particular Prof. Amber Malik and Prof. Saulat Siddique, for their hospitality and inspiration for providing the impetus behind this book.

I would also like to acknowledge the contribution of our cath lab staff and my colleagues, in particular Michael Lusignan, RT and David Downes, RT, our fellows and the many patients who are the subjects of the cases discussed in this book. I am indebted to them for teaching me and giving me the opportunity to pass it on.

I would like to thank my co-authors for their contribution to this project, for taking the time away from their families. I owe special gratitude to Dr. Azemi for his painstaking proofreading and attention to detail.

Special thanks is also owed to Jessica Ryan, PA for her assistance with this project.

Last but not the least, I would like to thank my family for their endless love and patience with me and allowing me the time away from them to complete this project.

Immad Sadiq

1

BASICS OF ANGIOGRAPHY

ANGIOGRAPHY SUITE

The foundation of a successful endovascular procedure is based on the ability to perform high quality angiography. The angiography suite where such procedures are performed should fulfill certain basic requirements to ensure patient safety and procedural success.

This topic thus deserves at least basic attention from the operator so certain fundamental requirements are kept into consideration.

BASICS

Procedure Room:

The procedure room must be of adequate size to accommodate the procedures that are expected to be performed in the suite.

- The room should be at least 500 sq ft with a control room space of approximately 150 -200 sq ft.
- As procedures become more complex and require more ancillary equipment, a larger space will be required. A growing trend has been to house angiography suites in operating room environments (hybrid suites) and this further increases the space requirement.
- There must be an adequate design in the procedure room to allow access to the patient from multiple approaches. Therefore, an island table is used with access from all four sides.
- There must be space for ancillary equipment that is used in the performance of procedures such as power injectors along with consoles to power devices such as lasers, ultrasound machines and atherectomy devices.
- Some patients may also require additional support with ventilators or cardiac support devices.
- There must be multiple sources of electrical power and medical gases so that equipment can be arranged so as to not interfere with the operator positions and so that there is a free range of motion of the imaging equipment.

- There must be adequate space for emergency equipment and storage of medication that will be used in the procedure.
- Monitoring equipment with hemodynamic, oximetry and EKG capabilities should be located so that it can be viewed by the operators regardless of the room configuration.
- There must be adequate workstations for support personnel to complete required documentation. A hybrid room or a room in which stent-grafts are performed should meet operating room standards for ventilation including laminar air flow with adequate air exchanges per hour in a positive pressure environment.
- The angiography suite must have adequate storage space for equipment and the facility should have pre-procedure holding and post-procedure recovery areas.

Imaging System:

- Modern imaging systems consist of a flat panel detector with digital subtraction capability.
- The system may be floor or ceiling mounted.
- The detector should be of adequate size to perform standard peripheral exams with a field of view that can be adjusted to meet the requirements of the region being evaluated.
- There should be a 'road-map' function so that a stored 'mask' image will be superimposed on the live fluoroscopic image to facilitate guide wire passage.
- The operator should have the ability to review images from table-side controls.
- Workstations should be available in the suite for image review and reporting purposes.
- The images should be stored on a digital archive system that has adequate redundancy and backup in order to prevent the loss of data.

> ***Road-map:***
> - This is a DSA sequence performed usually by using contrast injection during fluoroscopy. The frame with maximum vessel opacification is selected and becomes the mask 'road map' during subsequent live fluoroscopy.
> - This provides subtracted fluoroscopic images overlaid on a static image of the vessel. Devices such as wires etc can be easily seen and guided following the anatomy displayed on the road map.
> - Pre-requisites: The patient, table and detector should not move to prevent artifacts dues to misregistration of the images.

RADIATION SAFETY

Operators must have an an understanding of the principles of radiation safety and employ techniques to minimize exposure to the patient and to the medical personnel.

- Complex procedures have the potential to cause radiation induced injury (deterministic injury) in a dose related manner.
- Doses to a single area in excess of 3 - 5 Gy are likely to cause skin injury.
- At higher doses, permanent skin injury may occur.
- Operators and staff must be trained on the techniques to minimize radiation exposure including:
  - Minimizing fluoroscopy and cine acquisition time.
  - Using low fluoroscopy frame rates.
  - Avoiding imaging in the same projection for a prolonged time.
  - Using 'fluoroscopy save' modes instead of cine for documentation purposes.
  - Using collimators to avoid radiation to areas that are not of interest.
  - Positioning the detector as close to the skin as possible.
  - Maintaining an adequate distance from the x-ray source to the patient.
- Current imaging systems provide information on radiation dose. The

operator should be aware of how to interpret this data.

- The laboratory should be equipped with radiation protection devices for the staff including:
  - Lead aprons
  - Thyroid shields
  - Lead glasses
  - Movable radiation shields
- When properly used, a ceiling suspended radiation shield can reduce operator exposure by 90%.
- In addition, sterile radiation blocking pads can be placed on the field to further limit operator exposure.
- Staff exposure should be monitored with radiation badges and monitored by the radiation safety officer.

BACK TABLE SET UP

A sterile OR type table is set up to accommodate the supplies used during the procedure.

- This should include needles, sterile towels, drapes, 4 x 4 (gauze pads), sterile bowl with saline, medications in labeled syringes, and other equipment to be used such as inflation devices.
- A manifold system (figure 1) should be used that allows continuous measurement of pressure except during contrast injection.
- For peripheral angiography, we recommend the use of a 20 cc control syringe on the manifold instead of the 10-12 cc syringe used for coronary angiography.

PATIENT PREPARATION

PRE-OP MEASURES

Informed Consent:

- The patient should have informed consent that includes an explanation of the procedure, potential benefits, potential risks and reasonable alternatives.

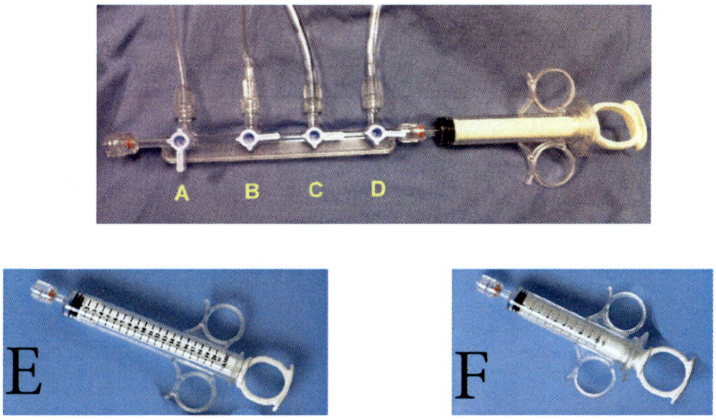

Figure 1.

A 4-port manifold.

A. Pressure line. B. Saline flush. C. Iodinated contrast.
D. Waste bag port.
E. A 20 cc control syringe.
F. A 10 cc control syringe.

History and Exam:

- The patient should have a recent history along with an updated physical examination on the day of the procedure to assess for any interval change.

Medication and Allergies:

- Medications should be reviewed and the patient should be instructed as to which medications to take or hold pre-procedure.
- Allergies should be reviewed with particular attention to prior contrast reactions.
- Generally warfarin is held so that the INR is less than 1.8-2.0.

Pre-op Blood Work:

- Appropriate blood work should be reviewed to include CBC, electrolytes, renal function and coagulation parameters if appropriate.

Fasting:

- The patient should be instructed as to how to fast prior to the procedure. Guidelines from the American Association of Anesthesiology recommend fasting from clear liquids for 2 hours before most procedures and refraining from a light meal for 6 hours prior to a procedure requiring anesthesia or sedation.

Risk of Contrast Exposure:

- The patient should be assessed for the potential risk of contrast-induced nephropathy. Potential risk factors for this condition include diabetes and pre-existing renal insufficiency.
- Preventative measures include adequate patient hydration and minimizing the amount of contrast.
- It does not appear that the use of N-acetylcysteine or sodium bicarbonate infusions offer more renal protection than hydration alone.
- When feasible, CO_2 angiography should be used in patients with chronic kidney disease and impaired renal function.

IV Access:

- Adequate IV access is mandatory for all patients undergoing an invasive procedure so that medication can be given readily.
- At least one peripheral IV site with a 20 gauge IV is required.

Documentation of Pulses:

- For arterial procedures, it is imperative that pertinent pulses (radial, ulnar, brachial and pedal pulses) should be documented 2+,1+, biphasic on Doppler, monophasic on Doppler or absent, as the case may be.

Universal Time-out:

- Prior to the commencement of the procedure (i.e. prior to sedation and anesthesia administration) a 'time-out' should be performed, verbalizing the patient identity, the intended procedure, site of access/incision, any allergies, pertinent laboratory data, mode of anesthesia and performing physician(s).

INTRA-OP MEASURES

Groin Shave and Access Area Prep:

- For patients requiring femoral artery access, the groins should be adequately shaved with an atraumatic razor without inflicting skin trauma (electric clippers recommended).
- Ideally, the area from below the umbilicus to five finger-breadths of the upper thigh should be shaved.
- These areas should then be prepped with antiseptic solution of choice. This should be applied using a sterile applicator in a spiral fashion, inward out (figure 2).

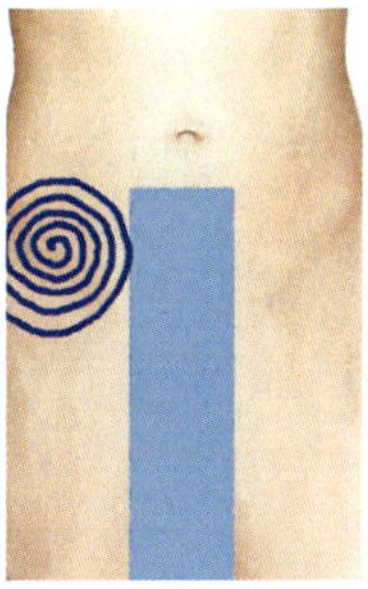

Figure 2.

Groin prep.

The groins are shaved and prepped from umbilicus down to proximal thigh. The antiseptic solution is applied from center outwards in a spiraling fashion.

Sedation:

- For conscious sedation, we routinely use IV fentanyl (25 -100 μg) and midazolam (0.5 -1 mg) increments.
- The patient should be able to breathe on their own. Supplemental O_2 may be required.

Anticoagulation:

- We consistently use IV unfractionated heparin (UFH) for anticoagulation during endovascular procedures.
- We administer 2000 U IV UFH after obtaining arterial access.
- Once diagnostic angiography is complete and an interventional plan is formulated, we administer supplemental IV UFH.

TABLE 1 - BED-REST POST-OP

- This assumes adequate recovery from anesthesia/sedation and secure hemostasis after removal of arterial/central venous sheaths.

ACCESS SITE	TYPE OF PROCEDURE	DURATION OF BED-REST
Radial	• Diagnostic	• 2-4 hours
	• Intervention	• 2-4 hours
Brachial	• Diagnostic	• 4-8 hours
	• Intervention	• 4-8 hours
Femoral	• Diagnostic	• 6 hours
	• Intervention	• 6-12 hours
Popliteal	• Diagnostic	• 12 hours
	• Intervention	• 12 hours
Pedal	• Diagnostic	• 4-8 hours
	• Intervention	• 4-8 hours

POST-OP MEASURES

- Once the procedure is complete, the patient should be monitored for a duration that ensures adequate recovery from sedation and secure hemostasis.
- After diagnostic angiography where therapeutic anticoagulation (ACT > 250 sec.) was not given, bed-rest is instituted for a shorter duration as compared to when therapeutic anticoagulation was given.
- Patients who undergo intervention require a longer duration of bed-rest as they have received therapeutic anticoagulation.
- In general, patients who undergo an upper extremity access can be am-

bulated earlier than those who had lower extremity access (table 1).

PRINCIPLES OF ANGIOGRAPHY

The lower extremity arteries are the most common blood vessels afflicted with atherosclerotic occlusive disease outside the heart. Satisfactory angiography of the lower extremities thus, is crucial to not only the definitive assessment of PAD in many cases, but also to plan and allow for the necessary endovascular or surgical intervention. In all elective and most emergent cases, this invasive assessment should be preceded by a non-invasive arterial study in addition to a sound history and thorough physical examination. This study usually consists of an ankle-brachial index (ABI) evaluation along with segmental Doppler or pulse volume recordings (PVRs). The information obtained from such a study facilitates the task of planning and performing angiography.

ACCESS

SITE

- Several factors are considered while selecting a vascular access site:
  - ▹ Region to be imaged.
  - ▹ Underlying vascular obstruction.
  - ▹ Site of proposed intervention.
  - ▹ Type of proposed intervention.
- For example, when considering femoral arterial access for carotid angiography, the operator must know if there is underlying common femoral or aorto-iliac occlusive disease that could potentially interfere with the procedure. In the case that such disease is present, alternative access via either contralateral CFA or brachial /radial artery should be considered prior to proceeding to angiography.
- Likewise, for venous access, lets say, for IVC filter placement, the operator must be aware of the presence or absence of underlying ileo-femoral DVT prior to gaining femoral venous access.
- We use a 4 Fr. micropuncture kit to obtain access.

Micropuncture kit:
This is an access kit that comes with a 21 G 7 cm needle, a 4 Fr. or 5 Fr. locking sheath (microcatheter) and dilator along with a 0.018" wire.

Supplied by several vendors such as Cook, Bard, etc.

SHEATH

- Our preference is to use a 5 Fr. system (including the sheath) for diagnostic angiography.
- Although a 4 Fr. system can also be used for this purpose without compromising image quality, as noted below, this precludes an accurate hemodynamic assessment of any borderline stenoses.
- A sheath with a radio-opaque tip is recommended as it can be easily visualized fluoroscopically. Several such sheaths are available and personal experience/preference, in addition to cost, usually determines the choice.
- We prefer a 5 Fr. /23 cm Cordis Brite-tip introducer as our initial straight sheath. The marker on the tip is easily visible under fluoroscopy. Additionally, at 23 cm long, the tip usually can be advanced into the abdominal aorta. This is helpful when performing hemodynamic assessment.

HEMODYNAMIC ASSESSMENT

RATIONALE

- Angiography is an imaging modality without any physiologic input.
- A two-dimensional image is generated. This does not always give an accurate estimate of the burden of obstructive disease and its physiologic significance in terms of underperfusion of the end-organ.
- Invasive hemodynamic assessment, which is analogous to non-invasive physiologic testing, should thus be carried out to determine the physiologic/clinical significance of this obstructive disease.
- This is not necessary for obviously severe/critical stenoses. However, it must be performed for moderately stenotic lesions prior to making a therapeutic decision.

Hemodynamics:
We use a 5 Fr. sheath and a 4 Fr. catheter.

TECHNIQUE

- This consists of recording blood pressure in a vessel and comparing it to a pressure in another vessel or another segment of the same vessel.
- Any difference in the two pressures implies the presence of obstructive pathology between the vessels/vessel segments.
- For arterial pressures, a difference of >/= 20 mm Hg is considered significant and is assumed to be due to an underlying significant stenosis.
- Usually one pressure is recorded through a catheter and the other through the arterial sheath. This requires two separate transducers.
- For reliable pressure transduction and measurement at least a 4 Fr. catheter is required.
- Since a 4 Fr. catheter introduced through a 4 Fr. sheath will cause the pressure in the sheath to be 'dampened' and appear falsely low, the sheath should be at least 5 Fr. in diameter. This is true for all catheters/sheaths – at least one Fr. difference should be present between them to allow accurate pressure measurement through both. Otherwise, the sheath pressure will be dampened (figure 3).

Hemodynamics:
For example, when assessing an EIA stenosis, the tennis racquet (TR) is kept in the aorta and the sheath tip below the diseased segment of EIA.
IA NTG is administered through the sidearm of the sheath (i.e. distal to the lesion) to provoke the gradient.

Hyperemia:
Vasodilators are always given downstream from the stenosis being interrogated.

Patient Set-up:

- Non-invasive blood pressure is checked in both arms prior to the patient being draped. If there is a significant difference (>/= 20 mm Hg) between the two sides, it indicates the presence of significant subclavian/innominate artery disease on the side with the lower pressure.
- While recording the 'opening' pressure, a simultaneous cuff pressure is

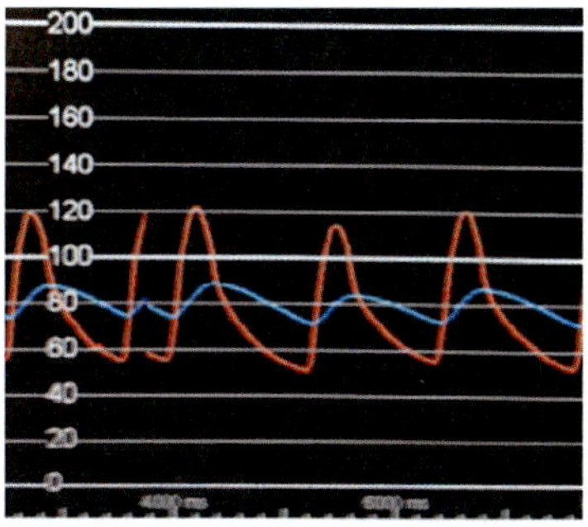

Figure 3.

A 'pseudo' gradient.

This is most commonly seen when the sheath and catheter (inserted through this sheath) have the same caliber. For example, both being 5 Fr. the sheath pressure (blue) is dampened.

Another common cause for this is thrombus in the sheath.

recorded on the side with the higher arm blood pressure. This may indicate the presence of significant subclavian/innominate artery (brachial BP < iliac BP) or aorto-iliac disease (iliac BP < brachial BP).

Two Transducer Method:

- Two transducers are set up. They are both leveled, flushed and 'zeroed'.
- One of them usually is a part of a manifold system.
- The other is a free-standing transducer that is connected through a high-pressure tubing to either the side-arm of the arterial sheath or the catheter.
- It does not matter which transducer is connected to which pressure source as long as they are recognized and labeled correctly.
- Simultaneous pressures are recorded at baseline. Any difference between the two is called a 'gradient'.
- Next hyperemia is induced with a vasodilator and peak gradient is recorded (figure 4).
- The gradient, if any, becomes exaggerated with hyperemia and simulates the impact of exercise, for example, in claudicating lower extremities.
- In other situations, such as renal artery stenosis, the impact of vasodilatation on the 'resting' gradient is irrelevant as no physiologic correlate of such phenomenon exists. A baseline gradient measurement is adequate to establish or rule out the diagnosis of a 'significant' stenosis.
- For example, a 4 Fr. TR catheter is placed in the abdominal aorta through a 5 Fr./23 cm Cordis Britetip sheath. Both are flushed and hooked to separate transducers.
- With the tip of the sheath in the common iliac artery, the recorded pressures reflect the gradient between the abdominal aorta and the common iliac artery. Both resting and hyperemic gradients are recorded.
- If the sheath is withdrawn so that its tip now rests in the external iliac artery, the recorded gradient would be between the aorta and the EIA. By laws of physics, this gradient cannot be any less than the gradient between the aorta and the CIA. If anything, it can be more.
- Again, resting and hyperemic gradients are recorded.

Single Transducer Method:

- With a single transducer only one pressure can be recorded at any instant. The transducer is connected to a catheter whose tip is downstream from a presumed 'lesion'.
- This setup is used for a 'pull-back' gradient assessment (figure 5).
- The pressure is recorded continuously at a slow 'paper speed' (10 mm / sec as opposed to the normal 25 mm / sec). This allows for a better appreciation of the pressure change as the

Figure 4.

A provoked gradient.

A. Two simultaneous pressures show no gradient across a lesion.

B. After administration of IA NTG, a significant gradient is noted.

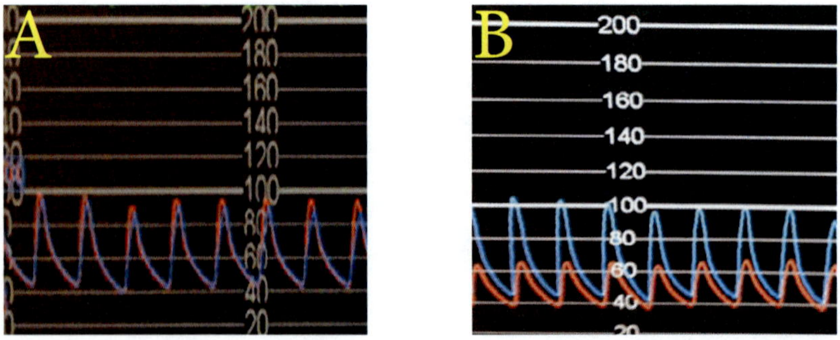

catheter is gradually withdrawn.

- The catheter should be withdrawn ('pulled back') at a speed no faster than 1 mm /sec.
- As the catheter is pulled upstream from a significant stenosis the pressure is seen to jump considerably above its previous level.

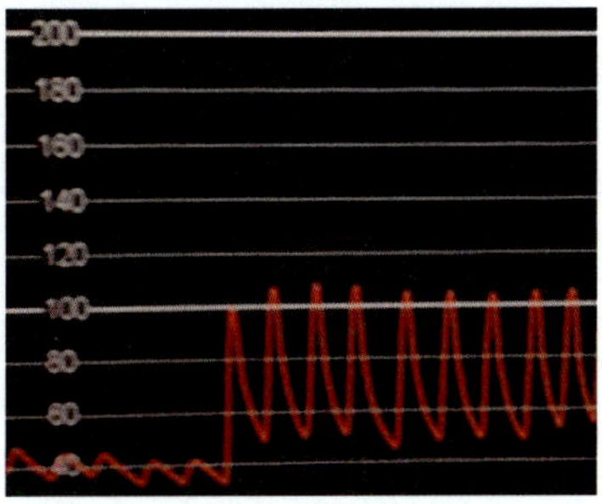

Figure 5.

Pullback gradient.

Using a single transducer, pressure is measured across a severe stenosis. On slow paper speed, the catheter is gradually withdrawn to demonstrate the gradient.

DIAGNOSTIC ANGIOGRAPHY

This consists of aortography and selective angiography.

AORTOGRAPHY

Flush Catheter:

- For thoracic (ascending, arch and descending) aortography, a non-angled pigtail catheter is used. From a radial/femoral approach a 5 Fr. /100 cm catheter is adequate. From a brachial approach, a shorter catheter can be used if necessary. Do not use CO_2 for angiography above the diaphragm due to the risk of stroke from gas embolism.
- For abdominal aortography, reasonable flush catheter choices include a 4 Fr./65 cm pigtail catheter, tennis racket catheter or an Omni-flush catheter.

Catheter Positioning:

- For ascending aortography, the catheter is positioned 2 cm above the aortic valve.
- If coronary bypass grafts are present and the idea is not to image them, then the catheter can be placed 2 cm caudal to the innominate origin. This will outline the arch, the great vessels and descending aorta.
- For abdominal aortography, regardless of the choice of the catheter, it is usually positioned so that it sits between the L1-L2 inter-space.

Imaging and Projection:

- Thoracic aortography is performed in an LAO projection to open the arch and reduce the overlap between the ascending and descending aorta.
- Initial abdominal aortography should be performed in the AP projection (figure 6).
- Digital subtraction angiography should be used whenever possible.
- A breath-hold is advisable to minimize motion artifact.
- For thoracic aortography, the field of view is set to include from the origin of the arch vessels to the diaphragm.

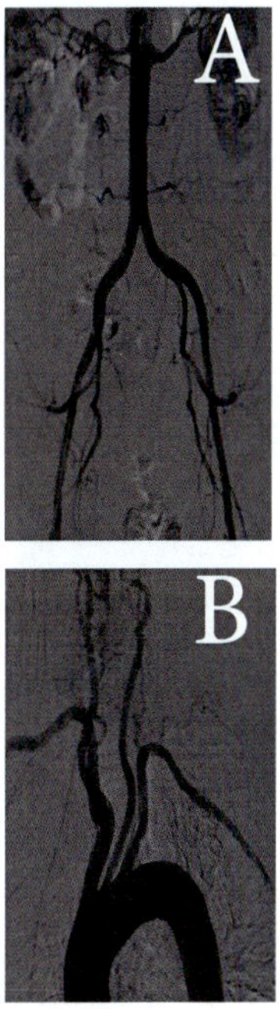

Figure 6.

Aortography.

A. Abdominal aortography.

B. Arch angiography.

- For abdominal aortography, the field of view is set to image from the top of the flush catheter to bilateral common femoral arteries.

Dye and Injection Rate:

- A power injector is used for aortography.
- The rate of power injection depends upon the purpose of the injection, the patient's body habitus, renal function and ability to comply with instructions of breath-hold during aortography.
- In patients with normal renal function, 70% dye with 30% heparinized normal saline (70:30 dilution) is usually adequate. If the creatinine clearance is less than 30 mL/min, then 50% dilution (50:50) should be used. Injection settings are 10-15 cc/sec for a total volume of 20-40 cc at 500-1000 psi. For abdominal aortography, 500-750 psi is usually adequate.
- CO_2 angiography should ***NEVER*** be used above the diaphragm in the arterial circulation or in the venous circulation if there is a known right to left shunt (see later).

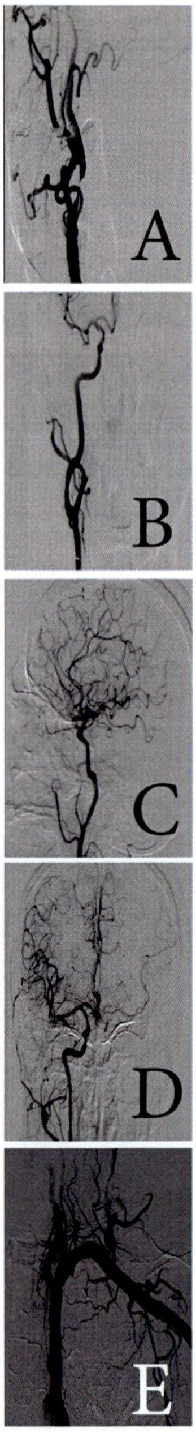

Figure 7.

Great Vessel Angiography.
A. Right CCA angiography in LAO.
B. Left CCA angiography in LAO.
C. Intracranial angiography in lateral projection.
D. Intracranial angiography in Towne's view.
E. Left subclavian artery angiography in LAO projection.

SELECTIVE GREAT VESSEL ANGIOGRAPHY

Catheter Selection:

- In order to perform selective great vessel angiography, a multitude of catheters are available. We prefer to use a 5 Fr./125 cm Vert catheter, anatomy permitting.
- If, however, the aortic arch is angulated and/or tortuous, a catheter such as the Vitek is more suitable.

Catheter Positioning:

- For selective great vessel angiography, the catheter should be placed at the origin of the vessel.

Imaging and Projection:

- Great vessel angiography is usually best performed in an LAO projection along with an additional AP or contralateral oblique views as necessary (figures 7A-E). For right subclavian artery, the ostial/proximal segments are better visualized in an RAO caudal projection. The same view may be necessary to see the ostium of the right CCA.

Dye and Injection Rate:

- Manual injection is usually adequate.
- DSA with breath-hold should be used.
- A single injection in each projection of nearly 5-8 cc is adequate.
- In patients with normal renal function, 70% dye with 30% heparinized normal saline (70:30 dilution) is usually adequate. If the creatinine clearance is less than 30 mL/min, then 50% dilution (50:50) should be used.
- CO_2 angiography should ***NEVER*** be used for great vessel angiography (see later) due to the risk of stroke.

SELECTIVE VISCERAL ARTERY ANGIOGRAPHY

Catheter Selection:

- Multiple catheters can be used to perform visceral angiography. The choice depends upon the diameter of aorta, any angulation within the aorta, and the angulation of the visceral arteries. Often a catheter like a JR4 serves well to perform celiac, SMA, renal and IMA angiography.

- More specialized catheters such a Cobra, Simmons 1, Contra 2, renal curve/double curve, hockeystick are also available.

Catheter Positioning:

- For selective visceral angiography, the catheter should be placed at the ostium of the vessel.
- For superselective angiography (e.g. gastroduodenal, splenic etc), the selective catheter is exchanged over a wire for a softer simple curve or straight catheter such as a straight/angled Glidecath®.

Imaging and Projection:

- Depending upon which vessel is being imaged, the projections may vary from AP to LAO 15-45° or complete lateral (figure 8).

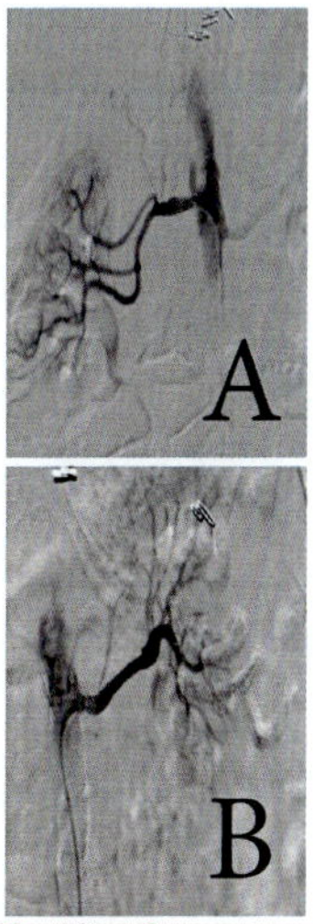

Figure 8.
Selective renal angiography.

A. Right renal artery angiography in LAO 15° projection.

B. Left renal artery angiography in LAO 20° projection.

Dye and Injection Rate:

- Manual injection is adequate.
- DSA with breath-hold should be used when possible.
- A single injection in each projection of nearly 5-8 cc is adequate.
- The injection should be performed that there is some reflux of dye into the aorta. This makes it easier to visualize the ostium of the vessel.
- In patients with normal renal function, 70% dye with 30% heparinized normal saline (70:30 dilution) is usually adequate. If the creatinine clearance is less than 30 mL/min, then 50% dilution (50:50) should be used.
- CO_2 may be used for this purpose and may be preferable in patients with renal dysfunction. Additionally, CO_2 refluxes easier than contrast dye from a vessel into the aorta. This may help delineate any ostial disease.

SELECTIVE ILIAC ARTERY ANGIOGRAPHY

Catheter Selection:

- In order to perform selective iliac angiography and go up-and-over the aortic bifurcation, multiple catheters can be used such as the Omniflush, IMA, Rim or Simmons catheter.

Catheter Positioning:

- For selective iliac angiography, the catheter should be placed at the aortic bifurcation engaging the CIA ostium.

Imaging and Projection:

- Iliac angiography is usually best performed in an AP projection along with an additional contralateral oblique view of nearly 20° (figure 9). The latter maneuver reduces the overlap between the internal and the external iliac arteries and is a must to rule out external & internal iliac artery (IIA) disease.

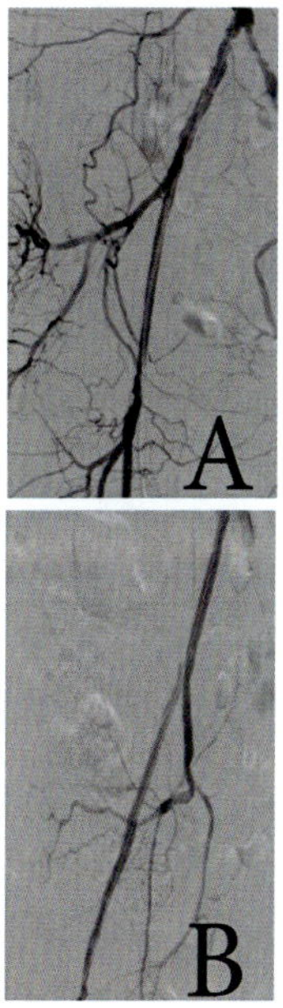

Figure 9.

Selective iliac artery angiography.

A. Selective angiography of the right CIA in AP projection and

B. LAO 20° projection.

Dye and Injection Rate:

- Manual injection is usually adequate.
- DSA with breath-hold should be used.
- A single injection in each projection of nearly 5-8 cc is adequate.
- In patients with normal renal function, 70% dye with 30% heparinized normal saline (70:30 dilution) is usually adequate. If the creatinine clearance is less than 30 mL/min, then 50% dilution (50:50) should be used.

- CO_2 may be used for this purpose (see later).

SELECTIVE INFRA-INGUINAL ANGIOGRAPHY

Once iliac angiography has been completed and if critical occlusive disease is not present, angiography of the common femoral (CFA) and infra-inguinal arteries should follow.

Catheter Selection:

- CFA angiography can be performed with the catheter in the CIA or the EIA. The latter is preferred as it allows better opacification.
- This is performed with a catheter that can be advanced smoothly up-and-over the aortic bifurcation to the distal EIA. Examples include the Omni-flush, Simmons and the Glidecath. Do not advance an IM or JR4 catheter over the bifurcation into the EIA/CFA as this may result in dissection of the vessel.
- Prior to this step, the wire should be advanced into the PFA (if patent) as this provides secure anchoring for this maneuver. The wire used for this purpose is a 0.035"steerable wire, such as an angled stiff Glide wire or a Wholey wire, and is advanced into the profunda (if patent) while the selective catheter is engaged at the ostium of the CIA. Over this 'rail', the catheter is then advanced to the distal EIA.
- If an Omni-flush was used for aortography that same catheter can be advanced into the EIA for selective angiography. However, if aortography was performed using a pigtail or tennis racket (TR) catheter then, as noted before, a separate catheter such as an IMA has to be employed to engage the CIA. While one can employ these flush catheters (pigtail or TR) by opening up their curve using a wire, to engage the CIA we caution against such a maneuver as it is unsafe and can often result in aortic or iliac dissection. Time and cost considerations should not get in the way of attention to detail for patient safety.

Catheter Positioning:

- For optimal imaging and to enable hemodynamic assessment (of the iliac arteries), the catheter should be advanced to the level of the distal EIA. Caution should be used when measuring pressure gradients in these situations as the selective catheter traversing the aortic bifurcation can sometimes kink (e.g. a Glidecath in a complex aortic bifurcation) and have a dampened waveform. If suspected, zoom in on the catheter at the aortic bifurcation to rule out this issue.

- After performing angiography of the CFA and proximal SFA/PFA, the catheter can be further advanced over a wire into the proximal/mid SFA if this vessel is not severely diseased/occluded. If the proximal SFA is indeed severely diseased avoid crossing it. Perform angiography of the distal segments with the catheter still in the CFA.

Imaging and Projection:

Common femoral artery:

- CFA angiography should be performed in an AP projection as well as an ipsilateral oblique of nearly 30° (figure 10A). Optional contralateral oblique angulation of 20-40° may be used.
- As the SFA and PFA are closer together in the proximal thigh the latter view opens the CFA bifurcation widely and reduces the overlap between the ostial/proximal superficial femoral artery and the PFA.
- Due to the separation of these vessels in this view it also serves as an ideal view for the PFA which should be imaged in its entirety along with any collaterals it provides to the distal SFA/popliteal arteries.
- The mid/distal PFA, like the SFA, is best imaged in an AP projection as these two vessels have coursed apart naturally with minimal overlap if any.

Superficial femoral and popliteal arteries:

- SFA and popliteal artery angiography is best performed in AP projection (figure 10B).

- In cases where optimal imaging in AP projection is not feasible, e.g. due to a knee joint prosthesis, a lateral view may be helpful. Alternatively, the use of IVUS is mandatory in these cases to assess the burden of disease in the segment of the vessel that is hiding behind the prosthesis.

Infra-popliteal arteries:

- Below-the-knee angiography is optimally performed in an ipsilateral oblique projection to minimize the overlap of the tibial arteries (figure 10C). This may also be accomplished by asking the patient to 'roll in and hold' their legs for a few seconds.
- For better visualization, this angiography should be performed with the catheter as close to the popliteal artery as possible. If the SFA is free of significant disease, the catheter should be advanced in the popliteal artery for infra-popliteal angiography. This minimizes the amount of contrast used and fluoroscopy time as well.

Foot arteries:

- Foot angiography should be performed in AP cranial and/or ipsilateral oblique projection (figure 10D).
- Digital subtraction should be used whenever possible for initial angiography and the frame rate should be between 3-5 frames per second.

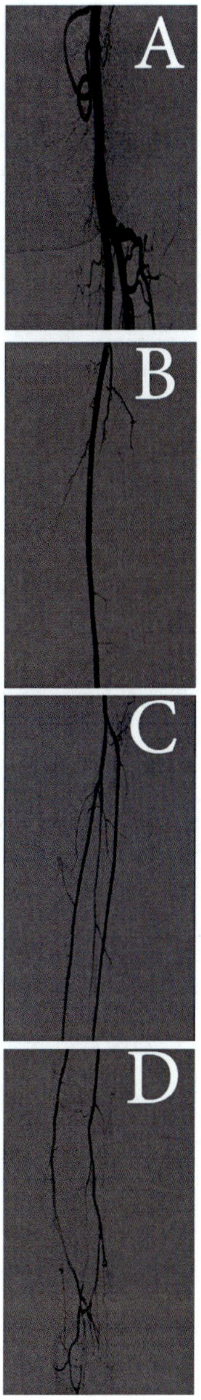

Figure 10.

Infra-inguinal angiography.

A. Left CFA angiography in LAO projection.

B. Left SFA angiography in AP projection.

C. Left infra-popliteal angiography in LAO 15° projection.

D. Left pedal angiography.

Dye and Injection Rate:

- Manual injection is usually adequate.
- Again DSA should be used when possible.
- A single injection in each projection of nearly 3-5 cc is adequate.
- In patients with normal renal function, 70% dye with 30% heparinized normal saline (70:30 dilution) is usually adequate. If the creatinine clearance is less than 30 mL/min, then 50% dilution (50:50) should be used.
- CO2 angiography may be used for this purpose.

CO_2 ANGIOGRAPHY

Carbon dioxide has been used safely and effectively for angiographic applications for over forty years. Despite its significantly low cost (up to 500 X cheaper compared to iodinated contrast agents) this modality has not gained mainstream popularity due to its somewhat complex setup and potential (figure 11) for operator induced error.

BASICS

Advantages of Using CO_2:

- Carbon dioxide is an endogenous gas and hence is non-allergenic and non-nephrotoxic.
- It is compressible, and unlike liquid contrast agents, does not mix with blood, rather displaces it due to its buoyancy.
- It is up to four hundred times less viscous than iodinated contrast.
- On DSA, CO_2 causes a negative image.

Disadvantages:

- Requires unique setup for delivery.
- Can cause some discomfort to the patient if 'trapping' occurs.
- Not suitable for cerebral angiography.

Figure 11.

CO_2 source.

A pressurized cylinder with medical grade CO_2.

INDICATIONS

- Renal insufficiency.
- Iodinated contrast agent allergy.
- To diagnose the site of bleeding (due to its low viscosity and tendency to expand, this gas is easily visible where iodinated contrast may not be very helpful).

APPLICATIONS

- Abdominal and lower extremity arterial angiography.
- Can be used for venous angiography and intervention such as IVC filter placement, venous PTA/stent (both above and below the diaphragm e.g. upper and lower extremity veins).

CONTRAINDICATIONS/CONCERNS

- Due to its potential for 'trapping', it should not be used for intracranial angiography as it can lead to irreversible cerebral ischemia.
- Supra-diaphragmatic angiography of the arterial system should not be performed with CO_2. It can be used safely above the diaphragm in the venous system(provided there is no right to left shunt).
- Not suitable for venography in patients with known right to left shunt and pulmonary hypertension.
- Do not use with nitrous oxide anesthesia (this gas can diffuse from the soft tissue into the existing CO_2 bubble and cause 'trapping'. This is 5-6 X more likely to occur with this anesthetic than without it).
- Large abdominal aneurysms should ideally not be imaged with CO_2 (the gas may not absorb fast enough in aneurysms due to their size and the surface area of the blood:gas interface). If this occurs, roll the patient side to side to expedite 'breaking' the surface tension and improving resorption).
- COPD (reduce the volume of CO_2 and increase the time between injections).
- Dialysis AV fistula: avoid reflux into the arterial system particularly intracerebral vessels).

DELIVERY SYSTEM

- For CO_2 angiography, the purified gas is delivered through a specialized, non-compressed closed system rather than a manifold (figure 13).

Source:

- A pressurized cylinder (up to 500 psi) with a controlled dispensing mechanism that contains medical grade CO_2 is used as the source for CO_2.
- A single cylinder which costs about $500 contains enough CO_2 for several dozen cases.
- The cylinder is used to fill up a disposable plastic bag which serves as a reservoir of the gas during angiography.
- The cylinder itself should NEVER be attached directly to a catheter as it can lead to accidental injections of excessive amounts of CO_2 at high pressure, leading to a harmful and potentially fatal outcome.
- The cylinder is non-sterile.

Components of the Delivery System:

The sterile components of the delivery system consist of the following:

A Reservoir bag:

- ▹ A 1200 -1500 mL flexible, plastic bag with a single port is used as the reservoir.
- ▹ In this bag the gas is subject only to atmospheric pressure. This avoids the risk of high-pressure injection of CO_2.

Filter:

- ▹ One micro-filter with a Luer-lok connector. This is used when CO_2 is collected from the cylinder into the bag. This filters air, microbes, methane, rust and other particulate contaminants.

Tubing:

- ▹ A 100 cm length, low pressure tubing to allow aspiration of the CO_2 from the bag. A 20 cm segment of this tubing connects the bag to a 30-60 cc syringe.
- ▹ This tubing has a one-way valve at the end which connects through a Leur-lok to the syringe.
- ▹ The other end of the tubing connects via a 4-way stopcock, that also has a one-way valve, to the catheter.

Check valves:

- ▹ Two built-in check-valves in the delivery tubing. These allow only one-way flow of the gas and prevent accidental introduction of air into the system.

Stopcock:

- ▹ Air contamination must be avoided during CO_2 angiography. Air is several times lighter than CO_2, does not resorb easily and is much less viscous. It is thus unsafe if injected into the circulatory system as it can be 'trapped' and can cause tissue ischemia.
- ▹ Stop-cocks make the system prone to operator error and allow inadvertent introduction of air into the system if a stop-cock is turned the wrong way.
- ▹ To reduce the likelihood of air-entry into the system, only one one-way stop-cock is used between the reservoir and the tubing and another (4-way stop-cock) is used between the tubing and the catheter.

A 30-60 cc delivery syringe:

- ▹ This syringe is used to inject CO_2.
- ▹ More than 50 mL should ***not*** be injected in a single bolus (20 - 30 mL is sufficient).
- ▹ Also, CO_2 is compressible. A 20 mL of compressed gas can contain up to twice that volume of non-compressed gas.

Connections of the Delivery System:

Step 1:

- Collecting CO_2 in the reservoir.
  - ▹ The reservoir bag is kept on the sterile field. The operator connects the bag through a sterile aerosol filter to the dispenser tubing of the cylinder (figure 12).
  - ▹ An assistant opens the gas into the bag.
  - ▹ The bag is filled to capacity. Gas valve is shut off, the filter left on the dispensing tubing. The bag is disconnected from the filter and manually deflated.
  - ▹ These steps are repeated three times to purge the bag of any residual air.
  - ▹ The bag is finally filled with CO_2 to capacity. It is detached from the filter and its port is locked.

Step 2:

- The bag is then connected via a one-way stop-cock to the 100 cm tubing (figure 12).

Step 3:

- A 30-60 cc syringe is connected to the tubing.

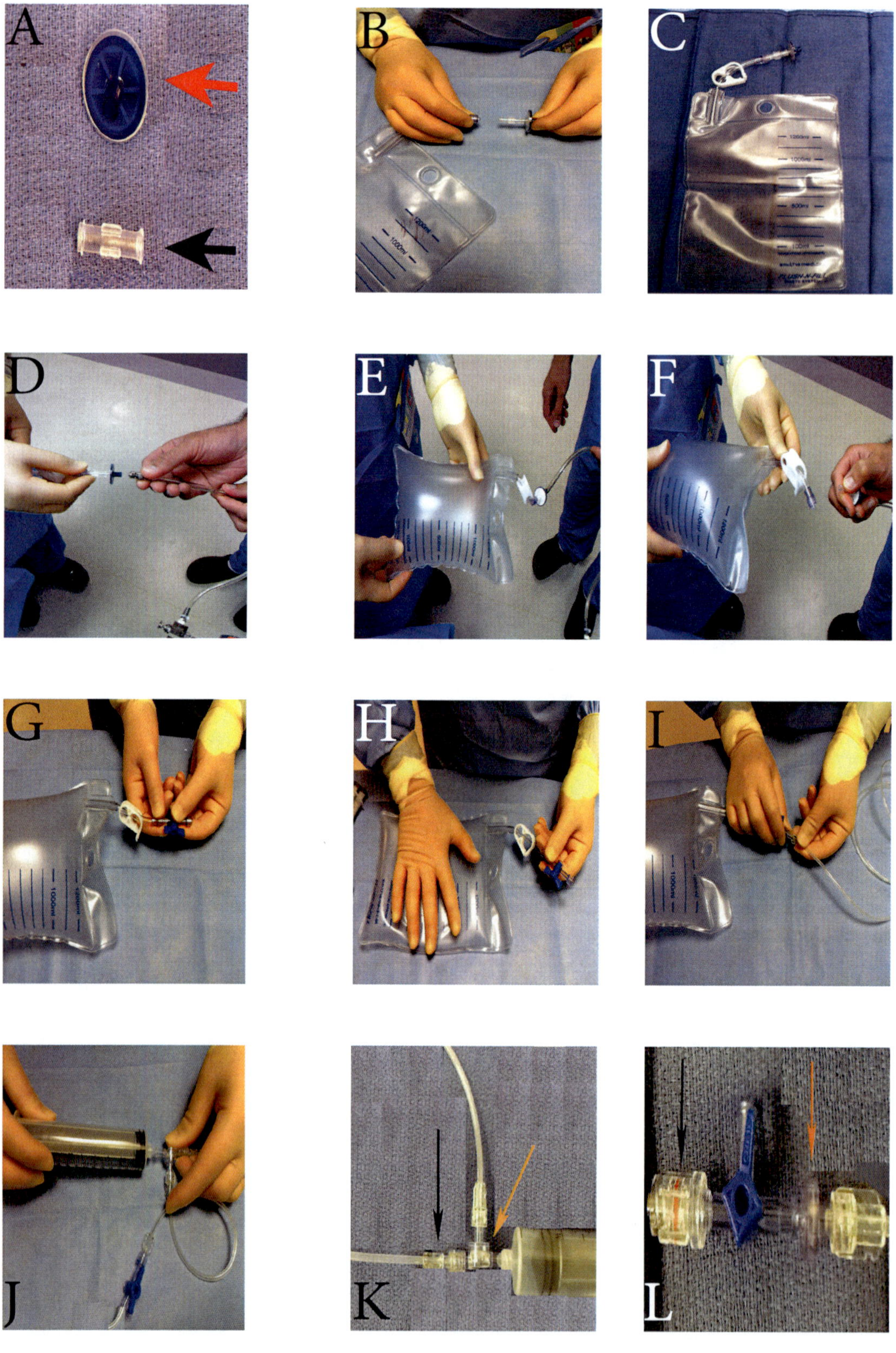
A
B
C
D
E
F
G
H
I
J
K
L

Figure 12 (opposite page).

Setup for CO_2 angiography.

A. An aerosol filter (red arrow) and connector (black arrow) are shown in close-up.

B & C. A flexible plastic bag with a single port is used as CO_2 reservoir. The filter and connector are connected to this bag.

D. An assistant helps connect the tubing of the CO_2 cylinder to the bag via the filter.

E. The bag is filled to capacity with CO_2.

F. The port on the reservoir is closed and the bag disconnected. The aerosol filter and connector, which are now non-sterile, are left connected to the cylinder tubing.

G. After removal of the filter, a one-way stop-cock is connected to the bag.

H. The bag is then deflated completely. The above steps are then repeated two more times.

I. After the bag has been filled the third time the one-way stop cock connects the bag to the delivery tubing.

J & K. A 60 cc syringe is used to draw CO2 from the bag and deliver forward to the outflow tubing (black arrow). The delivery tubing has a one-way valve (orange arrow) at both ends. The tubing is purged of air by injecting 20 cc of CO_2.

L. The final connection of the outflow tubing to a catheter (that is already in the body) is via a 4-way stop-cock with a one-way valve (orange arrow). The catheter connects via a Leur connector (black arrow).

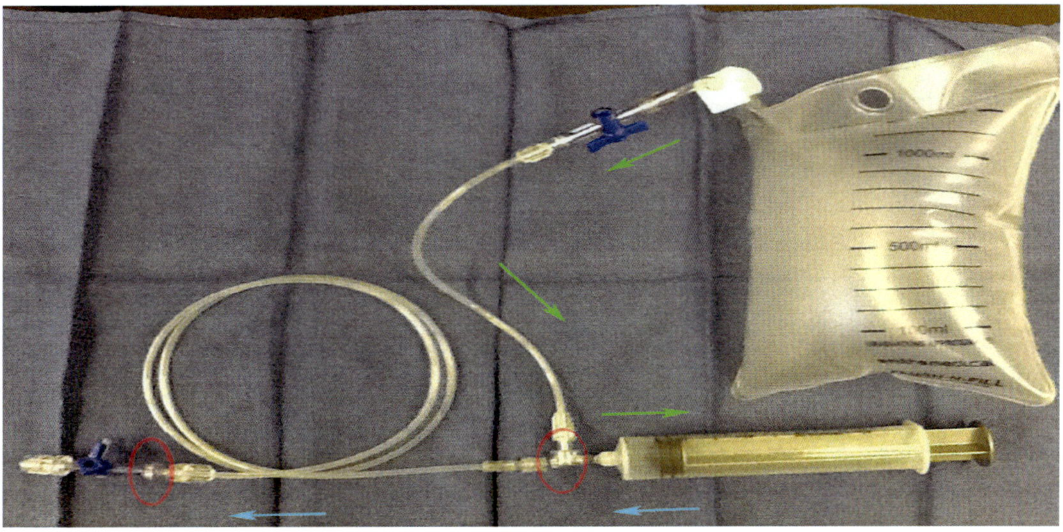

Figure 13.

Delivery set-up for CO_2 angiography.

The CO_2 reservoir bag has a (white) pinch-clip to close the port after CO_2 has been withdrawn. This reduces the chances of air contamination.
A one-way stop-cock then connects the port to the tubing. The green arrows indicate the flow of CO_2 from the bag into the syringe. The syringe is then used to deliver the CO_2 via the outflow tubing as indicated by the blue arrows. This tubing connects via a stop-cock assembly to the catheter. There are two one-way valves (outlined in red) that ensure only forward flow.

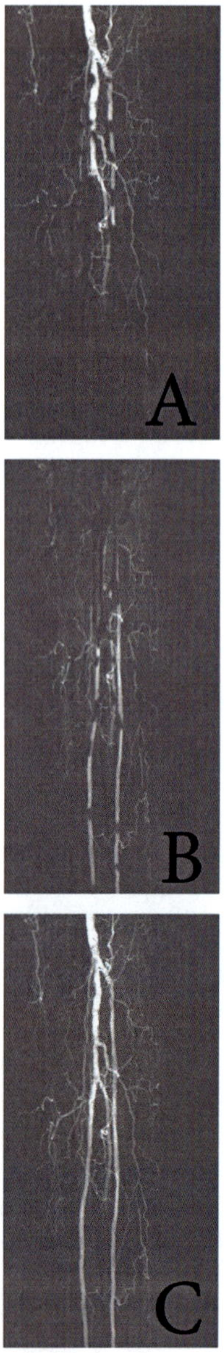

Figure 14.

'Stacked' CO_2 run.

A CO_2 angiogram of the lower leg ***(A and B)***. The stacked run ***(C)*** is a composite of the two and shows the anatomy in better detail.

Step 4:

- The tubing is purged of air by injecting 20 cc of CO_2 into it.

Step 5:

- The tubing is then connected via the four-way stop-cock to the catheter (figure 13).

Preparation of the Delivery System:

- When CO_2 is injected in a blood vessel it expands. A rapid injection through a catheter that is filled with fluid or blood causes sudden expansion of the gas that can be traumatic to the vessel.
- To avoid this 'explosive' impact of CO_2, flush the entire catheter with CO_2 prior to an angiogram.

Injection of CO_2:

- Manual injections are recommended. Using power injectors can result in delivery of excessive amounts of gas.
- Volume per injection is 1.5 times that of iodinated contrast.
- Since CO_2 is buoyant and tends to tend to flow better to non-dependant regions, we recommend elevating the area of interest 20 degrees or so off the table.
- In lower extremity arteries, injecting 200-300 mcg of intra-arterial NTG improves flow and CO_2 runoff.
- An interval of 1-2 minutes is recommended between injections.
- 'Explosive' delivery may cause rapid vascular distension and lead to nausea and tenesmus (when use in the abdomen).

IMAGING

- Dedicated DSA is required for CO_2 imaging (figure 14). Most imaging systems have a 'stacking' software that is required for processing of images. This allows creation of a composite 'run'.
- Imaging is performed at 3-5 frames/second (60 ms pulse width).
- If all the blood in a vessel has been displaced by CO_2, increasing the volume of injected gas will not increase the quality of the image.
- A vessel < 10 mm in diameter has 100% correlation with imaging accuracy compared to iodinated contrast.
- In larger vessels, there may be incomplete displacement of blood, the buoyant CO_2 rises and may give the impression of smaller size of the vessel and likely overestimation of the degree of stenosis.

AVOID

- Administration of excessive volume (large, single bolus or multiple, small boluses back to back) should be avoided as it can cause 'trapping'/ vapor lock.
  - Maximum single bolus volume should not exceed 50 mL (usually 20-30 mL is sufficient).
  - If an excess dose is injected, the operator must aspirate as necessary.
  - Changing patient position on the table may also help eliminate any 'vapor lock'.

2

DIAGNOSTIC ENDOVASCULAR EQUIPMENT

Equipment selection is one of the most important steps to performing a successful intervention. Planning a strategy prior to the intervention will allow for appropriate selection of equipment and ultimately a desirable outcome. This chapter will discuss and illustrate the different endovascular equipment available to the interventionalist.

ACCESS KITS

Percutaneous vascular access can be obtained using a variety of access devices or 'kits'. In contemporary clinical practice, modified Seldinger technique is used for access in most vascular beds except the radial artery where the original Seldinger technique is still commonly used.

Access is most commonly obtained using one of the following kits:

- 7 cm/18 Gauge (G) needle; accommodates a 0.035" wire.
- 7 cm/21 G needle; accommodates a 0.018" wire.

We prefer to use the latter, called a micropuncture kit (figure 1), as it allows the placement of a microcatheter in the vessel before an actual 'sheath' can be placed. If on angiography, the position of the microcatheter is found to be undesirable, it can be removed, manual compression held briefly and the access re-attempted in a more favorable segment of the artery.

For obese patients, a 7 cm long needle may not be able to reach the CFA. Use a 9 cm needle instead. A needle this length is available in 18 G and can be obtained by special order in 21 G.

Specially designed access kits are available for radial artery access. These package an access needle(21 G/4 cm), a 0.018"-0.021" wire and a 5-6 Fr./11 cm hydrophilic sheath with a dilator tip tapered to the caliber of the wire.

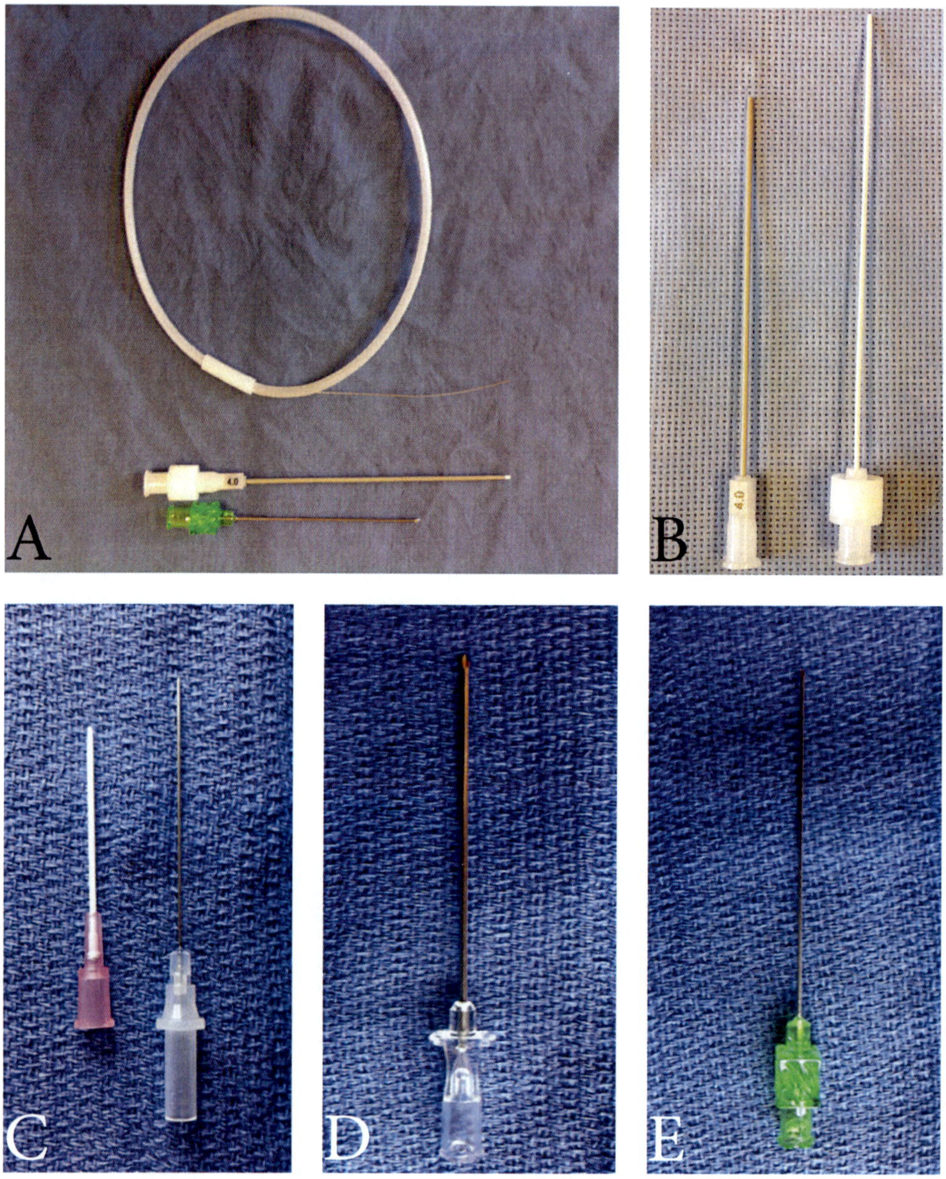

Figure 1.

Access kits and needles.

A. A 4 Fr. micropuncture kit with a 0.018" wire, a 21 G needle and a microcatheter.
B. The microcatheter used in a micropuncture kit, separated from its dilator.
C. A 20 G Jelco needle along with its outer catheter.
D. An 18 G/7 cm needle.
E. A 21 G/7 cm needle.

With an increase in infra-popliteal interventions, dedicated pedal artery access kits have been developed. These are similar to radial artery access kits with the usual exception of a 21 G/7 cm needle and a non-hydrophilic sheath.

SHEATHS

Introducer sheaths (fiugre 2) are used for endovascular procedures to establish a point of access to the vascular system for catheter insertion and manipulation without ongoing blood loss or vessel wall damage.

Diagnostic and interventional sheaths are available in multiple shapes and sizes. Table 1 lists different types of straight and pre-shaped sheaths.

Selection of the sheath depends upon the location of vascular access and the area of intervention (figure 3). The size of the sheath is dependent upon the type of procedure and the size of the catheters or equipment (i.e. stent or balloon size, atherectomy device, etc) used for the intervention.

- Hydrophilic sheaths are used for radial artery access.
- Cross-over sheaths are used for crossing over the aortic bifurcation. They sheaths are usually hydrophilic as well and kink-resistant.
- Guiding sheaths (e.g. Destination®, Terumo) are usually pre-shaped into specific shapes such as IM, HS, MP, JR4 and RDC.
- Specially designed sheaths are available for pedal access.
- Dryseal® (Gore) sheaths have a specially designed hemostatic valve that consists of an inflatable balloon.
- SoloPath™ (Terumo) is a balloon-inflatable sheath that is used for specific procedures such as transcatheter aortic valve replacement (TAVR), EVAR or TEVAR.

WIRES

Wire guides, or 'wires' as they are commonly referred to, are an integral part of the equipment used for any endovascular procedure.

Depending on the patient's anatomy and the nature of the procedure being performed, one type of wire may be preferred over others.

ACCESS/BASIC WIRES

These wires (table 2) are most commonly employed for sheath placement/removal and advancement of diagnostic and basic interventional equipment (e.g. balloons and stents). In patients with non-tortuous aorto-iliac arteries that are not severely diseased, a basic J-tip wire can be used for sheath placement and advancement/exchange of diagnostic catheters.

Sheath/Catheter Placement (0.035"):

- Starter® J wire.
- Emerald® J wire.
- Bentson® wire.
- Wholey® Hi-Torque Floppy wire.

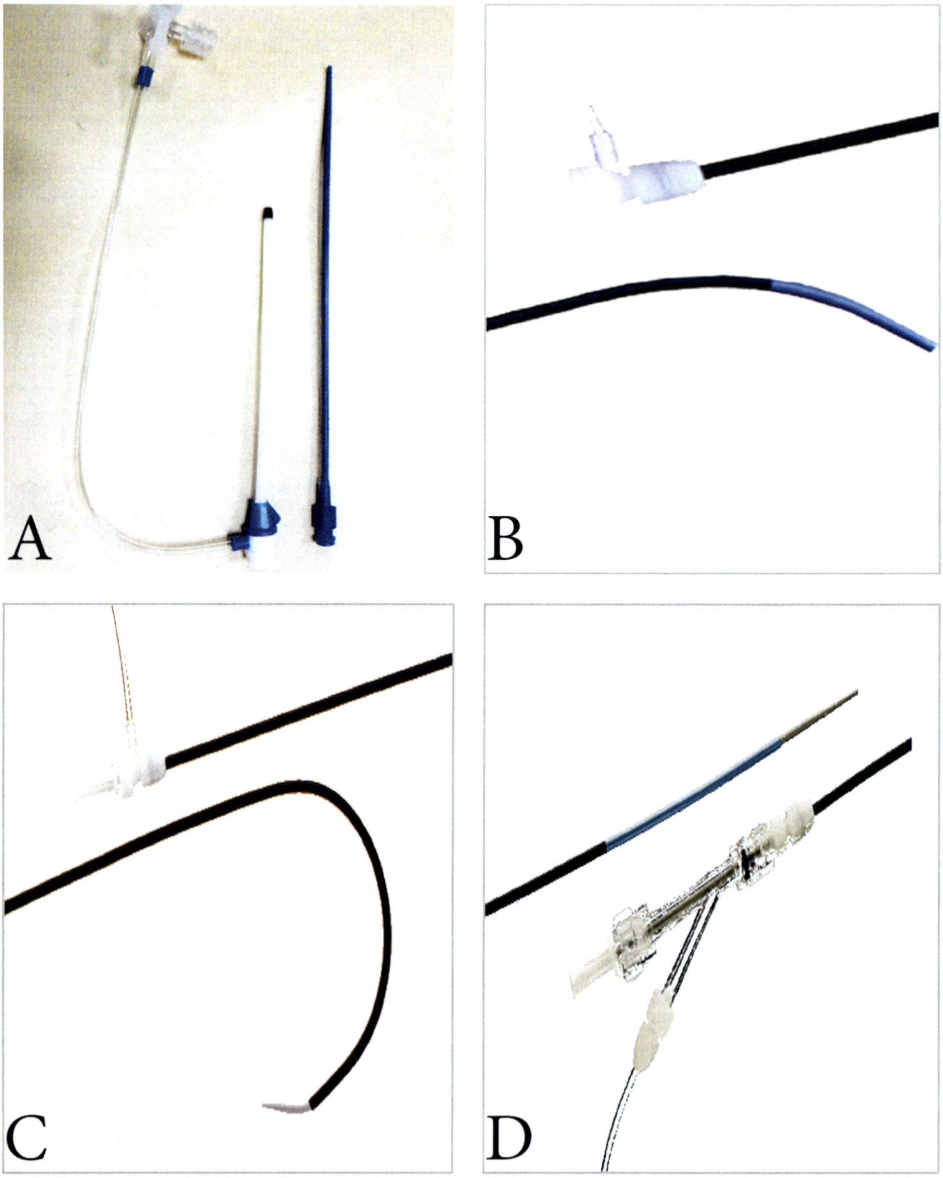

Figure 2.

Vascular access sheaths.

A. An 8 Fr. Britetip sheath.
B. Ansel 1 sheath. This is our preferred cross-over sheath with the diaphragm hub.
C. Balkan sheath can also be used for cross-over.
D. Shuttle sheath with a Tuohy-Borst adapter is used commonly for carotid stenting (6 Fr./90 cm) and for infra-popliteal interventions (5 Fr./90 cm) in our practice.

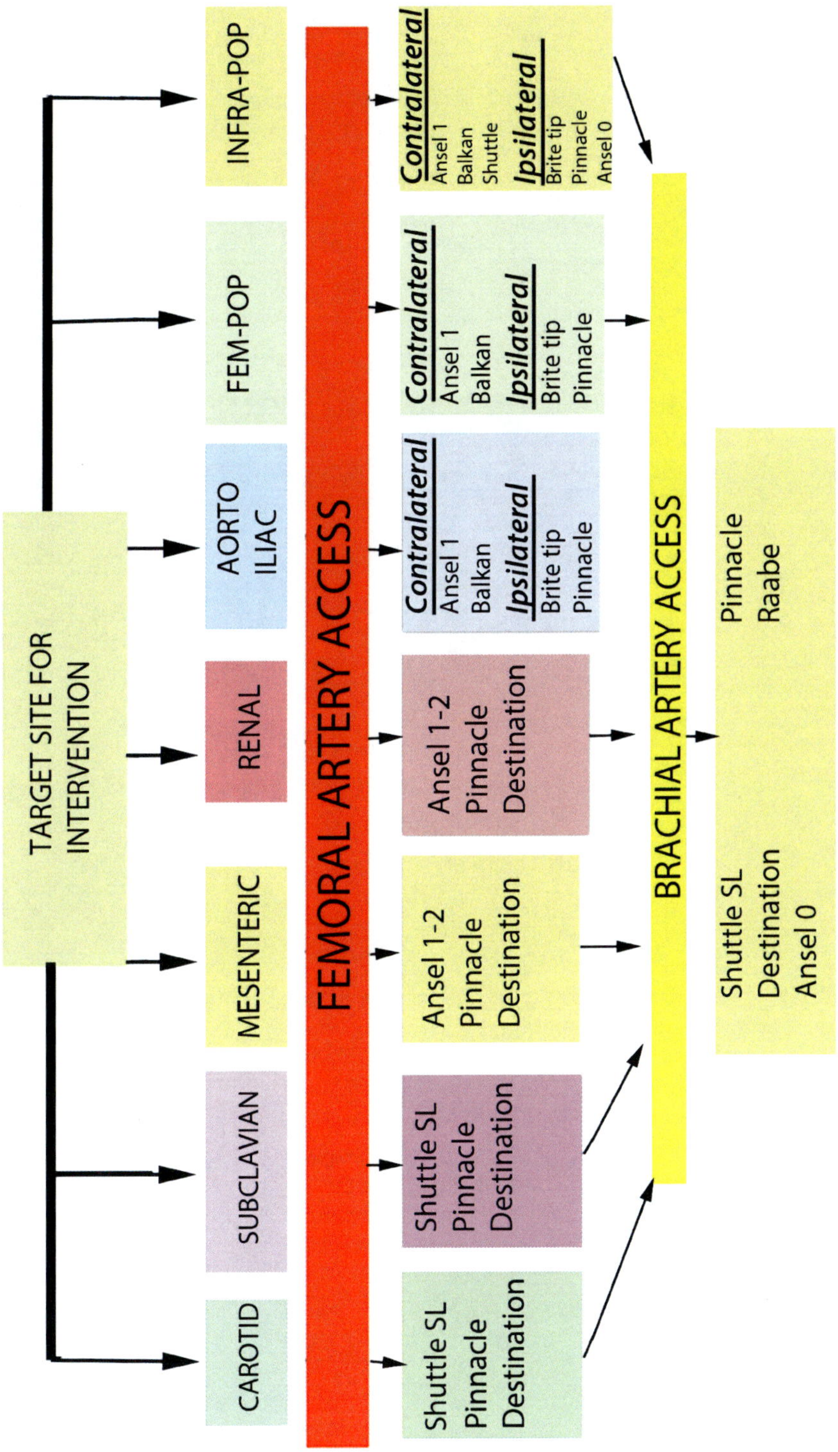

Figure 3.

Sheath selection flow chart.

(Ansel, Balkan and Shuttle SL - Cook Medical. Pinnacle, Destination - Terumo. Britetip - Cordis)

TABLE 1 - SHEATHS			
	Diameter (Fr.)	Length (cm)	Guidewire compatibility
STRAIGHT			
Ansel® 0 (Cook)	4-9	45, 90, 110	0.035", 0.038"
Britetip® (Cordis)	4-11	11, 23, 35, 45, 55, 90	0.035", 0.038"
Pinnacle® (Terumo)	4-11	10, 25	0.035", 0.038"
Pinnacle Destination® (Terumo)	5-8	45 ,65, 90	0.038"
Raabe® (Cook)	4-9	40	0.038"
Shuttle® (Cook)	4-8	80, 90, 110	0.038"
PRE-SHAPED / SELECTIVE			
Ansel® 1 (Cook)	5-12	45, 90	0.018", 0.038"
Ansel® 2 (Cook)	5-7	45	0.018", 0.038"
Ansel® 3 (Cook)	6	45	0.018", 0.038"
Balkin® (Cook)	5.5-8	40	0.038"
Pinnacle Destination® (Straight, IM, RDC, HS and MP curves) (Terumo)	5-8	45, 65, 90	0.038"
SPECIALITY SHEATHS			
Pedal Micropuncture® kit (Cook)	4 & 5 Fr.	7 cm	0.018" wire compatible. This is designed for dedicated pedal artery access. The needle is 4 cm long and has a corrugated echogenic tip for better visualization under ultrasound.
DrySeal® (GORE®)	12-26 Fr.	28 cm	0.035" wire compatible. Has a pressurized hemostatic valve with a built-in balloon. This is inflated with 2.5 mL of saline using supplied syringe. Provides excellent hemostatic control during delivery of multiple/large-bore devices e.g. aortic endografts, Impella, etc.
SoloPath™ (Terumo)	14,18,19, 21	25 cm.	0.038". Balloon-expandable transfemoral sheath. Inserted in the deflated state, once beyond the aortic bifurcation the 'Expander' (balloon) is inflated and deflated. The sheath reaches its intended diameter. The 'Expander' is removed and the sheath used in the usual fashion.

Tortuous Aorto Iliac Arteries (0.035"):

- Wholey® Hi-Torque Floppy wire.
- Angled non-stiff Glidewire®.
- Angled stiff Glidewire®.

Cross-over Sheath Placement:

- Angled stiff Glidewire®.
- Wholey® Hi-Torque Floppy wire.

SUPPORT WIRES

Support wires (table 2) are used to deliver equipment where the basic wires would fail or be outright dangerous.

Support wires, other than the Glidewires®, are usually not advanced directly into the aorta or other vessels but rather through a catheter as they are very stiff and provide minimal tactile feedback about the force on the tip. As a result, they can easily dissect and/or perforate the vessels if they are advanced bareback and not through a catheter. Hence, in our practice we always introduce these wires through an existing catheter.

Support Wires:

- Amplatz (not Amplatzer!) wires
- Extra-Stiff
- SuperStiff
- Ultra-Stiff
- Lunderquist
- Rosen
- Meir wire
- These wires are used to place:
    - ▹ Sheaths in patients who may have excessive scar tissue in the groin.
    - ▹ Cross-over sheath placement in patients with complex aorto-iliac anatomy.
    - ▹ Bulky interventional devices such as AAA or TAA repair endo-grafts.
    - ▹ TAVR delivery system etc.

INTERVENTIONAL WIRES

These wires (table 3) are used for crossing lesions and delivering interventional equipment in specific vascular beds. Typically, the interventional wire is introduced in the target vascular bed after a guiding catheter or sheath has been successfully delivered proximal to the vessel. Some support wires, e.g. Glidewire®, can be used interchangeably as a basic/support/interventional wire.

Carotid:

- 0.014" bare wire for use with proximal embolic protection (EPD).
- 0.014" filter wire for use as distal EPD.

Innominate:

- 0.014" plus 0.018" when using a 0.014" distal EPD.
- 0.018" retrograde when not using an EPD.

Subclavian

- 0.018"
- 0.035"

Work horse Wires:

Our workhorse wires for most endovacular procedures are the Glidewires®:

- Angled non-stiff for crossing severe stenoses including distal cap of some CTO SFA/popliteal arteries.
- Angled stiff for delivering interventional equipment and crossing CTO SFA/popliteal arteries.
- Straight (non-stiff and stiff) to cross CTOs.

Glidewires have a nitinol core and a polyurethane jacket with Tungsten and a hydrophilic coating.

They are available in various diameters (0.011"-0.038") and are used for various applications.

Mesenteric/Renal:

- 0.014"
- 0.018"

Aortic:

- 0.018"
- 0.035"

Iliac:

- 0.018"
- 0.035"

Femoro-popliteal

- 0.014"-0.035"

Infra-popliteal:

- 0.014"-0.018"
- 0.035" wires may be used to cross lesions.

Venous/Pulmonary:

- 0.035"

CATHETERS

Angiographic catheters are available in several different shapes and sizes and are divided into three major categories:

Most procedures require the use of more than catheter. A flush catheter is routinely used to study the anatomy of aorta and its major branches. However, for better characterization of the anatomy and pathology of a particular vascular bed selective catheterization and imaging is usually necessary.

1-FLUSH CATHETERS

Flush catheters are usually designed with side-holes to accommodate large volume of contrast injections, provide high flow and to allow dispersion of contrast without injuring the vessel wall. These catheters are used for angiography of large vessels including the aorta or vena cava. Figure 4 and table 4 list a few different types of flush catheters.

For abdominal aorta, we prefer to use a tennis racquet catheter. For the aortic arch/thoracic aorta, we prefer a non-angled pigtail catheter.

In the vena cava, a TR or non-angled pigtail may work equally well.

2-SELECTIVE CATHETERS

Selective catheters are pre-shaped for engagement of vessels as well as contrast angiography. Specific catheters are used for specific locations and lesions to facilitate selective engagement of a vessel. Figures 5 - 10 and table 4 list some of the various types of selective catheters including their sizes and lengths.

For great vessel angiography via transfemoral approach, a selective catheter that is at least 100 cm in length is required.

For visceral and lower extremity angiography via transfemoral approach, a selective catheter 65 cm in length is usually adequate.

3-SUPER-SELECTIVE CATHETERS

Super-selective catheters are used for crossing tortuous anatomy, allow for exchange of guidewires, and provide a conduit for injection of contrast and saline/medication. Table 5 lists the different super-selective catheters.

Most super-selective catheters are at least 100 - 150 cm in length. Super-selective catheters can be introduced via selective catheters.

TABLE 2 - 0.035" BASIC & EXTRA SUPPORT WIRES

	Stiffness	Length (cm)	Tip Shape	Comments
ACCESS/NON-STIFF WIRES				
Starter® wire	Soft	150-260	J	Not for crossing severe stenoses
Emerald®	Soft	300	J	Not for crossing severe stenoses
Wholey®	Soft	145, 260, 300	Floppy straight	Not for crossing severe stenoses
Bentson	Soft	145, 180, 260	Straight	Not for crossing severe stenoses
Glidewire®	Soft	80, 150, 180, 260	Straight, angled	Can be used for crossing severe stenoses(angled tip) or CTOs (straight tip)
SUPPORT/STIFF WIRES				
Glidewire®	Stiff	150-260	Straight, angled	Can be used for crossing CTOs
Amplatz Super Stiff®	Stiffer	150-260	Straight, J	Support wire for delivery of interventional devices
Amplatz Extra-Stiff	Stiffer than Super Stiff®	150-260	Straight, J	Support wire for delivery of interventional devices
Amplatz Ultra-Stiff	Stiffer than Extra-Stiff®	260	Straight, 3 cm curve	Support wire for delivery of interventional devices
Lunderquist®	Stiffer than Ultra Stiff®	260, 300	Straight, 7,7 mn curve	Support wire for delivery of interventional devices
Rosen®	Stiffer than Extra-Stiff®	80-260	1.5 mm curve	Support wire for delivery of interventional devices
Meier®wire	Stiffer than Extra-Stiff®	260	Straight	Support wire for delivery of interventional devices

TABLE 3 - CROSSING/INTERVENTIONAL WIRES			
	Tip load (g)/ Stiffness	Length (cm)	Comments
		0.014"	
Choice PT® Floppy (BSC)	Soft	300	Small vessel tight stenoses & CTOs
Fielder XT® (tip 0.009") (Abbott Vascular)	1.2	300	Small vessel tight stenoses
ProWater® (Abbott Vascular)	1.1	300	Small vessel tight stenoses
Regalia XS® (Asahi Intecc)	1.0	180, 300	Small vessel tight stenoses,
Confianza PRO® (9 & 12; tip 0.009") (Abbott Vascular)	9.3, 12.4	300	Small vessel CTOs
Astato 20® (Asahi Intecc)	20	180, 300	Small vessel CTOs
		0.018"	
Astato 30® (Asahi Intecc)	30	180, 300	Small vessel CTOs
Glidewire® (Terumo)	NA	80-260	Angled & straight 0.011" - 0.038"
Treasure® Floppy (Asahi Intecc)	4.0, 12	190, 300	0.014", 0.018"
		0.035"	
Glidewire® (Terumo)	NA	150-260	Workhorse wire for aorto-iliac and femoro-popliteal interventions. Angled and straight. Standard (non-stiff) and stiff.

TABLE 4 - FLUSH & SELECTIVE CATHETERS			
	Catheter size (Fr.)	Length (cm)	Guidewire compatibility
		FLUSH	
Straight	4-6	65-110	0.035", 0.038"
Pigtail	4-6	40-110	0.035", 0.038"
Tennis Racquet	4, 5	65	0.035"
Contra	4, 5	65, 90, 110	0.035", 0.038"
Omni-flush	4, 5	65, 90, 110	0.035", 0.038"
		VISCERAL	
Judkins Right (JR;3.5,4,5,6)	4-9	100, 125	0.038"
IM	5-7	45	0.018", 0.038"
Omni-Sos	4-6	45, 65	0.018", 0.038"
Simmons (1-4)	5.5-8	40	0.038"
Cobra (1,2)	5-8	45, 65, 90	0.038"
Yashiro	5	65	0.038"
		CEREBRAL	
Headhunter (1-3)	4-6	65-140	0.035", 0.038"
Hockeystick	4-7	90, 100	0.038"
JB (1-3)	4, 5	65-130	0.035", 0.038"
Simmons (1-4)	4, 5	65, 100	0.035", 0.038"
Vertebral	4-6	65-125	0.035", 0.038"
Vitek	5	100, 125	0.035", 0.038"
		MULTIPURPOSE	
DAV	4, 5	100	0.038"
KMP	4, 5	40-100	0.038"
MPA	4-6	50-100	0.035", 0.038"
MPB	4, 5	65, 100	0.035", 0.038"
TEGT	5	65	0.038"

TABLE-5 - SUPER-SELECTIVE CATHETERS			
	Catheter Size (Fr.)	Total length (cm)	Wire compatibility (in)
STRAIGHT			
Corsair® (Asahi Intecc)	2.6	135, 150	0.014”
Glidecath®(Terumo)	4, 5	65, 100, 120	0.038”
Quick-Cross® (Spectranetics)	2.6-5	65, 90, 135, 150	0.014”, 0.018”, 0.035”
Slip-Cath® (Cook)	4, 5	65, 100	0.038”
Slip-Cath XL® (Cook)	4	150	0.035”
FineCross MG® (Terumo)	2.6	130, 150	0.014”
Progreat® (Terumo)	2.4, 2.7	110, 130	0.018”, 0.021”
Progreat Ω (Terumo)	2.8	110, 130	0.021”
Stride® (Asahi Intecc)	2.2, 2.6	105-150	0.014”, 0.018”,
ANGLED			
Glidecath® (Terumo)	4, 5	65, 100, 120	0.038”
Quick-Cross Select (Spectranetics)	4/5	65, 90, 135, 150	0.014”, 0.018”, 0.035”
Slip-cath (Cook)	4, 5	40-125	0.038”

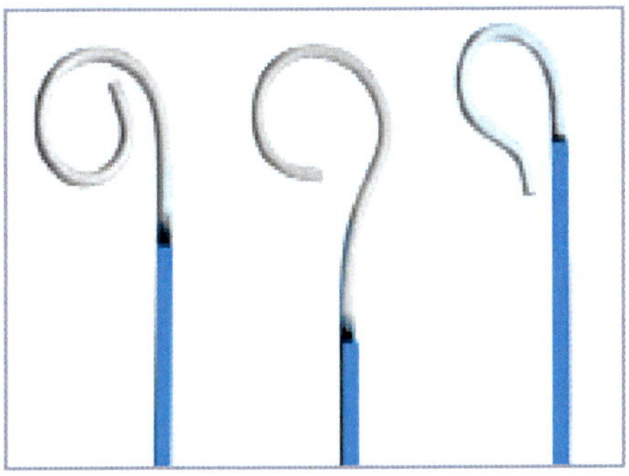

Figure 4.
Flush catheters.
Pigtail, tennis racquet (TR) and an Omni-flush.

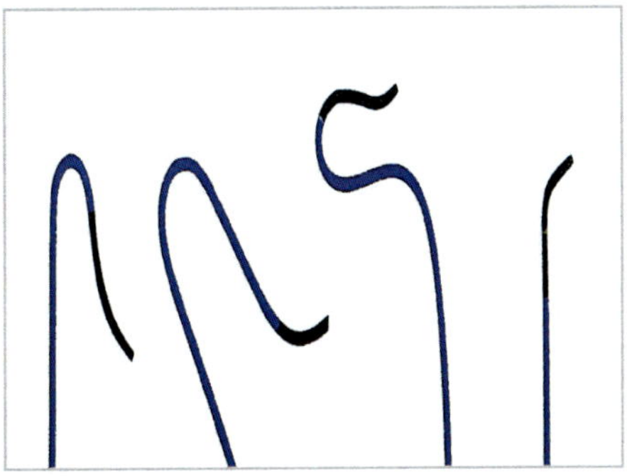

Figure 5.
Cerebral catheters.
Simmons 1, Simmons 2, Vitek and Vertebral catheter (left to right).

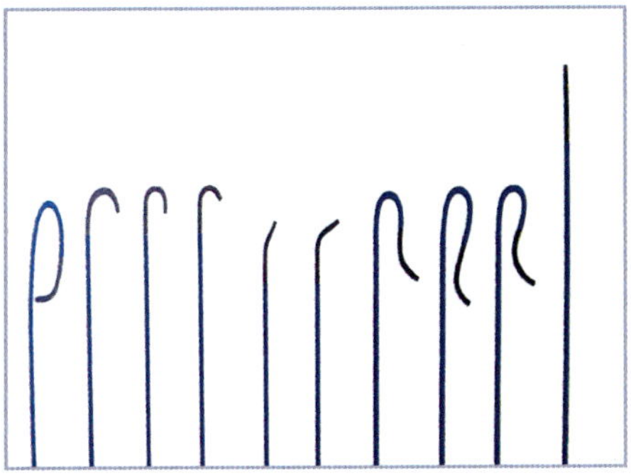

Figure 6.
Visceral catheters.
RH, Rim, TC, TC-BNK, TEGT, VERT, VS, VS1, VS2 and straight catheters (left to right).

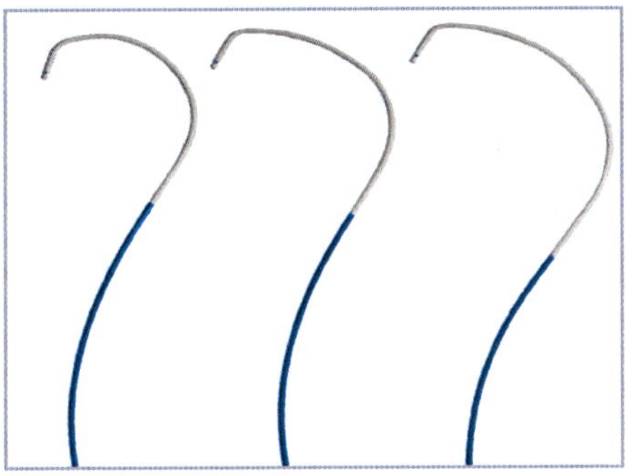

Figure 7.
Cobra catheters.
Cobra 1-3 (left to right).

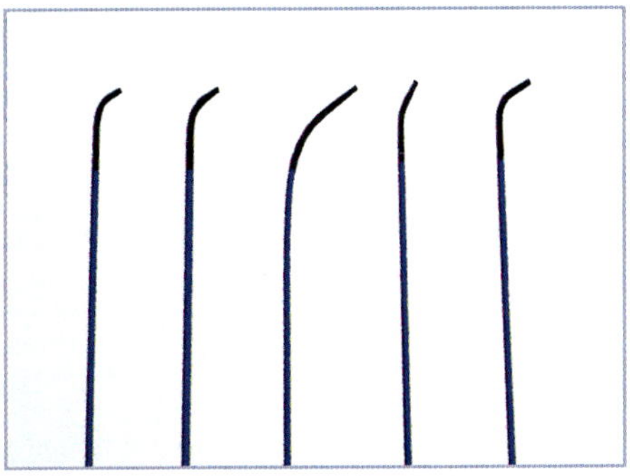

Figure 8.
Multipurpose catheters.
DAV, KMP, MPA, TEGT, VERT (left to right).

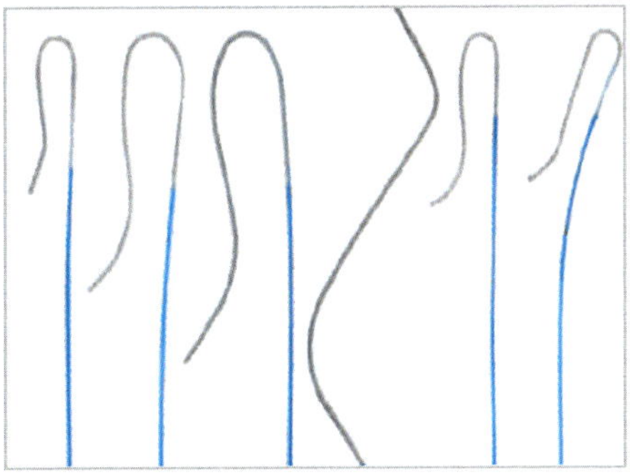

Figure 9.
Simmons catheters.
Simmons catheters 1-4,; modified Simmons 1 and modified Simmons 1.5 (left to right).

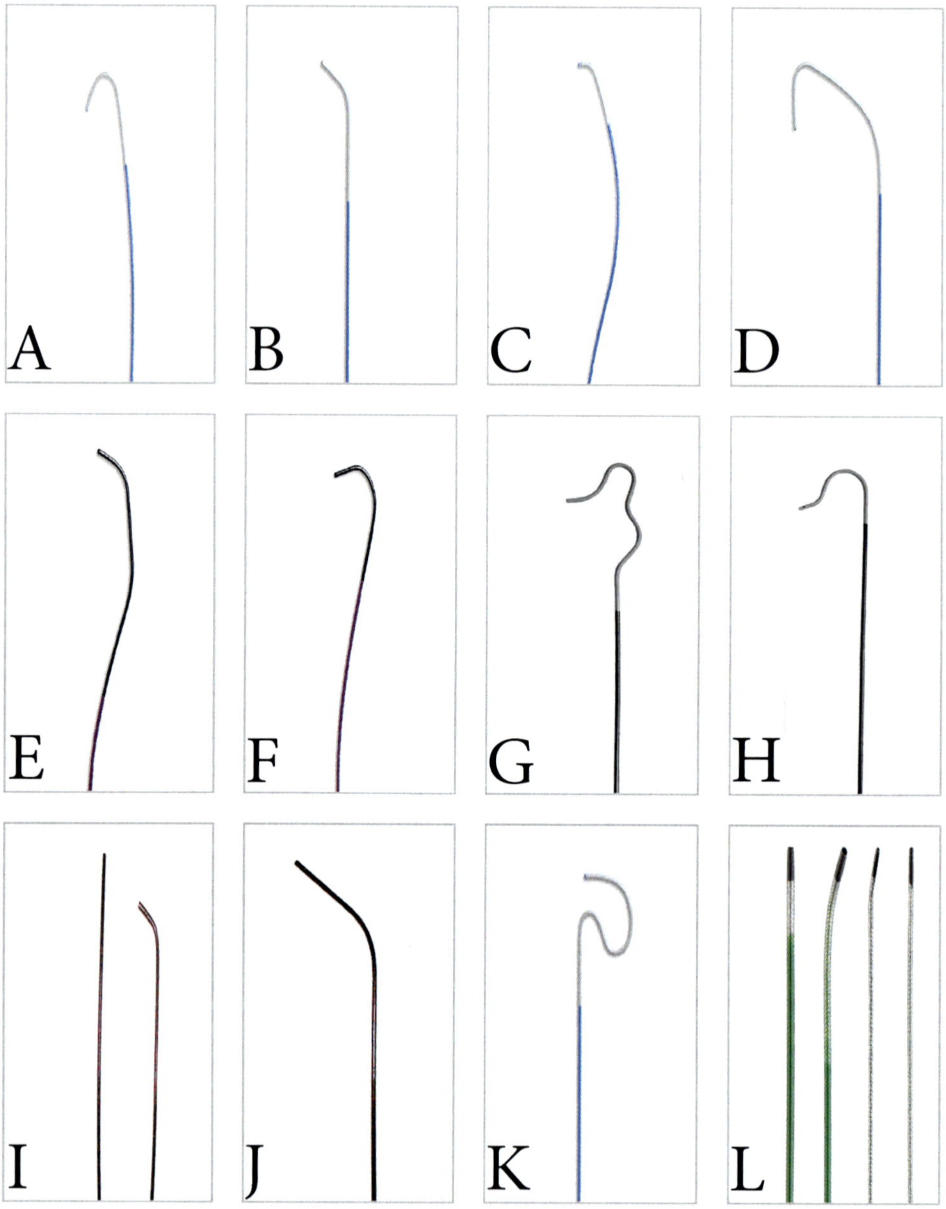

Figure 10.

Selective catheters.

A. IM. B. Vert. C. JR4. D. Renal double curve (RDC).
E. Headhunter. F. Hockeystick. G. Mickelson. H. Shepherd's Crook.
I. Glidecath straight and angled. J. Multipurpose Glidecath flush with a 3 cm tip. K. Mani.
L. CXI catheters.

3

INTERVENTIONAL EQUIPMENT

BALLOONS

For the sake of classification, the balloons used in endovascular therapy can be divided into the following categories:

- Angioplasty balloons
- Drug delivery balloons
- Specialized balloons
- Occlusion balloons (aortic)

ANGIOPLASTY BALLOONS

- Depending on the target site, angioplasty balloons used in endovascular interventions are available for delivery over a 0.014", 0.018" and 0.035" guide wire.
- Compliant balloons are used for predilatation.
- Non-compliant balloons are typically used for postdilatation or treatment of resilient stenoses such as fibrotic and heavily calcified lesions.
- Most 0.018" and 0.035" balloons are available only in over-the-wire (OTW) platforms.
- Most 0.014" balloons are rapid-exchange/monorail (RX) platforms.
- 0.014" platform balloons are used typically in the tibial vessels, renal/mesenteric and cerebrovascular (internal carotid and vertebral) arteries.
- 0.018" platform balloons are available in sizes that are mostly used in the tibial and femoropopliteal vasculature.
- 0.035" platform balloons are typically used in the larger vessels such as the aorta, venae cavae, ileo-femoral and arch vessels.

DRUG DELIVERY/IRRIGATING BALLOON

- ClearWay™ balloon (figure 1A; Atrium Medical, Hudson NH) is an example of an infusion balloon that allows local drug delivery.
- It is a micro-porous, non-compliant PTFE balloon that is designed for low-pressure (1-4 ATM) atraumatic inflations in an artery, transiently occluding flow, while a pharmacologic agent is delivered through the surface of the balloon.
- This procedure is referred to as OCI - *O*cclusion, *C*ontainment and *I*nfusion.
- These balloons are used for treatment of arteries with thrombotic burden, delivering potent

anticoagulants such as tPA or glycoprotein IIb/IIIa inhibitors locally.
- The OTW version of ClearWay requires a 7 Fr. sheath for 4.0 - 5.0 mm diameter balloons and an 8 Fr. for 6.0 - 8.0 mm diameter balloons.
- The Rx version of the balloon is 4 - 5 Fr. sheath (6 - 7 Fr. guiding catheter) compatible. The balloon ranges in 1.0- 4.0 mm diameter.

ClearWay:
This balloon is prepared with a 'positive prep' (i.e. by flushing the balloon inflation port with saline).

SPECIALIZED BALLOONS

- PolarCath® (Boston Scientific Corp., Natick MA) is a balloon catheter that is designed to deliver cryotherapy combined with angioplasty (CryoPlasty®) to stenotic lesions.
- This coaxial, dual balloon (figure 1B) fills up with liquid nitrous oxide that rapidly undergoes phase change into gas. This drops the temperature on the surface of the balloon to -10 °C for a 20 second duration.
- In theory, this induces apoptosis of the endothelium thus reducing the likelihood of a neo-intimal response to angioplasty.
- Available in 0.014" and 0.035" compatible sizes, this balloon is used for infra-inguinal intervention.

Angioplasty Balloons:
- Balloons used for PTA are prepared with at least a 50:50 mix of dye/saline to allow rapid inflation and deflation.
- These balloons are seldom inflated beyond nominal pressure, unlike coronary balloons.
- Peripheral balloons have longer shoulders.

AORTIC OCCLUSION BALLOONS

- These compliant balloons are used for achieving transient aortic occlusion during emergencies such as traumatic aortic injury (TAI) and/or aortic or iliac aneurysm rupture.
- They are also used for post-dilating or 'molding' abdominal and thoracic aortic endografts.
- Given their large volumes, these balloons should be prepared and inflated with diluted contrast (25:75 contrast to saline mix). This allows rapid inflation and deflation.
- For inflation/deflation, a 60 cc syringe is adequate and high-pressure inflation devices should be avoided.
- Examples include:
  - ▹ Coda balloon (Cook)
  - ▹ Equalizer (BSC)
  - ▹ Reliant (Medtronic)
  - ▹ Trilobe™ (Gore)

Aortic Occlusion Balloons:
These balloons are also called 'molding' balloons. They are very compliant and are intended for manual inflation only. Their rated burst pressure is usually less than 5 ATM.

Coda:

- This balloon is available in two diameters - 32 mm and 40 mm (figure 1C).
- Balloons of both sizes have a 10 Fr./120 cm shaft and are 0.035" wire compatible.
- A 14 Fr. sheath is required for their introduction.
- The maximum inflation volume is 34 mL for the 32 mm diameter balloon, and 40 mL for the 40 mm balloon.

Equalizer:

- This is a 7 Fr. catheter that requires a 14 Fr. sheath. It is 0.038" wire compatible (figure 1D).
- The balloon is available in two different shaft lengths: 65 cm and 100 cm and is 20-40 mm in diameter.

General Principles:

Do not continue advancing/withdrawing any device if resistance if felt.
Stop and assess the cause of the resistance.
Continued advancement/withdrawal under such circumstances may lead to vessel injury/perforation.

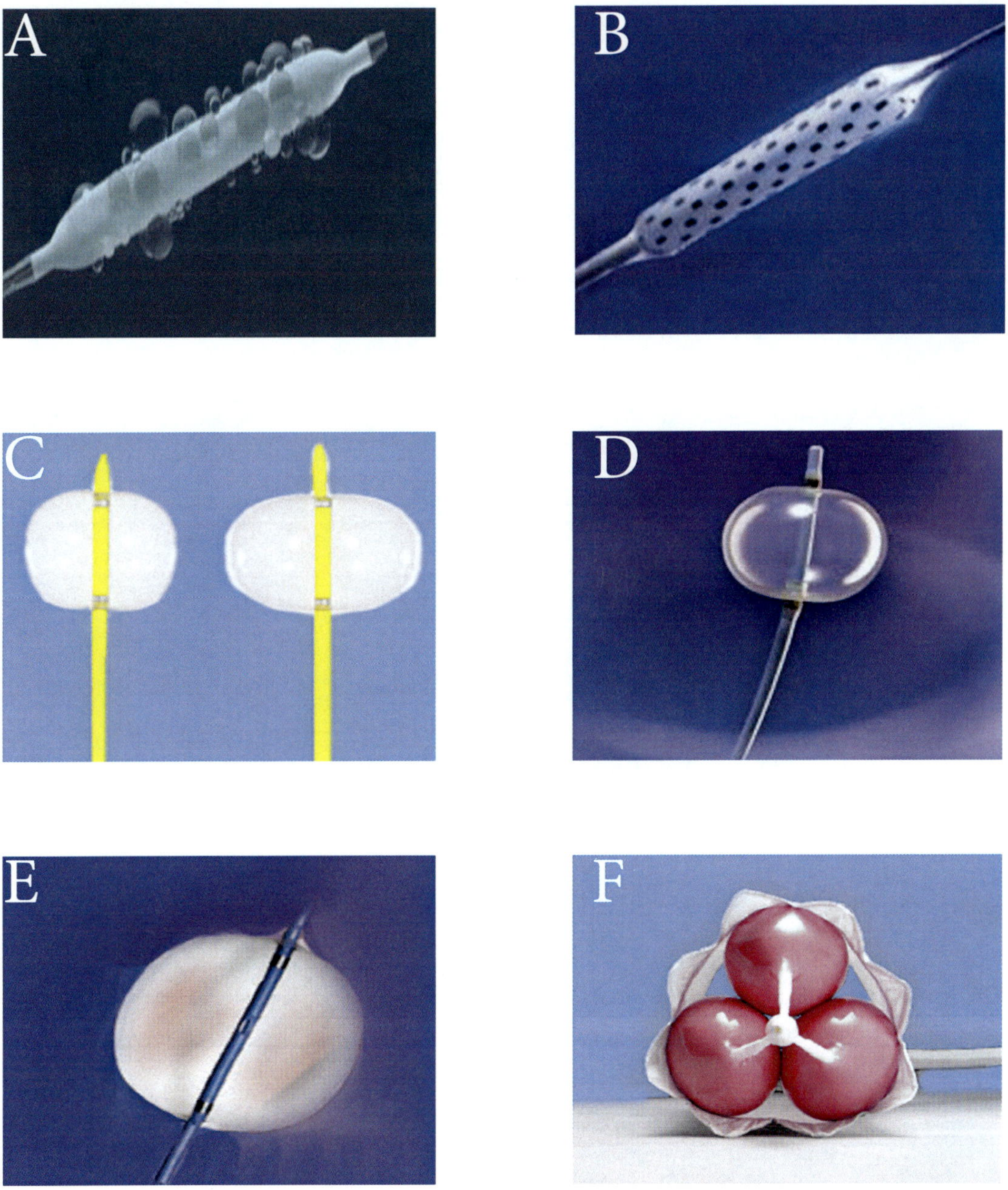

Figure 1.
Various balloon types used in endovascular intervention.

A. ClearWay™ irrigating balloon.
B. PolarCath™ cryoplasty balloon.
C. Coda™32 mm and 40 mm aortic occlusion balloons.
D. Equalizer® aortic occlusion balloon.
E. Reliant™ aortic occlusion balloon.
F. Tri-Lobe balloon (shown inflated inside an aortic endograft).

(Photos courtesy of respective manufacturers)

Reliant:

- This balloon has an 8 Fr./100 cm shaft and requires a 12 Fr. sheath for introduction (figure 1E).
- It is 0.038" wire compatible and its maximal diameter is 46 mm.

Tri-Lobe:

- This is a compliant, tri-lobed, polyurethane balloon (figure 1F).
- This balloon is used exclusively as a 'molding' balloon rather than an occlusion balloon.
- Its unique design does not cause aortic occlusion and in fact permits nearly 80% distal flow. This property allows distal perfusion.
- This balloon is available in two sizes, both 18 Fr. sheath compatible
    - ▹ Small for 16-34 mm luminal diameter vessels.
    - ▹ Large for 26-42 mm luminal diameter vessels.

STENTS

Stents are cylindrical, prosthetic devices with a mesh-like construct made of metal that are typically inserted in diseased (narrowed and sometimes aneurysmal) segments of blood vessels. Stents can, in general, be classified by their method of deployment as either balloon-expandable stents (BES) or self-expanding stents (SES) (figure 2).

BALLOON EXPANDABLE STENTS

- A BES is mounted on a 'carrier' balloon (figure 3B). Expansion of this balloon leads to the expansion of the stent.
- A BES usually reaches its 'nominal' diameter once its carrier balloon is fully inflated.
- BES are usually made of metals like 316 L stainless steel which is highly resistant to corrosion. Other metals such as nickel, cobalt and platinum are also used along with stainless steel.
- BES are usually created in the 'collapsed' state and then mounted on a balloon.
- These stents are more often reserved for locations where precision in placement and radial stiffness are needed and no significant bending or stretching of the vessel occurs on a regular basis, such as:
    - ▹ Ostial common iliac arteries
    - ▹ Ostial/proximal renal/visceral arteries
    - ▹ Ostial innominate and common carotid artery
    - ▹ Ostial/proximal subclavian arteries

Examples of BES:

- 0.035 " wire system:
    - ▹ Omnilink (Premounted)
    - ▹ Genesis (Premounted)
    - ▹ Palmaz XD (Hand-crimped)
    - ▹ Palmaz XL (Hand-crimped)

- 0.018 " wire system:
    - ▹ Palmaz Blue (Premounted)
    - ▹ Herculink (Premounted)

SELF EXPANDING STENTS

- An SES expands spontaneously due to the intrinsic properties of the metal from which it

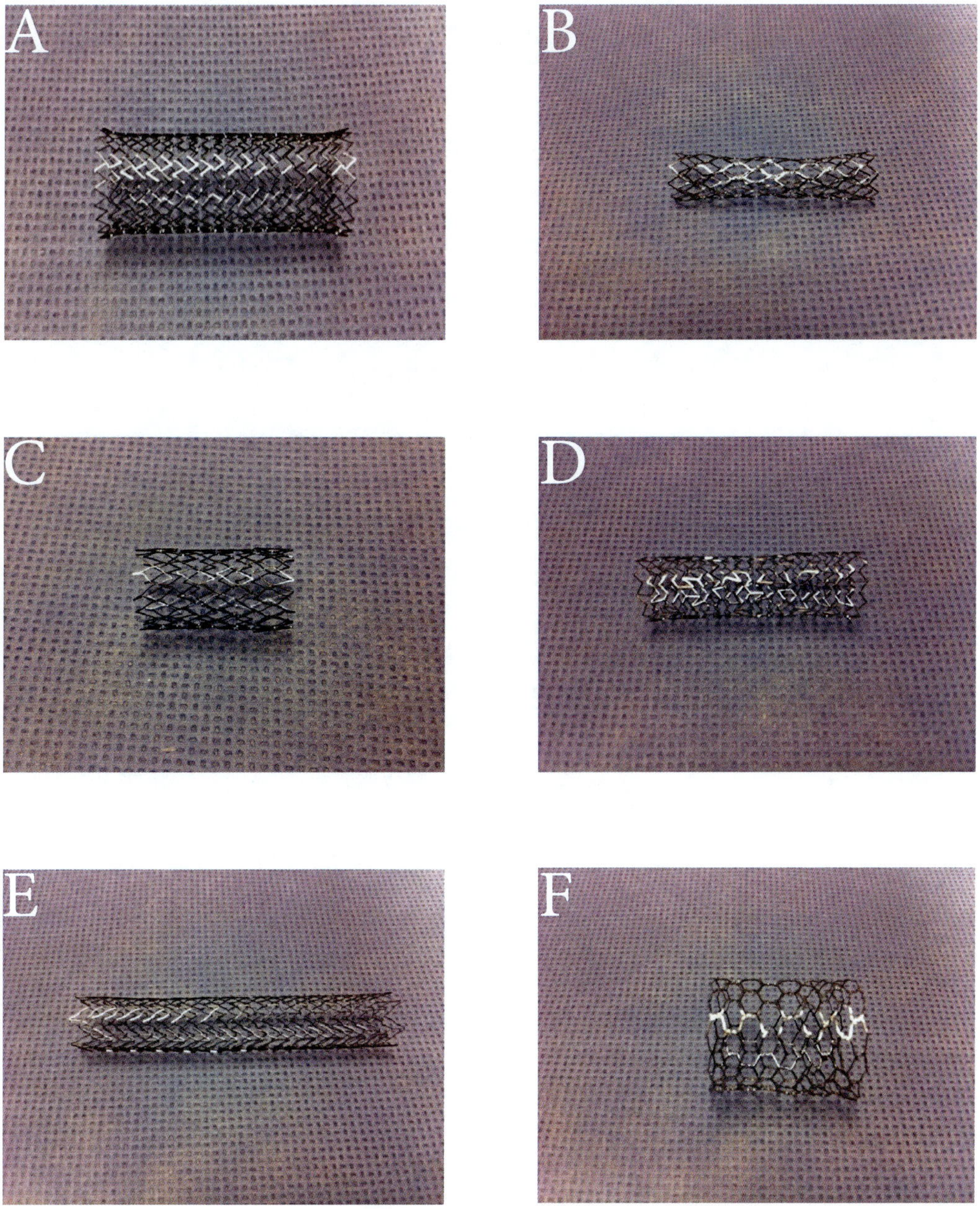

Figure 2.
Various types of stents used in endovascular intervention.

Self-expanding stents:
A. SmartControl
C. Precise
E. Absolute

Balloon-expandable stents:
B. Palmaz Blue
D. Express
F. Genesis XD

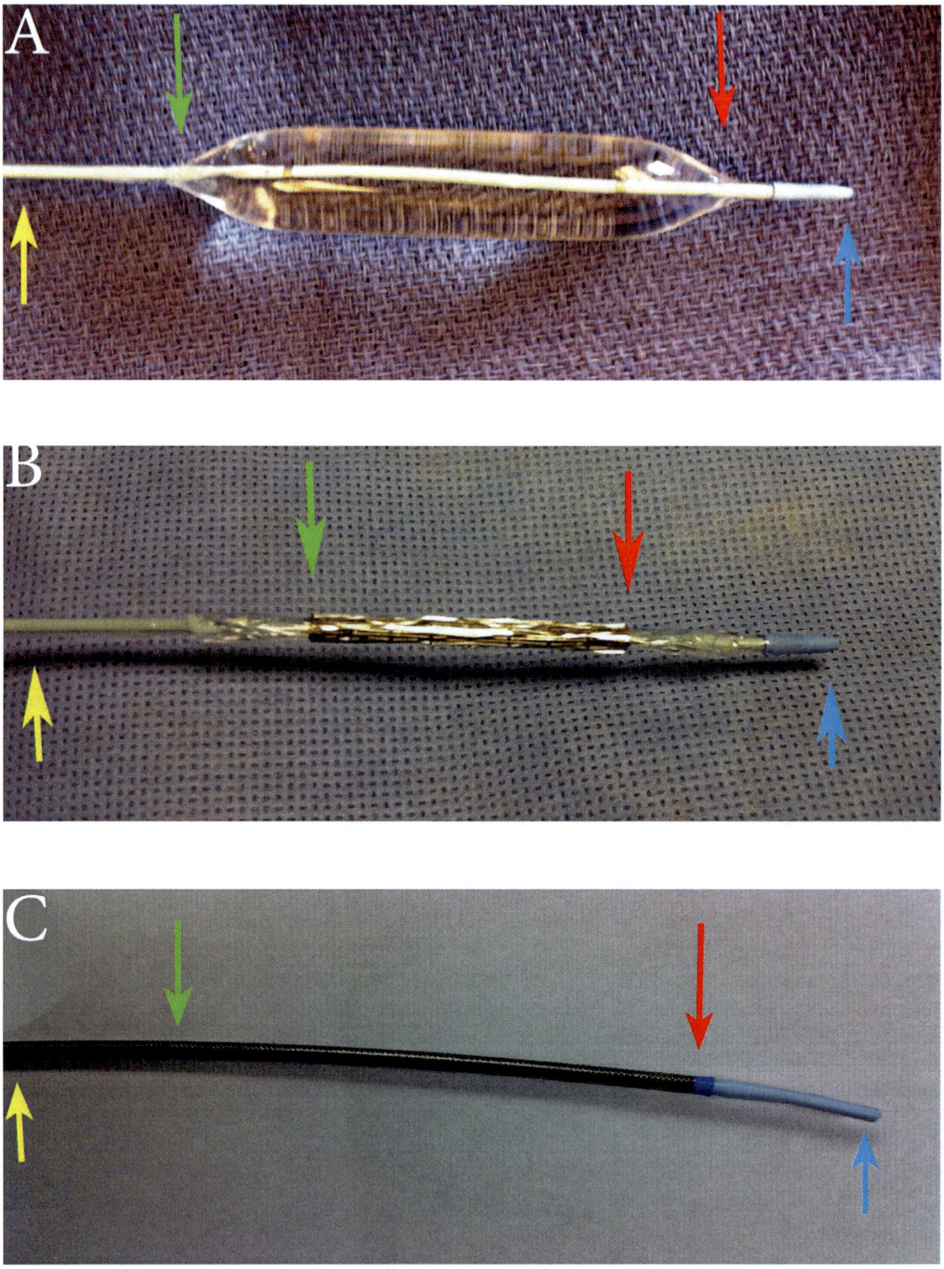

Figure 3.

A. An angioplasty balloon.
B. A balloon-expandable stent.
C. A self-expanding stent.

Yellow arrow: Hub end of delivery catheter.
Blue arrow: Tip end of delivery catheter.
Green arrow: Proximal end of balloon/stent.
Red arrow: Distal end of balloon/stent.

is made to assume its original size and shape. A balloon is not required to expand it (figure 3C).

- It is manufactured in its final expanded state and subsequently crimped inside a delivery catheter. For deployment, the delivery catheter is gradually pulled back to allow the stent to expand and assume its intended diameter.
- SES are almost universally made of nitinol (an alloy of nickel and titanium named for its contents and place of discovery - *Ni*ckel *Ti*tanium-*N*aval Ordnance *L*aboratory).
- Nitinol is well-known for its thermal shape-memory effect. This property allows nitinol materials to be pre-formed in a particular shape and size when heated, and into a transitional shape when cooled. When re-heated, the metal returns to its original shape and configuration.
- After deployment, a nitinol SES can be post-dilated with a balloon. However, the stent cannot be expanded beyond its nominal diameter due to nitinol's above properties.
- If a balloon larger in diameter than the stent is used for postdilatation the stent recoils to its nominal diameter after balloon deflation.
- SES should be oversized by 1-2 mm in diameter in comparison to the luminal diameter at the target site.
- These stents are usually used for more superficial arteries and those vessels that are prone to bending and stretching, such as the:
    - ▹ SFA/popliteal artery.
    - ▹ Internal carotid artery.
    - ▹ Mid/distal subclavian artery.
    - ▹ External iliac artery.

Examples of SES:

- 0.035" wire system:
    - ▹ Absolute (Abbot Vascular, Abbott Park, IL).
    - ▹ Smart Control (Cordis Corp., Bridgewater, NJ).
    - ▹ Zilver (Cook Medical Inc., Bloomington, IN).
    - ▹ Protégé (Covidien, Mansfield, MA).
    - ▹ LifeStent (C.R. Bard Inc., Murray Hill, NJ).

- 0.018" wire system:
    - ▹ Supera (IDEV Tech., Webster, TX).
    - ▹ Viabahn (W.L. Gore & Associates Inc., Flagstaff, AZ).
    - ▹ Zilver.

CHARACTERISTICS OF BES VS. SES

When selecting a stent for a particular lesion, a few qualities of BES vs. SES must be understood.

Radial Strength:

- The radial strength of a stent is defined as the external pressure that a stent is able to withstand without sustaining significant permanent deformation or reduction in lumen size.
- Contrary to common belief, SES have more radial strength than BES (figure 4), primarily due to the properties of nitinol mentioned above. An SES is very likely to return to its original size once the external force is removed.

Radial Stiffness:

- Radial stiffness of a stent is defined as the diameter reduction of a stent caused by the external application of force.
- BES have more radial stiffness than SES.
- BES are thus better suited for certain calcific lesions such as the common iliac and subclavian

arteries.

Delivery Profile:

- The delivery/'crossing' profile of a BES is almost always higher than that of a comparable size SES.
- This is primarily due to the fact that a BES is mounted on a balloon and the profile of the balloon adds to that of the BES.
- Additionally, the struts of an SES can be thinner compared to its BES counterpart due to the intrinsic properties of nitinol.
- The friction between the exposed metal of a BES and the hardened surface of a diseased artery can interfere with smooth delivery to the intended treatment site.

Placement Accuracy:

- BES are usually very precise in terms of their expansion at the site of deployment and tend not to move during this process.
- SES traditionally have some outward (hub-to-tip) motion in them at the time of deployment. This results from the stored potential energy in the stent that manifests as kinetic energy when the stent is 'unsheathed'.
- Hence, BES are preferred in locations where absolute precision is paramount (such as the coronary, ostial renal and ostial iliac arteries).

Stent Dislodgement:

- Stent dislodgement from the delivery catheter is unlikely with SES.
- It occurs only in BES, typically when a stent is advanced through a calcified/tortuous lesion without adequate predilatation or without being covered in a delivery sheath.

Direct Stenting:

- Direct stenting implies the placement of a stent in a lesion without predilatation with a balloon or another modality (e.g. atherectomy).
- This is not a *property* of a particular type of a stent, rather a technique of stent placement.
- Some of the drawbacks of this technique include:
    - ▹ Inability to fully visualize the proximal and distal ends of a severely narrowed lesion during stent placement as stent may become occlusive in the lesion. This may lead to imprecise stent positioning within the lesion.
    - ▹ Inability to deliver the stent successfully across the entire length of a severely stenotic lesion.
- A BES as well as an SES can be placed using direct stenting. However, this technique should be reserved only for BES and that too in select situations.
- SES placement directly in a stenotic lesion does not allow the stent to fully expand at deployment and there is minimal lesion modification. This poorly expanded stent in a lesion with a narrow lumen can 'trap' the delivery catheter of the stent.

Flexibility:

- This attribute refers to the ability of a stent to be delivered past tortuous segments of a vessel and to conform to the shape and configuration of a vessel wall, once deployed.
- The flexibility of a stent depends upon its design, its axial stiffness, bending compliance and strut thickness.
- Since the strut thickness of nitinol stents is usually less compared to BES, the latter are much less flexible than their SES counterparts.

Growth in Stent Diameter:

- BES are generally sized 1:1 to the luminal diameter of the vessel segment where they are

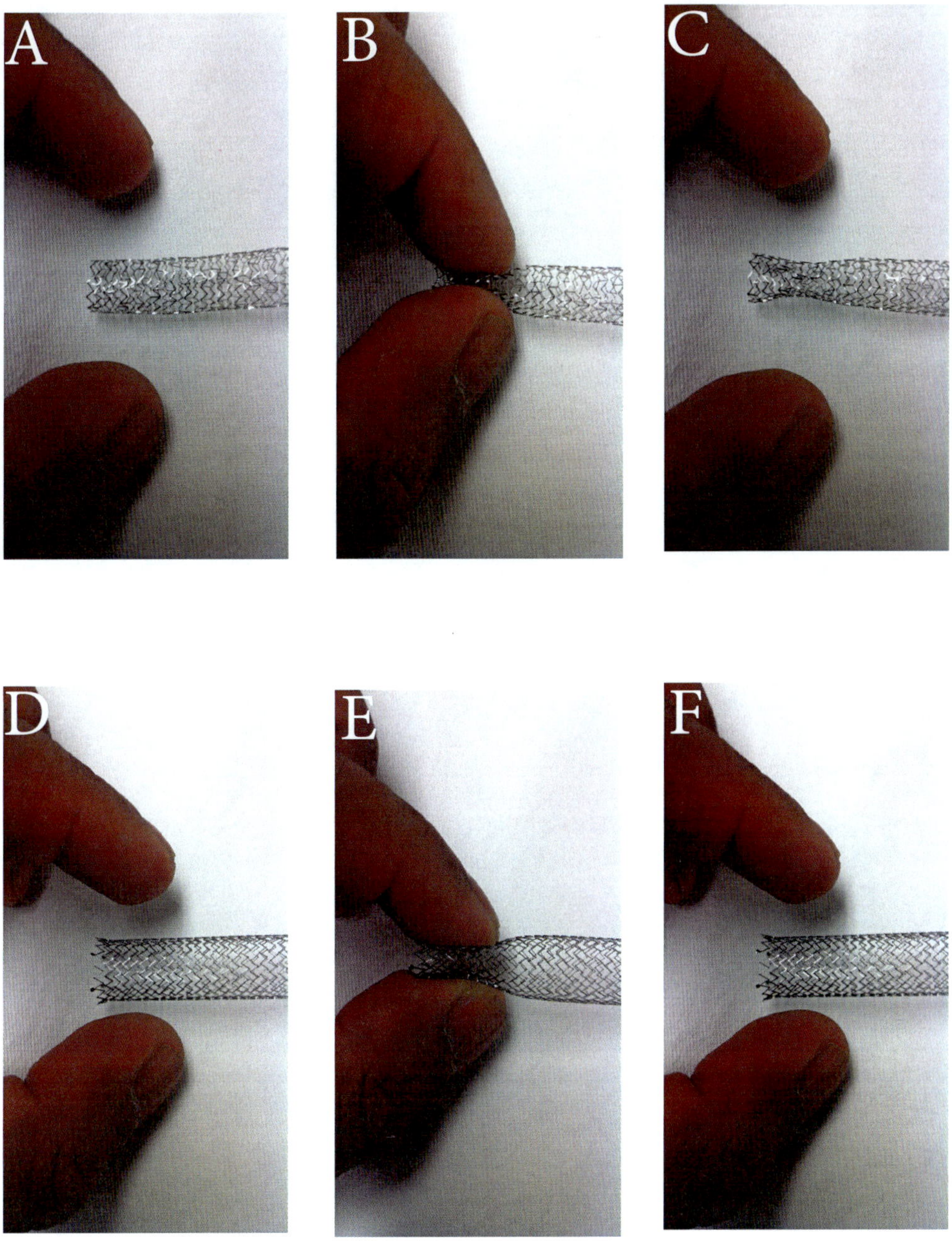

Figure 4.
Comparison of BES vs. SES radial strength in response to extrinsic compression.

A. A BES at baseline (after deployment to nominal diameter).
B. The BES is manually compressed.
C. After the compression is discontinued, the stent is noted to be deformed permanently.
D. A nitinol SES at baseline (after deployment to nominal diameter).
E. Manual compression of the SES.
F. After the compression is discontinued, the SES reverts to its original shape and diameter without any hint of permanent deformation.

deployed. SES are traditionally over-sized by 1-2 mm to allow for better apposition and expansion since a balloon is not used at deployment.
- Once deployed, a BES does not spontaneously 'grow' beyond its final balloon diameter. An SES, however, continues to expand in an attempt to reach its nominal diameter. This process may continue for 4-8 weeks post deployment.

COVERED STENTS/ STENT-GRAFTS

Covered stents or stent-grafts, as they are sometimes called, are stents with a hybrid structure consisting of a metal stent that is lined on the inside, outside or both with a graft material such as ePTFE (figure 5).

BALLOON-EXPANDABLE COVERED STENTS

iCast®:

- This is the only balloon-expandable covered stent (figure 5A) available in the US for use outside the coronary arteries.
- It is made of laser-cut 316L stainless steel that is encapsulated within an ePTFE film (inside and outside).
- It is available in 5 - 10 mm diameters.
- The 5.0 and 6.0 mm diameters require a 6 Fr. delivery sheath whereas the rest require a 7 Fr. delivery sheath.
- All sizes of this stent are 0.035" wire compatible.
- The placement of this stent should follow the same guidelines as those for other BES.

SELF-EXPANDING COVERED STENTS

Viabahn®:

- This is a nitinol stent with an inner ePTFE graft lining (figure 5B).
- This stent is available in 5 - 13 mm diameters and in 25, 50, 100 and 150 mm lengths in the US.
- It is available in a 0.018" (5 - 8 mm diameters) and a 0.035" (all diameters) wire compatible system.
- Viabahn (W. L. Gore & Associates, Inc., Flagstaff, AZ) is available in heparin-bonded and non-heparin-bonded constructs. The former should be avoided in patients who have a known hypersensitivity to heparin including previous heparin-induced thrombocytopenia (HIT) type II.

For femoro-popliteal arteries, Viabahn® is the only FDA-approved stent-graft. The following principles should be observed in cases where these stents are used (in the femoro-popliteal arteries):

- Aggressive debulking of the arteries should be performed with excisional atherectomy (see below) to allow proper stent expansion. This is particularly true for ostial SFA.
- Avoid leaving uncovered any balloon-/atherectomy-treated segments. Restenosis at these sites will compromise stent-graft patency.
- If there is ostial SFA disease, debulk and cover it with the Viabahn. If the ostium restenoses significantly, it can compromise the patency of the entire vessel/graft.
- Avoid over-sizing the stent-grafts. This can lead to 'pleating' of the fabric within the stent, promoting flow retardation and thrombosis.
- Avoid using in the mid/distal popliteal artery (even in cases of popliteal artery aneurysm) if the distal runoff is sub-optimal (less than 1 continuous vessel).

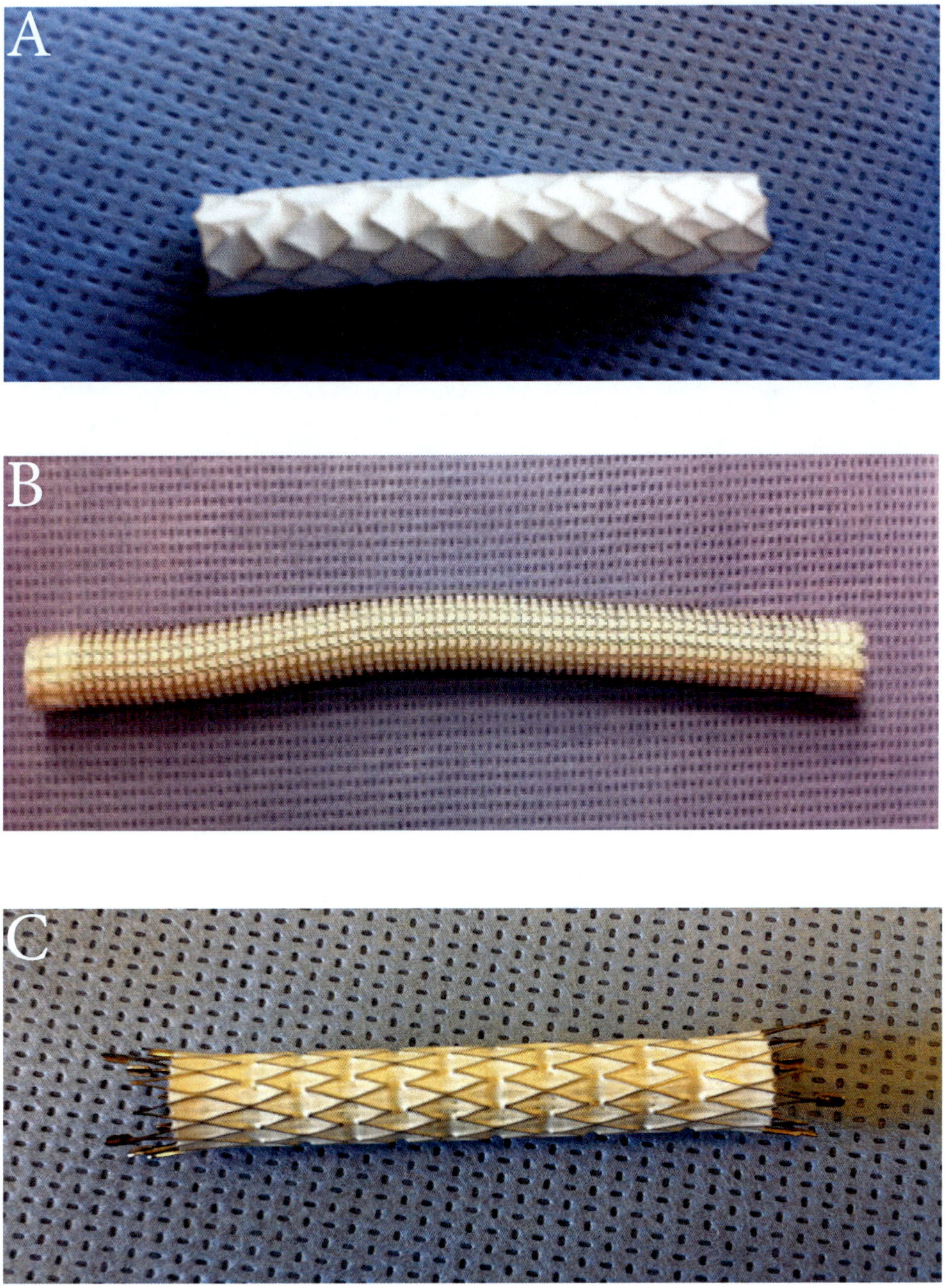

Figure 5.
Covered Stents.

A. An iCast balloon-expandable covered stent. The stainless steel stent is encapsulated within a film of ePTFE.
B. A Viabahn self-expanding covered stent. There is a nitinol stent on the outside of an ePTFE lining.
C. A Fluency Plus self-expanding covered stent. The terminal struts are exposed and the rest of the nitinol stent is encapsulated within two layers of ePTFE.

Fluency® Plus:

- This is a nitinol stent (based on Luminexx design; C.R. Bard Inc., Murray Hill, NJ) encapsulated within two layers of ePTFE (figure 5C).
- It is available in 40 - 80 mm lengths and 6 - 10 mm diameter.
- This stent requires an 8 - 9 Fr. delivery sheath.

THE PRINCIPLES OF COVERED STENT DEPLOYMENT

- Covered stents have a PTFE lining. Oversizing of these stents can lead to pleating/bunching of this lining and predispose to device thrombosis.
- The luminal diameter of the reference segment of the vessel should, therefore, be measured carefully and the stents sized almost 1:1 to the vessel. Use of a predilatation balloon is usually adequate for accurate estimation. Quantitative angiography, marker wires, marker catheters or IVUS can all be used for this purpose if necessary.
- Moisten the stents in saline prior to loading on the delivery wire.
- To avoid contamination, avoid touching the stent until just prior to advancing it through the diaphragm of the sheath.
- The length of the delivery wire should be at least twice that of the shaft of the stent delivery system. As a rule, a 260 cm long wire is a safe bet.
- Do NOT perform a negative prep on the iCast stent prior to it being positioned in the target segment of the vessel. This reduces the chances of stent dislodgement during delivery.
- If possible (avoid positioning across any side branches), position the proximal marker of the stent 1 cm proximal to the diseased segment to be treated.
- For an iCast keep the balloon inflated at least for 30 seconds prior to deflation.
- When two or more overlapping /telescoping stents are placed, the smaller diameter stent should be deployed first. The larger stent should be placed within this smaller stent, overlapping by at least 1 cm.
- Before placing the second stent, the first stent should be adequately post-dilated.

AORTIC STENT GRAFTS

- Aortic stent grafts are self-expanding covered stents that are designed primarily to treat aortic aneurysmal disease.
- For the abdominal aorta, these grafts primarily consist of a bifurcated modular graft system (figure 6A-C) with certain exceptions (e.g. in case of an aorto-uni-iliac configuration).
- For thoracic aorta, these grafts consist of a non-bifurcated tube graft (figure 6D).
- Fenestrated grafts (with perfusion capability to side branches) are not yet widely available in clinical practice in the US.
- These grafts are available in different diameters and lengths and their selection is based on careful review of the patient's anatomy on pre-operative imaging in the form of a CTA typically.

ATHERECTOMY DEVICES

Atherectomy involves percutaneous debulking, and in some cases removal, of atherosclerotic plaque from the arteries of interest. In non-coronary intervention, this modality is used almost exclusively for the arteries of the lower extremities.

Four types of atherectomy devices are currently used for vascular intervention:

- Orbital atherectomy.

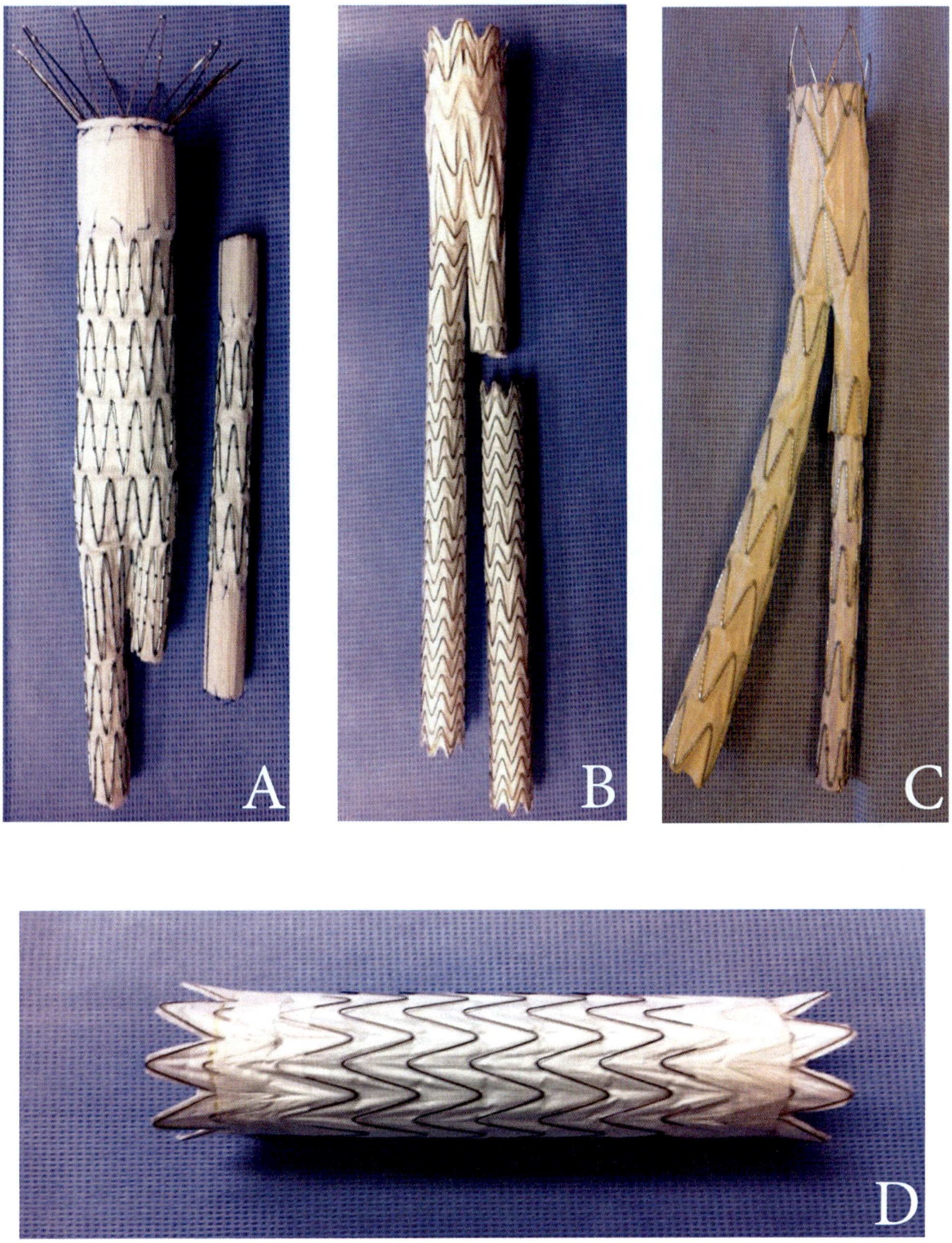

Figure 6.

Aortic endografts.

A. Zenith® (Cook) abdominal aortic bifurcated endograft and contra-lateral limb.
B. Excluder® (Gore) abdominal aortic bifurcated endograft and contra-lateral limb.
C. Endurant® (Medtronic) abdominal aortic bifurcated endograft and contra-lateral limb.
D. Gore TAG thoracic aortic endograft.

(Photos courtesy of respective manufacturers)

- Rotational atherectomy.
- Laser atherectomy.
- Directional atherectomy.

ORBITAL ATHERECTOMY

- This refers to the use of an atherectomy device in which the 'bur' (referred to as a 'crown' in this case) rotates in an 'orbital' fashion along with the wire rather than just rotating axially around the wire (as in rotational atherectomy).
- The diameter of the 'orbit' is variable and depends upon the diameter of the 'crown' and the speed at which it rotates.
- With increasing speed, more centrifugal force is generated that leads to a bigger orbit and therefore an increased luminal gain.
- This system is used for treatment of obstructive disease in infra-inguinal arteries.

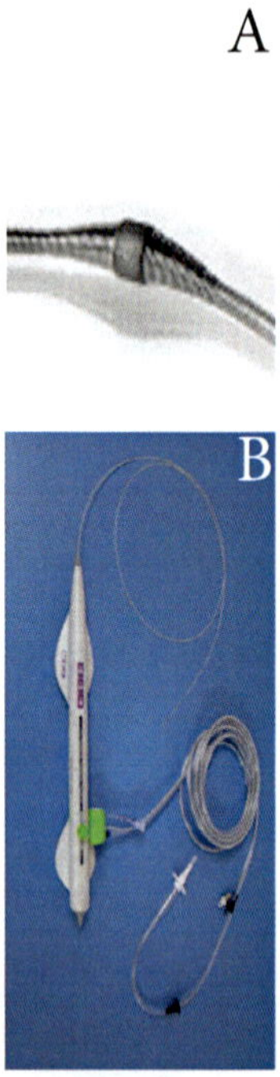

Figure 7. **Orbital atherectomy.**

A. Classic crown.
B. Driveshaft of Diamondback device.

The device (figure 7) used for this intervention is referred to as Diamondback 360° (DB 360) and/or Predator 360° (Pred 360). Its characteristics include:

- An eccentrically mounted 'crown' (figure 7A) that is diamond-coated.
- This crown rotates at speeds varying from 60,000 to 200,000 rotations per minute (rpm) creating centrifugal forces that press the crown against the lesion or plaque.
- The diamond-grit coating on the orbital atherectomy crown "sands" the plaque as the device comes into contact with the vessel wall.
- Microscopic particulate matter is generated from sanding the plaque and is continuously washed away in the blood stream.
- The system operates over a proprietary 0.014" guidewire (ViperWire™) with either a 0.014"/0.017" tip.
- This is a stainless steel wire with a silicone coating and a shapeable, radio-opaque spring tip.
- The device cannot be used over a regular 0.014" wire.

ROTATIONAL ATHERECTOMY

Rotational atherectomy in the peripheral arterial tree is carried out using a system known as Jetstream® Navitus and Jetstream® G3 SF.

- This device consists of rotating, expandable cutting blades/"flutes" at the tip of the catheter used to debulk the lesion.
- The device has two separate lumens that allow simultaneous infusion of saline as well as aspiration of debris from within the arterial lumen.
- This active aspiration is a unique feature of this atherectomy device and is carried out through a port in the catheter tip, reducing the potential for distal embolization. The aspirate is collected in an external bag on the console.
- The device allows for two-stage 'cutting' with expandable blade technology - 'blades down' and 'blades up' (figure 8).
- This system is used for treatment of obstructive disease (plaque, calcium, thrombus) in infra-inguinal arteries.

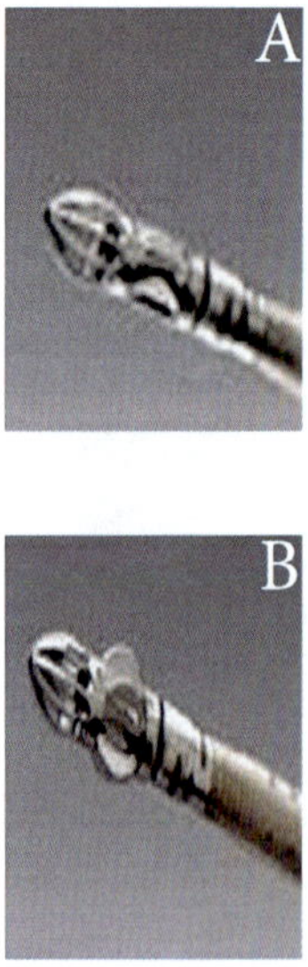

Figure 8.

Jetstream

with blades 'down' (A) and 'up' (B).

EXCIMER LASER ATHERECTOMY

The only excimer laser system for use in arterial photoablative therapy is the

CVX-300 system marketed by Spectranetics (Colorado Springs, CO).

- The system works by imparting a UV light pulse for 135 billionth of a second (135 nanosecond).
- In theory this breaks billions of molecular bonds in the targeted tissue and has photothermal effects leading to ablation of the target tissue.
- Tissue ablation occurs on direct contact without a rise in temperature of the surrounding tissue.
- Each pulse of energy can remove a tissue layer of 10 μm.
- This system is used for treatment of obstructive disease in infra-inguinal arteries.

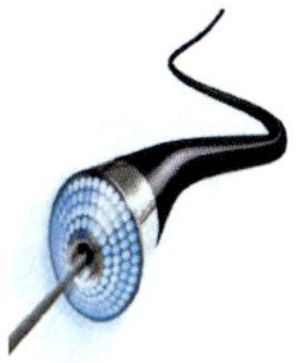

Figure 9. **Spectranetics TurboElite™ catheter.**

The catheter used to deliver this therapy is named Turbo Elite™ (figure 9).

DIRECTIONAL/EXCISIONAL ATHERECTOMY

This technique of plaque excision is carried out by a device that has an eccentric blade, coming in contact with only a segment of the arterial wall in a single longitudinal plane.

- The cutting assembly consists of a circular, rotating, high-speed blade, at the eccentric contact point, that is contained within a tubular housing.
- The blade rotates when the cutter is actuated and excises plaque, 'packing' it in a nose-cone at the distal end of the catheter, so that it can be removed.
- A handheld, battery-driven unit, the Cutter Drive, powers the unit.
- High speed cutting blades 'shave' a ribbon of plaque that is collected in the catheter nose-cone with each pass of the catheter through the lesion.
- The blade is redirected manually in all quadrants of the vessel lumen.
- It is available in two iterations i.e. SilverHawk™ and TurboHawk™ (Covidien, Mansfield MA) (Figure 10).

CHRONIC TOTAL OCCLUSION DEVICES

PROXIMAL CAP ENTRY DEVICES

Frontrunner:

- This is a device that is designed to cross CTO's with the intent of staying intraluminal, by penetrating the proximal cap using blunt dissection.
- The device has a blunted actuating tip that is opened and closed by manipulating a lever at the proximal end.
- This repetitive opening and closing of the tip results in micro-dissection.
- This device is introduced through a 4 - 4.5 Fr. catheter (the proprietary MicroGuide catheter is 4.5 Fr. but the device can be used through a 4 Fr. Glidecath). The catheter supports advancement of the device.
- The Frontrunner does not have a wire lumen.
- The device is available in two shaft lengths- 90 cm and 140 cm.
- To use this device effectively:
  - ▹ Advance through a catheter with the tip closed, to the proximal cap of a CTO and push against the cap.
  - ▹ Open the device and push forward. Advance the catheter over the device.
  - ▹ Close the device and retract nearly 1 cm. Repeat the above steps until across the distal cap.
  - ▹ Remove the device, confirm intra-luminal presence and substitute the device with a

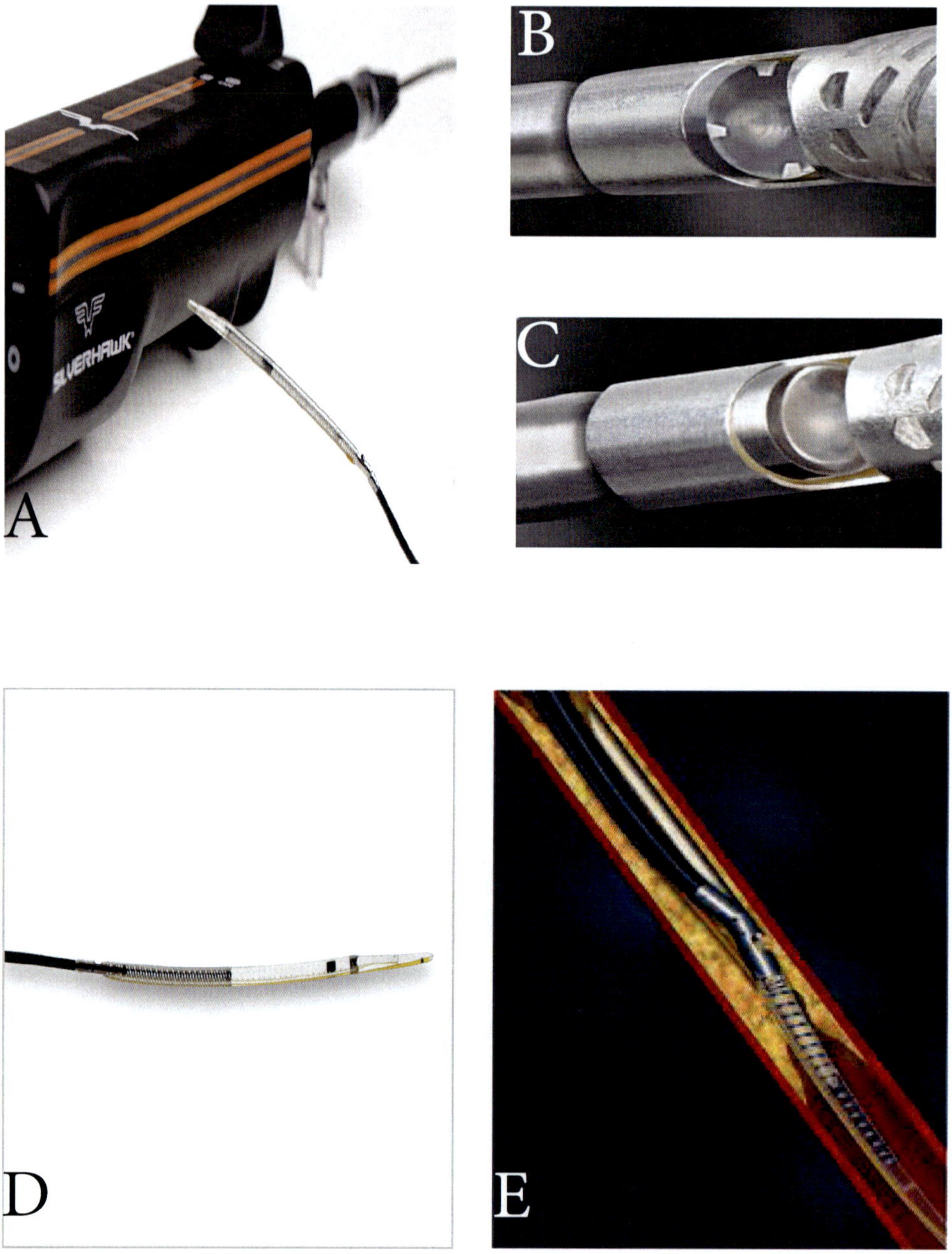

Figure 10.
Directional/Excisional atherectomy.

A. The Cutter Drive with the SilverHawk® catheter 'plugged' into it.
B. A zoomed in image of the SilverHawk Calcium cutter.
C. A zoomed in image of the SilverHawk smooth cutter.
D. The distal end of the SilverHawk catheter consists of the cutter assembly and a nosecone to collect the plaque that has been 'cut'.
E. A schematic of the SilverHawk in action. A 'jog' in the catheter deflects the cutter window to contact the vessel wall. The plaque that is excised is collected passively in the nosecone of the catheter.
(Photos courtesy of the manufacturer)

guide wire such as the ASGW.
- ▹ Proceed with final intervention.

TruePath™:

- The TruePath™ device (Boston Scientific Corp.) consists of a wire assembly with a pre-loaded torquer, a motor housing and an electronic control unit.
- The wire assembly consists of a 0.018" wire with a diamond-coated tip, the Active Tip.
- When the device is activated, the Active Tip rotates at 13,000 rpm to enable 'drilling' through CTOs.
- The device is 165 cm long and can be connected to an extension wire that increases the working length to over 300 cm, enabling catheter exchange.
- The distal tip is shapeable and transmits a 1:1 torque.
- The electronic control unit has a speaker for audible feedback and three LED indicators for visual feedback.
- To use this device, a wire is first advanced through an appropriate size catheter to the proximal cap of the occlusion. The wire is removed and the TruePath is inserted so that the tip comes in contact with the proximal cap.

Crosser™:

- The Crosser™ (C.R.Bard) is a device that converts AC power into mechanical vibrational force transmitted to the tip of the device, to help penetrate and cross a CTO.
- The device is available as Crosser catheter 14 (Rx, requires a 0.014" wire, 146 cm length, 6 Fr. guiding catheter compatible, 1.1 mm tip profile) or the Crosser catheter LP (does not require a wire, 146 cm length, 5 Fr. guiding catheter compatible, 0.6 mm tip profile).
- The device is connected to an electronic 'Generator' through a high-frequency 'Transducer.'
- It is activated with a 'Foot-switch.'
- Continuous irrigation with chilled saline (45 F) is required for device activation.
- It is advanced to the proximal cap through a support catheter. The proprietary catheter is called the Steward™.

Viance™:

- The Viance™ (Covidien, Mansfield MA) is a CTO crossing catheter with a hollow-core for a 0.014" wire.
- The catheter has an atraumatic, rounded distal tip with a 3 Fr. (0.038")profile.
- It is compatible with a 5 Fr. sheath.
- The shaft has a 150 cm working length.
- The catheter is advanced while being spun using bi-directional rotation.
- Once across the occlusion, the catheter is exchanged out for a 0.014" wire and the stenosis treated in the desired fashion.

Wildcat:

- The Wildcat (Avinger, Redwood City, CA) is a 0.035" wire compatible, over-the-wire device that has a 2 mm crossing profile and requires a 6 Fr. sheath.
- It has a shaft length of 110 cm.
- The tip of the catheter is blunt but has deployable 'Wedges' with spiral grooves.
- The proximal end has a 'Slider' that controls the deployment and retraction of the Wedges and deflection of the tip.
- A Tip indicator shows the position of the Wedges and whether the tip is deflected.
- A 'Rotator' controls the rotation of the tip independent of the catheter.
- It is designed for recanalization of femoro-popliteal arteries.
- Smaller 0.014" wire (5 Fr. sheath) compatible iterations of this device are also available as the Kittycat (1.5 mm crossing profile, 140 cm length) and Kittycat2 (1.7 mm crossing profile, 150

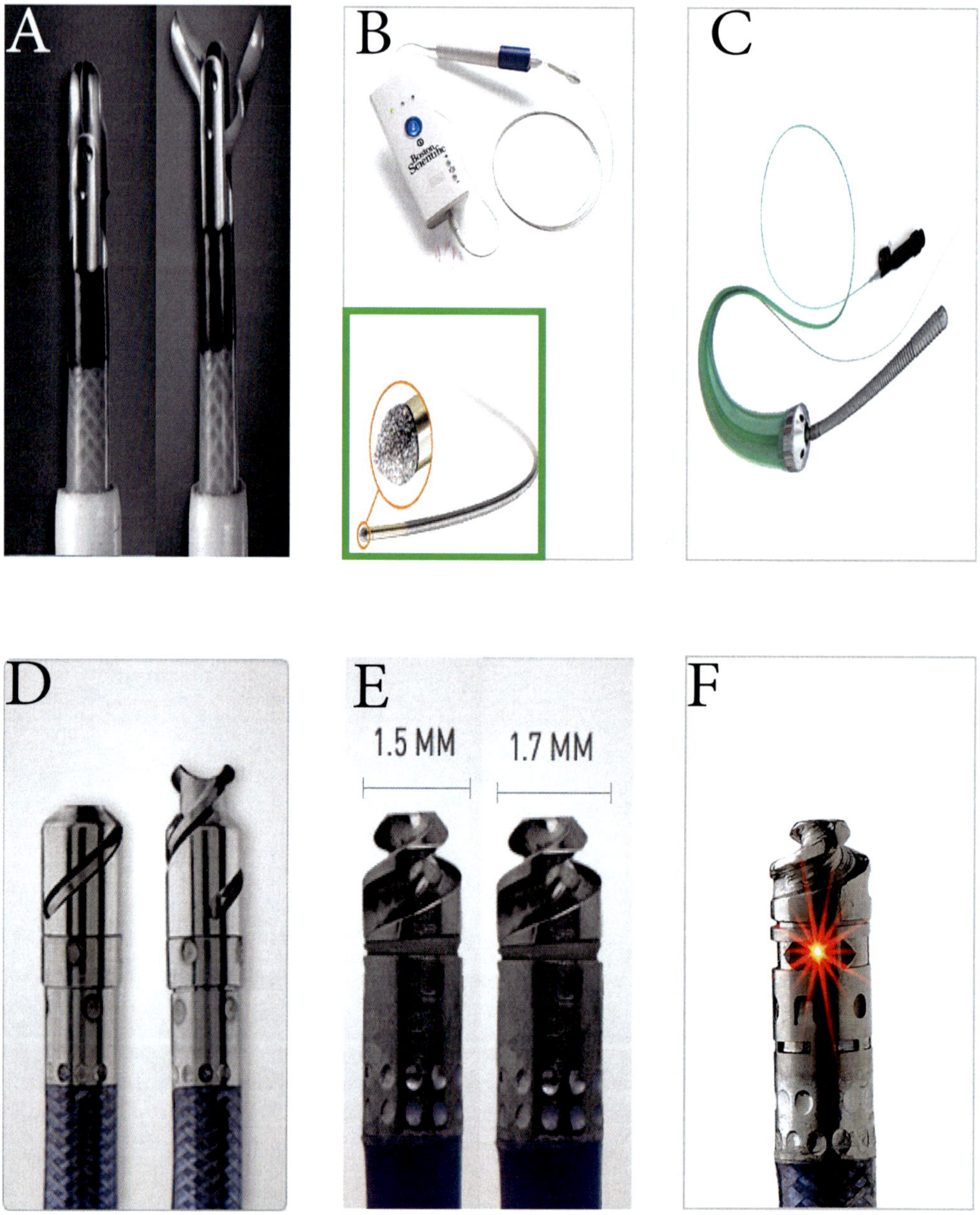

Figure 11.
Proximal Entry CTO Crossing devices.

A. The FrontRunner® with closed and open 'jaws'.
B. The TruePath™ device. The tip of the wire is shown upclose in the green box.
C. The Crosser™.
D. The Wildcat, without and with deployed 'wedges'. The diameter of the WildCat tip is 2 mm with undeployed 'wedges'.
E. The Kittycat and the KittyCat2.
F. The Ocelot (with OCT technology) with a 2 mm crossing profile.

(Photos courtesy of respective manufacturers)

cm length).

- The Ocelot is the first-ever CTO crossing catheter with a built-in real-time optical coherence tomography (OCT) capability (figure 11F).

DISTAL RE-ENTRY DEVICES

These devices are used to cross a CTO when attempts at crossing the distal cap and re-entry into the true lumen with conventional equipment (i.e. wires/support catheters) have failed.

Outback:

- This re-entry device (Cordis Corporation) is 6 Fr. sheath compatible and has no integrated IVUS.
- It is requires a single 0.014" guidewire.
- The shaft length is 120 cm with hydrophilic coating along the entire length.
- The device has a 22 gauge hollow, curved, nitinol needle near the distal tip.
- A 0.014" wire is advanced in the sub-intimal space, beyond the proximal cap, parallel to the distal reconstituted lumen.
- The device is then advanced under fluoroscopic guidance, over the 0.014" wire.
- When the catheter is noted to be adjacent to the distal lumen two orthogonal angiographic views are obtained to confirm proximity to the vessel lumen.
- An L -shaped marker is present at the distal tip, visible under fluoroscopy. In an orthogonal view it appears as an upside down T.
- A proprietary technique of 'L'ocate, 'T'une and deployment is used.
- Once the catheter is confirmed to be in a desirable location, the wire is withdrawn into the catheter, and the needle is actuated using a lever on the proximal end of the catheter.
- This curved hollow needle exits the side of the catheter, just proximal to the distal tip.
- Once the needle appears to have pierced across a flap into the true lumen, the 0.014" wire is advanced to enter the distal vessel and confirmed under fluoroscopy.
- The catheter is then removed without pulling the wire out.
- Intervention proceeds as usual thereafter.

Pioneer™:

- The Pioneer catheter (Medtronic Corp.) is another distal re-entry device.
- It is fairly similar to the Outback, except it has an integrated 64-element Volcano (Volcano Corp., Rancho Cordova CA)IVUS, requires two separate 0.014" wires and is 7 Fr. sheath compatible.
- The IVUS provides a cross-sectional view of vascular anatomy at the level of the tip of the catheter.
- The device has a hollow, curved 24 gauge nitinol needle, that is used to puncture into the distal true lumen.
- Over one wire, the catheter is advanced into the sub-intimal space and the other is used for advancing through the needle.
- Using IVUS guidance, the flow lumen of the reconstituted vessel is located.
- The catheter is rotated so that the needle is oriented at 12 O'clock position on the IVUS image.
- Once the needle is actuated and the second wire is advanced through it into the distal lumen, the device is exchanged out without removing the wire.
- Intervention proceeds in the usual fashion.

Enteer™ CTO Re-entry System:

- The Enteer™ system (Covidien, Mansfield MA) consists of the Enteer™ guidewire and Enteer™ re-entry catheter.
- The catheter is available in ATK and BTK treatment sizes, both with a crossing profile of 0.066", compatible with a 5 Fr. sheath.

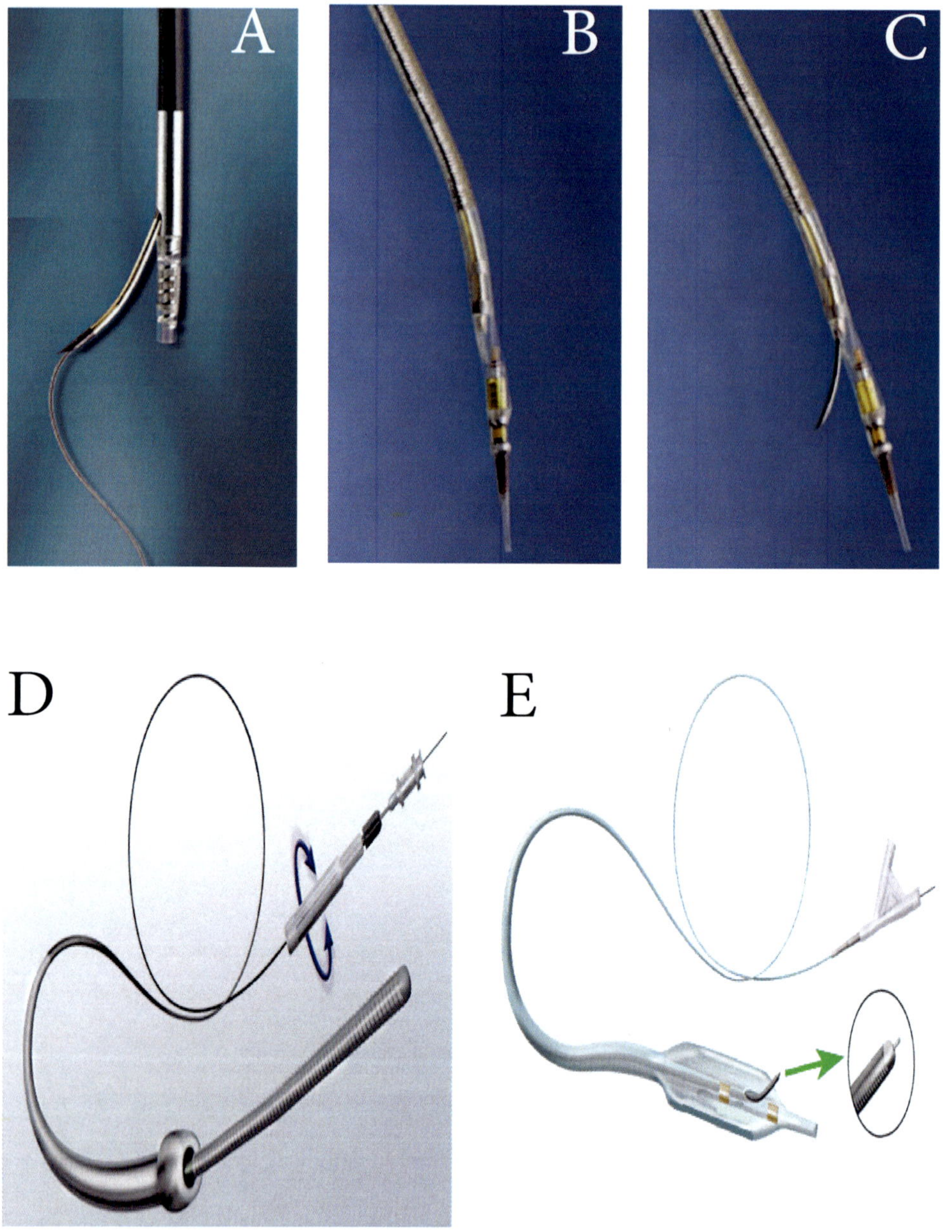

Figure 12.

Distal Re-entry CTO Crossing devices.

A. The Outback catheter with the nitinol needle actuated.
B. The Pioneer™ catheter without nitinol needle deployment. The IVUS transducer is noted proximal to the distal tip.
C. The Pioneer™ catheter with nitinol needle deployment.
D. The Viance™.
E. The Enteer™catheter and wire. The inset shows a close-up of the wire tip.

(Photos courtesy of respective manufacturers)..

- The catheters have an OTW balloon that can use a 0.014" - 0.018" wire.
- The balloon is 2 cm in length and is flattened side-to-side.
- It is intended for placement in the sub-intimal plane where it 'self-orients' itself in a circumferential manner, adjacent to the true lumen.
- The balloon has two radio-opaque markers and there is one exit port proximal to each of these markers. One hole exits on one side of the balloon, while the other on the opposite side. These holes are offset and allow the Enteer wire to exit through and poke into the true lumen.
- The Enteer wire has an angled tip and a fine probe (0.003" in diameter and 0.007" in length) for penetration and entry into the distal true lumen.
- Once distal entry has been made successfully, the Enteer wire is removed and a different 0.0.14" wire is inserted through the balloon into the distal lumen.
- The ATK catheter has a working length of 135 cm and its dimensions are 3.75 mm x 1.5 mm x 20 mm (width x height x length).
- The BTK catheter has a working length of 150 cm and its dimensions are 2.75 mm x 1.0 mm x 20 mm (width x height x length).
- Intervention then proceeds in the desired fashion.

SNARES

- A snare is a loop of a wire that is intended to remove, re-position or re-align foreign objects in the body.
- Snares have also been used to guide obtaining venous access in challenging cases.
- There are several different types of snares available in different diameters based on the caliber of the target vessel/chamber (figure 13).
- The snare kit typically consists of a snare loop, snare catheter and a loading tool.
- The basic technique of using a snare effectively remains the same regardless of the type of snare used (figure 14):
    - The snare is loaded onto a wire into a snare catheter.
    - The snare catheter is then advanced to the target site.
    - The snare is advanced out of the snare catheter and opened fully.
    - The desired object is grasped with the loop(s) of the snare.
    - Holding the snare steady, the snare catheter is gradually advanced over it, closing the loop.
    - The snare and snare catheter are then withdrawn or moved in the desired direction as a unit.

EMBOLIZATION DEVICES

Several materials are used for therapeutic embolization of arterial and venous circulation. By convention, the substance used to cause such embolization is termed an 'embolus' (figure 15).

Indications for embolization treatment include AV malformations, AV fistulas, micro-perforations, vascular trauma and aneurysm occlusion. Other indications targeted end-organ ablation and include treatment of neoplasms (TACE - transcatheter arterial chemoembolization), hypersplenism, uterine fibroids and varicocele among others.

COILS

- Coils are loops of metal wire (e.g. platinum) inter-twined with thrombogenic material such

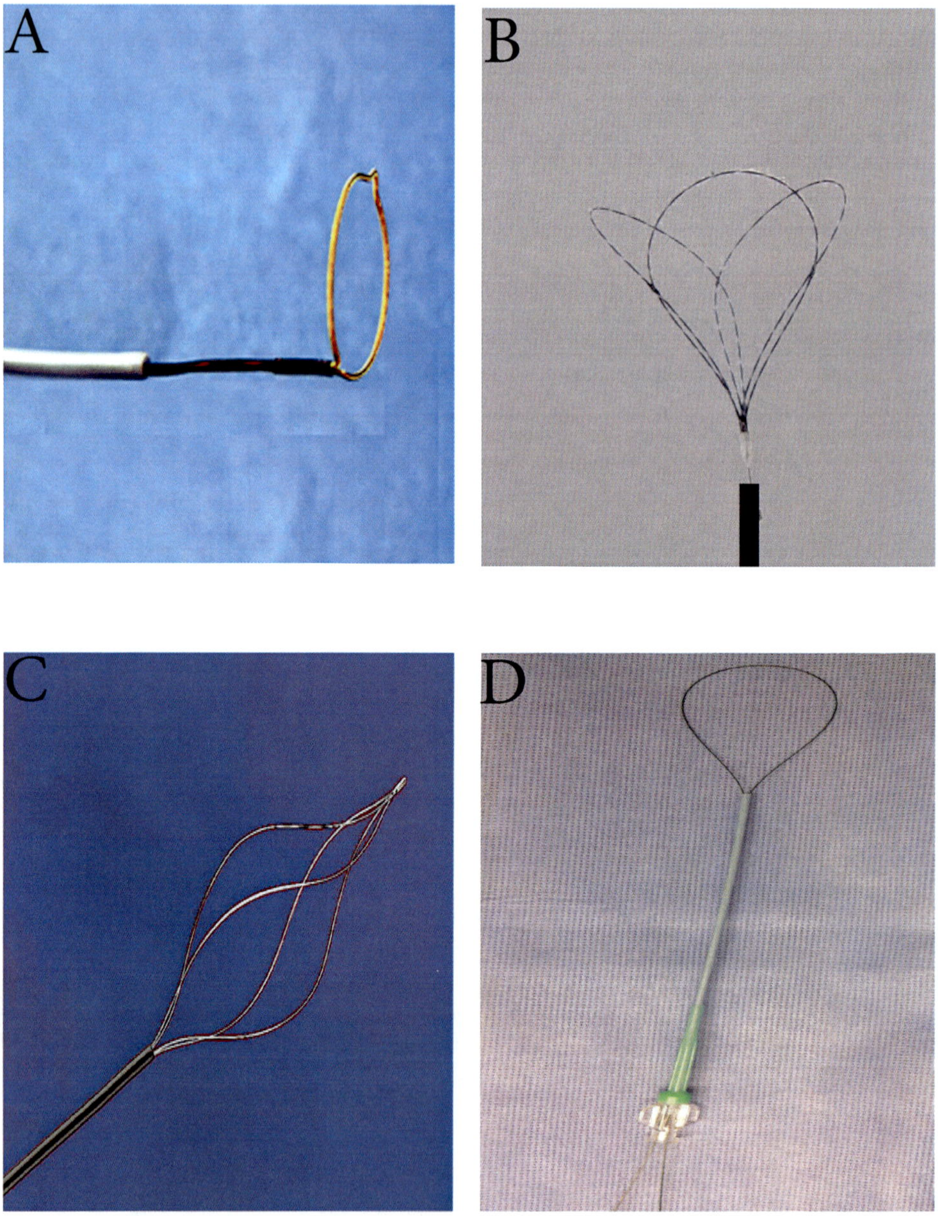

Figure 13.

Various types of snares.

A. Amplatz Gooseneck snare.
B. EnSnare™.
C. Expro Elite™ snare catheter.
D. A home-made snare constructed by advancing a 0.014"-0.018" wire through a catheter and looping it back through the tip of the catheter.

(Photos courtesy of respective manufacturers)

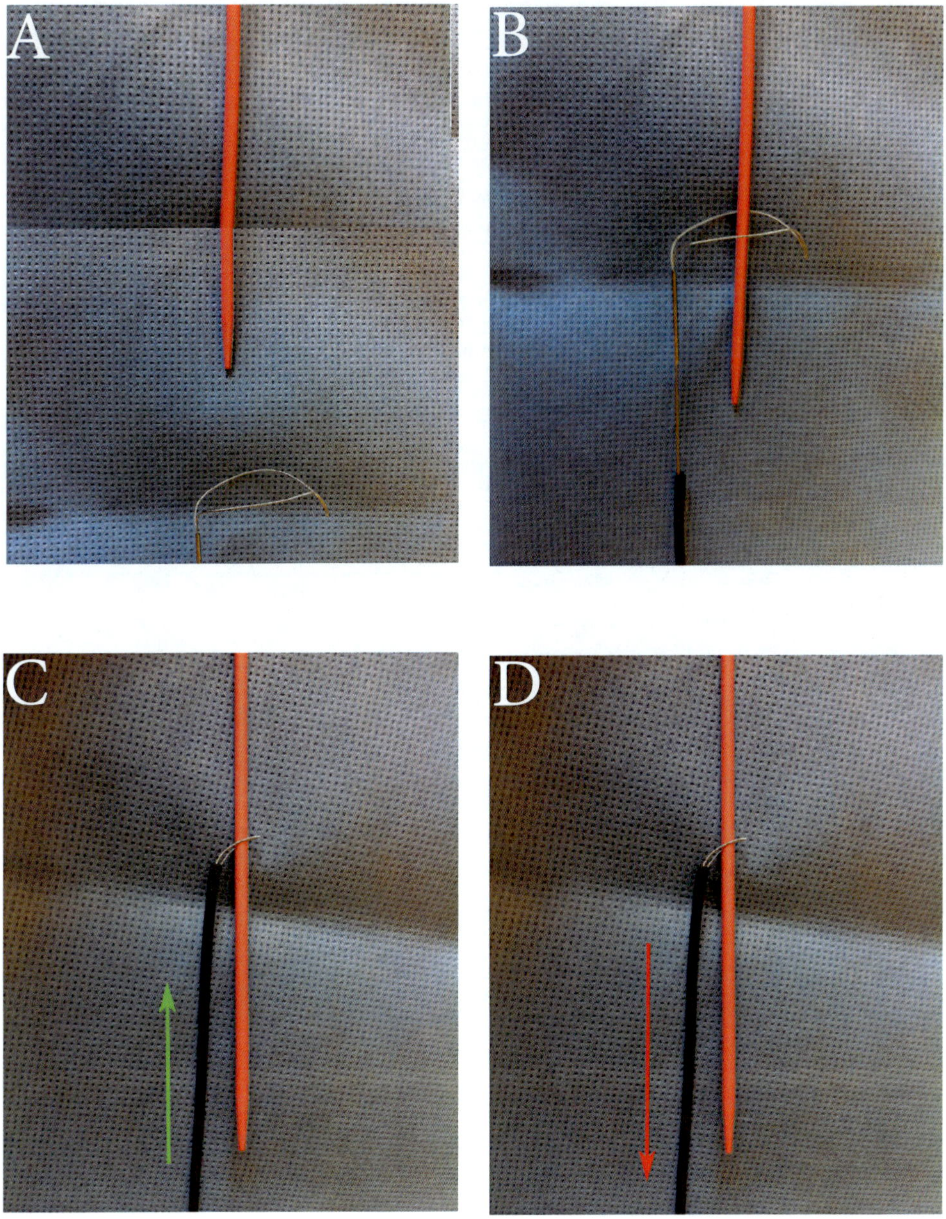

Figure 14.

How to use a snare.

A. The snare is positioned at the free end of the device to be removed/grabbed.
B. It is then advanced around the device.
C. While maintaining the snare's position, the snare-catheter is advanced until the loop(s) of the snare tightly closed around the device.
D. The snare and the snare-catheter are withdrawn as a unit.

as synthetic fibers.

- The coils are available in a variety of sizes and configurations.
- Coils are available as 'detachable' coils and 'pushable' coils.
- Most coils are provided pre-loaded in a loading cartridge.
- In the pre-loaded state, the coils are held in a more or less straight, strand-like configuration.
- In the deployed state, the coils tend to assume their unconstrained shape (usually a spiral loop).
- Coils are over-sized 30-50% in diameter to the target vessel.
- Most configurations are available in multiple coil diameters (0.018", 0.035" and 0.038") and lengths.
- Polyurethane catheters should not be used for coil delivery as the embolus can get lodged and thrombosed within these devices.
- Side-hole delivery catheters should be avoided as they can allow the coils to escape and embolize to non-target territories.

VASCULAR PLUGS

- Vascular plugs are made of a mesh of nitinol. These may be uni-, bi- or tri-lobed.
- These devices are very effective and reduce the number of coils required to successfully thrombose a larger vessel (due to their caliber).
- Due to their nickel content, it is purported that there is some release of this metal from the device for at least the first 60 days.

GLUES

- Glues used for embolization purposes are rapidly polymerizing liquid adhesives that transform into a solid. Examples include cyanoacrylate, or N-butyl-2-cyanoacrylate (NBCA).
- This polymerization occurs promptly on contact with blood.

MICRO-PARTICLES

- Particles such as PVA particles (polyvinyl alcohol) and acrylic gelatin microspheres can also be used for arteriovenous malformation (AVM) lesions.
- Microspheres (Embosphere; Biosphere Medical, Rockland, Mass) are biocompatible, hydrophilic, nonresorbable, and precisely calibrated particles produced from an acrylic polymer and impregnated with porcine gelatin.

GELFOAM

- Gelfoam is another agent, however it is not a permanent embolic agent and is generally used to temporarily occlude the abnormal vessels (e.g., bleeding or presurgical devascularization). Gelfoam is a water-insoluble, off-white, nonelastic, porous, and pliable material. Gelfoam may be cut without fraying, and it can absorb and hold many times its weight in blood and other fluids.

DETACHABLE BALLOON

- A detachable balloon that can be used to treat arteriovenous fistula (AVF) lesions.

PROXIMAL PROTECTION DEVICES

Proximal EPDs (table 1) are placed upstream from a target lesion. An 'occlusion' balloon is inflated (proximal to this lesion), causing flow interruption in the treatment segment. A stagnant column of blood is thus created. This state is maintained during the intervention. After the intervention is over,

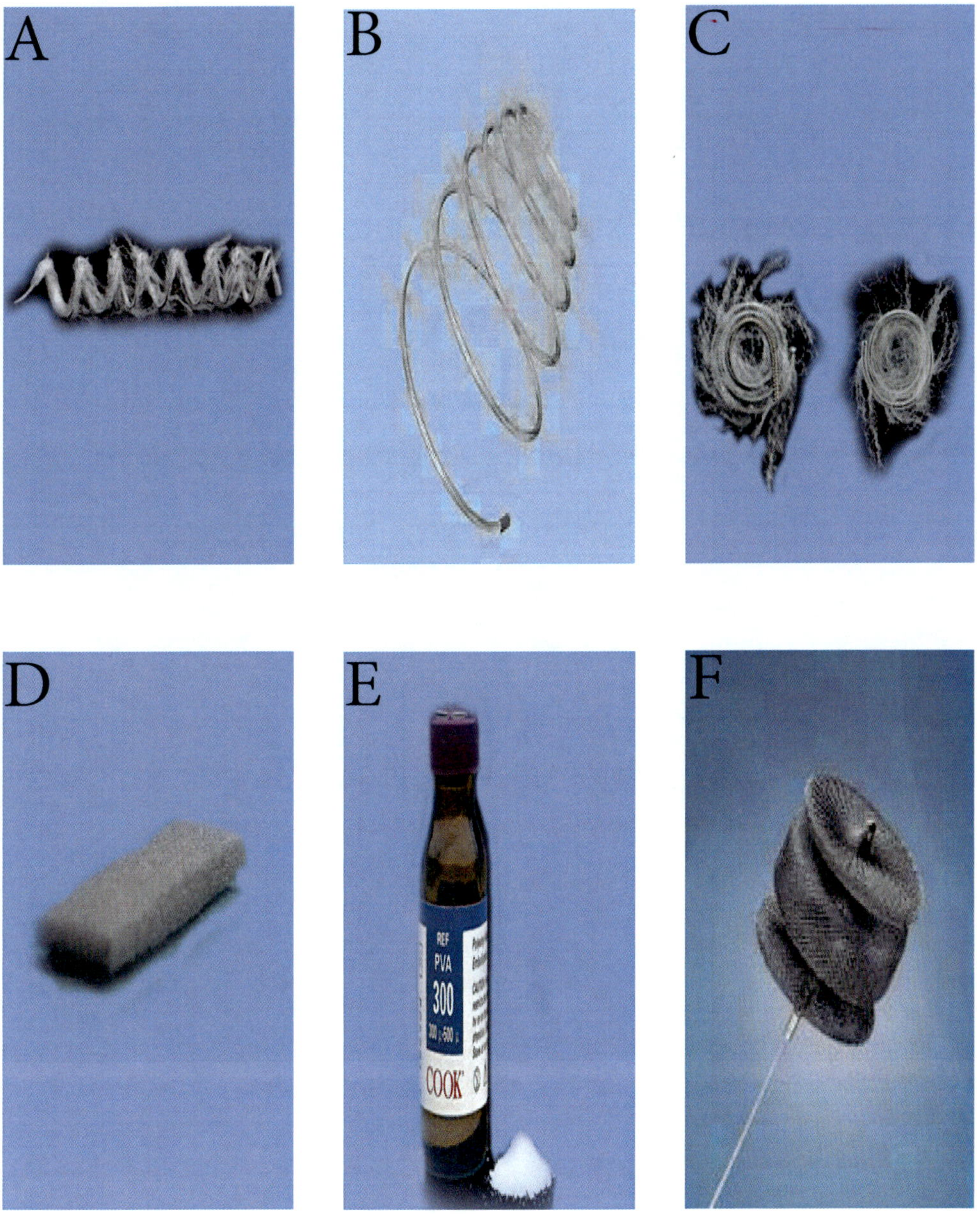

Figure 15.

Various devices used for embolization purposes.

A. Nestor coil.
B. Tornado coil.
C. Tornado coils.
D. Foam used for 'foam gel' creation.
E. PVA Glue.
F. Amplatzer vascular plug II.

(Photos courtesy of respective manufacturers)

TABLE 1 - PROXIMAL EMBOLIC PROTECTION DEVICES	
Gore Flow Reversal	• Gore. • Proximal occlusion with active flow reversal via an external circuit and back into the femoral vein. • 9.5 Fr. Gore balloon sheath (with dilator)/Gore balloon wire for ECA and Gore External Filter. • Creates an external AV shunt via 6 Fr. femoral venous sheath. • ECA diameter < 6 mm and CCA diameter 6-12 mm. • Used with 9 Fr./10 cm Terumo Pinnacle® or Terumo RadioFocus®sheaths. Non-Terumo sheaths > 10 Fr. required.
Mo.Ma Ultra	• Invatec Corp. • Proximal occlusion without active flow reversal. • 9 Fr. sheath/0.035" wire compatible/ 95 cm shaft. • ECA diameter < 6 mm and CCA diameter 6-12 mm. • Aspirate 60-100 cc of blood prior to balloon deflation.

TABLE 2 - DISTAL EMBOLIC PROTECTION DEVICES	
Accunet	• Abbott Vascular. • Fixed-wire filter. • 0.014" wire Rx- system 190 cm/300 cm lengths ; 4.5 mm - 7.5 mm diameter (for vessels 3.25 mm - 7.0 mm in diameter). • Maximum pore size of filter 150 um.
AngioGuard	• Cordis. • Fixed-wire filter. • 0.014" wire Rx- system 180 cm length; 4.0 mm - 8.0 mm diameter (for vessels 3.0 mm - 7.5 mm in diameter). • Maximum pore size of filter 100 um.
Emboshield Nav6	• Abbott Vascular. • BareWire technology; 0.014" PTFE-coated stainless steel wire with a 3 cm distal platinum/nickel tip; 190 cm/315 cm lengths. • Proximal end has Zebra-stripes; distal wire has a 0.019" stop. • Filter available in two sizes: Small/5.0 mm (2.5 mm - 4.8 mm vessels) and Large/7.2 mm (4.0 mm - 7.0 mm vessel diameter).
Filterwire® EZ	• Boston Scientific. • Fixed-wire filter. 0.014" wire. • 4 and 5 Fr./0.035" wire compatible; 45 cm/90 cm/ 135 cm length.
Gore Embolic Filter	• Gore. • Fixed-wire ePTFE filter. 0.014" PTFE coated wire 185 cm/300 cm length. • 0.014" wire/ 6 Fr. Guiding catheter compatible. • Two sizes 5 mm and 7 mm (for vessel diameter 2.5 mm - 5.5 mm). • Maximum pore size of filter 100 um.
Spider FX	• Covidien. • Fixed-wire Nitinol mesh filter. Cross lesion with guidewire of choice (0.014" or 0.018"). • 0.014" wire system; 190 cm/320 cm length. • 3 mm - 7 mm diameter (for coronary SVGs/carotids arteries 3.0 mm - 7.0 mm in diameter and lower extremity peripheral arteries 5.0 mm -7.0 mm in diameter).

aspiration is performed prior to restoration of antegrade flow.

Flow reversal, combined with proximal EPD, diverts the arterial blood in the CCA/ICA into an external catheter and through that into a femoral vein.

MO.MA ULTRA

- Mo.Ma® Ultra (Invatec S.p.A., Roncadelle, Italy) is a proximal cerebral protection device without flow reversal.
- It consists of a 9 Fr. system.
- At the conclusion of the intervention, manual aspiration of 60 - 100 mL of blood is performed.

GORE FLOW REVERSAL SYSTEM

- As the name implies, this is a proximal EPD with flow reversal.
- The device creates an external AV shunt leading to flow reversal and diversion of ICA blood into the femoral vein. This protects the neuro-circulation.
- It consists of a 9.5 Fr. system.

DISTAL PROTECTION DEVICES

Distal protection implies placement of an embolic protection device (EPD) downstream from a target lesion (i.e. between a target lesion and the organ being protected). Almost all distal EPDs are wire mounted and 0.014" wire systems. Two broad categories of EPDs exist:

FILTERS

Filter EPDs are fairly common in interventional practice. Most commonly used in carotid intervention for cerebral protection and coronary saphenous vein graft intervention, they have largely replaced their predecessor, the balloon EPDs and are now also being used for lower extremity and renal artery intervention.

Examples of filters in current practice include Filterwire EZ, Spider FX, Accunet, Emboshield NAV6, etc (see table 1).

BALLOONS

Balloon EPDs were the first generation of devices available to minimize the embolic risk of intervention, e.g. Guardwire® (Medtronic). Most, if not all, of these devices have been replaced by other alternatives in the form of filters or proximal EPDs.

THROMBECTOMY CATHETERS

ASPIRATION CATHETERS

Aspiration catheters are designed to remove thrombus primarily from the arterial circulation with the use of manual suction.

Several different types of catheters are available (table 3). These catheters are mostly 4 Fr. sheath compatible. Some larger bore catheters are 5-6 Fr. sheath compatible. One catheter, Pronto 0.035, is designed specifically for large vessels and requires a 10 Fr. sheath.

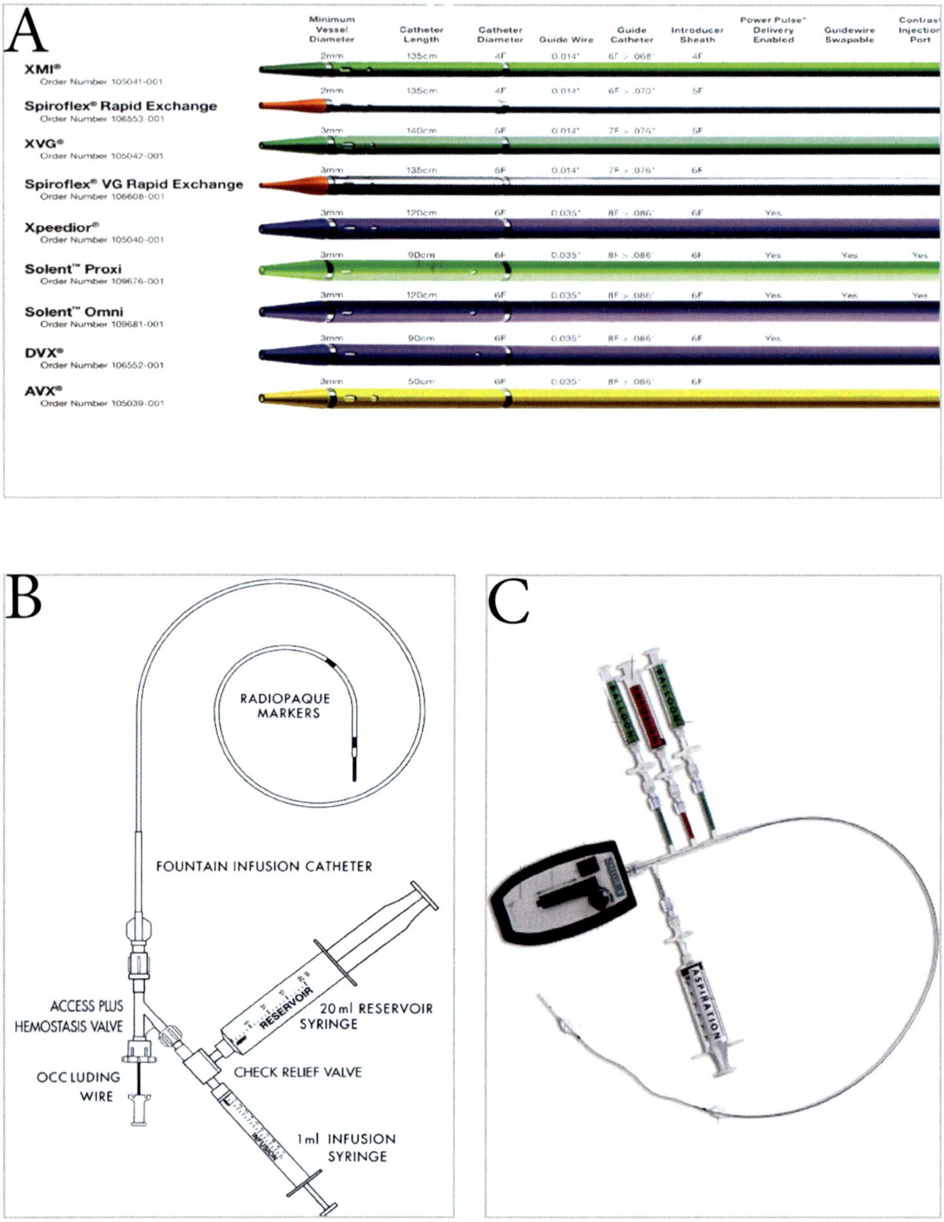

Figure 16.
Thrombectomy and infusion catheters.

A. Various selections of the AngioJet catheter.
B. The Fountain® catheter.
C. The Trellis-8 catheter.

(Photos courtesy of respective manufacturers)

- Fetch 2
- Pronto (V3, LP, etc)
- Pronto 0.035 Extraction Catheter
- Export
- Xtract

RHEOLYTIC THROMBECTOMY CATHETER

- The Angiojet® (Medrad Inc.,Indianola PA) rheolytic thrombectomy system functions by introducing pressurized jets of saline, utilizing Cross-Stream® technology, from orifices near the distal tip of the catheter.
- This generates a localized low pressure zone due to Bernoulli effect, entraining and macerating thrombus.
- The saline and thrombus are then aspirated back into the catheter and discarded outside the body.
- Multiple catheters are available for use in various vascular beds including coronary arteries (both native and saphenous vein grafts, peripheral arteries, deep venous system and dialysis AV-grafts/fistulas) (figure 16A & table 4).
- The catheters used in peripheral arteries (Solent™ Proxi) and deep veins (Solent™ Omni/DVX) are 6 Fr. sheath/0.035" wire compatible.
- Both of the above catheters are capable of delivering Power Pulse® spray (infusion of thrombolytic/anticoagulant agent via the AngioJet into a thrombus laden segment of the vessel for a duration of 20-30 minutes followed by aspiration).

TABLE 3 - ASPIRATION THROMBECTOMY CATHETERS	
ASAP	• Merit Medical Systems, Inc. • 6 Fr. Guiding catheter compatible /140 cm shaft
Diver C.E.	• Medtronic, Inc. • 6 Fr. Guiding catheter compatible /140 cm shaft
Export	• Medtronic, Inc. • 6/7 Fr. Guiding catheter compatible /145 cm shaft
Fetch2	• Medrad, Inc. • 6 Fr. Guiding catheter compatible /135 cm shaft
Pronto V3 Pronto LP	• Vascular Solutions, Inc. • 6 Fr. Guiding catheter compatible /140 cm shaft
Pronto 0.035	• Vascular Solutions, Inc. • 10 Fr. sheath compatible /115 cm shaft
QuickCat	• Spectranetics Corp. • 6 Fr. Guiding catheter compatible /145 cm shaft
Xpress-Way Rx	• Atrium Medical Corp. • 6 Fr. Guiding catheter compatible /140 cm shaft

OTHER MECHANICAL THROMBECTOMY CATHETERS

Trellis:

This catheter (figure 16C & table 4) is available in two different sizes Trellis™-6 (6 Fr.) and Trellis™-8 (8 Fr.) (Covidien, Mansfield MA).

- This thrombectomy/infusion system has two latex-free compliant balloons that allow isolation of the thrombus laden segment.
- The catheter is deployed over a 0.035" wire.

- The balloons are inflated (using the green syringes). Caution should be exercised to not over inflate the balloon as this may result in vascular trauma.
- The shaft of the catheter has an intrinsic sinusoidal wave (wire) that oscillates between 500-3000 rpm, to macerate the thrombus and to increase the surface area of the thrombus for the action of the thrombolytic agent.
- The thrombolytic agent (rtPA) is infused using the red syringe.
- After a brief dwell time of 5-15 minutes, aspiration is performed using the large white syringe.
- The balloons are then deflated.
- It may be beneficial to deployed a Spider EPD through the wire lumen of the catheter, distal to the distal balloon, prior to deflation.
- Angiography is then performed. The culprit lesion(s) is located.
- Once the device is retrieved, intervention on the culprit lesion is performed as appropriate.
- The device may be used within native vessels as well as stented vessels/bypass conduits.

TABLE 4 - MECHANICAL THROMBECTOMY CATHETERS	
AngioJet	• Medrad Interventional. • Multiple catheters available (XMI®, XVG®, Xpeedior®,Solent Omni®, etc). • 0.014"/0.035" wire compatible.
Trellis 6	• Covidien. • 6 Fr. sheath and 0.035" wire compatible. • 80 cm/120 cm length/ treatment length 10 mm/30 mm.
Trellis 8	• Covidien. • 6 Fr. sheath and 0.035" wire compatible. • 80 cm/120 cm length/ treatment length 15 mm/30 mm.

THROMBOLYTIC INFUSION CATHETERS

- These devices are designed for localized delivery of thrombolytic agents to enhance the fibrinolytic process.
- This also reduces drug dosage in an attempt to minimize bleeding complications.

Several catheters are available (table 5). The common ones in use in the U.S. are listed below. See chapter 13 for further details.

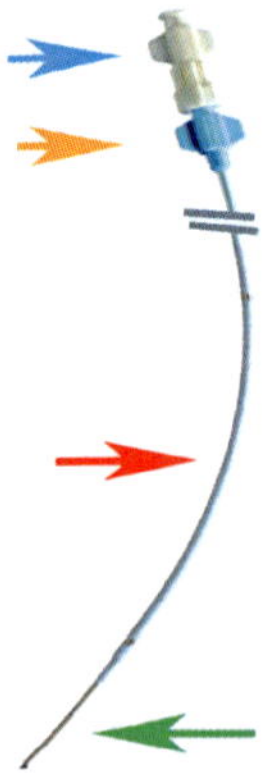

Figure 17.
A UniFuse catheter upclose.

THE FOUNTAIN® CATHETER

- The Fountain® catheter (Merit Medical, South Jordan, UT) is available in 4 and 5 Fr. sizes.
- This is a straight catheter with multiple side-holes in the distal part of the catheter, between two radio-opaque markers.
- When the distal end-hole is occluded with the accompanying 'occluder wire' the infused medication is directed out the side-holes thus entering the treatment site.
- When thrombolytic therapy is infused through this catheter the medication is infused typically for 12- 24 hours.

THE UNI*FUSE™

- The Uni*fuse™ system (AngioDynamics, Latham NY) consists of a PRO™ infusion catheter (orange arrow) and an Occluding ball guidewire (blue & green arrows) assembly. The latter has a useable hub (blue arrow) through which medication may be infused.

- Available in 4 and 5 Fr. profiles.

TABLE 5 - THROMBOLYTIC INFUSION CATHETERS	
Fountain catheter	• Merit Medical. • 4 and 5 Fr./0.035" wire compatible; 45 cm/90 cm/ 135 cm catheter length. • 5 cm/10 cm/20 cm/30 cm/40 cm/50 cm treatment lengths.
Pulse*Spray® Infusion system	• AngioDynamics. • 3-5 Fr./0.035" wire compatible; 45 cm/90 cm/135 cm length. • 2 cm/5 cm/10 cm/20 cm/30 cm/40 cm/50 cm treatment lengths.
SpeedLyser® catheter	• AngioDynamics. • 4 and 5 Fr./0.035" wire compatible; 45 cm/90 cm/ 135 cm length.
Uni*fuse™ catheter	• AngioDynamics. • 4 and 5 Fr./0.035" wire compatible; 45 cm/90 cm/ 135 cm length.
EkoSonic® MACH 4e Endovascular System	• EKOS® Corp. • 5.4 Fr./0.035" wire compatible;106 cm/135 cm catheter lengths. • 6 cm/12 cm/18 cm/24 cm/30 cm/40 cm/50 cm treatment lengths for 106 cm catheter. • 12 cm/30 cm/40 cm/50 cm treatment lengths for 135 cm catheter.

EKOSONIC™ MACH 4 ENDOVASCULAR SYSTEM

- EkoSonic Mach 4e endovascular system (EKOS Corp.) is designed to deliver controlled and selective infusion of thrombolytic (and other) agents into peripheral vasculature/pulmonary arteries.
- This catheter generates high-frequency (2-3 MHz), low power ultrasound waves in the treatment zone through the piezoelectric transduction of radio-frequency energy.
- The ultrasound emanates radially from the catheter into the surrounding thrombus/tissue to enhance dispersion and deeper penetration of the delivered drug.
- 6 Fr. sheath compatible device.
- Variable lengths of the catheter are available.

INTRAVASCULAR ULTRASOUND

- Intravascular ultrasound (IVUS) is a tomographic imaging modality that utilizes ultrasound images acquired through a transducer placed inside a blood vessel.
- Imaging via IVUS is complimentary to angiography and provides information not only about the vessel lumen but also its wall, atherosclerotic plaque, intra-luminal thrombus and adjacent vascular structures.
- Catheters are available in various sizes (table 6) appropriate for various vascular beds (coronary arteries, large vessels and aorta/IVC/SVC etc).
- This imaging modality is very useful for measurement of vessel diameter, identification of luminal abnormalities such as plaque, thrombus, dissection etc in cases where angiography is not definitive.

TABLE 6 - IVUS CATHETERS	
Atlantis 018	• Boston Scientific Corp. • 6 Fr. sheath/ 0.014" wire compatible/ 135 cm shaft. • Maximum image diameter 8 mm.
Atlantis PV	• Boston Scientific Corp. • 8 Fr. sheath/0.035" wire compatible/ 95 cm shaft. • Maximum image diameter 30 mm.
Eagle Eye Plati-num/ Gold	• Volcano Corp. • 5 Fr. sheath/0.014" wire compatible/ 150 cm shaft. • Maximum image diameter 20 mm. • 20 MHz frequency. • Rapid exchange catheter. • Has ChromaFlo®.
Vision PV 018	• Volcano Corp. • 6 Fr. sheath/0.018" wire compatible/ 135 cm shaft. • Maximum image diameter 24 mm. • 20 MHz frequency. • Rapid exchange catheter. • Has ChromaFlo®.
Vision PV 8.2	• Volcano Corp. • 9 Fr. sheath/0.038" wire compatible/ 90 cm shaft. • Maximum image diameter 60 mm. • 10 MHz frequency. • OTW catheter. • No ChromaFlo®.

4

VASCULAR ACCESS

RETROGRADE FEMORAL ARTERY ACCESS

- Femoral artery access is obtained in the common femoral artery (CFA).

ANATOMY OF CFA

- This artery is the continuation of the external iliac artery (EIA).
- It starts after the origin of the inferior epigastric artery and terminates, below the lower margin of the femoral head, by bifurcating into the superficial femoral artery (SFA) and profunda femoris artery (PFA).
- The CFA is 3-5 cm in length.
- Its luminal diameter is 6 - 8 mm in an adult male and 5 - 7 mm in an adult female.

INDICATIONS

- This site serves as the primary access for most diagnostic and interventional (arterial) procedures.
- Access in alternate sites is obtained only for specific indications as discussed later in the chapter.

PATIENT AND ROOM SETUP

- The patient is placed in supine position.
- Both groin sites are prepared in a standard fashion and cleaned with antiseptic solution prior to draping the patient.
- For most supra-inguinal procedures (where no intervention on infra-inguinal vessels is planned) the operator stands on the right side of the patient. This is true not only for right CFA access but also for the left (figure 1).
- However, if simultaneous access in the left brachial artery is being considered, operator position to the left side of the patient is a natural choice, regardless of which CFA is used.
- For infra-inguinal angiography and intervention, retrograde CFA access in the contralateral CFA is used (right CFA for left leg angiography and vice versa). In this case, the operator always stands on the side where CFA access is being obtained and opposite from the leg that

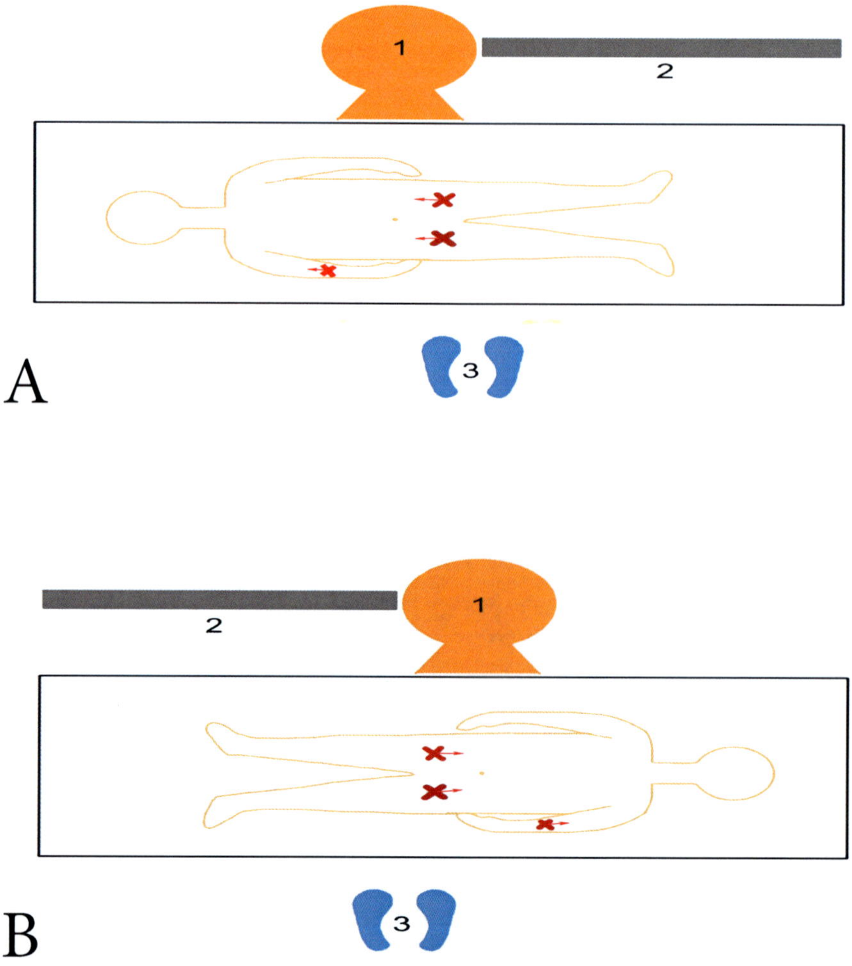

Figure 1.

Room set-up for right or left retrograde CFA for supra-inguinal angiography.

A. Right or left CFA retrograde access with or without concurrent right brachial artery access.
B. Right or left CFA retrograde access with concurrent left brachial artery access.

- 1. Image intensifier.
- 2. Image display and hemodynamic monitors.
- 3. Operator position.
- Red crosses denote arterial access sites.
- Red arrows denote direction of the access.

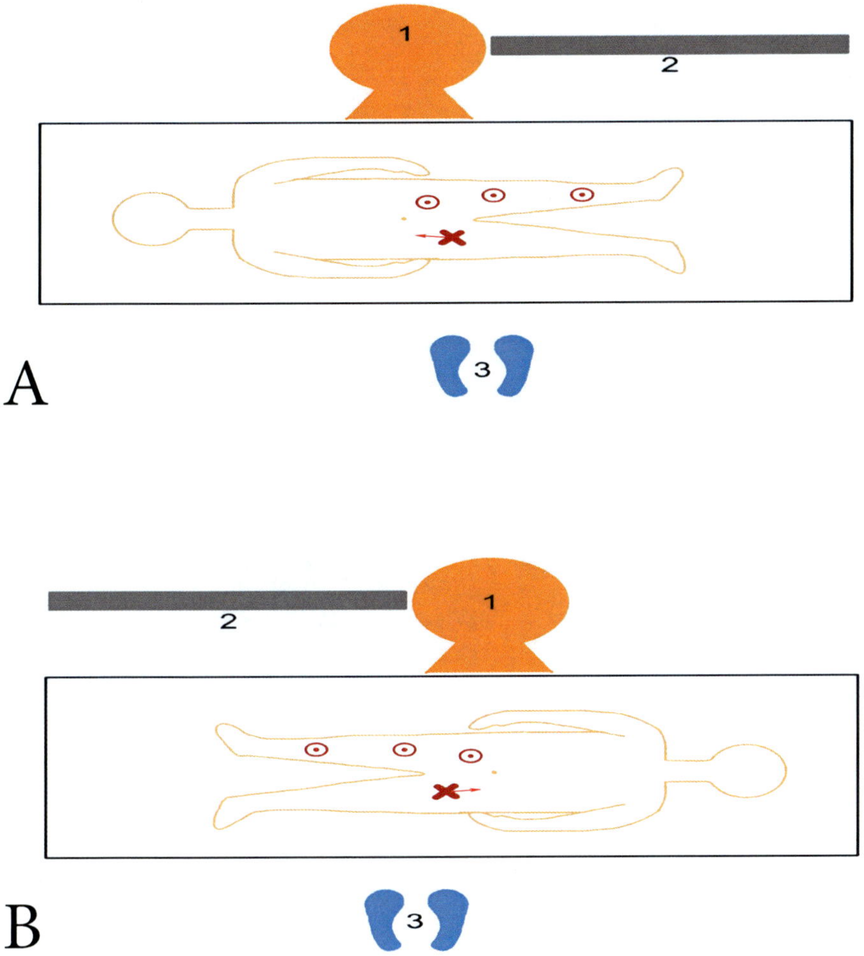

Figure 2.

Room set-up for leg angiography and intervention using retrograde CFA access.

A. Left leg angiography and intervention.
B. Right left angiography and intervention.

- 1. Image intensifier.
- 2. Image display and hemodynamic monitors.
- 3. Operator position.
- Red crosses denote arterial access sites.
- Red arrows denote direction of the access.
- Red circles denote target vessels for intervention.

is being imaged and intervened upon (right groin access and right side position of the operator for left leg angiography and vice versa). Though this may not always be the most obvious and 'user-friendly' position for the operator (e.g. left CFA access and left position of the operator for right leg angiography), it not only provides optimal imaging but better radiation safety (figures 1 & 2).

LANDMARKS

- Femoral artery access is used when the pulse in this artery is normal or near normal and easily palpable.
- Prior to obtaining access in this artery the following landmarks are palpated (figure 3):
    - ▹ The anterior superior iliac spine
    - ▹ The pubic tubercle
- A straight line is imagined to connect these two bony landmarks. This line is bisected at the junction of its medial third with the lateral two thirds.
- The CFA is located at this point and is accessed nearly two finger-breadths below this level.
- Once a femoral pulse is felt at this point, it is marked with a radio-opaque marker such as a hemostat/"pickups" and visualized under fluoroscopy to confirm this position.

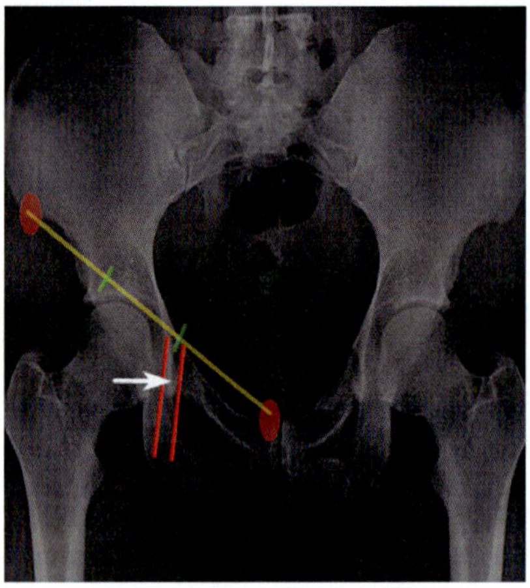

Figure 3. **Bony landmarks.**

Bony landmarks for obtaining CFA access. See text for details.

EQUIPMENT

- We use a 4 Fr. micropuncture kit for obtaining femoral artery access (figure 4 & table 1).
- This kit consists of :
    - ▹ A 0.018" wire.
    - ▹ A 21 Gauge, 7 cm long, hollow bore needle.
    - ▹ A 4 Fr. microcatheter with dilator.
- A 0.035" 150 cm J-tip wire.
- A 5 Fr./23 cm sheath with a radio-opaque marker on the tip (e.g. Cordis Britetip®).

TECHNIQUE

Skin Incision:

- We prefer to make a small incision in the skin at the site of the punc-

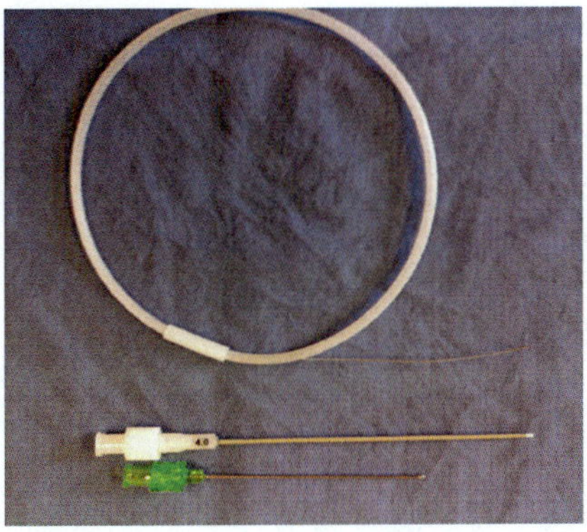

Figure 4.
A 4 Fr. micropuncture kit.

The three contents (needle, wire, microcatheter/dilator) come in one package.

Sheath size:
- Infra-popliteal PTA / stenting - 5 Fr.
- Infra-popliteal atherectomy - 5/6 Fr.
- Femoro-popliteal PTA - 5 Fr.
- Femoro-popliteal stenting- 5/6 Fr.
- Femoro-popliteal atherectomy - 6/7 Fr.

ture.

- Ideally, this should be done before the needle is inserted in the skin, so that a subcutaneous 'track' can be made.
- This incision is usually made at the level of the lower border of the femoral head which is marked under fluoroscopy by placing a hemostat. The arteriotomy is made at the level of the center of the femoral head (figures 5A & B).

Needle Alignment:

- The needle should be held at a 45° angle to the skin, bevel up (at 12 O' Clock position).
- The needle should be aligned in a lateral to medial direction, from the skin incision to the target arteriotomy site.

Needle Advancement:

- The needle should be advanced slowly and smoothly in a lateral to medial direction.
- It should be aimed towards the most prominent pulse so that it enters the artery at the level of the center of the femoral head (figure 5C).
- Blood return through a 21 Gauge needle in the micropuncture kit is not as brisk and pulsatile as is usually seen with an 18 Gauge thin wall needle.
- Instead, the flow is more like a constant drip (roughly a drop or two per second). Venous blood will be darker and drip much slower (1 drop every 3 - 5 seconds, in patients with normal central venous pressure).

Wire Advancement:

- The 0.018" wire has a soft straight tip that can be easily damaged, hence it is advanced very gently.
- Additionally, tactile feedback is suboptimal with this wire.
- The wire and needle should be coaxial to each other (figure 5D). This reduces the friction in advancement and withdrawal of the wire, generated by angulation between the two.
- Caution should also be exercised when attempting to withdraw this wire through the needle. If the wire is not coaxial to the needle, its distal end can get unraveled and trapped in the tissues, cause vessel dissection or be severed from the shaft.

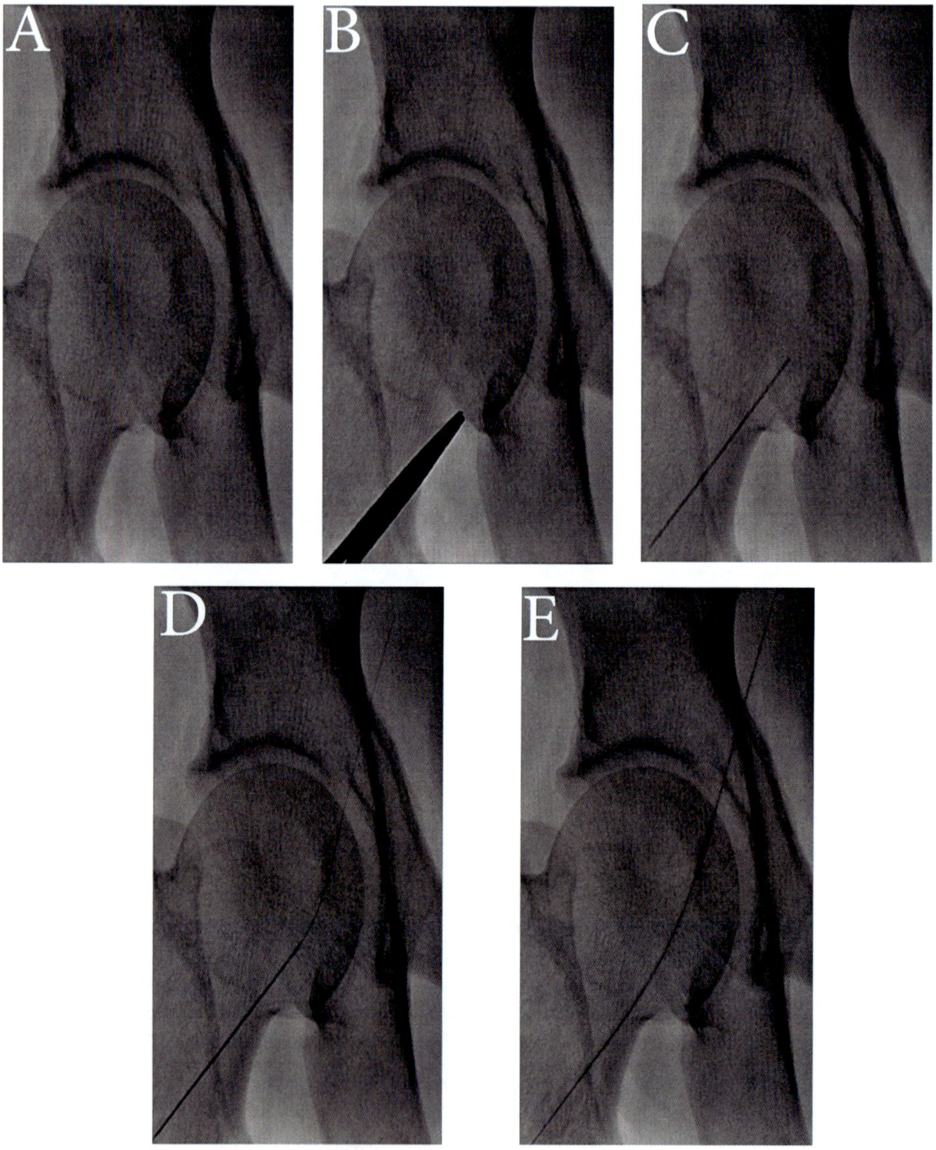

Figure 5.

Fluoroscopic correlates of retrograde CFA access.

A. Right femoral head.
B. A hemostat is placed at the level of the lower margin of the right femoral head. This is done to mark the location of the skin incision.
C. A 21 G needle is advanced through the skin incision to the right CFA.
D. The 0.018" wire supplied with the micropuncture kit is advanced through the needle into the right EIA.
E. After removal of the needle, the microcatheter is advanced over the wire into the CFA. This catheter is advanced till its hub is at the skin level.

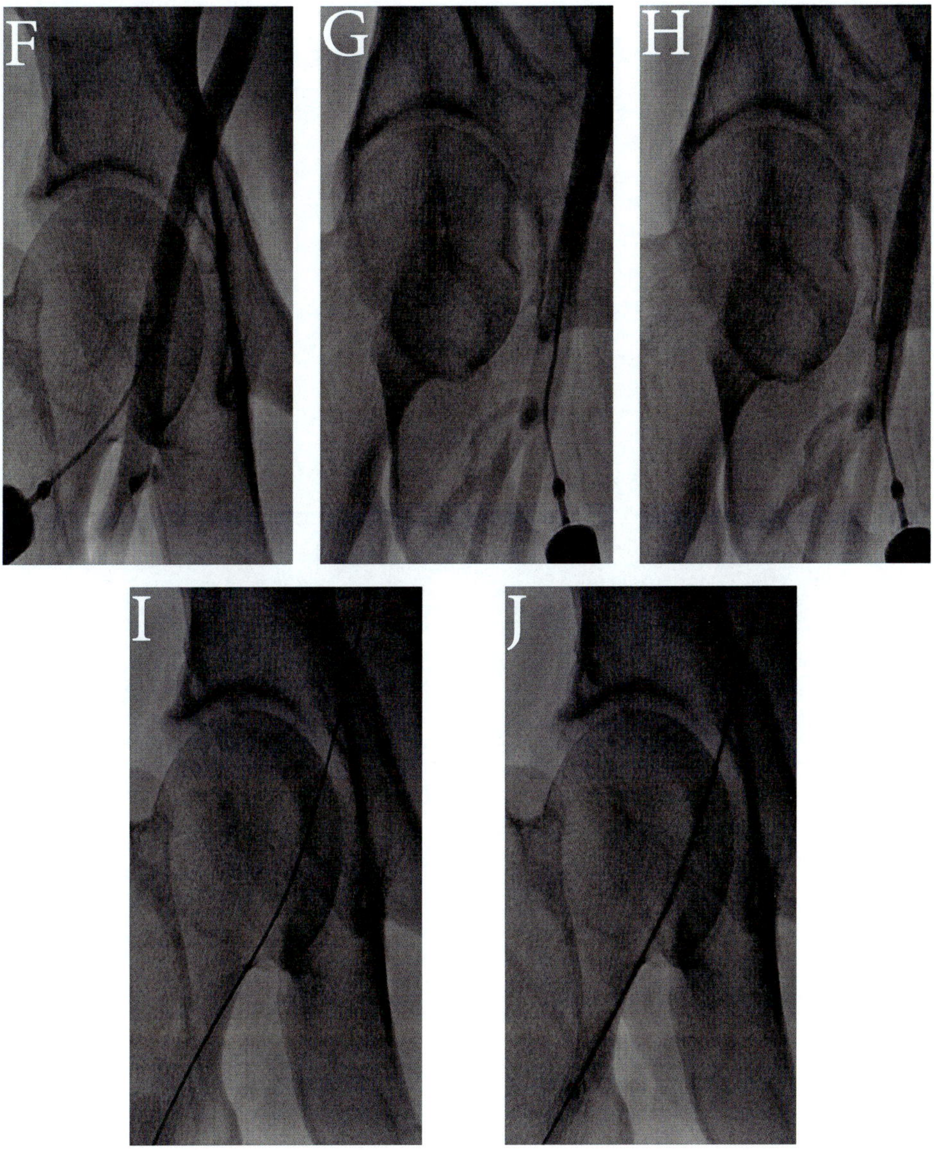

Figure 5 (continued).

F - H. After removal of the wire and dilator from the microcatheter, an angiogram is obtained in various projections to confirm accurate arteriotomy positioning.

I. Once it is confirmed that the microcatheter is in a desirable position in the CFA, a 0.035" wire is advanced into the aorta. The microcatheter is then removed.

J. A 5 Fr. 'working' sheath is then inserted over the 0.035" wire into the CFA. After removal of the dilator, the sheath is flushed with heparinized saline.

Use of Fluoroscopy:

- As this wire provides minimal feedback with regards to the tip's interaction with the vessel wall, plaque or branch vessels, fluoroscopy should be used when the wire is advanced.
- If, for example, the wire encounters plaque in the vessel at the tip of the needle the operator may not feel any resistance while advancing it.
- On fluoroscopy, however, the wire may be seen getting curled and 'bunched up' at the entry site. Continued advancement of the wire may result in vessel dissection.
- Likewise, on fluoroscopy the wire may be seen to enter a side-branch and can thus be appropriately redirected prior to advancement of the microcatheter into the vessel.

Microcatheter:

- Once the wire is in the desired location, the needle is removed and the microcatheter is advanced to its hub, into the vessel (figure 5E).

Angiography:

- At this point, the catheter is fixed in the artery using the left hand, while the dilator and 0.018" wire are removed.
- An angiogram is performed to confirm that the arteriotomy is in the desired spot (figures 5F-H).
- A 3 cc syringe with a 50:50 mix of dye is connected to the microcatheter for this purpose.
- An ipsilateral 30° projection is optimal as it displays the CFA bifurcation well.
- If necessary, contralateral 20° and AP projections may be used.
- These steps prevent 'up sizing' the sheath if the arteriotomy is in an undesired location in the CFA or in the PFA/SFA. In such a case, the microcatheter is removed, manual compression is applied to the arteriotomy site for 3 - 5 minutes and access re-attempted.

0.035" Wire:

- A 0.035" wire (J-tip, angled or other wire of choice) is introduced through the microcatheter into the vessel (figure 5I). This wire is advanced into the abdominal aorta. If resistance is encountered prior to reaching the aorta, wire advancement should be halted at that point.

Working Sheath:

- The microcatheter is removed and the working sheath (e.g. a 5 Fr. /23 cm) is inserted over the 0.035" wire up to a point where it is clearly intraluminal, yet the tip of the dilator is not beyond the wire tip (figure 5J).
- The dilator is removed and the sheath flushed.

TROUBLESHOOTING

Inability to 'Stick' the Artery:

- Rarely, despite a reasonable femoral pulse the operator is unable to make a satisfactory arteriotomy. This is indicated by either no or poor flow through the needle.
- In this case, the first thing to do is to re-assess the anatomic landmarks and re-palpate the CFA pulse. Make sure the skin incision and the track created for the arteriotomy guide the needle in the correct direction.

- Fluoroscopy should be used to identify CFA calcification, if any, and localize the vessel. In our experience, linear calcification of the CFA is present more commonly in the medial wall of the vessel than the lateral. This can help direct the needle lateral and parallel to the calcification ridge.
- If fluoroscopy in an antero-posterior (AP) projection remains unsuccessful in finding the artery, a 20-30° ipsilateral oblique projection should be used. In this view, the femoral vessels lie medial to the femoral head. CFA calcification is easier to visualize in this projection. This view is also helpful in highlighting an incorrect needle track to the vessel (medial to lateral track vs. lateral to medial).
- Another option is to use ultrasound guidance to identify the CFA and obtain access under direct duplex visualization. This may be particularly helpful in patients with an increased risk of bleeding complications (e.g. before or after thrombolytic therapy administration, obese patients, etc).

Inability to Advance the Wire:

- Sometimes the needle appears to be in the correct place in the CFA and a reasonable blood return is seen. However, the wire cannot be advanced successfully into the vessel.

Wire cannot be advanced beyond needle tip:

- If the wire cannot be advanced freely beyond the needle tip it should be removed (figure 6A). Good blood return should be confirmed.
- If satisfactory flow is present, the needle should be withdrawn 1-2 mm (figure 6B) and the bevel of the needle should be rotated 30-90° (counterclockwise for right CFA retrograde approach, and clockwise for left CFA). If the flow remains reasonable, under fluoroscopic guidance the wire should be re-advanced.

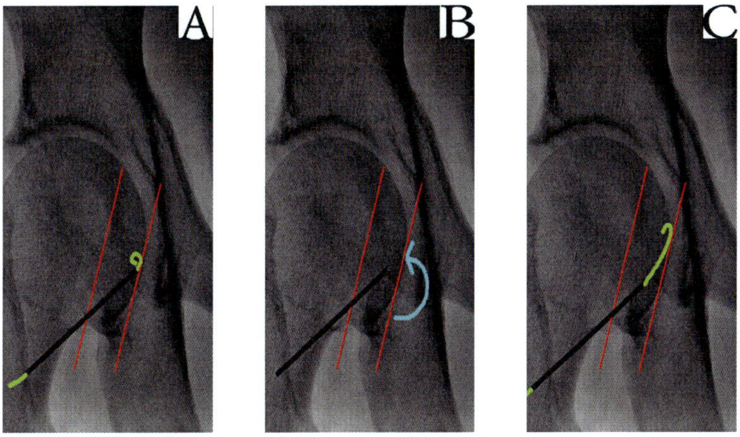

Figure 6.

Inability to advance wire beyond needle tip.

The needle is withdrawn 1-2 mm, rotated and the wire re-advanced after confirming normal flow.

Wire cannot be advanced beyond the CFA without resistance:

- Assuming that the blood return was reasonable through the needle and the wire had no difficulty exiting the needle into the vessel, it should be determined if the resistance is encountered by the tip of the wire or the shaft.
- If the wire tip seems to be moving freely but there is resistance upon advancing it, the track of the needle with respect to the wire should be checked under fluoroscopy.
- Often this resistance is a result of the friction between the wire shaft and the needle tip, due to an inaccurate needle entry angle. This would be obvious under fluoroscopy with the needle not being coaxial with the wire.
- Without withdrawing the needle tip, the hub of the needle is directed towards the vessel plane, the wire and needle become more and less coaxial and much of the friction disappears.
- The wire can then be advanced without much trouble.

Inability to Advance the Working Sheath:

- One situation that occurs not infrequently and demands attention is difficulty in advancing a sheath over a 0.035" wire into the CFA.
- This may be due to one of the following reasons:
    - ▹ Scar tissue from previous surgery.
    - ▹ Calcification in the artery.
    - ▹ Long tortuous sheath track through the subcutaneous tissue and skin.
- In this situation advancing a dilator (without the sheath) over the wire and 'making the track wider' may not be helpful either.
- The simplest way to overcome this problem is to switch the wire out for a stiffer wire such as a 0.035" SuperStiff® Amplatz wire.
- The following steps should be taken:
    - ▹ Ensuring an adequate incision in the skin and dilatation of the track.
    - ▹ A straight 4 Fr. /65 cm Glidecath® is advanced bareback (i.e. without a sheath in place) under fluoroscopic guidance over the initial 0.035" wire, into the aorta. Since the catheter has no marker(s) on the tip, it may be difficult to visualize without adequate zoom.
    - ▹ Once the catheter is in the proximal abdominal aorta, the wire is exchanged out for a J-tip, 180 cm SuperStiff® Amplatz wire. This step is performed under fluoroscopic guidance as well.
    - ▹ The Glidecath® is removed and the sheath (along with its dilator) is advanced over this wire into the CFA.
    - ▹ With this method usually no significant resistance is encountered and the sheath reaches its destination.
    - ▹ If, however, the sheath cannot be advanced despite the above steps, the following should be considered:
        - ♦ Using a sheath with braiding and/or hydrophilic coating (e.g. Arrow®, Pinnacle®, Raabe® etc)
        - ♦ Dilatation of the track with serial dilators of progressively larger size until the desired sheath French size is reached.
        - ♦ Use of alternate access site.

Tips:

- Caution should be used to not advance the SuperStiff® Amplatz wire without fluoroscopy. Major vascular perforation may occur due to the stiffness of the shaft.

TABLE 1 - EQUIPMENT FOR ARTERIAL ACCESS	
Needle	• 21 G needle preferably. • 18 G thin wall needle (for prosthetic graft access).
Sheaths	• 4 Fr. micropuncture kit (21 G needle). • 5 Fr./11 cm or any sheath in the desired size.
Wires	• 0.018"/30 cm stainless steel or nitinol wire. • 0.035" ▹ Standard J-tip wire. ▹ Angled stiff Glidewire. ▹ Angled non-stiff Glidewire. ▹ J-tip SuperStiff® Amplatz wire.
Ancillary Equipment	• 4 Fr./65 cm straight Glidecath®. • Sterile stockinette/sleeve. • Aperture drape. • Vascular ultrasound.

ANTEGRADE FEMORAL ARTERY ACCESS

The goal of antegrade CFA access is to allow placement of an arterial sheath into the CFA with the tip pointing cranio-caudad (i.e. towards the foot). The sheath invariably is advanced from the CFA into the proximal SFA.

INDICATIONS

Antegrade CFA access is more challenging and has a steeper learning curve. This access is used for ipsilateral infra-inguinal intervention. Most often this access is used when an 'up-and over' approach across the aortic bifurcation from the contralateral CFA is not feasible. Examples of scenarios that necessitate such access include:

- Previous AAA repair (endovascular or surgical with an aorto-bifemoral or aorto-bi-iliac graft). The newly created 'aortic bifurcation' in these cases is steep with a 180° bend making it very challenging to track equipment from one CFA to the other leg.
- 'Kissing' aorto-iliac stents with crisscross deployment.
- Challenging aortic bifurcation with excessive aorto-iliac tortuosity.
- Infra-popliteal intervention may sometimes be facilitated by this access as it provides a more direct path for the wire and enhances torque control. Additionally, it allows more distal delivery of devices due to the shorter distance between the access point and target sites.

CAUTION/CONTRAINDICATION

- Due to the presence of excess soft tissue between the skin and CFA in obese patients, the risk of bleeding goes up from an antegrade access site. Thus caution should be used when using this access method in obese patients.
- Morbid obesity is a contraindication to antegrade access.

PREREQUISITES

- Antegrade CFA access is feasible only in those patients who have minimal disease in the proximal SFA, as the sheath is 'parked' in this vessel.
- Ideally, the PFA should be patent as well, otherwise there is a risk of peri-operative limb ischemia as the CFA sheath, which enters the SFA, may be at least partially occlusive.

PATIENT AND ROOM SETUP

- The patient is placed in supine position (figure 9).

- Both groin sites, from umbilicus down, are prepared in a standard fashion and cleaned with antiseptic solution.
- The patient is draped from head to toes: the face is left exposed but a 'shroud' is created - sterile towels are placed around the face so that the likelihood of sterile equipment coming in contact with the skin and getting contaminated is minimum.
- The 'back' (supplies) table is setup in tandem with the operating table, at the head end.
- Since the skin puncture site on the side of the antegrade stick is cephalad to the inguinal ligament, the groin aperture in the drape is centered higher than usual.
- We recommend obtaining retrograde CFA access on the contralateral side so that a catheter can be advanced over the aortic bifurcation for either 'road-mapping' the ipsilateral CFA or placing a 'target wire' in it.
- Alternatively, although not ideal, antegrade access may be obtained under ultrasound guidance.
- Standing on the left side of the patient affords a right-handed operator a suitable position for forehand access (to either CFA). We favor this set up even in the case of left lower extremity intervention.
- The operator, therefore, stands on the left side of the patient, regardless of the extremity being intervened upon (figure 9).

LANDMARKS

- The target entry site into the CFA remains the same in antegrade femoral access as in retrograde access (figure 7). The path from skin to CFA is different, however.
- Anatomical landmarks for location and palpation of femoral pulse remain the same as well:
  - ▹ The anterior superior iliac spine
  - ▹ The pubic tubercle
- An imaginary straight line is used to connect these two bony landmarks. This line is bisected at the junction of its medial third with the lateral two thirds.
- The skin incision for antegrade CFA access is made two finger-breadths *above* this point of intersection.
- The CFA is located, nearly two finger-breadths below this point of intersection.

Figure 7.

Landmarks for antegrade CFA access.

- Bony landmarks for CFA pulse palpation are the same. However, the skin incision is made 2 finger-breadths cephalad of the imaginary inguinal ligament line.
- Green arrow denotes the site of skin incision.
- White arrow denotes the site of CFA puncture.

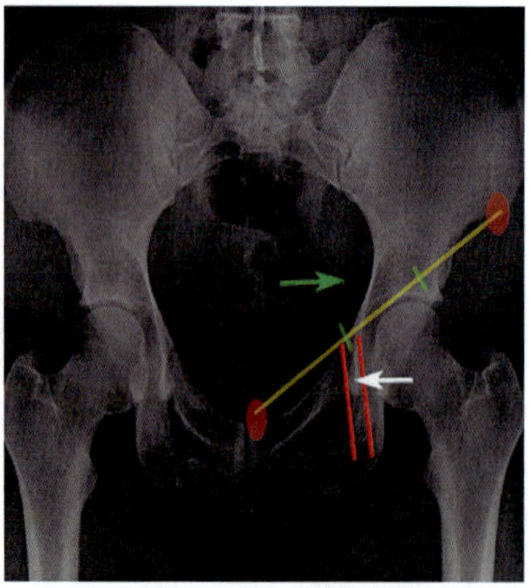

EQUIPMENT

- We use a 4 Fr. micropuncture kit for obtaining femoral artery access.
- Alternatively, an 18 Gauge needle may be used with a steerable 0.035" wire such as an angled stiff Glidewire® or the Wholey wire.
- If this latter needle is used, we prefer a 9 cm long needle particularly in mild to moderately obese patients.
- A 5 - 7 Fr. working sheath is used depending upon the site and nature of the intended intervention.

TECHNIQUE

Retrograde Access in Contralateral CFA:

- We always obtain access in the contralateral (to the side where antegrade access is planned) CFA and insert a 4 Fr./11 cm sheath. We secure this with 2-0 Silk. After flushing this sheath, we advance a selective catheter to engage the ostium of the (ipsilateral) common iliac artery. We obtain a 'road-map' or advance a 0.035" wire through this catheter into the ipsilateral SFA. This 'target' wire highlights the CFA and is used to direct the antegrade needle into the CFA.
- We give 2000 units of IV UFH at this point but do not fully anticoagulate the patient until the antegrade sheath has been successfully placed.

Skin Incision:

- Any tissue pannus should be taped cranially and away from the intended track of the needle.
- We prefer to make a small incision in the skin at the site of the puncture.
- Ideally, this should be done before the needle is inserted in the skin, so that a subcutaneous 'track' can be made.
- This incision is usually made two finger-breadths above the level of inguinal ligament.

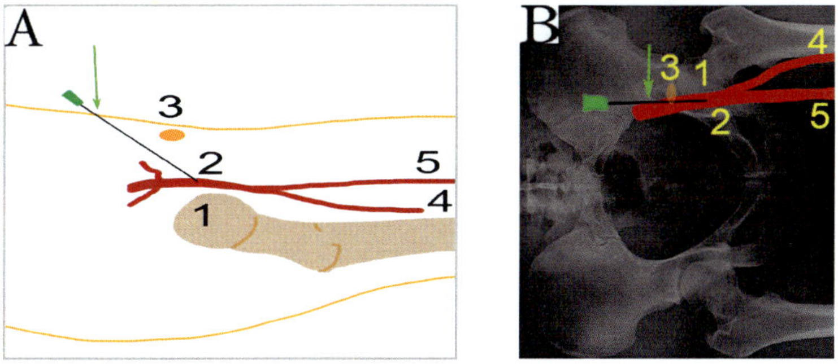

Figure 8.

Needle track for antegrade CFA access.

A. Schematic of lateral view of the supine patient to show landmarks and relations.

B. AP view of the supine patient to show landmarks and relations.
1. Femoral head
2. CFA
3. Inguinal ligament
4. PFA
5. SFA

Green arrow denotes the site of skin incision.

Needle Alignment:

- The needle should be held at an angle of < 45° to the skin, bevel up (at 12 O' Clock position).
- An angle steeper than this may cause difficulty in steerability of the wire in the CFA and into SFA.
- The needle track should be coaxial to the SFA (figures 8A & B).
- A medial to lateral track may direct the wire into the PFA.

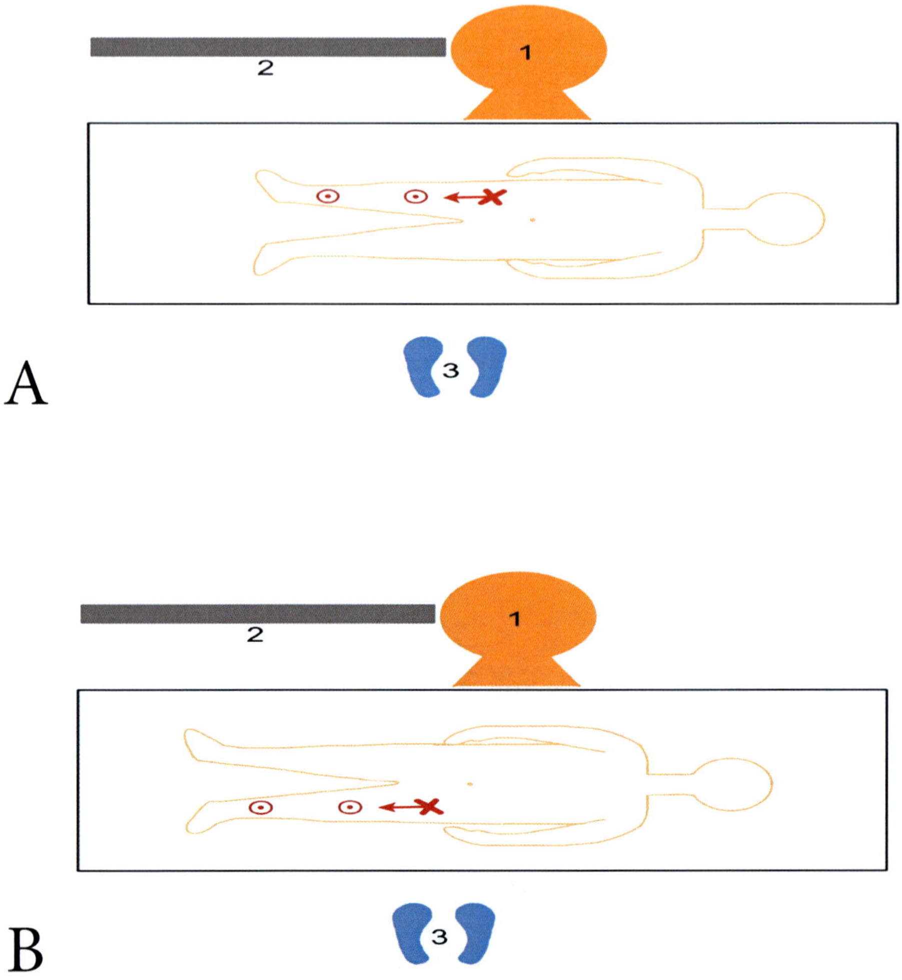

Figure 9.

Room set-up for lower extremity angiography and intervention using antegrade CFA access.

A. Room set-up for right lower extremity antegrade CFA access.
B. Room set-up for left lower extremity antegrade CFA access.

- 1. Image intensifier.
- 2. Image display and hemodynamic monitors.
- 3. Operator position.
- Red crosses denote arterial access sites.
- Red arrows denote direction of the access.
- Red circles denote target vessels for intervention.

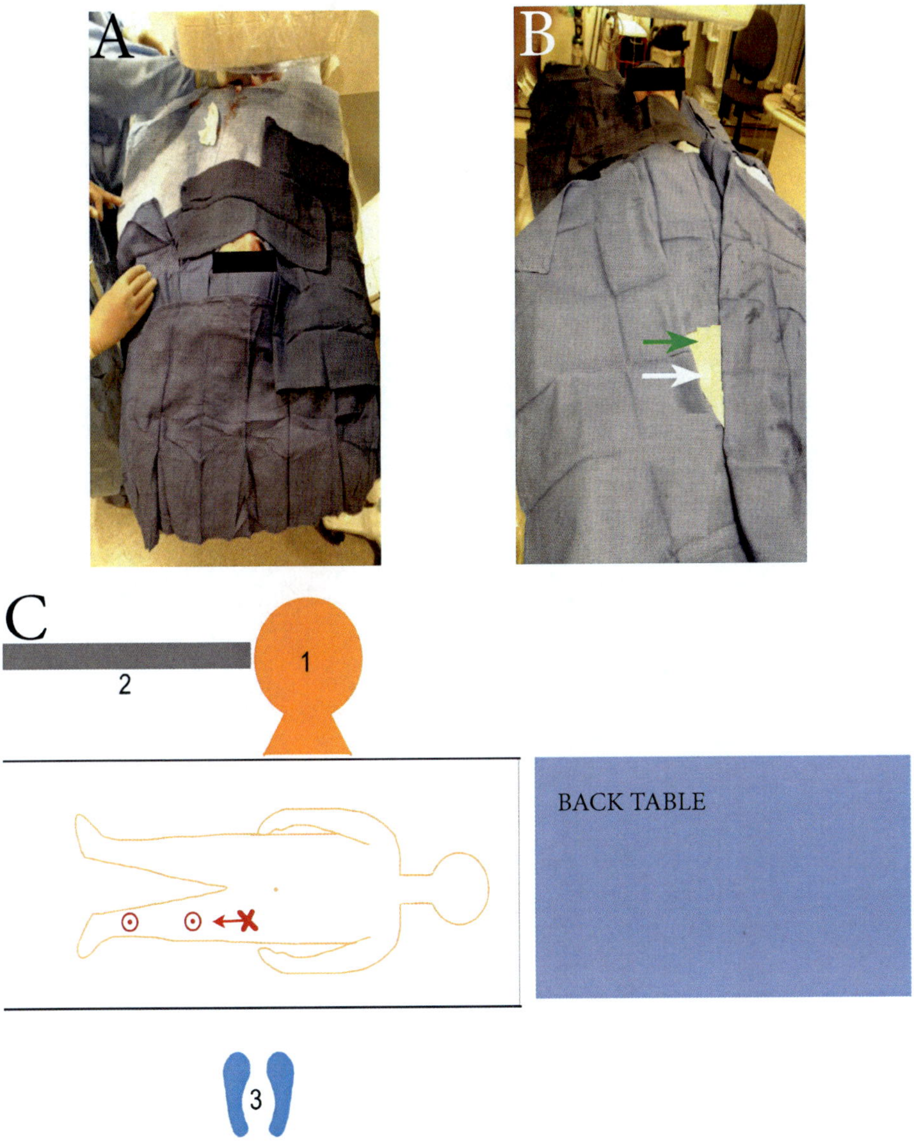

Figure 10.

Patient preparation for antegrade CFA access.

A. The patient is draped from head to toes, leaving the face exposed.
B. The green arrow denotes the skin incision site and the white arrow indicates the position of the desired femoral artery stick site, at the center of the femoral head.
C. The back table is placed behind the head of the patient. This provides an ergonomic work surface that is in line with the patient.

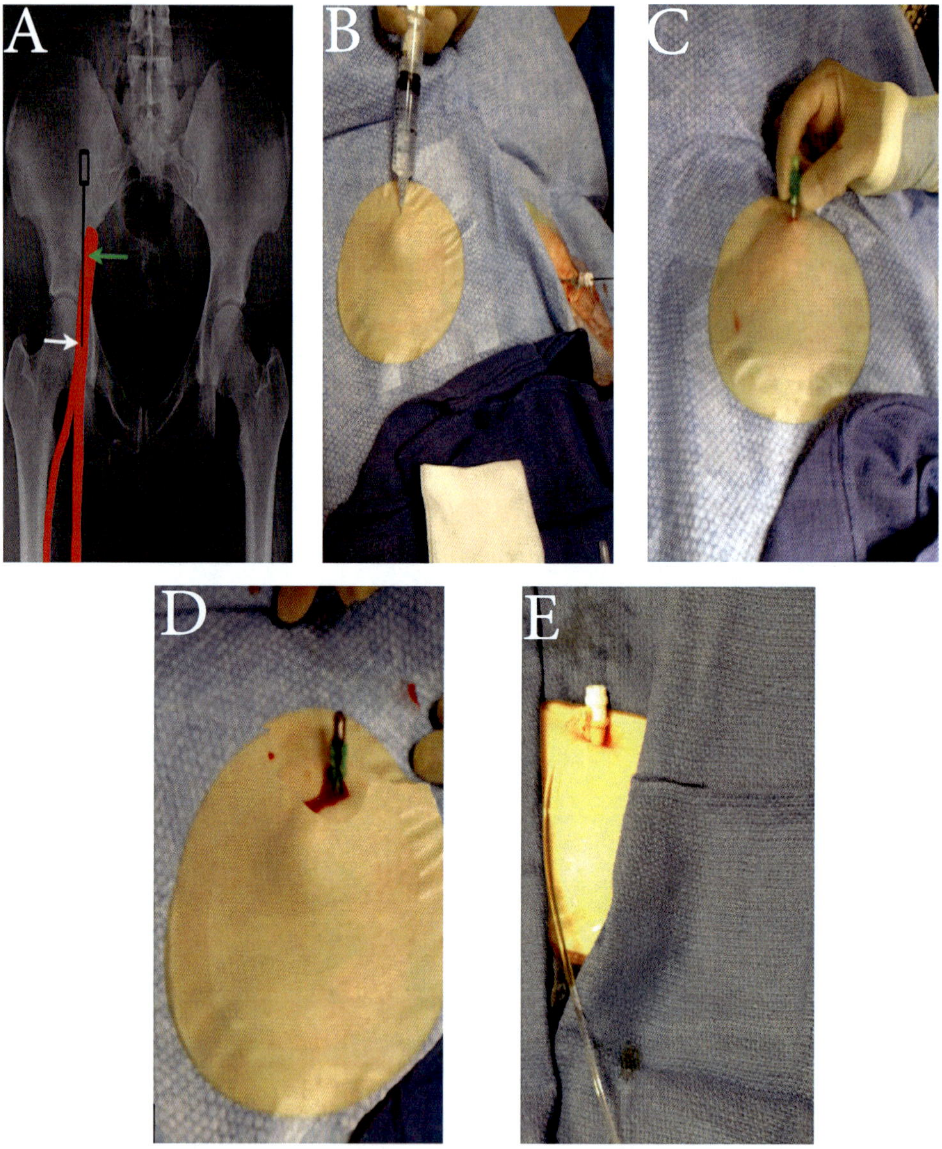

Figure 11.

Fluoroscopic correlates of antegrade CFA access.

A. Needle entry site into the skin (green arrow) and the CFA (white arrow).
B. Local anesthetic infiltration at the proposed skin entry site after confirmation under fluoroscopy/road map.
C. A 21 G needle is advanced through the skin incision to right CFA. This step should be performed under fluoroscopy as well.
D. The needle enters the artery. Despite the thin body habitus of this patient, there is abundant tissue between the skin entry site and the CFA, as indicated by the fact that needle is in almost to the hub.
E. A 7 Fr. sheath has been advanced into the CFA. This sheath is 23 cm long and it has been advanced to its hub. Most of the sheath thus sits in the right SFA.

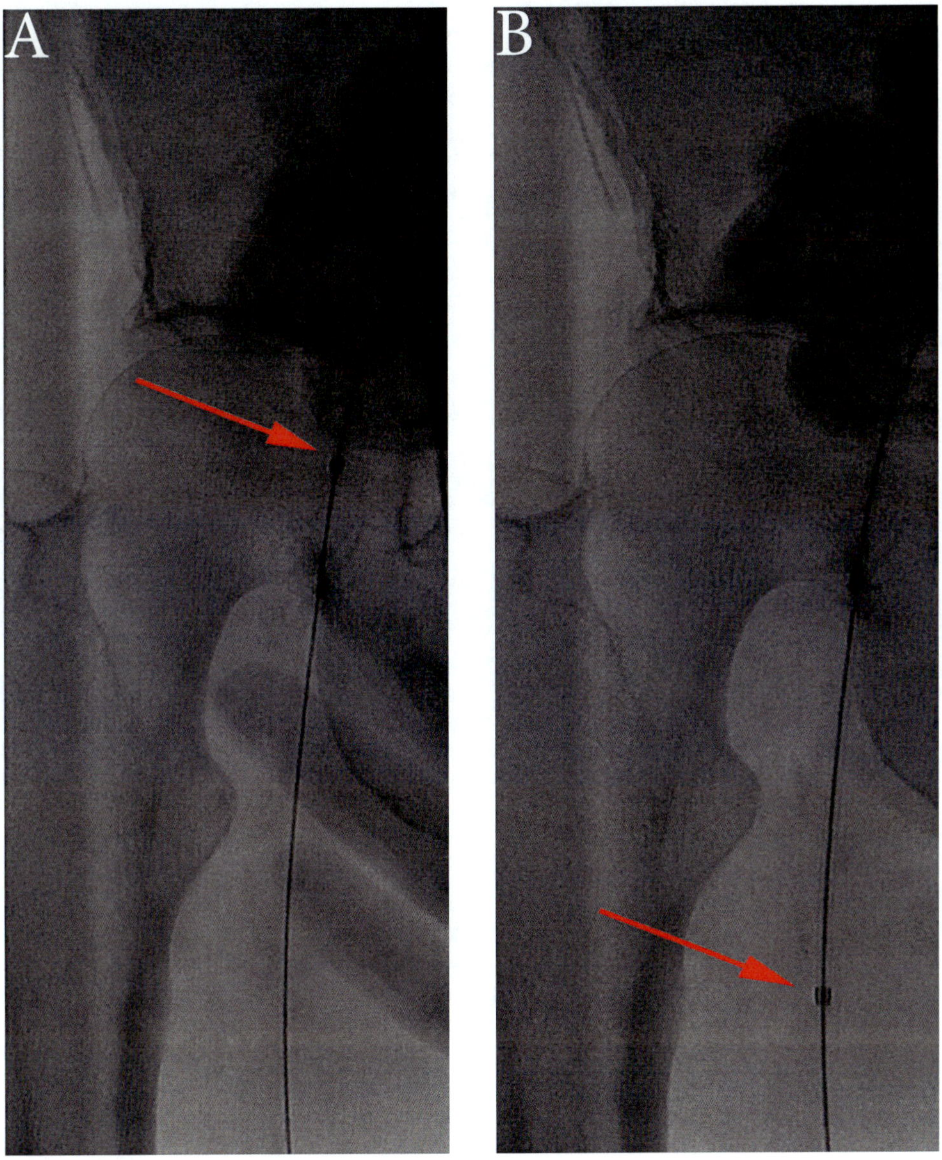

Figure 12.

Insertion of an antegrade CFA sheath.

A. The sheath tip (red arrow) is seen at the level of the arteriotomy.

B. The sheath has been advanced in the proximal SFA.

Needle Advancement:

- The needle should be advanced slowly and smoothly, cranial to caudal, in the direction of the CFA with the aim to enter the artery at or just above the center of the femoral head.
- If the artery is entered above this level, in the EIA, it may result in retroperitoneal hemorrhage that may be catastrophic after removal of the sheath.
- If the artery is entered below this point, in the vicinity of the CFA bifurcation, there may not be enough room to maneuver the wire into the SFA.

Micropuncture wire:
- The micropuncture wire can be safely withdrawn before it exits the needle tip.

Use of Fluoroscopy & Wire Advancement:

- The arteriotomy should be made under fluoroscopic guidance, either using 'road-map' or aiming for a 'target' wire in the CFA, as noted above.
- The wire should always be advanced under fluoroscopy, to prevent injury to the vessel and to avoid entry into branches like the PFA or lateral circumflex artery.
- The wire should be advanced into the SFA.
- Ideally, the wire should not be withdrawn through the needle. It can get unraveled and be trapped in the tissues, cause vessel dissection or be severed. The needle should be removed over the wire.
- This is also true of the Glidewire® which should be withdrawn through an 18 Gauge needle only if necessary and with extreme caution, first making the needle and wire perfectly coaxial.

Stiff wire:
- A 0.035" Super-Stiff®Amplatz wire is adequate.
- Use the wire with a soft J-tip.
- Always advance through a catheter e.g. Glidecath®.
- Always advance under fluoroscopy.

Microcatheter:

- Once the wire is in the desired location, the needle is removed and the microcatheter is advanced to its hub, into the vessel.

Angiography:

- The microcatheter is fixed in the artery, while the dilator and 0.018" wire are withdrawn.
- An angiogram is performed to confirm that the arteriotomy is in the desired location prior to up sizing the sheath.
- This can also be performed through the contralateral access catheter but may not be as accurate as through the microcatheter itself.

0.035" Wire:

- Once the arteriotomy is confirmed to be safe, a torqueable 0.035" wire (e.g. angled stiff Glidewire® or Wholey wire) is advanced through the microcatheter and steered into the SFA.
- If the SFA is diseased do not advance this wire beyond the mid segment as this may result in a dissection.

Sheath size:
- Infra-popliteal PTA / stenting - 5 Fr.
- Infra-popliteal atherectomy - 5/6 Fr.
- Femoro-popliteal PTA - 5 Fr.
- Femoro-popliteal stenting- 5/6 Fr.
- Femoro-popliteal atherectomy - 6/7 Fr.

Working Sheath:

- After removal of the microcatheter, a 5 - 7 Fr./23 cm sheath with a radio-opaque tip marker (e.g. Cordis Britetip®) should be advanced under fluoroscopic guidance.
- After this sheath is in place the patient can be fully anticoagulated for the rest of the procedure.

Arteriotomy Care:

- At the conclusion of the procedure, the arteriotomy is dealt with in more or less the same fashion as a retrograde CFA access.

- Closure devices such as the Angioseal® and Perclose Proglide® have been safely used with this access as well. Their adequacy can be confirmed on angiography via the contralateral catheter.
- One major departure from standard practice is that if manual compression is to be used for arteriotomy closure, we choose to reverse anticoagulation at the end of the case. The sheath is then removed while the patient is on the table.
- Compression is held for at least 20 minutes.
- Again, satisfactory hemostasis can be confirmed on angiography via the contralateral catheter.

TROUBLESHOOTING

Inability to Advance the Wire:

- This may be due to one of several reasons:
  - ▹ A steep angle between the needle and artery.
  - ▹ The needle tip may have advanced too deep into the vessel wall.
  - ▹ The needle tip abuts a plaque.
- The remedial steps consist of the following:
  - ▹ Remove the wire.
  - ▹ Withdraw the needle 1-2 mm.
  - ▹ Confirm adequate back flow.
  - ▹ Attempt re-wiring.
  - ▹ If continued resistance is encountered at the tip of the needle, withdraw the wire, confirm good back flow, rotate the needle 30-90° clockwise or counterclockwise and re-wire.
  - ▹ If unsuccessful, remove the needle and hold pressure on the CFA for 3-5 minutes.
  - ▹ Change the angle of approach and make it less steep (if already not < 45°).
  - ▹ Change the fluoroscopic view.
  - ▹ Reattempt.

Inability to Advance the Working Sheath:

- This happens more frequently with antegrade than retrograde CFA access.
- The most common reason is a long, tortuous subcutaneous track created by a steep angle of approach.
- The solution consists of either removing the wire and starting anew with the access, or switching the wire out for a stiffer wire (as outlined in the previous section).
- The latter is usually successful.

RETROGRADE POPLITEAL ARTERY ACCESS

INDICATIONS

- Retrograde popliteal artery access is usually obtained for crossing an SFA CTO in a retrograde direction that otherwise could not be crossed using an antegrade approach.
- The goal is to place a 4 Fr. microcatheter/Glidecath® to assist in advanc-

ing a wire retrograde and crossing occlusive disease in the SFA.

PREREQUISITE

- The popliteal artery must be patent and relatively healthy in the proposed access area.
- An assistant is required to do an angiogram as access is being obtained.

PATIENT AND ROOM SETUP

- The patient is placed in supine position.
- Both groin sites, from umbilicus down, are prepared in a standard fashion and cleaned with antiseptic solution (in case antegrade CFA access is also needed).
- Additionally, for popliteal artery access the leg must be generously cleaned with antiseptic solution from upper thigh to the toes.
- The leg and foot are rested on a sterile sheet.
- The patient is draped in a standard fashion.
- A 12" x 12" H-shaped cut is made in the drape, in the section overlying the proximal thigh.
- The leg, still sterile, is pulled through this hole in the drape to the top of the table.
- If desired, a sterile stockinette can be put on the leg and foot.

LANDMARKS

- The target entry site into the popliteal artery is determined by angiographic findings.
- There are no clear cut bony landmarks that apply to all cases.

EQUIPMENT

- We use a 4 Fr. micropuncture kit for obtaining access.
- A 4 Fr./100 cm straight Glidecath®.
- A 0.035" straight or angled stiff Glidewire®.
- A short 4 Fr. sheath may be inserted over a 0.035" wire, after removal of the microcatheter.

TECHNIQUE

Retrograde Access in Contralateral CFA:

- We always obtain access in the contralateral CFA and insert a 6 Fr./45 cm Ansel sheath, which is advanced over the aortic bifurcation into the ipsilateral EIA.
- An angiogram is obtained to identify the popliteal artery entry site.
- A healthy segment of the popliteal artery should be selected.
- We give 2000 units of IV UFH at this point. We do not fully anticoagulate the patient until the popliteal artery sheath has been successfully placed.

Leg Positioning:

- The leg is bentslightly at the hip and knee joints so that the popliteal fossa is exposed (figures 13A & 13B).
- The micro-puncture kit is prepared.

Skin Incision:

- We do not make an incision in the skin prior to making an arteriotomy.

- With the leg elevated, an angiogram is obtained to determine the site of popliteal artery puncture. Alternatively, ultrasound with sterile technique can be used for this purpose.
- Accordingly, the corresponding skin puncture site is selected.
- Local anesthetic is given in a standard fashion.
- The leg is rested back on the table for a couple of minutes.
- Once the needle has entered the artery, a short incision is made in the overlying skin at the needle entry site.

Needle Alignment:

- The needle should be held at an angle of < 45° to the skin, bevel down (at 6 O' Clock position).
- The direction of the track is not as important here as in CFA access. If it is possible to do an intra-op ultrasound, the relationship of the artery to the adjacent vein can be identified and venipuncture avoided.

Needle Advancement:

- The arteriotomy should be made under fluoroscopic guidance, with dye injection from above to guide it.
- The leg is elevated as above and, under fluoroscopy or ultrasound guidance (either using calcium in the artery or dye injection from the CFA sheath for guidance), the popliteal artery is punctured with the 21 G needle (figures 13B & 13C).
- The needle should be advanced slowly and smoothly, caudal to cranial, in the direction of the popliteal artery at or above the level of the femoral condyles.

Use of Fluoroscopy & Wire Advancement:

- Once reasonable (arterial) back bleeding is noted, the wire is advanced. An assistant may have to hold the leg as this step is performed.
- The wire should be introduced under fluoroscopy.
- The wire should be advanced as close to the distal cap of the SFA occlusion as safely possible.

Microcatheter:

- Prior to advancing the microcatheter, we make a small, transverse incision in the skin.
- Once the wire is in the desired location, the needle is removed and the microcatheter is advanced into the vessel.
- If the artery has been entered close to the distal cap of the occlusion or diseased segment, do not advance the tip of the microcatheter beyond the visible lumen.
- At this point, the leg can be placed back on the drape. A roll of towels or a small pillow placed under the drape, behind the calf, may allow access to the microcatheter in the popliteal artery, without repeatedly elevating the leg off the table.

Angiography:

- The microcatheter is fixed in the artery, while the dilator and 0.018" wire are withdrawn.
- Using a 3 cc syringe, an angiogram is performed to confirm that the arteriotomy is in the desired location prior to advancing a 0.035" wire.

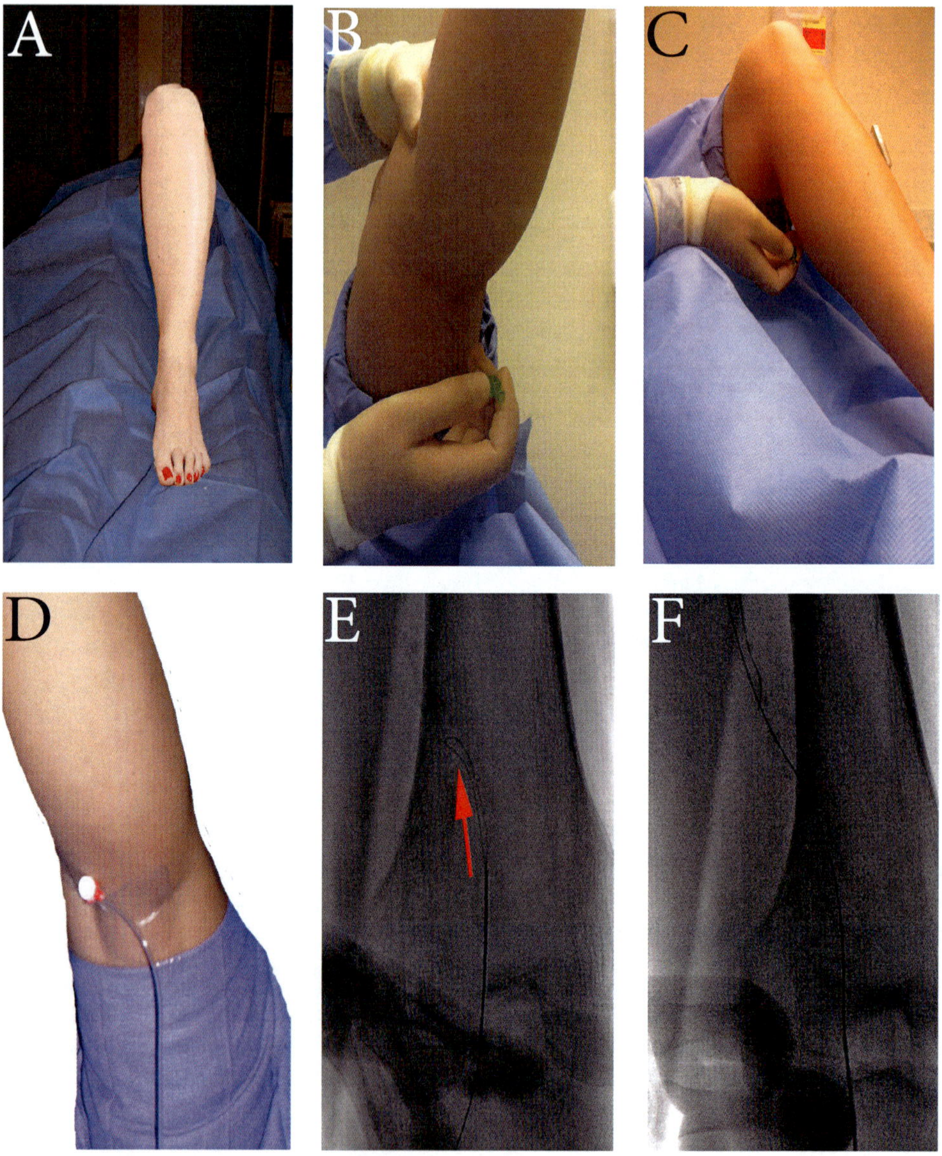

Figure 13.

Supine retrograde popliteal artery access.

A. The left leg is placed on top of the drape after being cleansed and prepped.
B. The patient's leg is bent at the knee and hip. It is elevated off the table. After administration of local anesthetic and under fluoroscopic guidance, the operator advances a 21 G needle into the popliteal artery.
C. Access may also be obtained if the leg is elevated only partially, using a steeper angle of approach from behind the popliteal fossa.
D. A 4 Fr. sheath may be inserted for easier re-entry of devices.
E. The 0.018" wire (red arrow) is being advanced through the 21 G needle in a retrograde direction.
F. Through the microcatheter a 0.035" ASGW is then advanced across the distal cap in a retrograde direction.

- This step should be done through the popliteal artery microcatheter rather than indirectly confirming through CFA sheath injection. The operator has to be certain that the catheter is present in a relatively healthy part of the popliteal artery and that it is not in a dissection plane or in the popliteal vein.

0.035" Wire:

- Once the arteriotomy is confirmed to be desirable, a 0.035" straight stiff Glidewire® is advanced through the microcatheter, up to the distal cap of the occlusion.

Working Sheath:

- The microcatheter is removed and replaced with either a short 4 Fr. Britetip® sheath or a bareback 4 Fr./100 cm straight Glidecath®.
  - ▹ The sheath should be advanced under fluoroscopic guidance, making sure the tip of the dilator does not extend beyond the wire .
- The patient may be fully anticoagulated at this point.

> ***CAUTION:***
> - Do NOT withdraw the wire through the needle. The wire can get unraveled and be trapped in the tissues, cause vessel dissection or be severed.

Intervention:

- Once the occlusion has been successfully crossed, the rest of the intervention proceeds in an antegrade fashion, via the CFA sheath.
- This is done after 'externalizing' the popliteal wire through the femoral sheath.

Arteriotomy Care:

- Once antegrade wire placement across the occlusion has been secured and confirmed, the Glidecath®/sheath from the popliteal artery can be removed.
- A prolonged balloon inflation at the arteriotomy site may be sufficient to seal it. At times it may be necessary to put a stent in the popliteal artery to seal an ongoing brisk extravasation or a prominent AV fistula with the popliteal vein.
- Although closure devices such as the Angioseal® and Perclose Proglide® have been safely used with this access, we do not advocate the use of these devices.
- Satisfactory hemostasis can be confirmed on angiography via the CFA sheath.

TROUBLESHOOTING

Inability to 'Stick' the Artery:

- This may be due to one of several reasons:
  - ▹ The artery rolls under the tip of the needle.
  - ▹ The artery is heavily calcified.
  - ▹ The vein is entered instead of the artery.
  - ▹ A steep angle between the needle and artery.
  - ▹ The needle tip may have advanced too deep into the vessel wall.
  - ▹ The needle tip abuts a plaque.
- The remedial steps consist of the following:
  - ▹ Remove the wire.
  - ▹ Withdraw the needle 1-2 mm.
  - ▹ Confirm adequate back flow.
  - ▹ Attempt re-wiring.

- ▹ If continued resistance is encountered at the tip of the needle, withdraw the wire, confirm good back flow, rotate the needle 30-90° clockwise or counterclockwise and re-wire.
- ▹ If unsuccessful, remove the needle and hold pressure for 3-5 minutes.
- ▹ Try sticking the artery at a different level.
- ▹ Reattempt.
- ▹ If still unsuccessful, consider putting the patient in the prone position and obtaining arterial access under ultrasound guidance.

DORSALIS PEDIS ARTERY ACCESS

INDICATIONS

- Retrograde dorsalis pedis (DP) artery access is usually obtained for crossing femoro-popliteal or tibial CTOs that cannot be crossed in an antegrade fashion.
- The goal is to place a 4 Fr. microcatheter/Glidecath®/sheath to assist in advancing a wire retrograde and crossing occlusive disease.

PREREQUISITE

- The DP (and ideally the distal AT) must be patent and relatively healthy in the proposed access area.
- An assistant is required to do an angiogram/ultrasound as access is being obtained.

PATIENT AND ROOM SETUP

- The patient is placed in supine position.
- Both groin sites, from umbilicus down, are prepared in a standard fashion and cleaned with antiseptic solution.
- Additionally, for DP access the foot must be generously cleaned with antiseptic solution from upper leg to the toes.
- The leg and foot are rested on a sterile sheet.
- The patient is draped in a standard fashion.
- A 6" x 12" H-shaped cut is made in the drape, in the section overlying the lower leg.
- A sterile stockinette is put on the leg and foot and they are pulled through this hole in the drape to the top of the table.
- A hole is cut in the stockinette, in the area overlying the dorsum of the foot and ankle joint.

LANDMARKS

- Pulse in the DP is usually weak or not palpable due to upstream PAD.
- Access is typically obtained in the segment between the dorsum of the foot and ankle joint.
- The target entry site into the DP is determined by angiographic findings.
- There are no clear cut bony landmarks that apply to all cases.

EQUIPMENT

- We use a 4 Fr. micropuncture kit for obtaining access.

- A 4 Fr./100 cm straight Glidecath®.
- A 0.035" straight or angled stiff Glidewire® (for femoro-popliteal CTO) or a 0.014" interventional wire (for tibial CTO).

TECHNIQUE

CFA Access:

- We almost always obtain access in the CFA (contralateral retrograde or ipsilateral antegrade) and insert at least a 5 Fr. sheath.
- If contralateral CFA access is obtained, a cross-over sheath is inserted over the aortic bifurcation, into the ipsilateral EIA.
- An angiogram is obtained to identify the DP entry site.
- A healthy segment of the artery should be selected.
- The patient may be fully anticoagulated, if necessary, prior to obtaining DP access.

Foot/Leg Positioning:

- The leg is kept straight and internally rotated so the toes point to the ceiling, exposing the dorsum of the foot to the operator.
- The micropuncture kit is prepared.
- Under fluoroscopy or ultrasound guidance, the DP is punctured with the 21 G needle.
- If using ultrasound, a 15 MHz hockey-stick probe is used. The artery is identified in short axis and accessed in this view.
- Gray-scale imaging is used.

Skin Incision:

- We do not make an incision in the skin prior to the arteriotomy.
- An angiogram is obtained through the CFA sheath to determine the site of puncture. Alternatively, ultrasound with sterile technique can be used for this purpose.
- Accordingly, the corresponding skin puncture site is selected.
- Approximately 1-2 cc of local anesthetic is administered in a standard fashion.
- Once the needle has entered the artery, a very superficial and short incision is made in the overlying skin at the needle entry site.

Needle Alignment:

- The needle should be held at an angle of < 90° to the skin, bevel up (at 12 O' Clock position).
- However, when using ultrasound guidance sometimes, for better needle visualization, a 90° angle has to be used. Caution should be used when introducing the wire in this case.

Needle Advancement:

- The arteriotomy should be made under fluoroscopic guidance, with dye injection from above or ultrasound to guide it.
- The needle should be advanced slowly and smoothly, caudal to cranial, in the direction of the artery at or below the level of the ankle joint.

Use of Fluoroscopy & Wire Advancement:

- Once reasonable back bleeding is noted, the wire is introduced.
- The wire should also be always visualized under fluoroscopy when being advanced.

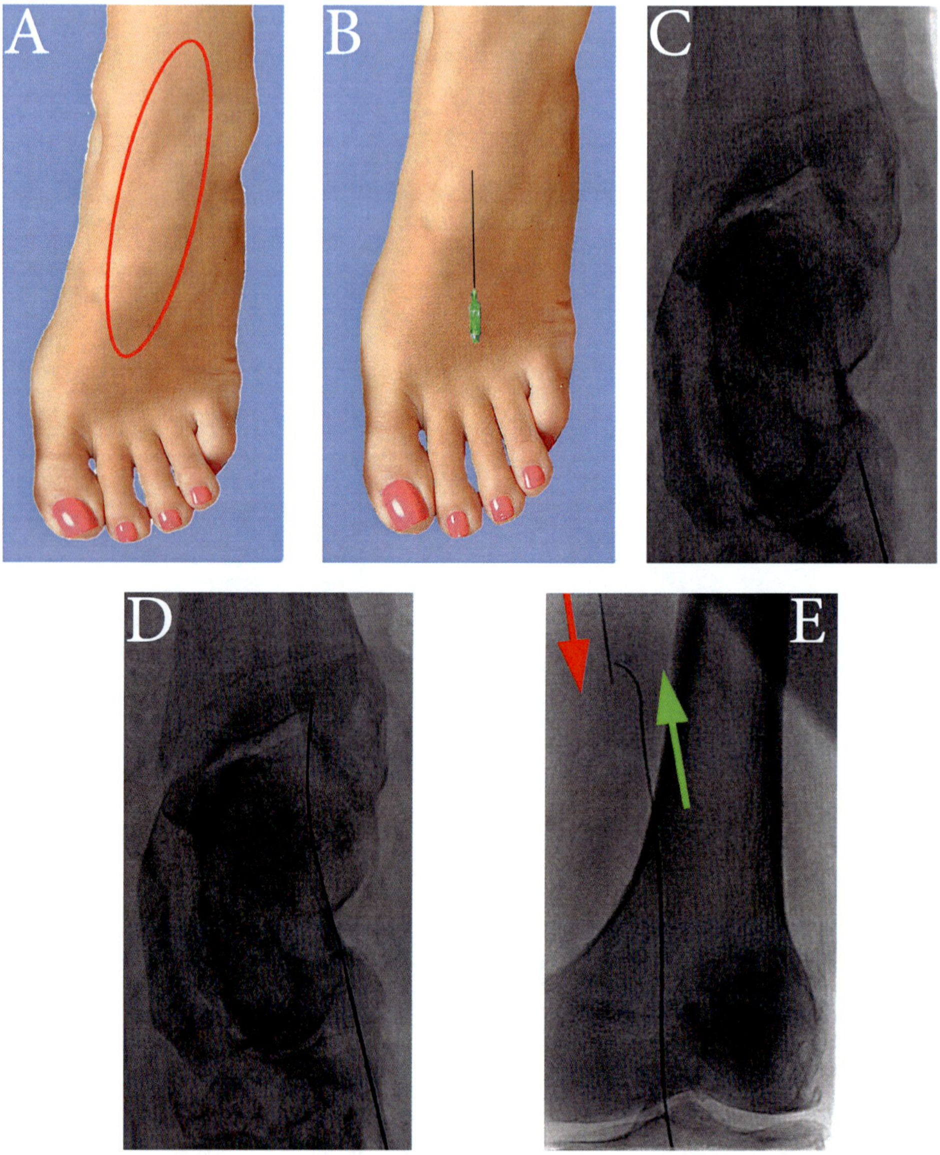

Figure 14.

Dorsalis pedis artery access.

A. The target site for DP access is in the proximal foot.
B & C. A 21 G needle is advanced using either US or fluoroscopic guidance to enter the DP.
D. The 0.018" wire supplied with the micropuncture kit is advanced through the needle into the AT.
E. A microcatheter is inserted over the wire. Once the dilator is removed, a 0.035" ASGW is used for femoro-popliteal occlusive disease. Here an antegrade wire (red arrow) marks the proximal cap while the retograde wire has crossed the occludsion (green arrow).

Microcatheter:

- Once the wire is in the desired location, the needle is removed and the microcatheter is advanced into the vessel.
- If the artery has been entered close to the distal cap of the occlusion or diseased segment, do not advance the tip of the microcatheter beyond the visible lumen.

Angiography:

- The microcatheter is fixed in the artery, while the dilator and 0.018" wire are withdrawn.
- Using a 3 cc syringe, an angiogram is done to confirm that the arteriotomy is in the desired location prior to advancing a 0.035" wire.
- This step should be done through the microcatheter.

Wire:

- For patients with AT occlusion, a 0.014" wire is usually used for retrograde crossing of the diseased segment.
- However, for patients with femoro-popliteal CTO, once the arteriotomy is confirmed to be safe, a 0.035" angled stiff Glidewire® is advanced through the microcatheter, up to the distal cap of the occlusion.

Glidecath®:

- The microcatheter is removed and replaced with a bareback 4 Fr./100 cm straight Glidecath®.
- This should be advanced under fluoroscopic guidance.

Intervention:

- Once the occlusion has been successfully crossed, the rest of the intervention proceeds in an antegrade fashion, via the CFA sheath.
- This is done after 'externalizing' the retrograde wire through the femoral sheath.

Arteriotomy Care:

- Once antegrade wire placement across the occlusion has been secured and confirmed, the microcatheter/Glidecath® from the DP can be removed and manual compression applied.
- Satisfactory hemostasis is usually achieved in minutes without difficulty, even in patients who are fully anticoagulated.
- A sterile dressing is applied and the site is checked intermittently during the rest of the procedure, for bleeding or hematoma formation.

TROUBLESHOOTING

Inability to 'Stick' the Artery:

- This may be due to one of several reasons:
  - ▹ The artery rolls under the tip of the needle.
  - ▹ The artery is heavily calcified.
  - ▹ The vein is entered instead of the artery.
  - ▹ A steep angle between the needle and artery.
  - ▹ The needle tip may have advanced too deep into the vessel wall.

POSTERIOR TIBIAL ARTERY ACCESS

INDICATIONS

- Retrograde posterior tibial (PT) artery access, like DP access, is usually obtained for crossing femoro-popliteal or tibial CTOs that cannot be crossed in an antegrade fashion.
- The goal is to place a 4 Fr. microcatheter/4 Fr. sheath (e.g. Precision™ sheath, Terumo)/Glidecath® to assist in advancing a wire retrograde and crossing occlusive disease.

PREREQUISITE

- The PT must be patent and relatively healthy in the proposed access area.
- An assistant is required to do an angiogram/US as access is being obtained.

PATIENT AND ROOM SETUP

- The patient is placed in supine position.
- Both groin sites, from umbilicus down, are prepared in a standard fashion and cleaned with antiseptic solution.
- Additionally, for tibial access the foot must be generously cleaned with antiseptic solution from upper leg to the toes. The leg and foot are rested on a sterile sheet.
- The patient is draped in a standard fashion.
- A 6" x 12" H-shaped cut is made in the drape, in the section overlying the lower thigh.
- A sterile stockinette is put on the leg and foot and they are pulled through this hole in the drape to the top of the table.
- Cut out a 4"x 9" hole in the stockinette, in the area overlying the medial aspect of the leg, from mid calf to ankle.
- The patient's leg is bent slightly at the knee joint and then externally rotated at the hip. This exposes the medial aspect of the lower leg (figure 15A).

LANDMARKS

- Pulse in the PT is usually weak or not palpable due to upstream PAD.
- Access is typically obtained in the segment above the medial malleolus, although access in the medial/lateral plantar arteries has been described as well.
- The target entry site into the PT is determined by angiographic/US findings.
- There are no clear cut bony landmarks that apply to all cases.
- Often calcium in the artery may be seen.
- Ultrasound guidance is a very effective method of obtaining access in the tibial arteries.
    - ▹ A hockey-stick probe (covered in a sterile probe cover)is used and the artery is visualized in short axis.

EQUIPMENT

- We use a 4 Fr. micropuncture kit for obtaining access.
- A 4 Fr./100 cm straight Glidecath®.

- A 0.035" straight or angled stiff Glidewire® (for femoro-popliteal CTO) or a 0.014" interventional wire (for tibial CTO).

TECHNIQUE

CFA Access:

- We always obtain access in the CFA (contralateral retrograde or ipsilateral antegrade) and insert at least a 5 Fr. sheath.
- If contralateral CFA access is obtained, a cross-over sheath is inserted over the aortic bifurcation, into the ipsilateral EIA.
- An angiogram is obtained to identify the entry site.
- A healthy segment of the artery should be selected.
- The patient may be fully anticoagulated, if necessary, prior to obtaining PT access.

Foot/Leg Positioning:

- The leg is bent slightly at the knee and the foot externally rotated to expose the medial aspect to the operator.
- The micropuncture kit is prepared.
- Under fluoroscopy or ultrasound guidance, the PT is punctured with the 21 G needle.

Skin Incision:

- We do not make an incision in the skin prior to the arteriotomy.
- An angiogram is obtained through the CFA sheath to determine the site of puncture. Alternatively, ultrasound with sterile technique can be used for this purpose.
- Accordingly, the corresponding skin puncture site is selected.
- Approximately 1-2 cc of local anesthetic is administered in a standard fashion.
- Once the needle has entered the artery, a very superficial and short incision is made in the overlying skin at the needle entry site.

Needle Alignment:

- The needle should be held at an angle of <45° to the skin, bevel up (at 12 O' Clock position).

Needle Advancement:

- The arteriotomy should be made under fluoroscopic guidance, with dye injection from above or ultrasound to guide it.
- The needle should be advanced slowly and smoothly, caudal to cranial, in the direction of the artery at or below the level of the medial malleolus.

Use of Fluoroscopy & Wire Advancement:

- Once reasonable back bleeding is noted, advance the wire.
- The wire should also be always visualized under fluoroscopy when being advanced.

Microcatheter:

- Once the wire is in the desired location, the needle is removed and the microcatheter is advanced into the vessel.
- If the artery has been entered close to the distal cap of the occlusion or diseased segment, do not advance the tip of the microcatheter beyond

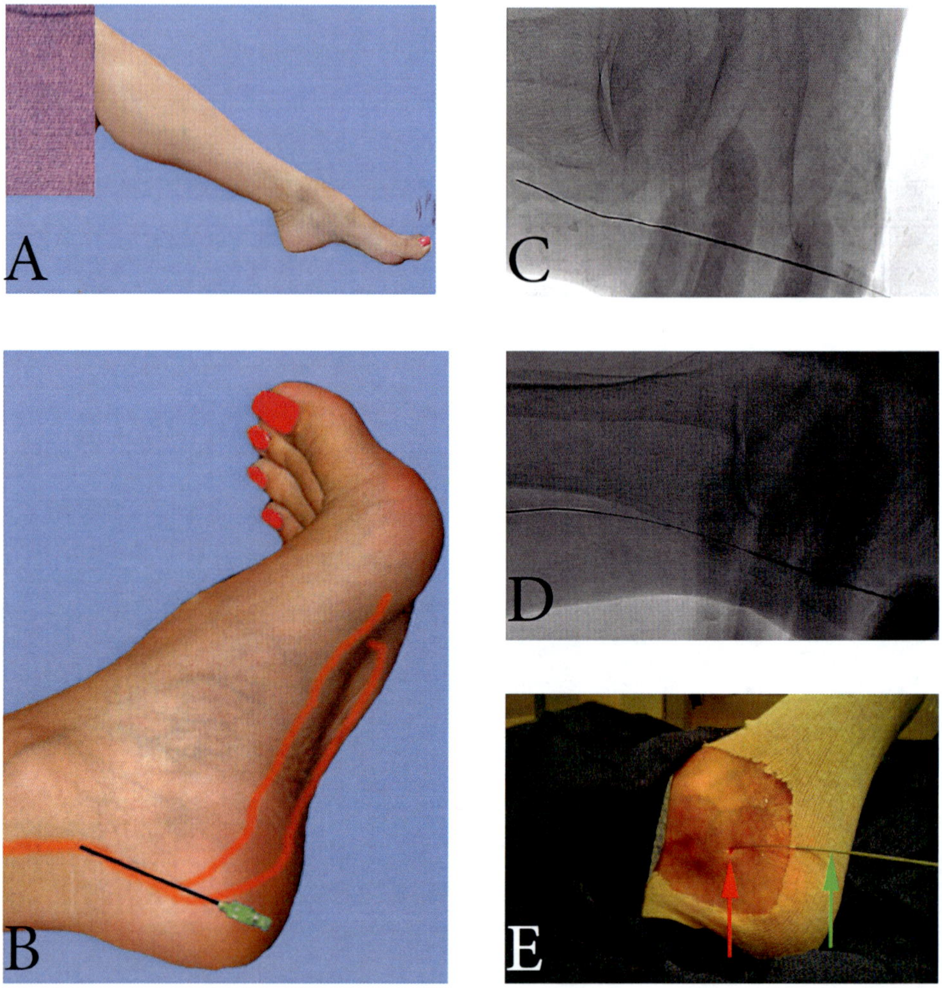

Figure 15.

Posterior tibial artery access.

A. Patient positioning for PT access.
B. Direction of the needle puncture for PT access. A 21 G needle is advanced through the skin parallel to the PT.
C. An 0.018" wire supplied with the micropuncture kit is advanced through the needle into the PT.
D. As the wire is advanced further it is seen under fluoroscopy.
E. The foot has been draped. The red arrow marks the skin entry site and the green arrow shows the microcatheter being inserted into the PT.

the visible lumen.

Angiography:

- The microcatheter is fixed in the artery, while the dilator and 0.018" wire are withdrawn.
- Using a 3 cc syringe an angiogram is done to confirm that the arteriotomy is in the desired location prior to advancing a 0.035" wire.
- This step should be done through the microcatheter.

Wire:

- For patients with tibial occlusion, a 0.014" wire is usually used for retrograde crossing of the diseased segment.
- However, for patients with femoro-popliteal CTO, once the arteriotomy is confirmed to be safe, a 0.035" angled stiff Glidewire® is advanced through the microcatheter, up to the distal cap of the occlusion.

Glidecath®:

- The microcatheter is removed and replaced with a bareback 4 Fr./100 cm straight Glidecath®.
- This should be advanced under fluoroscopic guidance.

Intervention:

- Once the occlusion has been successfully crossed, the rest of the intervention proceeds in an antegrade fashion, via the CFA sheath.
- This is done after 'externalizing' the retrograde wire through the femoral sheath.

Arteriotomy Care:

- Once antegrade wire placement across the occlusion has been secured and confirmed, the microcatheter/Glidecath® from the PT can be removed and manual compression applied.
- Satisfactory hemostasis is usually achieved in minutes without difficulty, even in patients who are fully anticoagulated.
- A sterile dressing is applied and the site is checked intermittently during the rest of the procedure, for bleeding or hematoma formation.

TROUBLESHOOTING

Inability to 'Stick' the Artery:

- This may be due to one of several reasons:
  - ▷ The artery rolls under the tip of the needle.
  - ▷ The artery is heavily calcified.
  - ▷ The vein is entered instead of the artery.
  - ▷ A steep angle between the needle and artery.
  - ▷ The needle tip may have advanced too deep into the vessel wall.
  - ▷ If the artery cannot be entered in the distal (perimalleolar) segment, it may be accessed more proximally if an adequate target entry site is present.

FEM-FEM BYPASS GRAFT ACCESS

INDICATIONS

- A femorofemoral or cross-femoral bypass graft is an extra-anatomic bypass that consists of a 6 - 10 mm diameter Dacron® or ePTFE/xPTFE conduit, connecting the right CFA to the left.
- It is used for the treatment of severe inflow disease on one side. This conduit is tunneled through the anterior abdominal wall, superior to the pubis.
- Access in this graft is used to treat either aorto-iliac disease on the donor side or infra-ingui-

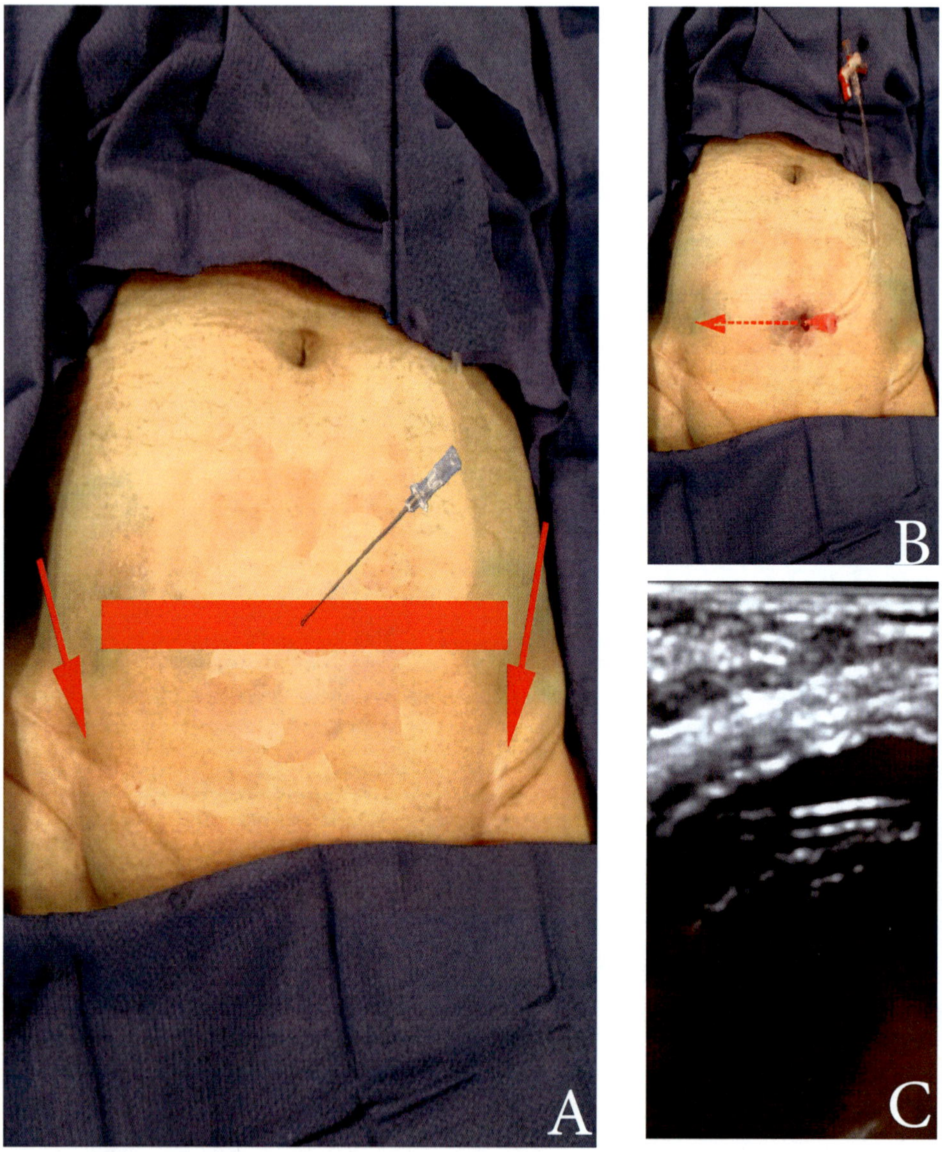

Figure 16.

Femorofemoral bypass graft access.

A. A patient with fem-fem bypass graft is prepped and draped. The red arrows indicate the scars over each femoral artery. The bypass graft is roughly outlined by the red rectangle. An 18 G needle is shown aiming at the center of the graft, pointing towards the right leg (target for intervention).
B. A 4 Fr. sheath has been inserted in the bypass graft. The dotted red arrow indicates the direction of sheath insertion.
C. An ultrasound image shows the sheath ('tram line') in the bypass graft in the long axis.

nal disease on either side.

- The goal is to place a sheath that points towards the side of interest, and can be used to perform angiography and intervention on the affected leg.

PREREQUISITE

- The graft must be patent.

PATIENT AND ROOM SETUP

- The patient is placed in supine position.
- Both groin sites and the entire lower abdomen from umbilicus down are prepared in a standard fashion and cleaned with antiseptic solution.
- A standard drape can be used by placing one of the groin apertures on top of the site where access in the graft is planned.
- Alternatively, a hole can be devised in the drape over the site of interest.

LANDMARKS

- The graft is identified by palpation of the supra-pubic region. Usually the corrugations on the surface of the graft are easily felt (figure 16A).
- Moreover, a pulse may be palpable.
- The graft may also be visualized under fluoroscopy.
- Alternatively, intra-op ultrasound can be used to localize the graft.
- The arteriotomy is typically made in the straight (horizontal) portion of the graft, in the midline (figure 16A) or closer to the contralateral third.

EQUIPMENT

- As there is usually some scar tissue above the graft and the graft itself is fairly stiff, we use an 18 G needle and a 0.035" angled stiff Glidewire®.

TECHNIQUE

Skin Incision:

- A small superficial incision is made in the skin prior to the arteriotomy.

Needle Alignment:

- The needle should be held coaxial to the graft (directed from one CFA to the other), pointing towards the side being considered for treatment, at an angle of < 45° to the skin, bevel up (at 12 O' Clock position relative to the long axis of the graft).

Needle Advancement:

- The arteriotomy may require fluoroscopic or ultrasound guidance. Usually this is not necessary as the graft is easily palpable.
- The needle should be advanced slowly lateral to medial towards the ipsilateral CFA.

Use of Fluoroscopy & Wire Advancement:

- Once reasonable back bleeding is noted, the wire is advanced.
- The wire should also be always visualized under fluoroscopy when being advanced.

Wire:

- As noted above, a 0.035" angled stiff Glidewire® is used due to its steer-

ability and stiffness.
- Enough length of the wire should be advanced either into the SFA/PFA or the iliac artery to provide adequate anchoring to insert the sheath in the graft.

Arteriotomy Care:
- At the conclusion of the procedure, the sheath is removed when the ACT is less than 180 seconds.
- Manual compression is used for hemostasis.
- Excessive pressure should be avoided as the graft can be deformed and may thrombose.
- A sterile dressing is applied and the site is checked intermittently for bleeding or hematoma formation.

TROUBLESHOOTING

Inability to Advance the Working Sheath:
- This may be encountered in old grafts.
- The solution is to switch the wire out for a stiffer wire (as outlined in the retrograde CFA access section).

AXILLOFEMORAL BYPASS GRAFT ACCESS

INDICATIONS
- An axillofemoral bypass graft is an extra-anatomic bypass that consists of a 6 - 8 mm diameter Dacron® or ePTFE/xPTFE conduit, that connects the axillary artery to the CFA (usually ipsilateral and rarely contralateral).This conduit is tunneled through the flank /abdominal wall.
- This access is usually used for performing:
  - ▹ Thrombectomy of the axillofemoral bypass graft itself.
  - ▹ Endovascular treatment of ileofemoral disease.
  - ▹ Subclavian, carotid, or coronary angiography in patients with occluded aorta and contralateral subclavian artery.
- The goal is to place a sheath that points towards the side of interest, (axillary artery or the CFA) and can be used to perform angiography and intervention.

PREREQUISITE
- The graft must be patent.

PATIENT AND ROOM SETUP
- The patient is placed in supine position.
- The operator stands on the side of the table where the bypass graft is located.
- The ipsilateral arm is abducted at nearly 90° at the shoulder.
- Both groin sites and the entire chest and abdomen from chin down are prepared in a standard fashion and cleaned with antiseptic solution.
- A standard drape can be used by devising an aperture in it, over the graft site of interest.

LANDMARKS
- The graft is identified by palpation of the axillary/lateral abdominal wall region. Usually the corrugations on the surface of the graft are easily felt.
- If the graft is patent a pulse may be palpable.
- The graft may also be visualized under fluoroscopy.

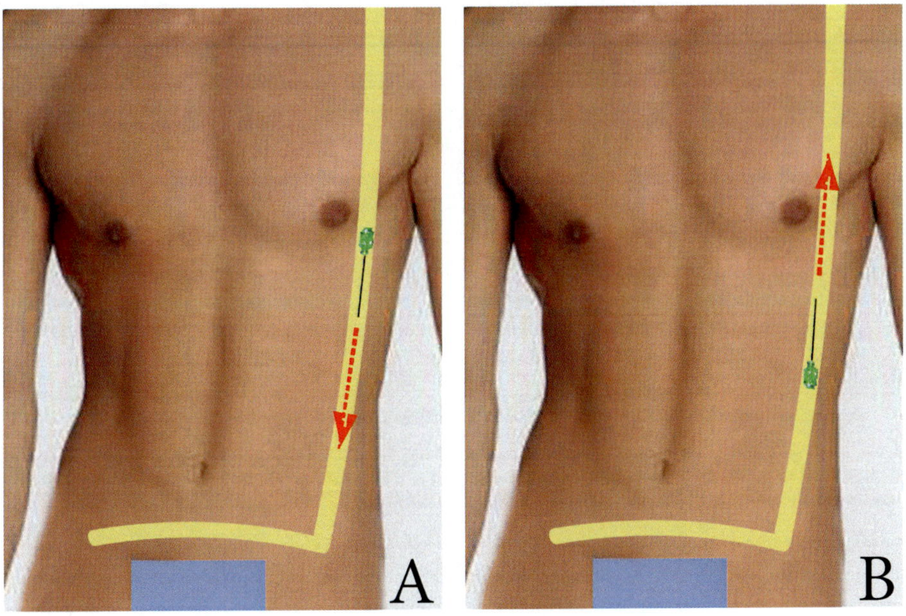

Figure 17.

Accessing an axillofemoral bypass graft.

The operator stands on the side of the bypass. After administering subcutaneous anesthesia, an 18 G needle is advanced into the graft. Also shown here is a left to right femorofemoral bypass graft.

A. For lower extremity angiography, a craniocaudal direction of the needle is used.

B. For access to subclavian/arch vessels the needle is directed caudal to cranial.

- Alternatively, intra-op ultrasound can be used to localize the graft.
- The arteriotomy is typically made in the straight (vertical) portion of the graft, closer to the middle third.

EQUIPMENT

- As there is usually some scar tissue above the graft and the graft itself is fairly stiff, we use an 18 G needle and a 0.035" angled stiff Glidewire®. Caution should be used when this wire is maneuvered through the needle to not sever it. Alternatively, a non-hydrophilic steerable wire such as the Wholey can be used.

TECHNIQUE

Skin Incision:

- A small superficial incision is made in the skin prior to the arteriotomy.

Needle Alignment:

- The needle should be held coaxial to the graft (directed cranio-caudad), pointing towards the CFA, at an angle of < 45° to the skin, bevel up (in 12 O' Clock position).

Axillofemoral BPG:
- This graft is usually created in patients with infected abdominal aneurysms/ grafts and/or aorto-iliac occlusive disease.

Needle Advancement:
- The arteriotomy may require fluoroscopic or ultrasound guidance. Usually this is not necessary as the graft is easily palpable.
- The needle should be advanced slowly cranial to caudal towards the target/recipient CFA.

Use of Fluoroscopy & Wire Advancement:
- Once reasonable back bleeding is noted, the wire is advanced.
- The wire should be visualized under fluoroscopy when being advanced.

Wire:
- As noted above, a 0.035" angled stiff Glidewire® is used due to its steerability and stiffness.
- Enough length of the wire should be advanced into the CFA or the iliac artery to provide adequate anchoring to insert the sheath in the graft.

Arteriotomy Care:
- At the conclusion of the procedure, the sheath is removed when the ACT is less than 180 seconds.
- Manual compression is used for hemostasis.
- Excessive pressure should be avoided as the graft can be deformed and may thrombose.
- A sterile dressing is applied and the site is checked intermittently for bleeding or hematoma formation.

TROUBLESHOOTING

Inability to Advance the Working Sheath:
- This may be encountered in old grafts.
- The solution is to switch the wire out for a stiffer wire (as outlined in a previous section).

BRACHIAL ARTERY ACCESS

- Brachial artery, once a popular site for access, is not commonly used in contemporary practice.
- In comparison to the radial artery, brachial artery access provides nearly a 30 cm shorter distance to the target site(s).This becomes an important consideration in endovascular procedures dealing with iliac and infra-inguinal arteries, as most devices are not available in shaft lengths greater than 150 cm.
- If the brachial artery pulse is normal on both sides, the left side is preferred over the right for access due to several reasons:
    - ▹ In theory, a lesser risk of embolic stroke as the catheter crosses the path of only the left vertebral artery. In contrast, a catheter inserted via the right brachial artery to the descending aorta crosses the path of bilateral vertebral arteries as well as common carotid arteries.
    - ▹ Less distance from left brachial artery to abdominal aorta.
    - ▹ No need to traverse the aortic arch.
- Some situations, however, make the right brachial artery the only feasible choice for initial arterial access. These include:

- ▹ Presence of a left arm AV fistula for hemodialysis (absolute contraindication).
- ▹ LIMA-LAD as the only patent coronary graft (relative contraindication).
- ▹ Severely diseased/occluded left subclavian artery.
- ▹ Innominate or right vertebral/subclavian/carotid artery intervention, particularly in the presence of a complex aortic arch.

ANATOMY

- The brachial artery is the continuation of the axillary artery.
- It starts at the lower border of the teres minor muscle and terminates about 1 cm distal to the antecubital crease by bifurcating into the radial and ulnar arteries.
- The artery is located midway between the lateral and medial humeral epicondyles.
- The artery is usually 4 - 5 mm in diameter.
- It lies medial to the biceps brachii muscle and at the level of the elbow it is medial to its palpable tendon.
- Medial to the artery lies the median nerve which, if inadvertently punctured with a needle, causes a painful sensation felt at the wrist level.

Branches of Brachial Artery:

- Profunda brachii artery.
- Superior and Inferior ulnar collateral arteries.
- Radial artery.
- Ulnar artery.

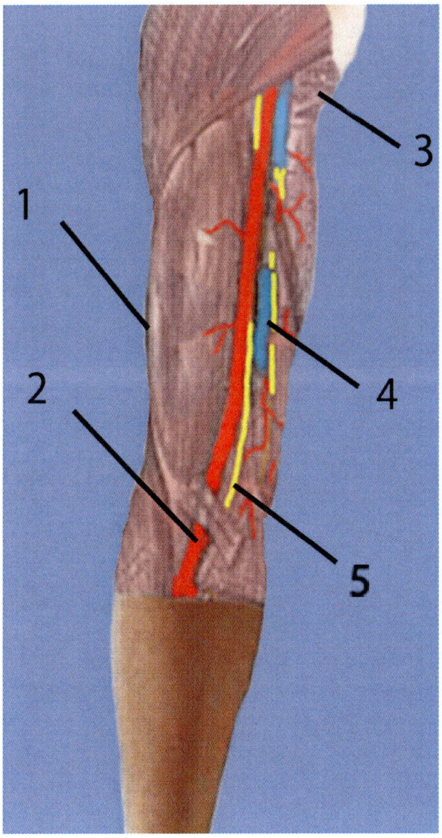

Figure 18.

Anatomy of the brachial artery and its relations to adjacent structures

1. Biceps brachii.
2. Brachial artery.
3. Teres Minor.
4. Basilic vein.
5. Median nerve.

INDICATIONS

- Diagnostic arteriography of any target site can be performed via brachial artery access.
- This access should be used when alternative access via femoral /radial arteries is not feasible.

PATIENT AND ROOM SETUP

- The patient is placed in supine position.
- For right brachial artery access, the arm is kept at the patient's side, with the forearm also supinated and somewhat externally rotated.
- A right-handed operator stands on the patient's right to obtain right brachial artery access.
- For the left brachial artery, the arm is placed on an arm board, abducted to 90° at the shoulder. The forearm is again supinated and somewhat externally rotated.
- The right-handed operator stands on the left side of the table between the table and the abducted arm.
- Both groin sites, from umbilicus down, are prepped in a standard fashion.
- The patient is draped in a standard fashion.
- A separate drape with an aperture (4"x 4") is applied to the arm, centered 1 - 2 cm above the antecubital fossa.

LANDMARKS

The artery is located slightly above the antecubital fossa, 1 - 2 cm above the crease. The size of the brachial artery makes it a favorable access site for peripheral intervention.

EQUIPMENT

- We use a 4 Fr. micropuncture kit for the access.
- This is then changed to a 4/5 Fr. sheath for diagnostic angiography and 6/7 Fr. sheath for interventional procedures.

TECHNIQUE

- The brachial artery is located 1-2 cm above the antecubital crease and punctured at that level after administration of local anesthetic in the subcutaneous tissue.
- Ultrasound guidance is advised if a successful arteriotomy is not made in the first three attempts or if the pulse is weak to begin with.

ARTERIOTOMY CARE

- At the conclusion of the procedure, the sheath is removed when the ACT is less than 180 seconds.
- Manual compression is used for hemostasis.
- A sterile dressing is applied and the site is checked intermittently for bleeding or hematoma formation.
- The arm is immobilized at the elbow with an 'armboard' for four hours to restrict unwanted patient movement.
- Distal pulses should be checked to assess perfusion to the hand, during/after hemostasis.

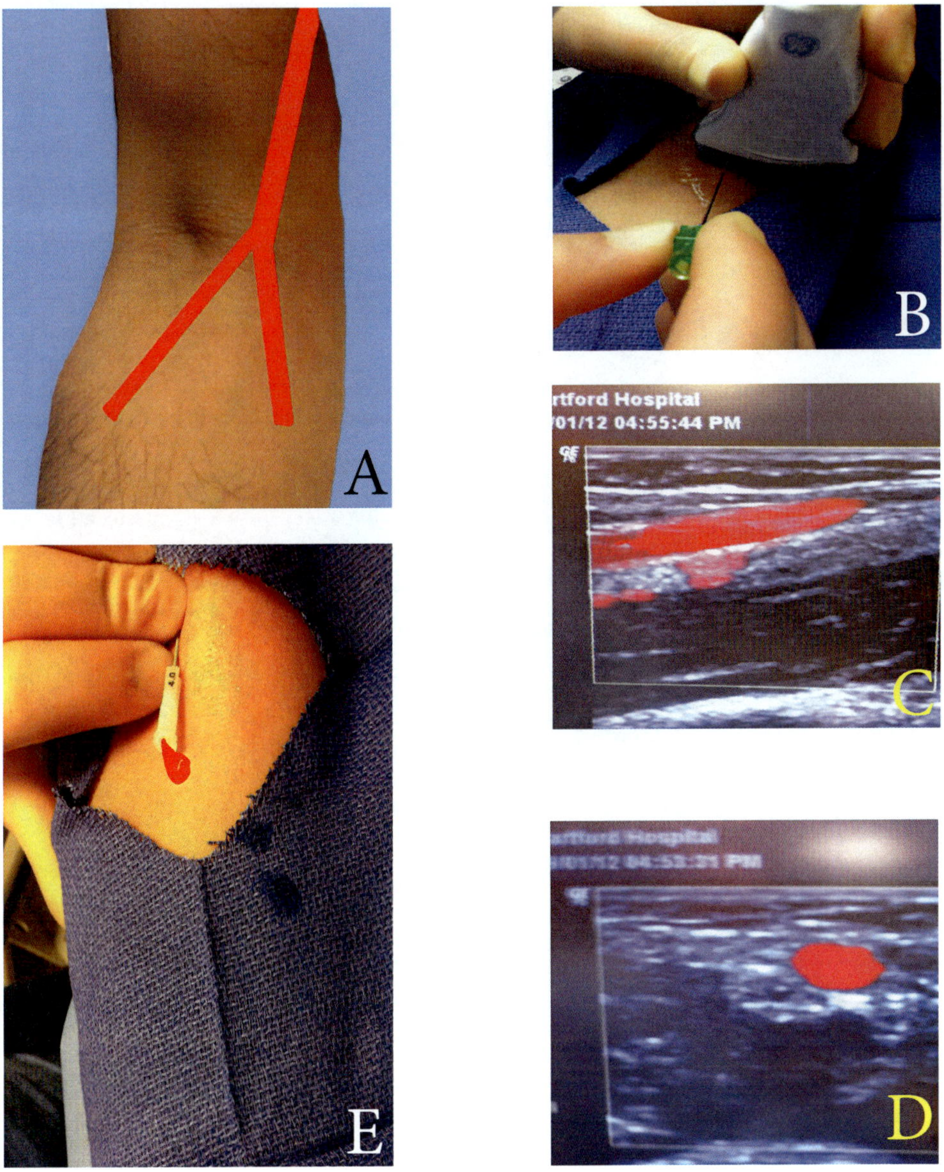

Figure 19.

Brachial artery access.

A. Surface anatomy of the right brachial artery.
B. Ultrasound guidance is used if the pulse is not brisk.
C. The brachial artery is identified by its pulsatile flow in the long axis view.
D. The brachial artery in the short axis view.
E. Brisk flow is seen after the microcatheter is in place.

RADIAL ARTERY ACCESS

- Radial artery access is becoming more popular in the recent years, particularly in coronary artery procedures.
- This is a result of significant improvement in the equipment and the introduction of specifically designed catheters, wires, guides and hemostatic devices.
- One of the advantages of using this access site is the lower risk of a vascular complication.
- However, the caliber of devices that can be safely introduced through the radial artery dictates the feasibility of this access for diagnostic and interventional procedures.
- Also, due the distance from the iliac and infra-inguinal arteries, this access is not ideal for infra-inguinal interventional procedures.

ANATOMY

- The radial artery is a main artery that arises from the bifurcation of the brachial artery in the antecubital fossa and runs along the lateral aspect of the forearm.
- It starts about 1 cm distal to the antecubital crease, and terminates in the hand by joining the respective branch of the ulnar artery to form the deep palmar arch.
- The radial artery is the smaller of the two terminal branches of the brachial artery.
- The artery is usually 3 - 4 mm in diameter.

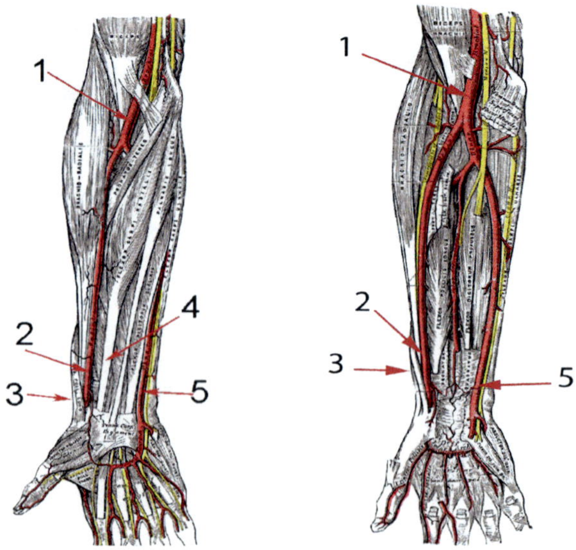

Figure 20.

Anatomy of the radial artery and its relations to adjacent structures.

1. Brachial artery
2. Radial artery
3. Styloid process of the radius
4. Tendon of flexor carpi radialis
5. Ulnar artery.

(Images courtesy of 20th U.S. Edition of Gray's Anatomy of the Human Body)

PATIENT AND ROOM SETUP

- Patient and room set-up for radial artery access is similar to that for brachial artery access.
- The arm is abducted slightly at the shoulder and stabilized on a radiolucent arm board.
- The wrist is hyperextended and placed over a roll of (non-sterile) towels under the wrist for support (figure 21A).
- Adhesive tape is placed across the palm of the hand to prevent flexion of the wrist during the procedure.

LANDMARKS

- The artery is located medial to the styloid process of radius and lateral to the tendon of flexor carpi radialis.

PREREQUISITES

- The first critical step in getting the access is to make sure that there is adequate dual blood supply to the hand and a patent palmar arch is present.
- This is determined by performing a modified Allen's test.
  - ▹ The patient's arm is elevated and the patient is asked to make a fist for 30 seconds.
  - ▹ Simultaneous pressure is applied to both the radial and ulnar arteries so as to occlude them.
  - ▹ With the hand still elevated, the patient is asked to open the hand. It should appear pale.
  - ▹ The pressure applied to the ulnar artery is released and the palm observed for return of color within 7 seconds.
  - ▹ The test is considered normal if the color in the palm returns within 7 seconds indicating that the ulnar artery provides adequate circulation to the deep palmar arch.
- A more sensitive test to evaluate the patency of the palmar arch collaterals is the Barbeau test (figure 22) using waveform plethysmography with a pulse oximeter placed on the thumb.
- The technique is similar to the modified Allen's test.
- Plethysmography responses are categorized into 4 types:
  - ▹ Type A = no dampening.
  - ▹ Type B = slight dampening of the pulse tracing.
  - ▹ Type C = loss of the pulse tracing followed by recovery.
  - ▹ Type D = no recovery of the pulse tracing within 2 minutes.
- Radial access is considered suitable with type A, B and C responses.

INDICATIONS/APPLICATIONS

- Most of the coronary diagnostic and interventional procedures including PCI, rotablation, complex PCI, even CTOs are being attempted from the radial access. The equipment and the guides have seen a great improvement over the last few years with dedicated catheters and wires.
- For peripheral intervention, radial artery access can be used for most target sites up to the aortic bifurcation.
- Although intervention on femoro-popliteal arteries can be performed as well, very few devices have adequate shaft length to reach from the radial artery to beyond the femoral bifurcation.

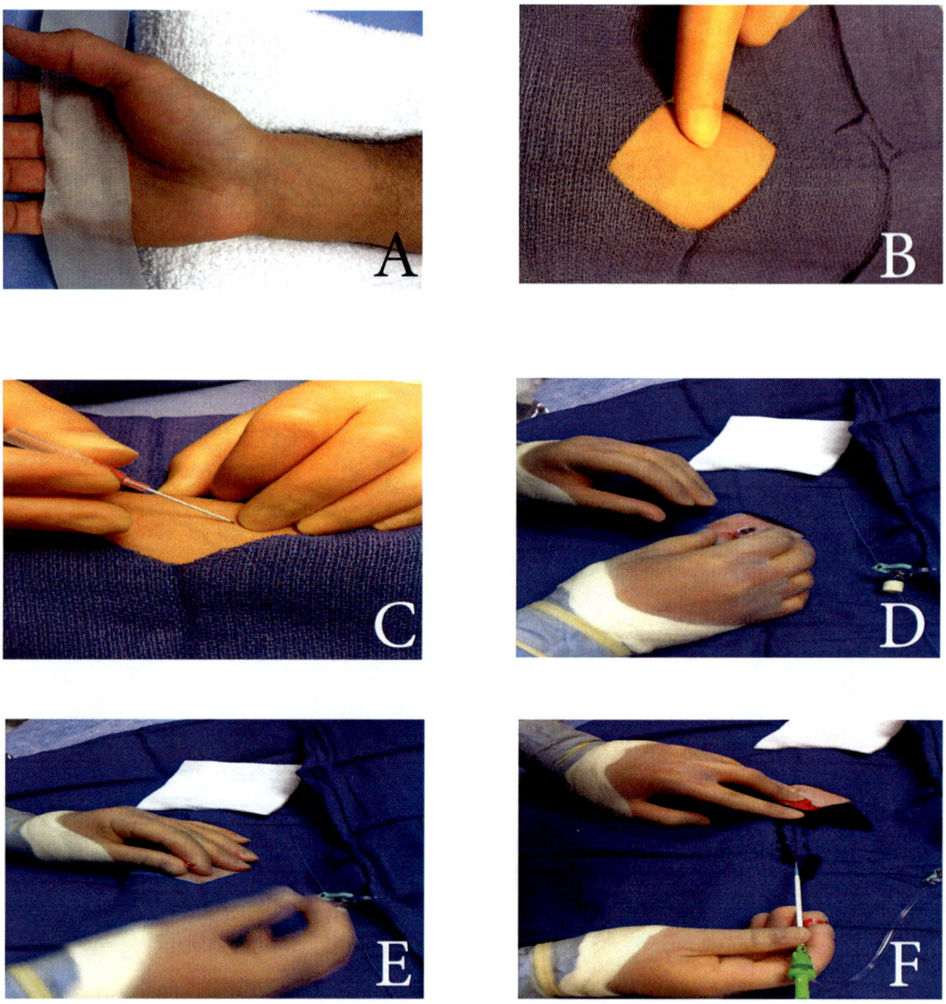

Figure 21.

Radial artery access.

A. The patient's hand is extended at the wrist and taped.
B. An aperture drape is placed over the sterile radial artery access site.
C. A 20 G Jelco needle is advanced through the skin to right radial artery.
D. After seeing a 'flashback', the needle is advanced (approximately 3-5 mm) through the artery (posterior wall). The needle is removed. The catheter is then gradually withdrawn until pulsatile flow is seen.
E. The 0.018" wire supplied with the radial sheath kit is advanced through the catheter into the right radial artery.
F. Over this wire, the sheath is advanced into the radial artery.

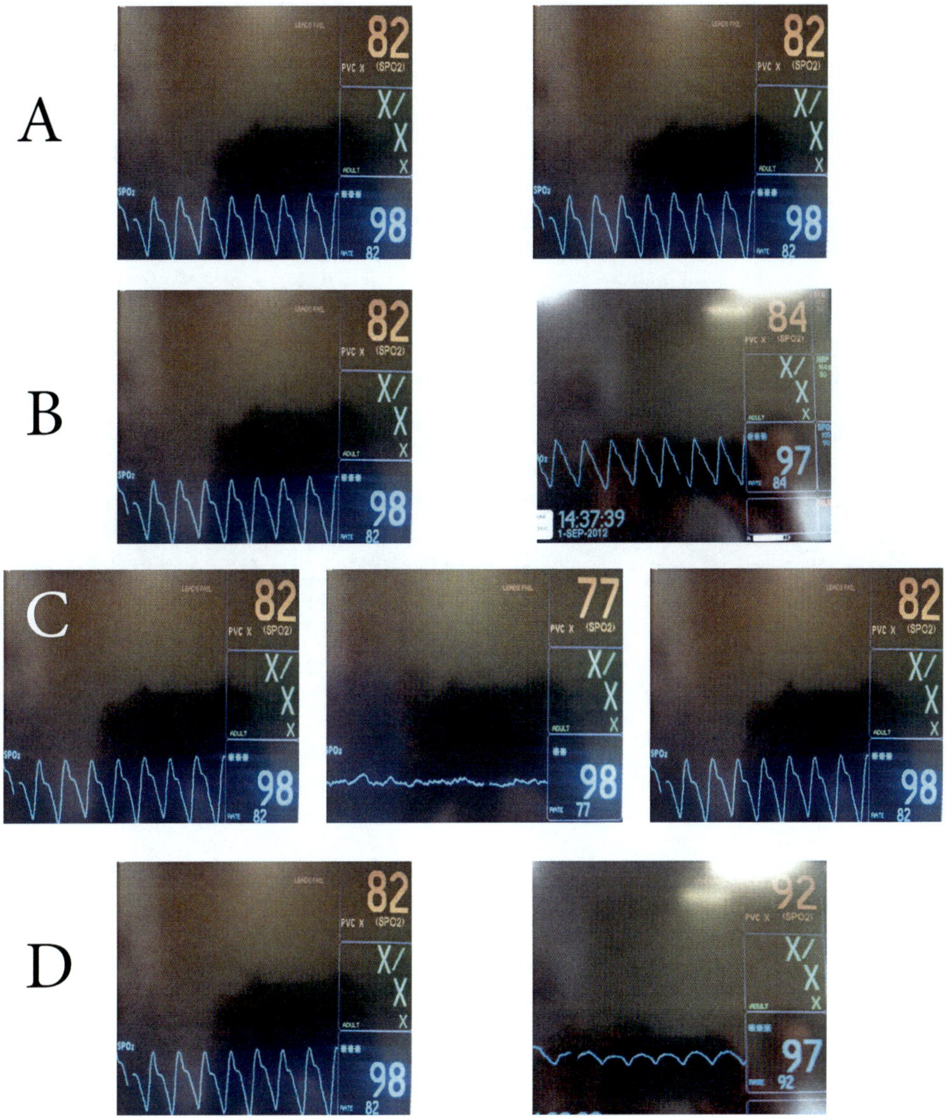

Figure 22.

Radial artery assessment - Barbeau test.

A. Type A response with no change in waveform.
B. Type B response with slight dampening of the waveform.
C. Type C response with loss of plethysmogram tracing followed by gradual recover (in < 2.0 min.).
D. Type D response with no recovery.
.

Jelco®
- Consists of a sharp needle inside a radiopaque FEP polymer catheter.
- It accommodates a 0.018" wire.

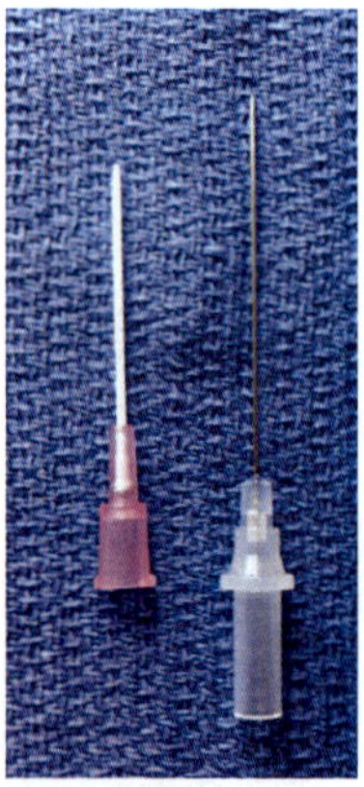

Figure 23.

Jelco needle.

A 20 G Jelco needle along with its outer catheter.

EQUIPMENT

- A 20 G Jelco® (Smiths Medical, Carlsbad CA) needle or a 4 Fr. micropuncture kit can be used to obtain radial artery access.
- Eventually, a hydrophilic sheath is placed in order to minimize intimal trauma, improve patient comfort and long-term arterial patency.
- A 'radial cocktail' is given to decrease the chance of radial artery spasm and/or thrombosis. The mixture consists of heparin 60 units/kg, Verapamil (2.5-5 mg) and nitroglycerin (0.1-0.3 mg).

TECHNIQUE

- The radial artery is palpated and following the administration of a small amount of subcutaneous lidocaine, the artery is punctured at a 45° angle to the skin, approximately 1 - 2 cm proximal to (above) the styloid process.
- Seldinger technique is employed. After seeing a 'flashback' of blood in the hub of the needle, the Jelco is advanced through the artery and the needle is removed.
- The catheter is slowly withdrawn and once blood flow is obtained, a guide wire is advanced into the radial artery followed by the sheath.
- If a micropuncture needle is used, a single anterior wall stick is performed and the guide wire is advanced into the artery.
- Once the sheath is inserted, a guide wire is advanced to the aorta.
- Tortuosity may be encountered particularly from the right radial approach as the innominate artery may enter the aortic arch at unfavorable angle.
- It may be helpful to use a steerable flexible guide wire such as a Wholey wire (Covidien, Mansfield, MA) or a Tiger wire (St Jude Medical Inc., St Paul, MN) to access the ascending aorta.
- In addition, asking the patient to 'take in and hold a deep breath' for a few seconds may help straighten the tortuous vasculature and may facilitate wire and catheter manipulation.
- Rarely, difficulty may be encountered as the guide wire is being advanced at the region of the elbow crease. This is often due to a loop in the radial artery. This may require the use of coronary steerable guidewires to negotiate this region or may require that this access be abandoned and an alternative site used.

ARTERIOTOMY CARE

- The radial artery is closed with manual compression.
- There are commercially available devices to assist in compression. These devices include the TR® band (Terumo Interventional Systems Inc., Somerset, NJ) and Bengal closure device. It is important to achieve hemostasis without complete occlusion of the vessel in order to decrease the chance of radial artery occlusion.
- A method that we frequently use is to apply a pulse oximeter to the thumb or index finger of the involved hand.
- A volume of air is placed in the TR band so that hemostasis is achieved along with evidence of some residual flow through the radial artery on the pulse oximeter while compressing the ulnar artery.

5

AORTO-ILIAC INTERVENTION

Patients with aorto-iliac occlusive disease (AIOD) may present with claudication, critical limb ischemia or acute limb ischemia. In these patients claudication typically affects the calf, although the thigh and buttock may also be affected. The latter is usually present when the common or internal iliac artery distribution has severe flow impairment. Additionally, men with chronic, bilateral common or internal iliac artery disease may also have erectile dysfunction.

ANATOMY

The abdominal aorta terminates at the level of L4-L5 vertebrae by bifurcating into the right and left common iliac arteries (CIA) (figure 1).

COMMON ILIAC ARTERY

- The CIA is usually 3-5 cm in length.
- Its luminal diameter is typically 8 -12 mm in adult males and 7-9 mm in females.
- Normally, the CIA gives no branches prior to its bifurcation into the external (EIA) and internal iliac arteries (IIA). Rarely, however, branches to the inferior pole of the kidney, the main renal artery for a pelvic kidney, ureteral, muscular and periosteal branches are present. Hence, prior to CIA intervention it is critical to assess it for origin of any anomalous vessel(s).

EXTERNAL ILIAC ARTERY

- This vessel continues in the path of the CIA. It is bigger than the IIA.
- The luminal diameter of this vessel in healthy adult males is 6-8 mm and in females 5-7 mm.
- EIA gives two branches normally, the deep circumflex iliac artery and inferior epigastric artery. The latter serves as the anatomical landmark for transition of the vessel into the CFA.

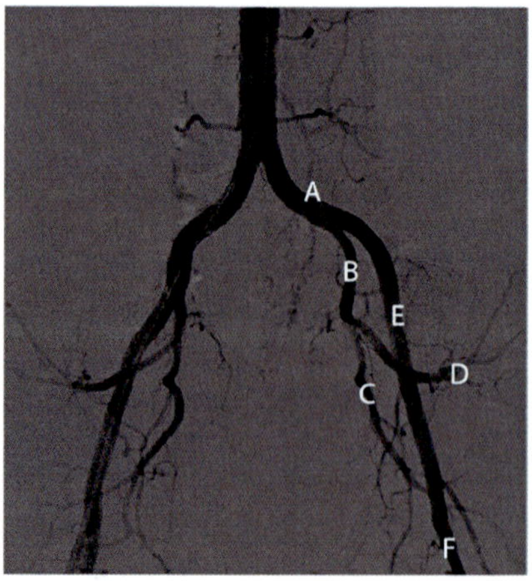

Figure 1.
Normal iliac arteries.

A. CIA.
B. IIA.
C. Anterior division of IIA.
D. Posterior division of IIA.
E. EIA.
F. CFA.

Branches of IIA:

Anterior Division:
- Superior vesical
- Inferior vesical
- Middle rectal
- Uterine/Ovarian/Vaginal
- Internal pudendal
- Inferior gluteal

Posterior Division:
- Iliolumbar
- Lateral sacral
- Superior gluteal

INTERNAL ILIAC ARTERY

- This is the smaller of the two terminal divisions of CIA.
- It has a luminal diameter of 4-8 mm in adult males and 4-6 mm in females.
- It is usually 4 cm in length.

DIAGNOSIS

On physical examination clues to the presence of aorto-iliac disease include diminished ipsilateral pulses, presence of a femoral or aortic bruit and other manifestations of chronic PAD.

The ankle brachial index is usually diminished but may be preserved at rest. Doppler waveforms and PVRs are usually abnormal and revealing. However, at times the ABI/PVR study may be normal at rest. If the history is strongly suggestive of PAD, an exercise test should be performed in these patients. If the exercise ABIs are also normal, one must rule out internal iliac artery disease.

In cases of severe and longstanding symptoms, a total occlusion is often the underlying cause. When suspected, and particularly in cases of severe bilateral symptoms, it is prudent to obtain an MRA or CTA prior to conventional angiography. This allows better planning of diagnostic angiography and endovascular intervention.

DIAGNOSTIC ANGIOGRAPHY

ACCESS

Site:

In cases of suspected severe unilateral AIOD, retrograde arterial access can usually be safely obtained in the contralateral CFA. However, if bilateral severe inflow disease is suspected then the site of access is best determined by the information obtained from a preoperative MRA or CTA. Otherwise, brachial access should be considered for initial angiography. Although radial artery access can be used for diagnostic angiography, this access may not be feasible for iliac intervention due to the extended distance.

Micropuncture kit:
This is an access kit that comes with a 21 G 7cm needle, a 4F or 5F locking sheath (micro-catheter) and dilator along with a 0.018" wire.

Supplied by several vendors such as Cook, Bard etc

Sheath Size:

We prefer to obtain initial access with a 21 gauge needle and 0.018" wire (e.g. 4Fr. micropuncture kit), regardless of the site of access. This is then upsized over a 0.035" wire to a 5Fr. sheath that is at least 20 cm long. Advance the tip of the this sheath to the distal CIA.

ABDOMINAL AORTOGRAPHY

Catheter:

- Use any of the several 4 Fr. flush catheters such as a pigtail, tennis racquet (TR), or Omni flush catheter.

Imaging:

- Start in an AP projection. Routine oblique projections are not necessary.
- However, these may be beneficial (see below) in delineating disease in the distal CIA, EIA and IIA.
- Abdominal aortography should allow visualization of juxta/infra-renal aorta to the common femoral arteries bilaterally.
  - ▹ Place the catheter at L1/L2 level and include bilateral femoral heads in the field of view.
- Use digital subtraction angiography (DSA) as this will reduce the volume of dye administration without compromising image quality.
- Remember to use 'breath hold' with DSA.

Dye Injection:

- A power injector is used for aortography.
- The rate of power injection depends upon the patient's body habitus, renal function and ability to comply with instructions of breath-hold during aortography.
- In patients with normal renal function, use 70% dye with 30% heparinized normal saline dilution (70:30). If the creatinine clearance is less than 30 mL/min, then use 50% dilution (50:50). CO_2 may also be used for this purpose (manual injection only).
- Injection settings for iodinated contrast are 10-20 cc/sec for a total volume of 20-40 cc at 500-750 psi.

SELECTIVE ANGIOGRAPHY

Catheter:

- Via contralateral femoral artery access a selective catheter such as an IM, Omni flush, Rim or Simmons catheter is used and placed at the aortic bifurcation to engage the CIA ostium.
  - ▹ Via brachial access a multi-purpose, angled hydrophilic catheter (e.g. angled Glidecath), JR4, or Vert catheter can be used.

Imaging:

- AP and particularly contralateral oblique 20° projection (RAO 20° for the left iliac and LAO 20° for the right iliac) with nearly 20° caudal angulation is helpful in delineating the EIA and IIA with respect to the CIA.

Dye Injection:

- DSA with breath holding is used with manual injections of 5-8 cc of dye through the selective catheter.
- In the setting of an occluded iliac artery, obtain a prolonged run to visualize collaterals to the distal reconstituted segment.

HEMODYNAMICS

Hemodynamic assessment is an essential component of the invasive evaluation of AIOD. In cases of angiographically severe stenoses, the measurement of a hemodynamic gradient across these lesions may be less important. However, for lesions that appear to be <70% narrowed on angiography, it is imperative to measure the trans-stenotic gradient.

Ideally, the gradient is measured by using simultaneous pressure recordings proximal and distal to the diseased segment of interest. In order to obtain a reliable measurement, the caliber of the measuring catheters should be at least 1 French size apart. Otherwise, artifactual dampening of the pressure transduced through the sheath may be noted and misinterpreted as a significant gradient. For example, if using a 5 Fr. sheath in the common femoral artery (distal to an external iliac artery stenosis) the aortic catheter should be 4Fr. in caliber. Obviously if two separate access points are used for measuring the gradient (e.g. the right CFA and the left brachial artery), the difference in the caliber of the catheters is irrelevant.

> ***Hemodynamics***
> For example, when assessing an EIA stenosis the TR is kept in the aorta and the sheath tip below the diseased segment of EIA. IA NTG is administered through the sidearm of the sheath (i.e. distal to the lesion) to provoke the gradient.

Measurements should be made before and after hyperemia. Induce hyperemia by administering 200 mcg of intra-arterial (IA) NTG distal to the stenosis. Some operators use Papaverine for the same purpose.

If the patient has bilateral iliac artery disease in the ostial/proximal segments, the most reliable way to perform hemodynamics with two transducers is to use bilateral femoral artery access. Simultaneous aortic and iliac pressures are measured first on one side then the other.

> ***Hemodynamics with bilateral iliac disease:***
> Ideally, access should be obtained on both sides and the pressure in each iliac artery should be compared to the aortic pressure.

INTERVENTION

ACCESS

The choice of access site for intervention is made after diagnostic angiography. Often the same access site as for diagnostic angiography can be used for intervention after the sheath is appropriately upsized. However, this decision depends upon the following anatomical factors:

Location of the Lesion:

- **Ostial CIA stenosis:**
  - Lesions within the first 5 mm of the origin of the CIA are considered ostial.
  - These lesions are best treated from ipsilateral femoral or (left) brachial artery access as this allows for a more coaxial route and offers the necessary support. As an example, if diagnostic angiography via right CFA revealed a severe left CIA ostial stenosis, it is best treated via left femoral access.

- **Non-ostial CIA stenosis:**
  - Lesions in the CIA that are non-ostial can be treated from contralateral femoral access provided there is enough room proximally to keep the stent coaxial.

> ***Brachial artery access:***
> - Ideally, a 6 Fr. sheath is used.
> - Although up to an 8 Fr. sheath may be used if necessary, the risk of complications including brachial artery thrombosis is increased with the larger sheath size.

Severity of the Lesion:

- **Total Occlusion:**
  - ▹ These lesions are best approached in an antegrade fashion.

- **Non-total occlusion:**
  - ▹ These lesions can be approached in an antegrade or retrograde fashion depending upon their location.

Burden of Disease in the Iliac Arteries:

- In case of bilateral ostial iliac disease, 'kissing' ostial stenting is usually performed keeping stent protrusion in the aorta to a minimum. This intervention is performed via ipsilateral femoral access.

EQUIPMENT

GUIDING SHEATH

Size:

- A 6 Fr. sheath is adequate for delivering non-covered stents in iliac artery diameters. For covered stents appropriately sized for the iliac arteries (e.g. iCast, Viabahn or Fluency) at least a 7 Fr. sheath is required.

Shape:

Straight:

- When using ipsilateral femoral or brachial artery access for intervention, a straight sheath which is long enough to reach the aortic bifurcation should be used.
  - ▹ Brachial: 90 cm
  - ▹ Ipsilateral femoral:>20 cm

Curved:

- With contralateral femoral access for intervention a 'cross-over' sheath is used.
- Cross-over sheaths can be straight (e.g. Raabe) or pre-shaped (e.g. Ansel).

> ***Cross-over sheaths:***
> These sheaths are designed to be advanced from CFA access up and across the aortic bifurcation and are usually 45-55 cm in length.
>
> Examples include:
> - Ansel
> - Raabe
> - Destination etc.

> ***Stiff vs Non-stiff wire:***
> Avoid using a non-stiff wire for device delivery (including sheaths).

WIRES

- Guidewires used for iliac artery interventions are commonly 0.035" in diameter.
- Polymer-coated hydrophilic wires (e.g. Glidewire) are the preferred work-horse wires for crossing these lesions. Once across the lesion(s), these wires can be exchanged out for a J-tipped non-hydrophilic 180-260 cm long wire depending upon the shaft length of the available balloons and stents.
- The shape of the 'crossing' hydrophilic wire is chosen based on lesion severity and morphology.

> ***Hydrophilic Wires:***
> We use 0.035" Glidewire from Terumo as the workhorse hydrophilic wire for aorto-iliac intervention.

Total Occlusion:

- A straight non-stiff or stiff hydrophilic wire (the latter is preferred) is the wire of choice supported by a 4Fr. catheter. After initial penetration of the proximal cap, if this wire does not successfully cross the length of the occlusion (i.e. unable to cross the distal cap), it may be switched out for an angled stiff hydrophilic wire to allow for better steerability. The two wires supported by a catheter may have to be used interchangeably until the occlusion is successfully crossed.

Balloon inflation:
We use the following general principles:
- Balloons should be inflated until the waist is eliminated or nominal pressure reached.
- Predilatation:10 sec.
- Stent deployment:10 sec.
- Postdilatation: 10 sec.

Avoid continued dilatation in the aorta, iliac/visceral arteries if the patient is experiencing pain at the site of inflation.

Non-total Occlusion:

- An angled non-stiff hydrophilic wire (e.g. Glidewire, Terumo) is our preferred wire to traverse a severe/critical stenosis. The ASGW is not as torqueable as the angled non-stiff hydrophilic wire and can dissect the artery while crossing a tight lesion.

BALLOONS

Predilatation:

- Predilatation (prior to deploying a stent) is recommended as it creates a channel for smooth delivery and precise positioning of the stent. Additionally, this step allows for appropriate stent sizing.
  - ▹ Choose a balloon that is at least 2 - 3mm smaller in diameter than the luminal diameter of the reference vessel.

Postdilatation:

- This step is desirable as it allows for appropriate stent expansion and apposition. If the stent was undersized, postdilatation allows for stent expansion to a 1:1 ratio.
  - ▹ Choose a balloon with a diameter no greater than the luminal diameter of the reference vessel.

BES 'Negative' prep:
Do not perform a 'negative prep' on the BES until it is in correct position within the lesion as it may increase the chances of stent dislodgement during delivery.

STENTS

Balloon Expandable Stents (BES):

- BES are recommended for treatment of ostial/proximal CIA and IIA lesions.
  - ▹ These are sized 1:1 to the luminal diameter of the reference vessel.
  - ▹ They allow for precise placement.
  - ▹ They provide higher radial force that may be necessary for heavily calcified lesions.

Self Expandable Stents (SES):

- SES are used for EIA and may also be used for mid/distal CIA.
  - ▹ These stents are 'over-sized' by at least 1 mm in diameter compared to the reference vessel.

Covered vs Non-covered stents:
Covered stents are used in lesions prone to perforation:
- CTO
- Calcification
- Older patients
- Females

They may also be used for the treatment of ectatic or aneurysmal segments.

Covered BES are used in the CIA and covered SES in EIA.

ANTICOAGULATION

- IV unfractionated heparin is the preferred anticoagulant.
- Goal ACT > 250 sec.
- Bivalirudin or GP IIb/IIIa inhibitors should be avoided because of the inability to reverse the effect in case of perforation.

TECHNIQUE

As noted above, diagnostic angiography is usually performed via the common femoral artery on the less diseased side. Based on the findings of initial imaging different access points may have to be utilized for intervening upon iliac artery disease.

Summarized below are the recommended steps for intervention in each scenario based on anatomical findings.

Figure 2.

CIA TOTAL OCCLUSION

Aortography showing common iliac artery occlusion is depicted schematically in figure 2. The following steps are recommended for treating this lesion:

Access:

- Right CFA access for aortogram.
- Left brachial artery access after aortography, advancing a 6 Fr. 90 cm Shuttle sheath to the level of the aortic bifurcation over a 0.035" 260 cm angled stiff Glidewire (ASGW).
- A road map is obtained.

Guiding Sheath/Catheter & Wire:

- A 0.035" compatible crossing catheter (e.g. Glidecath, QuickCross etc) is loaded in a "telescoping" fashion through a 6 Fr.'Multipurpose' (MPA) guiding catheter. This crossing catheter should be long enough to cross the occlusion and this usually requires a length of at least 120 cm from the brachial approach.
  - ▹ The longest Glidecath/QuickCross catheter is 150 cm.
- These catheters are advanced over the wire through the Shuttle sheath to the aortic bifurcation.
- A straight stiff Glidewire (SSGW) is chosen to cross the CTO. The ASGW is exchanged out for a SSGW.

Technique:

- The crossing catheter is advanced to the tip of the MPA guide which is directed at the proximal cap. The wire tip is kept inside the catheter at this time.
- If a 'micro-channel'/'beak' exists at the proximal cap the catheter is used to 'engage' it. The SSGW is then advanced with a sharp 'poke' at the cap.
- The SSGW is advanced across the proximal cap of the left CIA occlusion.
- Once the wire is across the occlusion, the crossing catheter (Glidecath) is advanced over it beyond the occlusion. Do not advance the MPA guiding catheter across the occlusion!
- The wire is removed, the crossing catheter is flushed and a confirmatory angiogram is performed.
- If the plan is to use a non-covered stent then the 6 Fr. brachial sheath can be used for this purpose without obtaining left CFA access.
- If, however, a covered stent is planned a 7 Fr. sheath will be needed. To avoid up-sizing the brachial sheath to 7 Fr. (for covered stent placement), left CFA retrograde access is obtained.
  - ▹ After obtaining left CFA access, an ASGW is advanced through the crossing catheter (e.g. GC) from above and 'externalized' through the left CFA sheath. This can be done either by maneuvering the wire into the sheath or by snaring it from below.
  - ▹ Once the wire is externalized, clamp the end outside the CFA sheath with a hemostat and remove the crossing catheter from above.
  - ▹ Apply a 'torquer' on the wire outside the brachial sheath to avoid the wire from being pulled in. It is also important to ensure that the Tuohy-Borst adapter of the Shuttle sheath is tight around the wire to prevent the wire from being pushed out from below.

> ***"Telescoping":***
> This implies the introduction of one catheter through another catheter.
> - Requires at least 2Fr. size difference between the catheters
> - The inner catheter should be sufficiently longer than the outer to allow it to be delivered to its destination.
>
> The purpose of this technique is to obtain better 'guide' support from the outer catheter while the inner catheter is used to cross the lesion once the wire is across it.

"Externalization":
This implies a wire introduced into the body through one access point is retrieved (e.g. snared out) through another access point.

This, for example, facilitates intervention across CTOs by allowing a wire that has crossed an occlusion to serve as a rail to deliver devices.

"Shepherding":
Advancing a sheath/catheter into a lesion over a wire/balloon combination.

This maneuver allows smooth delivery and precise placement of the stent within the lesion minimizing the risk of dislodgement of the stent.

'Road map':
A road map can be obtained by simultaneous injection via right and left sheaths.

 - ▹ The rest of the intervention will then be performed via the CFA sheath
- This technique is also referred to as 'body flossing'.
- Load a pre-dilatation balloon on the wire and advance it across the lesion.
- Perform PTA of the CIA lesion and gradually upsize the balloon as necessary from a small diameter (e.g. 4-6mm) to a larger balloon (7-8 mm).
- Pull the deflated balloon halfway into the sheath and gradually advance the sheath and the balloon across the lesion. This step is called 'shepherding'.
  - ▹ The use of the balloon allows smoother transition and a 'tapered' tip to the sheath during this maneuver reducing the likelihood of vessel dissection or perforation.
  - ▹ Alternatively, the dilator of the sheath can be reintroduced into the sheath and used for this purpose.
  - ▹ If this maneuver is unsuccessful, the wire must be switched out for a stiffer wire with more support. Advance the balloon beyond the lesion (to the CFA if coming from brachial access; to the juxta-renal aorta if coming from below), the 0.035" hydrophilic wire is removed and replaced with a stiffer 0.035" 260 cm wire (e.g. Amplatz Superstiff). Reattempt this maneuver with either the balloon or the sheath dilator in place.
- Obtain an angiogram to reassess the lesion, its response to PTA and any dissection or perforation.
- Obtain a fresh road map.
- Select an appropriate size BES, load it on the wire, advance across the lesion and position it precisely to cover the lesion.
- Withdraw the sheath while maintaining the stent in position.
- Deploy the BES. 'Shepherd' the sheath back into the stent.
- Remove the stent balloon and obtain another angiogram to assess the result.
- Post-dilate the stent with an appropriate size balloon and reshepherd. Perform another angiogram.
- If satisfied with the result, perform final hemodynamics and abdominal aortography (if renal function allows).

CIA OSTIAL/PROXIMAL STENOSIS

(Figure 3)

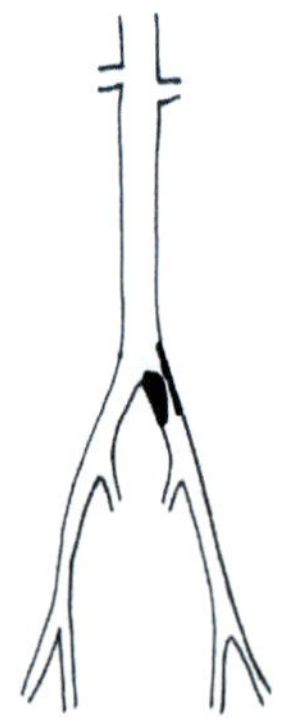

Figure 3.

Access:

- Right CFA access for aortogram. After the aortogram remove the TR. Advance the sheath close to the aortic bifurcation or above it to assist in further imaging to guide precise stent placement on the left side (via the left CFA).
  - ▹ It also allows delivery of a wire across the right CIA in case of plaque shift during left CIA stenting.
- Left CFA access is used for the intervention (6 Fr. for non-covered stent; 7 Fr. for covered stents). A straight sheath long enough to reach the aortic bifurcation (>20 cm) is used. Initially, it is inserted only halfway.
  - ▹ During this access keep the ASGW below and out of the lesion to avoid dissection of the artery.
  - ▹ Advance the sheath under fluoroscopic guidance and keep the dilator tip out of the lesion.

Technique:

- Check the gradient between the aorta and the left CFA.
- Obtain a road map. Use an angled non-stiff hydrophilic wire to cross the lesion.
- Advance a crossing catheter across the lesion and exchange the wire for an ASGW or another stiffer wire.
- Load a pre-dilatation balloon on the wire and advance it across the lesion.
- Perform PTA of the CIA lesion and gradually upsize the balloon from a small diameter (e.g. 4-6 mm) to a larger balloon (7-8 mm).
- Prior to withdrawal of the balloon, shephed the left CFA sheath over the balloon and across the lesion.
- Withdraw the balloon. Repeat an angiogram to assess the result of PTA.
- Obtain a new road map.
- Select an appropriate size BES, load it on the wire, advance across the lesion and position in its precise location.
- Withdraw the sheath while maintaining the stent in position.
- Deploy the BES. 'Shepherd' the sheath back into the stent.
- Remove the stent balloon and obtain another angiogram to assess the result.
- Post-dilate the stent with an appropriate size balloon and reshepherd. Perform another angiogram.
- If satisfied with the result, perform final hemodynamics and abdominal aortography (if renal function allows).

CIA MID/DISTAL STENOSIS

(Figure 4)

Access:

- Right CFA access for aortogram.

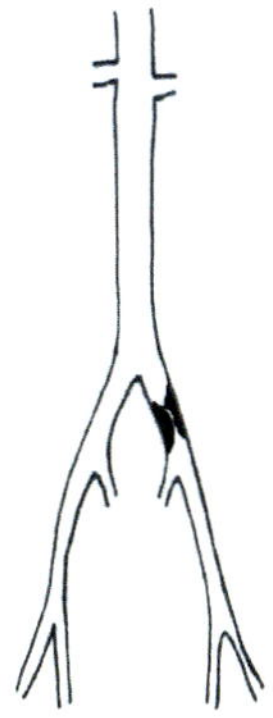

Figure 4.

Technique:

- Use an IM catheter to engage the ostium of the left CIA, obtain a road map and direct an angled non-stiff hydrophilic wire across the lesion. Advance this wire into the distal PFA/SFA.
- Before a 'cross-over' sheath can be advanced over the aortic bifurcation a stiffer wire such as the ASGW should be in place. Remove the IM catheter.
- Advance a straight 4 Fr. hydrophilic catheter (e.g. GC) over the angled non-stiff hydrophilic wire in the PFA and exchange the wire out for an ASGW.
- Exchange out the hydrophilic catheter and the right CFA sheath for a cross-over sheath (6 Fr. for non-covered BES;7 Fr. for covered BES). Advance this sheath over the bifurcation, under fluoroscopic guidance, into the proximal contralateral CIA. As the sheath is being advanced over the bifurcation, it may be necessary to partially withdraw its dilator to facilitate easier advancement over the bifurcation.
- Do not allow the dilator to enter the lesion.
- Once the sheath is in the proximal left CIA, remove the dilator and flush the sheath.
- Obtain a new road map.
- Advance a pre-dilatation balloon across the lesion.
- Perform PTA of the CIA lesion and gradually upsize the balloon as necessary from a small diameter (e.g. 4-6mm) to a larger balloon (7-8 mm).
- Prior to withdrawal of the balloon, shephed the sheath over the balloon and across the lesion.

- Withdraw the balloon.
- Obtain a new road map.
- Select an appropriate size BES, load it on the wire and position within the lesion.
- While maintaining the stent in position withdraw the cross-over sheath proximal to the lesion, still keeping it over the aortic bifurcation.
- Deploy the BES. 'Shepherd' the sheath back into the stent.
- Remove the stent balloon and obtain another angiogram to assess the result.
- Post-dilate the stent with an appropriate size balloon and reshepherd. Perform another angiogram.
- If satisfied with the result, perform final hemodynamics.

EIA STENOSIS / TOTAL OCCLUSION

(Figures 5A/5B)

Access:

- Right CFA access for aortogram.

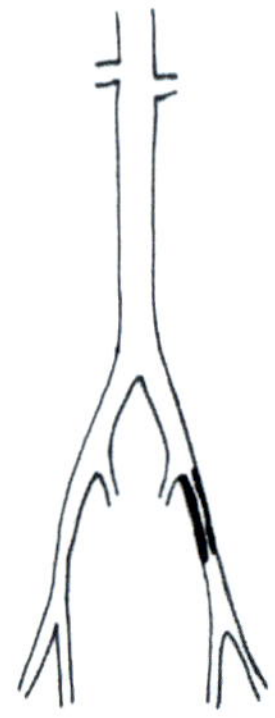

Figure 5A.

Technique:

- Use an IM catheter to direct an angled non-stiff hydrophilic wire into the anterior or posterior division of the patent hypogastric artery, to act as an anchor.
- Exchange this wire out for the angled stiff hydrophilic wire through a 4 Fr. straight hydrophilic catheter. Remove the catheter.
- Exchange the right CFA sheath for a cross-over sheath and advance it over the bifurcation into the mid left CIA under fluoroscopic guidance.
- Remove the dilator from the sheath and the wire from the hypogastric artery.
  - ▹ **For EIA stenosis: (Figure 5A)**
    - ♦ Use an angled non-stiff hydrophilic wire to cross the lesion.
  - ▹ **For EIA occlusion: (Figure 5B)**
    - ♦ Use a SSGW supported by either an angled or straight tip hydrophilic catheter to cross. Once the wire is across, follow with the catheter. Remove the wire, flush the catheter and confirm that the wire is intraluminal by measuring arterial pressure and taking an angiogram.
- Insert a 0.035" angled stiff hydrophilic wire and advance deep into the PFA/SFA.
- Remove the hydrophilic catheter.
- Load a pre-dilatation balloon on the wire and place across the lesion.
- Perform PTA of the lesion (balloon diameter 4 - 6 mm). Shepherd the sheath across the lesion. Withdraw the balloon. Repeat the angiogram.
- Obtain a fresh road map. Select an appropriate size SES (for a covered SES > 6.0 mm in diameter a 7 Fr. sheath is required).
- Withdraw the cross-over sheath while maintaining the stent in position.
- Deploy the stent and reshepherd the sheath into the stent with caution. The transition between the sheath and the stent catheter may not allow for a smooth delivery. If resistance is encountered, remove the stent catheter without shepherding. Try delivering the postdilatation balloon into the stent. If resistance is encountered, remove the balloon and attempt to shepherd with either the pre-dilatation balloon or the dilator of the sheath.
- Post-dilate the stent with an appropriate size balloon. Repeat angiography.
- If satisfied with the result, perform final hemodynamics.

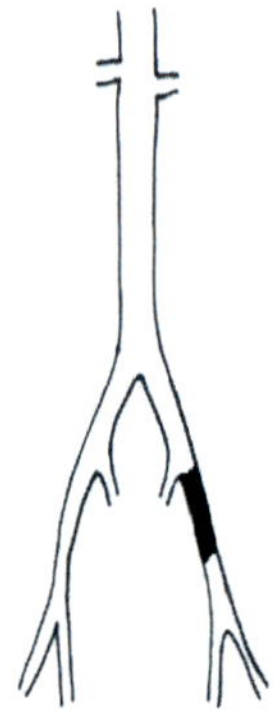

Figure 5B.

BILATERAL OSTIAL CIA SEVERE DISEASE
(Figure 6)

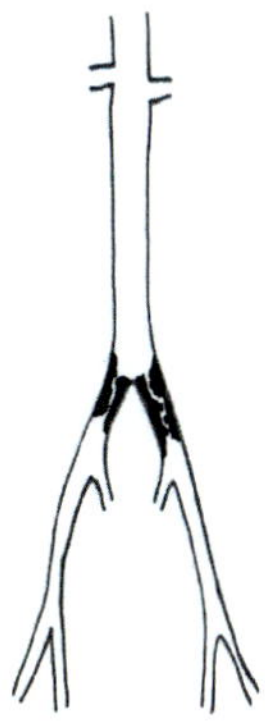

Figure 6.

Access:

- Right CFA access for aortogram. For the intervention this sheath will have to be upsized to a 6 or 7 Fr. depending upon the need for a covered stent.
- Left CFA access for retrograde intervention of the left CIA (6 or 7 Fr.).
- Use straight sheaths (>20 cm length) with radio-opaque tips for bilateral access.

Technique:

- Cross the lesions on each side with a 0.035” angled non-stiff hydrophilic wire.
- Next, exchange this wire through a catheter on each side for a stiffer wire.
- With stiff wires across both CIA lesions advance a pre-dilatation balloon across the right CIA lesion.
- Perform PTA of the right CIA lesion and gradually upsize the balloon from a small diameter (e.g. 4-6 mm) to a larger balloon (7 - 8 mm) as necessary.
- Prior to withdrawal of the balloon, shephed the right CFA sheath over the balloon and across the lesion. Withdraw the balloon. Repeat angiogram and assess the result.
- If no perforation is noted repeat the same steps on the left side.
- Obtain a fresh road map through either sheath making sure to clearly outline the aortic carina and the CIA ostia.
- Select an appropriate size BES for each side. Do not perform a ‘negative prep’ on these stents until they are in correct position within the lesion.
- Withdraw the sheaths while maintaining the stents in position.
- Once the stents are correctly positioned, deploy them using a “kissing technique” (simultaneous inflations). Ideally the original aortic carina should be preserved. This allows future ‘cross-over’ access from one side to the other in case of future interventions.
- Shepherd each sheath into its respective stent. Remove the balloons. Repeat angiography and assess the results.
- Post-dilate the stents with an appropriate size balloon in a “kissing” fashion. Repeat angiography.
- If satisfied with the results, perform final hemodynamics and abdominal aortography (if renal function allows).

HOW TO CROSS-OVER A CHALLENGING AORTIC BIFURCATION?

A challenging aortic bifurcation (figure 8A) has at least one of the following characteristics:

- Heavily calcified distal aorta, CIA and/or EIA.
- Right and left CIA run parallel to each other for some length.
- Exaggerated tortuousity of the aorta, CIA and/or EIA.

These features will prevent the wire (even a non-stiff hydrophilic wire) from being delivered distally and lead to poor catheter support at the aortic bifurcation.
Early recognition of the challenge (on an aortogram, CTA or MRA) should lead to the selection of appropriate equipment.

The sequence for crossing in this case should be (figures 7 & 8):

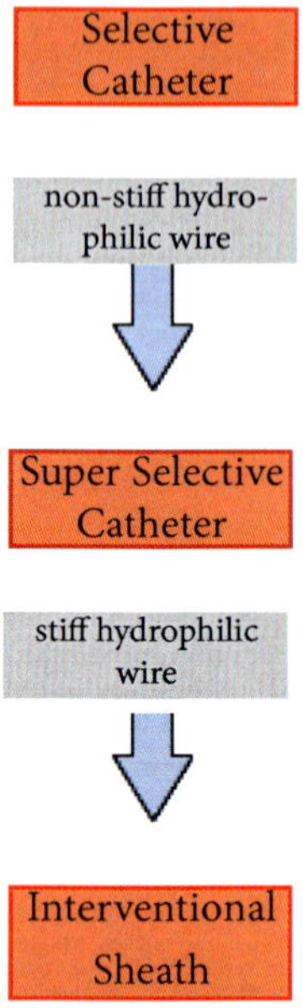

Figure 7.

Wire Stiffness:

- Non-stiff hydrophilic < stiff hydrophilic < Amplatz Superstiff < Amplatz Extrastiff < Lunderquist = Meir wire.
- To prevent vessel trauma, ALWAYS advance /remove an Amplatz, Lunderquist or other stiff wire *through a catheter*!

Crossing aortic bifurcation over a balloon:

- Advance the sheath to the aortic bifurcation.
- Bring the balloon halfway outside the sheath, into the contrlateral CIA.
- Inflate to 1-2 ATM.
- Keep traction on the wire.

Advance the sheath over the balloon. The two may be advanced together or in a stepwise fashion.

1. Selective catheters:

- Start with a selective catheter with a longer tip (e.g. Contra or Simmons) for 'deep seating' in the CIA.
- DO NOT waste time with catheters like the IMA, Rim or other similar short tip catheters.
- Use 5 Fr. selective catheters instead of 4 Fr. catheters.

2. Non-stiff steerable wire:

- Use a non-stiff angled hydrophilic wire (e.g. angled non-stiff hydrophilic wire) to traverse the iliac arteries and deliver it into the distal PFA or distal SFA (if patent).

3. Super-selective catheters:

- Exchange out the selective catheter for a super-selective catheter. Advance this catheter to the tip of the non-stiff wire if possible.
- Use a straight tip 4 Fr. hydrophilic catheter as it is easier to deliver than an angled catheter.

4. Stiff wire:

- Exchange out the non-stiff wire for a stiffer wire (e.g. angled stiff hydrophilic wire, Amplatz Superstiff, Amplatz ExtraStiff or a Lunderquist).
- Initial choice should be a steerable stiff wire such as the angled stiff hydrophilic wire. This can be delivered farther than the other stiffer wires mentioned above due to its hydrophilic coating and steering properties. Also, because it is not 'super' stiff it tends to dislodge the delivery catheter to a lesser degree and does not tend to 'pull it back' as the wire is advanced distally. If, however, this fails to allow the delivery of the cross-over sheath due to the lack of adequate support then one of the other wires should be used. Use 260 cm wires preferably.

5. Interventional sheath:

- Exchange for a 'cross-over' sheath of desired profile and deliver to its destination.
- If the sheath cannot be advanced despite the above measures, remove the dilator and instead use a balloon 4-6 mm (diameter) x 40 mm (length).
- Use a hydrophilic, braided sheath with a pre-formed curve (e.g. Ansel) for better delivery and less kinking at the bifurcation.
- After sheath delivery, the Amplatz and Lunderquist wires should be exchanged out through a catheter for another less traumatic wire. However, if the situation demands continued use of these wires, then confirm their tip position in a safe zone prior to proceeding.

MANAGING AN ILIAC PERFORATION

Equipment:
7 Fr. sheath

Balloons:
EIA:6 mm x 40 mm
CIA:8 mm x 30 mm

EIA stent:7 mm x 38 mm iCast
CIA stent: 8/9 mm x 59 mm iCast

Perforation of the iliac artery is a catastrophic complication that may be fatal if not treated promptly.

CAUSES

This is an iatrogenic complication that occurs more commonly in the EIA than the CIA. The following serve as risk factors:

- Advanced age.
- Female gender.

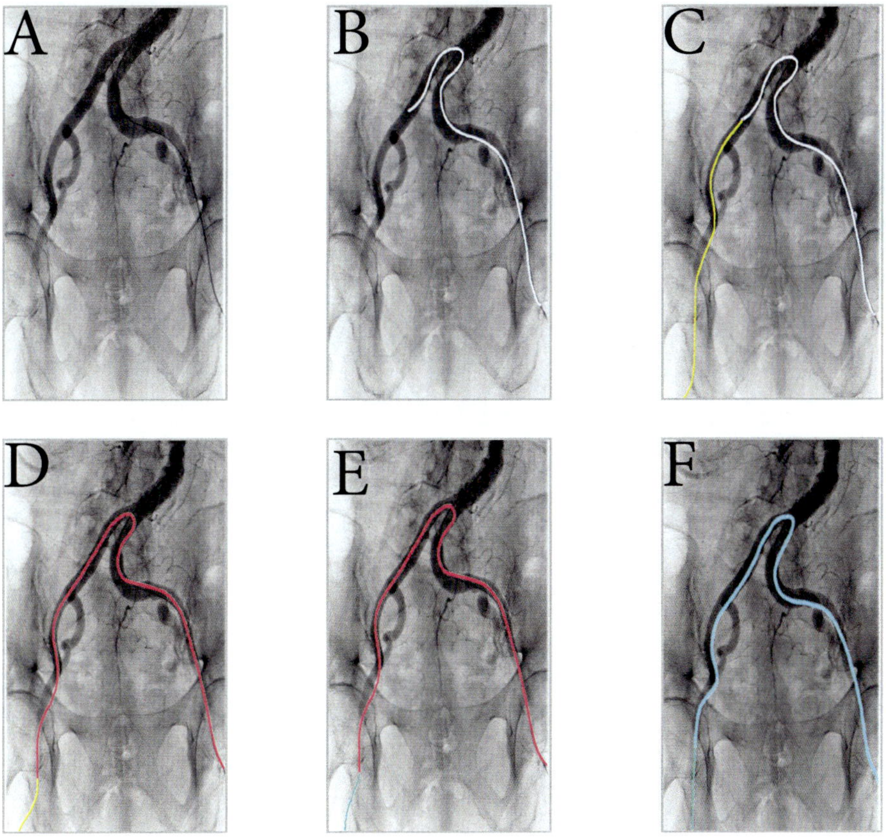

Figure 8.

How to cross a challenging aortic bifurcation.

A. Angulated and tortuous aortic bifurcation and iliac arteries.
B. A selective catheter (Simmons 1) is used to engage the bifurcation.
C. An ANSGW is advanced into PFA.
D. The selective catheter is exchanged out for a superselective catheter (Glidecath).
E. Once the Glidecath is in the PFA, the ANSGW is exchanged out for a stiff GW.
F. Over the stiff GW the superselective catheter is exchanged out and the interventional sheath is placed across the aortic bifurcation.

- Heavy calcification.
- Chronic total occlusion.
- Tortuous arteries.

Perforation occurs due to vessel trauma caused by:

- Angioplasty or stent placement.
- Sheath delivery (figure 9).
- Wire tip.

Illustration:

- An 87-year old male developed severe hypotension after obtaining initial access for aortic valvuloplasty.
- Exaggerated iliac tortuosity, ectasia and calcification was noted at the time of arterial access (8 Fr./25 cm sheath).
- Angiography revealed right EIA perforation
- The sheath was promptly advanced across the site, a 7.0 mm x 40 mm balloon was used for tamponade
- An 8.0 mm x 59 mm iCast was placed across the perforation site and overlapped with another 8.0 mm x 38 mm iCast due to incomplete coverage
- The stents were postdilated cautiously with a 10.0 mm balloon (as the arteries were ectatic)
- Prompt resolution of the hemorrhage was noted angiographically.
- Heparin did not have to be reversed.
- No blood transfusion was required required.
- The patient recovered completely and valvuloplasty was performed after 48 hours via the same access site.

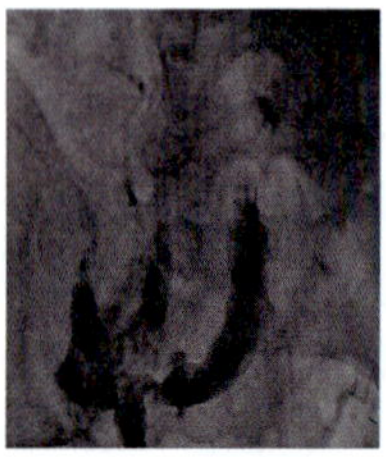

Figure 9.

DIAGNOSIS

If a perforation is suspected, prompt angiography at the site of the most recent intervention should be performed. Use DSA if possible. If no obvious angiographic perforation is identified withdraw the sheath partially (without pulling back the wire) and repeat the angiogram to rule out sheath-related trauma.

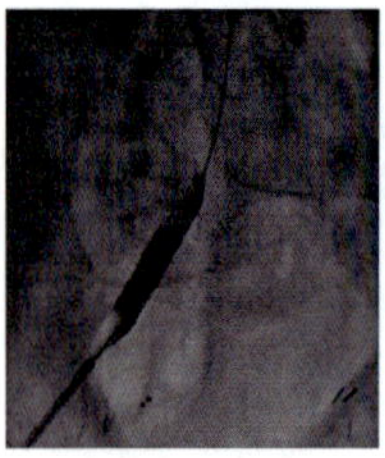

Figure 10.

TREATMENT

Once identified the treatment should be swift and methodical:

Balloon Tamponade:

- Obtain immediate balloon tamponade at the site of perforation with an appropriate size balloon (6-8 mm diameter x 40 mm for the EIA; 7-10 mm diameter x 20-30 mm for the CIA). Inflate at nominal pressure (figure 10).

Sheath Selection:

- If the sheath in place is < 7 Fr., prepare a 7 Fr. sheath right away
  - ▹ 7 Fr./20 cm straight sheath for ipsilateral iliac perforation.
  - ▹ 7 Fr./45 cm 'cross-over' sheath for a contralateral iliac perforation.

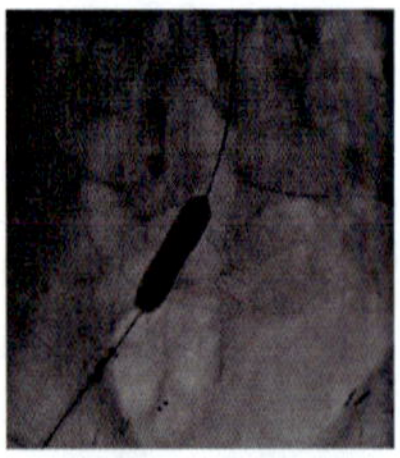

Figure 11.

Stent Selection:

- For most perforations in the iliac arteries, a 7.0 mm - 9.0 mm diameter balloon-expandable covered stent (e.g. iCast) is adequate.

Balloon Removal:

- Prior to balloon removal consider exchanging the interventional wire through it for a stiff wire if not in use already.
- Shepherd the sheath across the perforation.

Sheath Up Sizing:

- Minimum 7 Fr.

Obtain A Road Map:

- For precise stent placement.

Stent Delivery:

- Deliver the stent to the desired location (figure 11). Withdraw the sheath, confirm placement and deploy at nominal pressure.
- Re-shepherd the sheath into the stent.

Repeat Angiogram:

- Assess if postdilatation or placement of another stent is necessary (figures 12 &13).
- Use DSA.
- Appropriate imaging at this juncture is critical to exclude ongoing bleeding. For CIA perforations, consideration should be given to aortography with undiluted contrast.

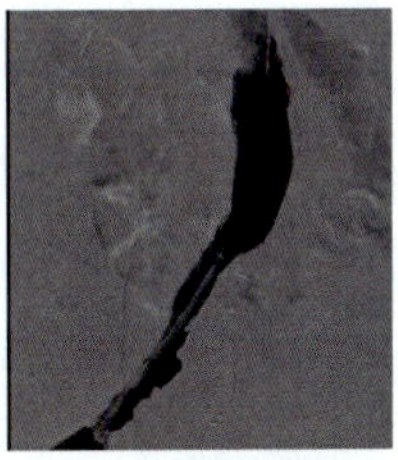

Figure 12.

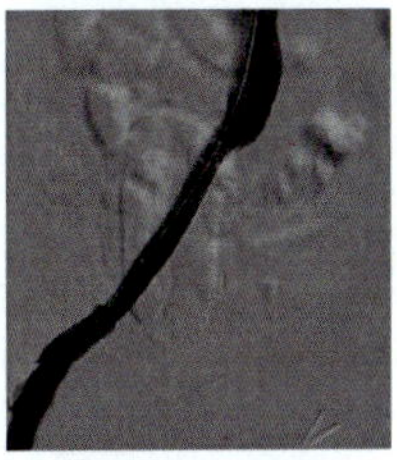

Figure 13.

ADDITIONAL MEASURES

Concurrent with the above steps the following measures should be considered:

- Volume resuscitation.
- Blood type and crossmatch.
- Anticoagulation reversal with protamine if not contraindicated.
- Vascular surgery consultation if endovascular treatment remains unsuccessful.

POST-PROCEDURE FOLLOW-UP

- Repeat Hb/Hct at two to four hour intervals until a stable trend is noted.
- Transfuse pRBCs as necessary.
- If significant retroperitoneal bleeding is suspected prior to successful treatment, a non-contrast CT of abdomen and pelvis may be helpful at 12-24 hours. This helps assess the extent of the hematoma and any mass effects from it (e.g. hydroureter etc).

Protamine:

- Give a test dose of 5 mg and observe for any allergic response.
- Dosing for reversal is typically 10-15 mg IV for every 1000 U of IV UFH given in the last 2 hours.
- Max. dose: 50 mg.

Precautions:

It may cause:

- Anaphylaxis
- Severe hypotension
- Non-cardiogenic pulmonary edema
- Pulmonary HTN

Contraindications:

- Prior protamine use.
- NPH insulin use (current or prior).
- S/p vasectomy.
- Fish allergy.

CASE 1

Ms. S. is a 79-year-old female with a known history of open AAA repair and lower extremity PAD. Approximately 20 years ago, she had undergone an open repair for AAA with a tube graft. Subsequently, she was diagnosed to have colon cancer and underwent total colectomy. Six years after the AAA repair, she was found to have severe iliac occlusive disease primarily on the right side. She had disabling intermittent claudication in both legs and over time developed a non-healing ulcer on the right foot. She underwent a right axillary artery to right femoral bypass graft and a cross femoral bypass graft. Post-operatively, she had complete resolution of her symptoms. However, approximately 1 year later, these symptoms recurred primarily affecting the right leg. At that time, noninvasive arterial testing revealed severe disease in the right leg with moderate disease on the left. PVRs suggested bilateral aorto-iliac disease, right worse than left. An MRA of her abdomen, pelvis and lower extremities was obtained prior to angiography.

Initial Diagnostic Equipment:

- 4 Fr. micropuncture kit.
- 5 Fr./23 cm long straight sheath.
- 0.035" 260 cm angled stiff hydrophilic wire (Glidewire).
- 4 Fr./65 cm tennis racket (TR) catheter.
- Power injector.

Access & Diagnostic Angiography:

- For initial diagnostic angiography, arterial access was obtained in the left common femoral artery using a 5 Fr./23 cm Cordis Brite-tip sheath.
- 2000 units of IV heparin was given.
- Over a 0.035" ASGW wire a 4 Fr. tennis racket catheter was advanced into the distal abdominal aorta.
- Abdominal aortography was performed at a rate of 10 cc per second for a total of 30 cc.

Interventional Technique:

Start up Interventional Equipment

- 4 Fr/65 cm TR catheter.
- 4 Fr./100 cm straight hydrophilic catheter.
- 7 Fr./23 cm straight sheath with marker tip (Cordis Brite-tip).
- 0.035" 260 cm angled non-stiff hydrophilic wire (Glidewire).
- 0.035" 260 cm angled stiff hydrophilic wire (Glidewire).
- 0.035" 260 cm straight stiff hydrophilic wire (Glidewire).
- 0.035" 150 cm straight catheter (QuickCross).
- 0.035" compatible PTA balloon.
- 0.035" compatible balloon-expandable covered stent for right CIA (9 mm x 59 mm iCast).

Engaging the RIGHT EIA Occlusion

- Due to distal aortic and proximal right common iliac artery disease, antegrade approach was not ideal.
- However, given the short stump of the right CFA (Figure 14 A&B), retrograde access was challenging. Hence, it was considered as a backup option for access.
- Using a 4 Fr. IMA catheter, right common iliac artery was cannulated via the left femoral access.
- In cases where the external iliac artery is occluded proximally, the ipsilateral hypogastric artery is used to advance the wire into in an anchoring position and then a cross-over sheath

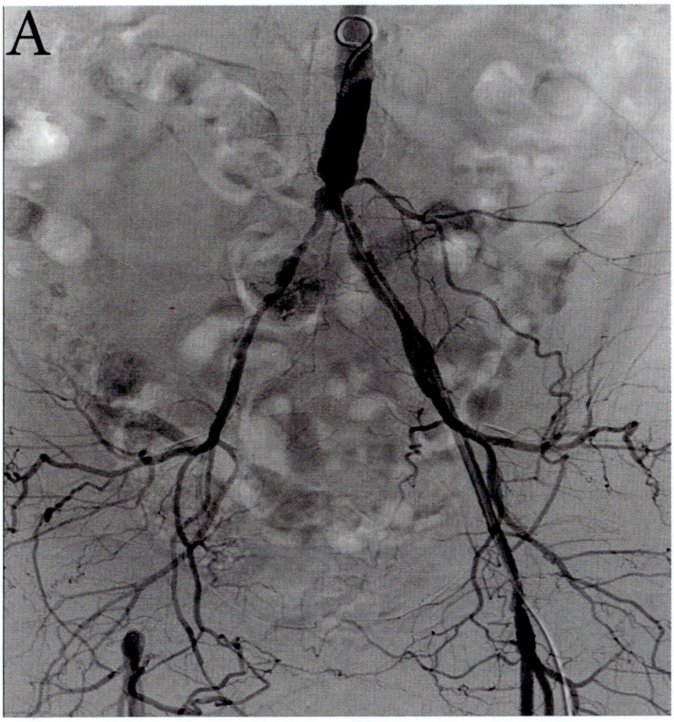

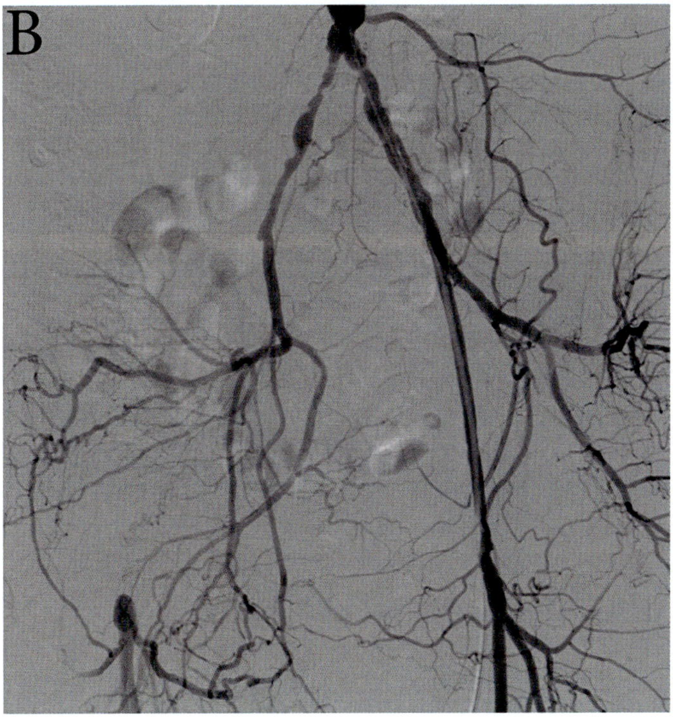

Figure 14A & B.

Findings:

1. Aortography revealed a patent distal aortic tube graft.
2. Diffuse severe disease of distal aorta was noted.
3. The right common iliac artery was diffusely diseased with a severe stenosis.
4. The right hypogastric artery was patent and providing abundant collaterals to the right common femoral and profunda femoris artery.
5. The right external iliac artery was occluded at its ostium with no reconstitution of this vessel.
6. Right common femoral artery was patent and filling via right hypogastric and right profunda. femoris collaterals.
7. Left common iliac artery was also diffusely diseased in the proximal half.
8. Distal left common iliac artery was patent and free of significant disease.
9. Left hypogastric artery was noted to be patent.
10. Moderate disease was noted in the proximal left external iliac artery.
11. Bilateral SFA and profunda femoris were patent proximally.

is advanced over the aortic bifurcation using this wire as a rail.

- Although stiff wires serve as a good rail, it is challenging and not always desirable or safe to advance a wire, such as an angled stiff Glidewire, into the hypogastric artery.
- Initially, an angled non-stiff Glidewire was advanced into the distal hypogastric artery (superior gluteal artery).
- This wire was then exchanged out through a 4 Fr. straight Glidecath for an angled stiff Glidewire.
- Over the angled stiff Glidewire, a cross-over sheath was advanced over the aortic bifurcation (Figure 15 A).
- This may not always be smooth because of the obstructive disease at the common iliac artery (as in our case).
- Thus, while dealing with the contralateral common iliac artery disease, the tip of the sheath is parked at the level of the aortic bifurcation.
- The dilator was removed and the sheath was flushed out; the wire was still kept in the hypogastric artery. PTA and stenting of the mid (non-ostial) common iliac artery was performed with an adequate size balloon/stent (figure 15 B).
- Appropriate length selection of the stent is critical here as extension of the stent into the hypogastric artery or the distal aorta can be very detrimental to the planned revascularization of the occluded external iliac artery.
- After the common iliac artery stent was postdilated to a desired diameter, the dilator of the sheath was reinserted over the angled stiff Glidewire, and the sheath was carefully advanced under fluoroscopic guidance through the stent to the distal edge of the common iliac artery. At this point the dilator was removed, the sheath was re-flushed.
- In LAO 20° projection, a road map of the right common and external iliac arteries was obtained.

Crossing the RIGHT EIA Occlusion

- A 4 Fr./65 cm straight Glidecath was advanced over the Glidewire into the sheath. The wire was removed from the right hypogastric artery and a straight stiff 260 cm Glidewire was inserted to the tip of the catheter.
- Through this catheter, the wire was advanced across the occluded right external iliac artery.
- The wire traversed the occlusion without significant difficulty and entered the distal right CFA as well as the SFA (Figure 15C).
- The straight Glidecath was followed over the wire and advanced into the right SFA (Figure 15D).
- The wire was removed, the catheter flushed and a confirmatory angiogram was performed (Figure 15D).
- After confirming the intraluminal position of the Glidecath, a 0.018" V. 18 wire was advanced through it into the SFA. A separate 0.014 inch choice PT floppy wire was also advanced alongside the V. 18 wire through the Glidecath.
    - ▹ This was done to confirm access to both the SFA and PFA and to rule out the presence of an intimal flap
- In an RAO 30° projection, a "road map" was obtained to identify the distal right CFA, profunda and SFA.
- The choice PT floppy wire was pulled back to the CFA bifurcation and unsuccessful attempts were made to maneuver it into the profunda.
- The profunda could not be visualized on selective angiography of the right CFA via the cross-over antegrade route suggesting reentry into the true lumen was into the SFA rather than the CFA.

Retrograde RIGHT CFA Access

- At this point, under fluoroscopic guidance retrograde access was obtained in the right CFA to salvage the profunda and SFA (Figure 16A).
- Using a 4 Fr. micropuncture kit, right CFA access was achieved. The 0.018 inch wire was

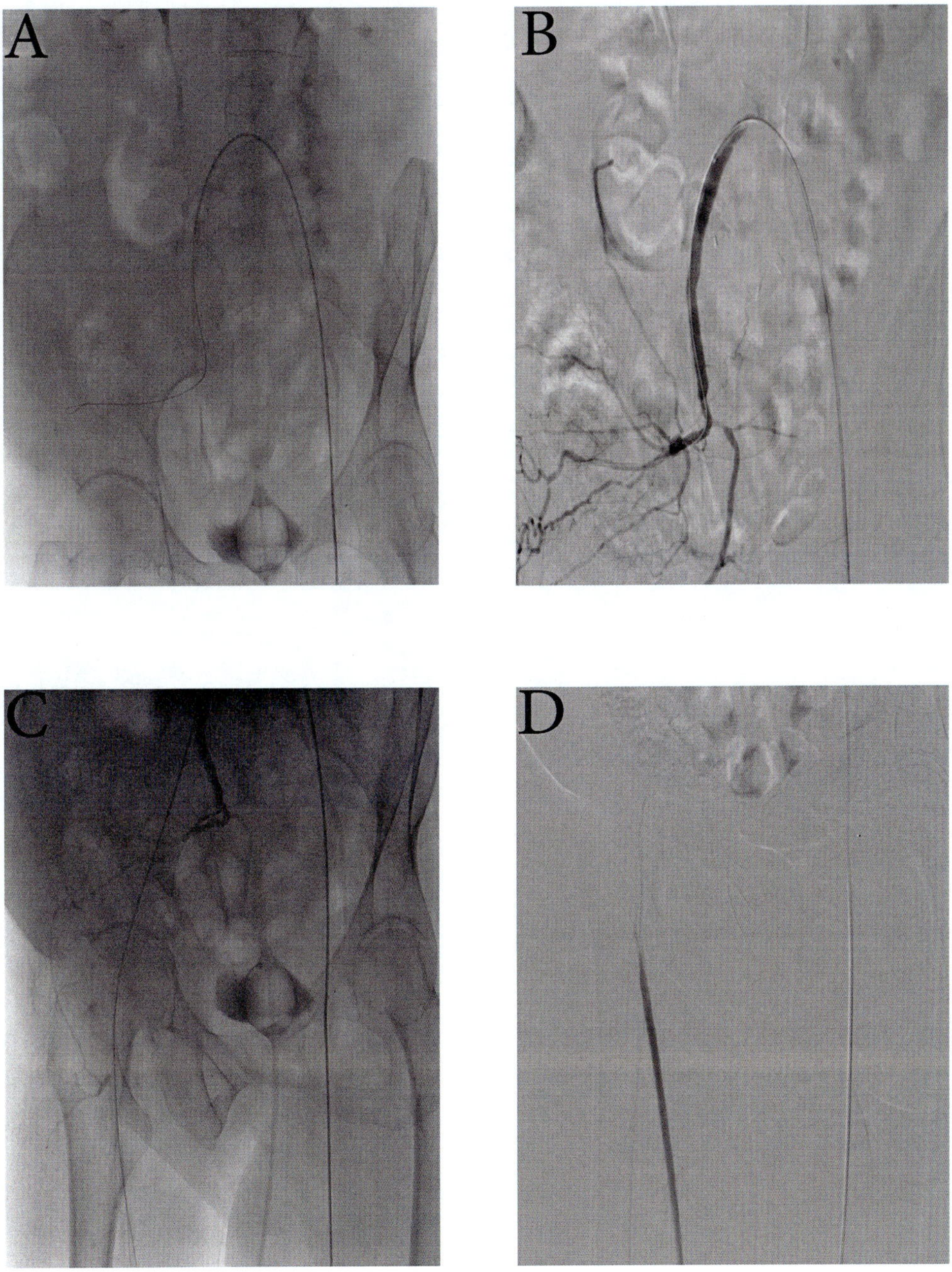

Figure 15.
A. The hydrophilic stiff wire is in the superior gluteal artery, a branch of the posterior division of the IIA. Note the loop
at the tip of the wire is not only a safety feature against perforation but also provides additional anchoring.
B. Angiogram via the cross-over sheath after right CIA PTA/stenting.
C. The straight stiff hydrophilic wire is advanced across the right EIA occlusion, supported by a hydrophilic catheter. The tip of the wire is in the SFA.
D. A confirmatory angiogram through the hydrophilic catheter once its advanced into the SFA and the wire is re moved. Note the PFA does not fill.

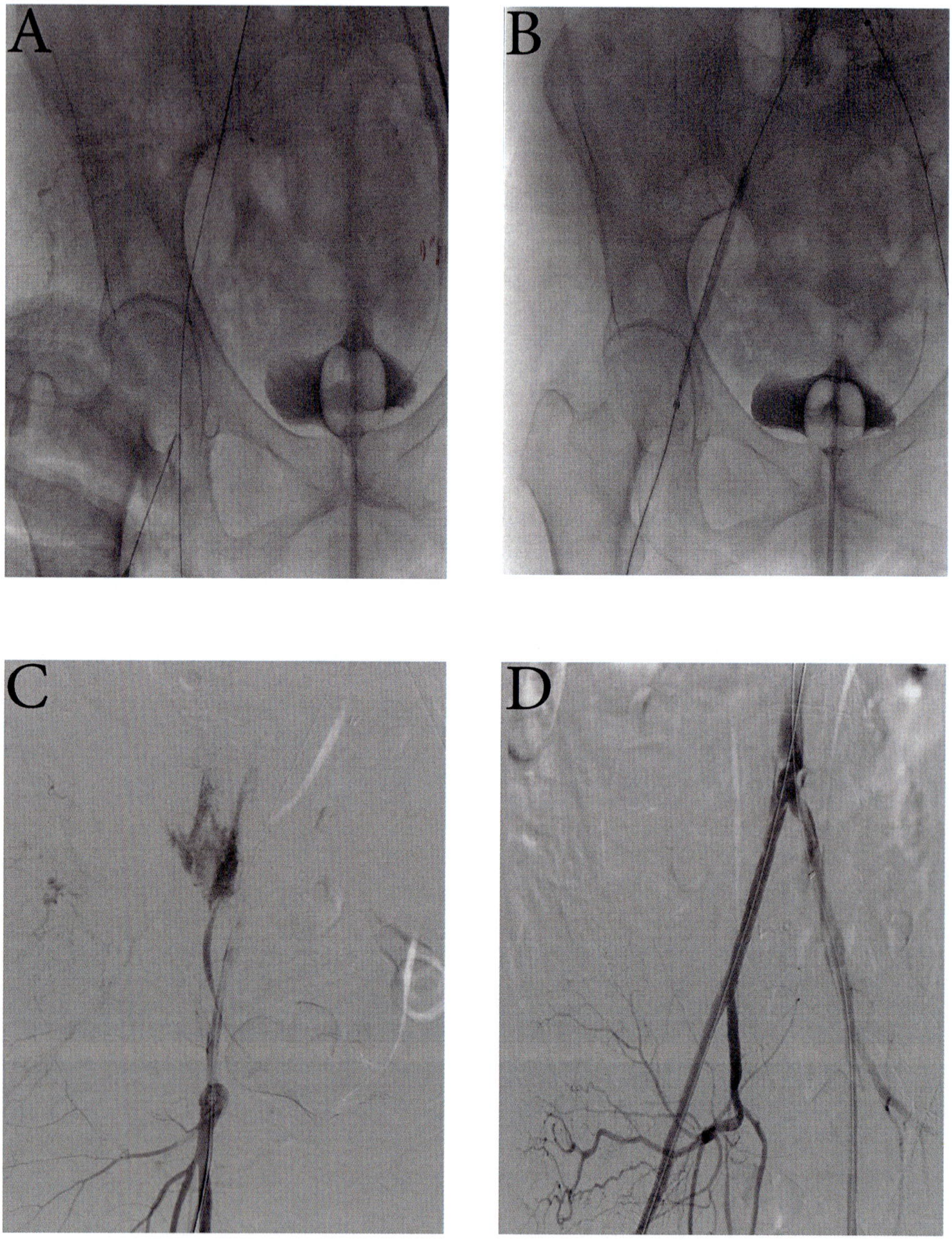

Figure 16.

A. Retrograde access being obtained in the right CFA with a 21G needle and a 3cc syringe (filled with contrast) to confirm precise entry.
B. PTA being performed via CFA access after the wire has been successfully advanced into the aorta via the right CFA.
C. Right EIA perforation as a result of PTA.
D. Angiogram of the right EIA after placement of a covered SES (6.0 mm x 100 mm Viabahn). Note that this stent was intentionally not extended across the hypogastric artery which would have been excluded otherwise. This vessel can be an important source of collaterals to the CFA/PFA in the future.

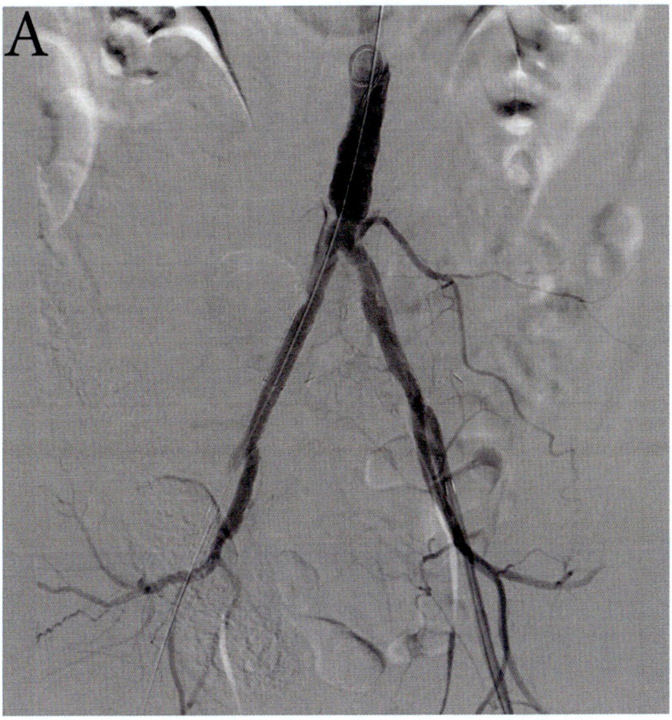

Figure 17A.

Dissection of the distal aorta into the CIA bilaterally.

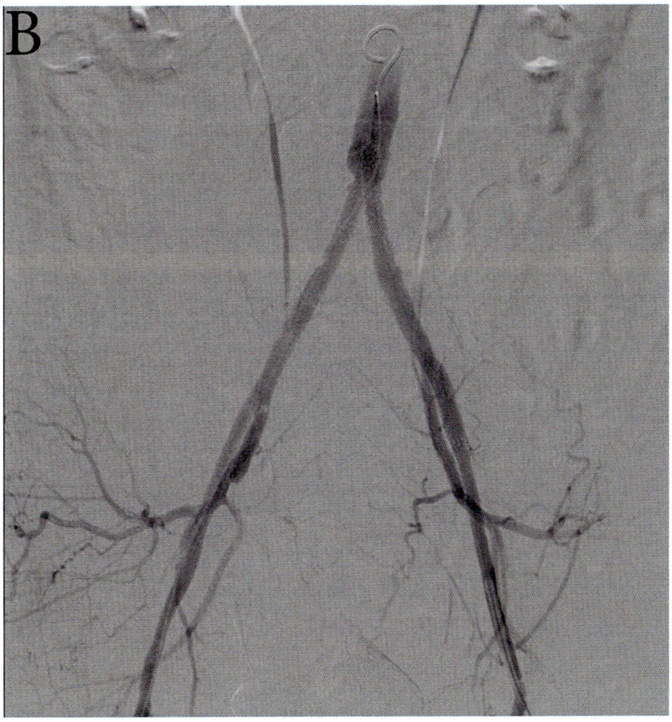

Figure 17B.

Final aortography after bilateral CIA kissing stent placement.

1. As noted, covered BES were used. No residual flow is noted around the tube graft. The dissections of bilateral CIA have resolved. No perforation is noted.
2. Bilateral internal iliac arteries are patent.
3. Satisfactory flow is noted in bilateral ileofemoral arteries.

advanced in a retrograde fashion without difficulty, through the needle and into the right external iliac artery.

- Once the 4 Fr. micro-catheter was in the right CFA, a straight stiff 260 cm Glidewire was advanced in a retrograde fashion into the right common iliac artery. At this point, the right femoral micro-catheter was removed and a 4 Fr./65 cm straight glide catheter was advanced over the straight stiff Glidewire into the right common iliac artery. The wire was removed, the catheter flushed and then used to perform a confirmatory angiogram.
- After the glide catheter tip position was confirmed in the right common iliac artery, a 0.035" 260 cm Wholey wire was advanced through the Glidecath into the distal abdominal aorta.
- The Glidecath was then removed and a 7 Fr./23 cm Cordis Brite-tip sheath was inserted into the right CFA and advanced into the distal right external iliac artery. The dilator was removed, the sheath was flushed and a confirmatory angiogram was performed.
- At this point, the cross-over sheath in the right common iliac artery was withdrawn to the left common iliac artery.

Balloon Angioplasty of the RIGHT EIA Occlusion

- Balloon angioplasty was performed with a 6.0 mm x 60 mm balloon of the occluded segment of the right external iliac artery (Figure 16B). This was successful but complicated by an immediate perforation of the proximal right external iliac artery (Figure 16C). The perforation was temporarily sealed with balloon inflation.

Stent Placement and Deployment

- The right external iliac artery was stented with a 6.0 mm x 100 mm Viabahn stent from the CFA up across the perforated segment (Figure 16D).
- The perforation was sealed with the stent.
- The distal segment of the right common iliac artery (adjacent to the right hypogastric artery) was treated with an 8.0 mm self-expanding Absolute stent which was overlapped proximally and distally with the previously placed external and common iliac artery stents (Figure 16D).
    - ▹ A non-covered stent was used here to allow perfusion to the hypogastric artery which may serve as a future source of collaterals to the leg.
- At this point, attention was focused on the distal aorta and bilateral ostial/proximal common iliac arteries.

Postdilatation and Final Aortography

- A 6.0 mm x 60 mm balloon was loaded on the wire through each femoral sheath and kissing angioplasty was performed at the aortic bifurcation.
- Angiography revealed a limited retrograde dissection of the distal abdomen aorta on both sides (Figure 17A).
- Due to a suboptimal angioplasty result and dissection of the distal aorta into the ostial iliac arteries, a decision was made to use covered balloon-expandable stents (iCast) bilaterally. Therefore, the sheaths in both groins had to be upsized to a 7 Fr. Cordis Brite-tip sheaths. Two 8.0 mm x 38 mm iCast stents were deployed using a kissing technique in the bilateral common iliac arteries and distal abdominal aorta just below the tube graft.
- Subsequent angiography revealed complete resolution of the stenoses involving the distal abdominal aorta as well as the bilateral common iliac arteries (Figure 17B).
- No residual dissection was noted in the distal abdominal aorta or the bilateral iliac arteries.
- No hemodynamic gradient was identified between the abdominal aorta and either common femoral artery.
- Final aortography revealed patent distal abdominal aorta and bilateral CIA as well as EIA.

CASE 2

Mr. W is a 61 year old male with a history of coronary artery disease, diabetes, hypertension and disabling intermittent claudication of the RIGHT lower extremity. Physical exam revealed a faint right and a normal left common femoral pulse. His non-invasive studies revealed an ABI of 0.56 on the RIGHT with a normal ABI on the left. The PVRs suggested severe right aorto-iliac occlusive disease.

Initial Diagnostic Equipment:

- 4 Fr. micropuncture kit.
- 5 Fr./23 cm long straight sheath with a radio-opaque tip.
- 0.035" 260 cm angled stiff hydrophilic wire (Glidewire).
- 4 Fr./65 cm tennis racket (TR) catheter.
- Power injector.

Access & Diagnostic Angiography:

- For initial diagnostic angiography, a 5 Fr./23 cm long sheath was inserted in the left common femoral artery.
- 2000 units of IV heparin was given.
- A 4 Fr. tennis racket catheter was advanced over a 0.035" ASGW wire into the distal abdominal aorta.
- Abdominal aortography was performed at a rate of 10 cc per second for a total of 30 cc using a power injector.

Goals of Intervention:

- Revascularization of right common iliac artery with a covered stent.
- Revascularization of right external iliac artery if necessary with PTA/stent.
- Revascularization of ostial left common iliac artery if necessary.

Interventional Technique:

Start up Interventional Equipment

- 6 Fr./90 cm long hydrophilic sheath (Shuttle).
- 6 Fr. MP guiding catheter.
- 7 Fr./23 cm straight sheath with marker tip (Cordis Brite-tip).
- 0.035" 260 cm angled stiff hydrophilic wire (Glidewire).
- 0.035" 260 cm straight stiff hydrophilic wire (Glidewire).
- 0.035" 150 cm straight catheter (QuickCross).
- 0.035" compatible PTA balloon.
- 0.035" compatible balloon-expandable covered stent for right CIA (9 mm x 59 mm iCast).

Engaging the Right CIA Ostium

- Given the ostial location of the occlusion (figure 18A), an antegrade approach via brachial access is ideal.
- Left brachial artery access was obtained and 6 Fr./90 cm straight sheath was advanced over the 0.035" angled stiff hydrophilic wire into the distal abdominal aorta.
- A road map was obtained through this sheath.
- The 0.035" QuickCross catheter was loaded into the 6 Fr. multipurpose guiding catheter in a 'telescoping' fashion. The two were then advanced through the brachial sheath over the 0.035" angled stiff hydrophilic wire to the occlusion (figure 18B). Once the guide was engaged into the ostium of the right CIA, the angled stiff hydrophilic wire was exchanged out

Figure 18A.
Findings:

1. Moderate atherosclerosis in the distal abdominal aorta with no evidence of a severe stenosis or an aneurysm.
2. Patent left common and external iliac arteries.
3. The left hypogastric artery appears occluded.
4. There is moderate disease at the ostium of the left CIA.
5. Occlusion of the right common iliac artery just beyond its origin with distal reconstitution at the level of the proximal external iliac artery.

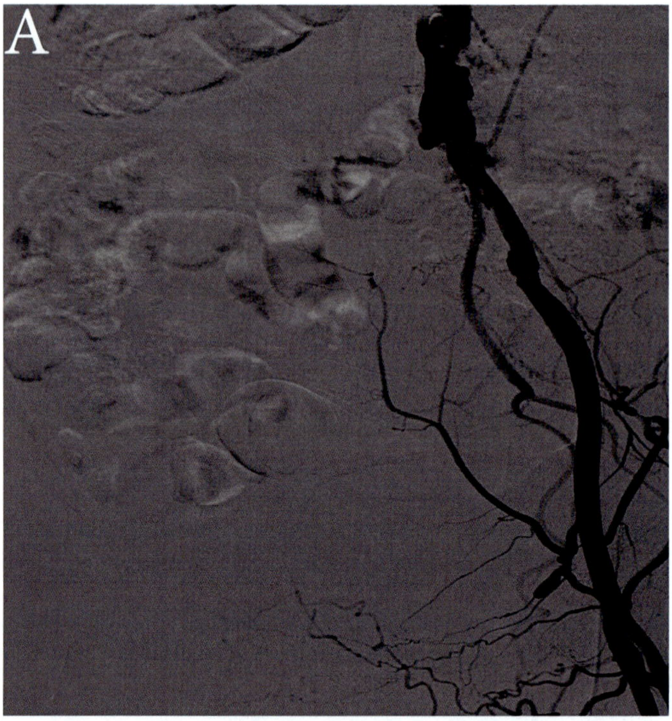

Figure 18B.

This figure shows a schematic of a 6 Fr. Shuttle sheath (black) through which a 6 Fr. MPA GC (blue) is advanced to the proximal cap of the occlusion.
A SSGW is then advanced in the direction of the right CIA/ EIA (red arrow).

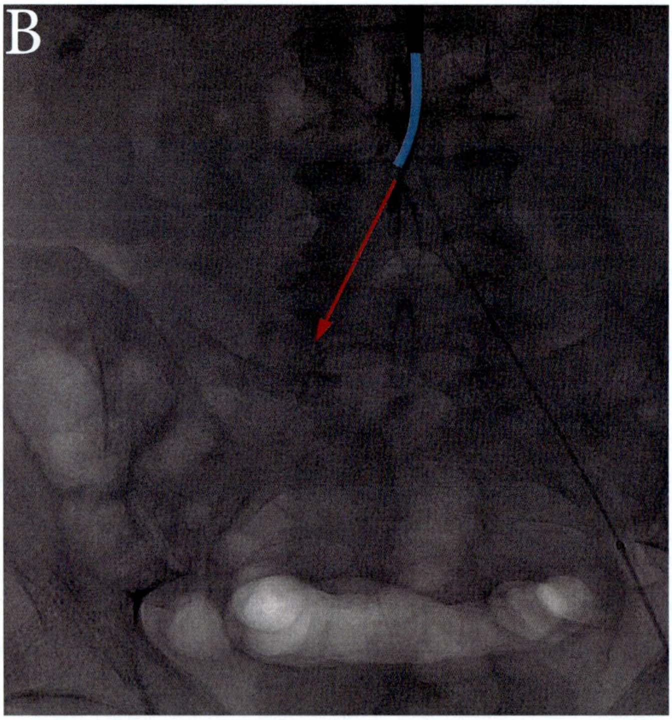

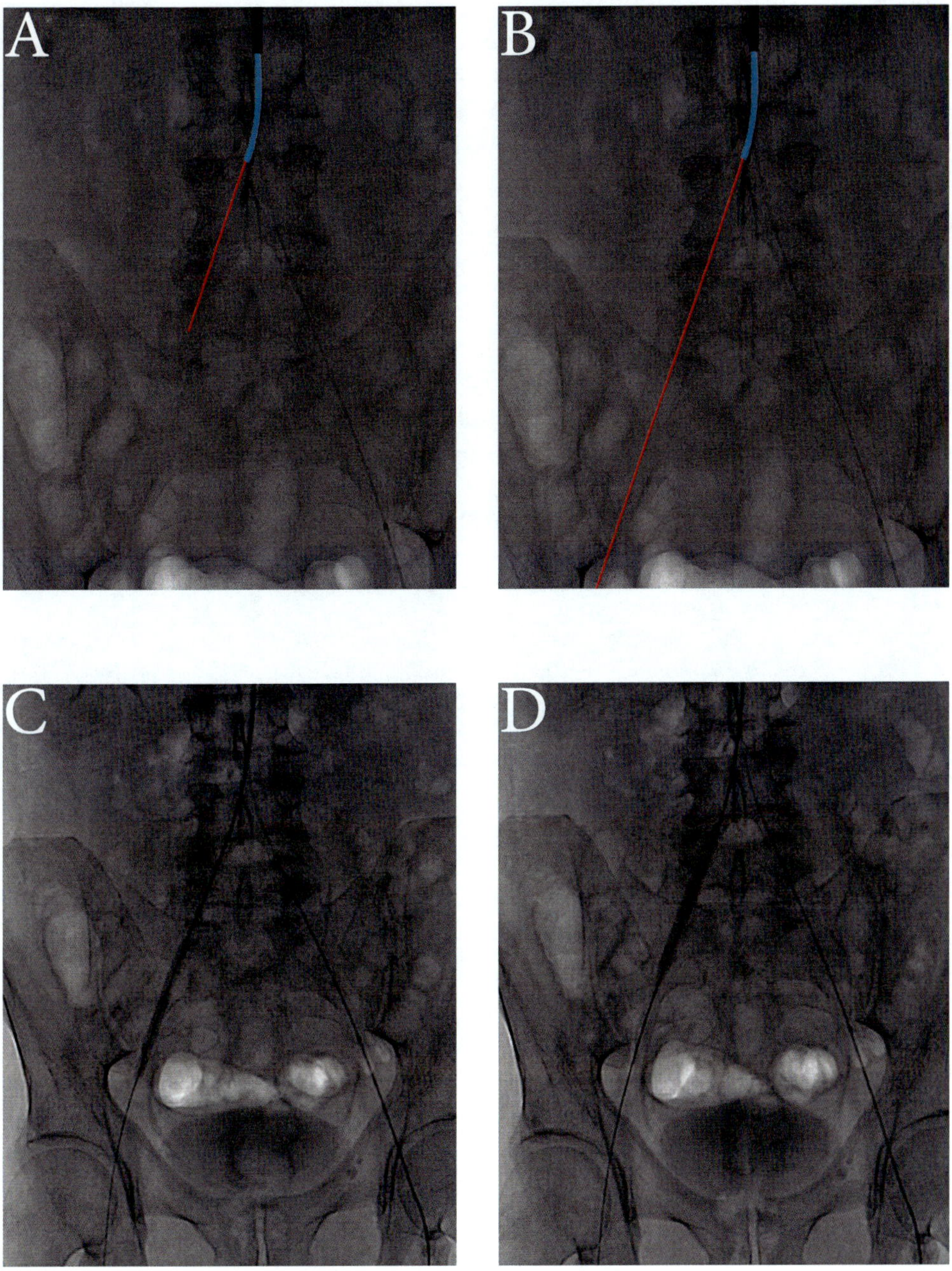

Figure 19.

A & B. The straight hydrophilic wire is being advanced across the occlusion in the right CIA and enters the EIA. The left femoral artery sheath and the TR catheter through it is noted in the aorta.

C & D. The wire has been advanced into the SFA. Angioplasty of the EIA and the CIA is performed.

for a 0.035" straight stiff hydrophilic wire

Crossing the Right CIA Occlusion

- Once the straight stiff hydrophilic wire was in position, it was advanced across the occlusion, supported by the QuickCross, without much difficulty. It was then advanced under road map beyond the EIA and into the CFA (figure 19A & 19B).
- A QuickCross catheter (0.035", 150 cm) was advanced over the wire into the CFA, the wire was removed and a confirmatory angiogram was performed through this catheter to document intraluminal position.
- An angled stiff hydrophilic wire was advanced through the QuickCross catheter into the CFA.

Balloon Angioplasty

- The catheters were removed and a 6.0 mm x 60 mm Mustang balloon was used to perform angioplasty of the EIA and the CIA. (Figure 19C & D, respectively).

Stent Placement and Deployment

- Due to heavy calcification in the CIA and chronicity of the lesion the risk of perforation due to intervention was considered high. Therefore, it was decided to use a covered stent (iCast) in the CIA.
- Since a 7 Fr. sheath is required for this stent, right CFA retrograde access was obtained rather than up sizing the brachial sheath.
- A 7 Fr./23 cm Brite tip sheath was inserted into the right CFA and the wire from above was externalized through this sheath.
- The sheath was then advanced along with the dilator over this "externalized" wire into the proximal right CIA, beyond the original occlusion. The dilator was removed and the sheath flushed.
- A road map was obtained.
- A 9.0 mm x 59 mm iCast stent (figure 20A) was advanced over the wire into the correct position. While maintaining the stent in position, the CFA sheath was pulled back to expose the stent.
- The stent was deployed.
- The sheath was shepherded over the stent balloon which was eventually removed. An angiogram was performed.
- An 8.0 mm x 30 mm Absolute stent was advanced into the EIA via the CFA sheath, overlapped with the above stent, and deployed in a standard fashion.

Postdilatation & Repeat Aortography

- The sheath was shepherded into the stent. An angiogram was obtained and using an 8.0 mm balloon the stent was postdilated.
- Follow-up angiography revealed patent right CIA and EIA stents. However, the ostium of the left CIA became narrowed because of plaque shift.
- Left CFA access was then up sized to a 6 Fr./23 cm long sheath which was advanced into the distal abdominal aorta.
- The dilator removed and the sheath flushed.
- An 8.0 mm x 39 mm balloon-expandable stent (non-covered; Genesis) was advanced into the sheath and positioned at the ostium.
- The sheath was withdrawn and the stent deployed in a 'kissing fashion' with a balloon in the right CIA.
- Final hemodynamics and aortography was performed after removal of both balloons (figure 20B). Widely patent right and left common iliac arteries were noted. Flow was also noted in bilateral hypogastric arteries. Moderate diffuse narrowing was noted in the right EIA, however, without any gradient across it. This segment was thus left untreated with the expectation that over time the caliber of the vessel would improve due to restored patency of the right common iliac artery. The goal (hemodynamic normalization but not necessarily angiographic perfection) was achieved.

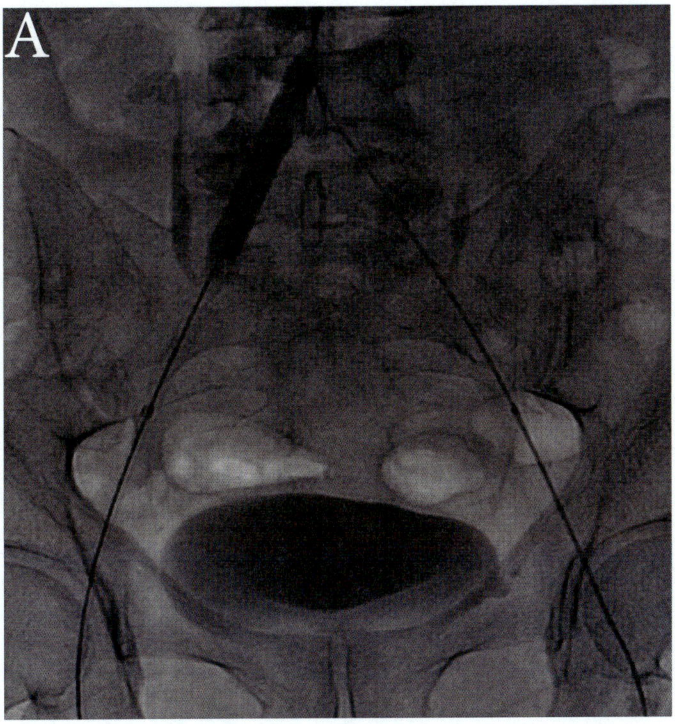

Figure 20A.
Covered stent placement at the ostium of the right CIA. Note a sheath in the right EIA that is inserted via the right CFA. This is a 7 Fr. sheath and allows delivery of the covered stent. The brachial sheath is 6 Fr.

The catheter across the left CIA is maintained during this intervention to allow 'bail out' in case of significant 'plaque shift' into this vessel. Additionally, this catheter allows the ability to image prior to stent deployment so the positioning is precise.

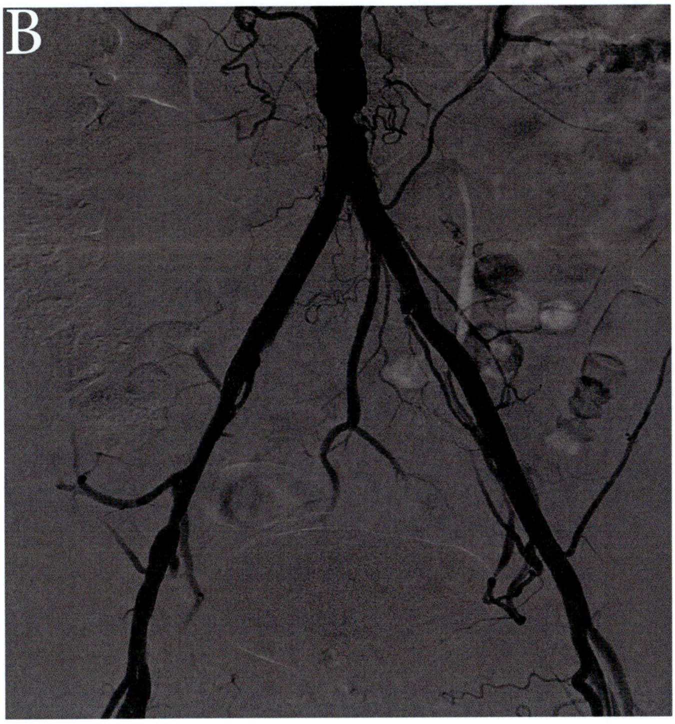

Figure 20B.

An aortogram is performed. This shows patent bilateral CIA osita. The stent at the ostium of the left CIA is a non-covered BES. Rationale for choosing a non-covered stent here is the relatively low risk of perforation in this artery which was compromised by plaque shift rather than native calcific plaque.

6

FEMORO-POPLITEAL ARTERY INTERVENTION

Femoro-popliteal obstructive disease is common in patients with intermittent claudication (IC) and is usually a significant factor in those with critical limb ischemia (CLI). In patients with IC, femoro-popliteal disease is typically the major burden of their PAD while in those with CLI it is usually accompanied by aorto-iliac (inflow) and/or infra-popliteal (outflow) disease. In cases of popliteal artery disease, a concurrent SFA stenosis or occlusion can significantly reduce collateral flow and cause CLI even in the absence of any inflow or outflow disease.

The most common site of atherosclerotic injury to the SFA is at the level of the Hunter's canal. This is believed to be a result of torsion, elongation, and other physical forces the vessel is exposed to in this region thus leading to endothelial injury and plaque formation.

ANATOMY

The femoro-popliteal arteries consist of the:

- Common femoral artery (CFA).
- Superficial femoral artery (SFA).
- Profunda femoris artery (PFA).
- Popliteal artery.

COMMON FEMORAL ARTERY

- This artery is the continuation of the EIA.
- Angiographically it starts after the origin of the inferior epigastric artery and terminates at the origin of the PFA and SFA (figure 1).
- Normally, its luminal diameter is 6 - 8 mm in adult males and 5 -7 mm in females.
- It is usually 3 cm in length. However, it may be as short as 1 cm and as long as 6 cm.
- It terminates by bifurcating into the SFA and PFA.
- Prior to its termination it gives four named branches (lateral to medial):
    - ▹ Superficial iliac circumflex artery
    - ▹ Superficial epigastric artery
    - ▹ Superficial external pudendal artery
    - ▹ Deep external pudendal artery

Femoral Triangle:
It is a triangular anatomical space in the proximal third of the thigh, apex pointing towards the foot. Its boundaries include:
- The inguinal ligament superiorly.
- The sartorius laterally and
- Adductor longus medially.

Femoral vessels course through its middle, base to tip.

Hunter's Canal:
This is an aponeurotic tunnel bound by:
- Vastus medialis anterolaterally
- Adductor longus and magnus postero-medially.

PROFUNDA FEMORIS ARTERY

- This is the smaller of the two terminal branches of the CFA.
- Its normal luminal diameter is 3 - 6 mm.
- It typically starts at the level of the femoral head, from the posterior and lateral aspect of the CFA. Rarely, it arises off the postero-medial aspect instead.
- It terminates in the distal third of the thigh as the 'fourth perforating artery'.
- Its branches include:
  - ▹ Lateral femoral circumflex artery.
  - ▹ Medial femoral circumflex artery.
  - ▹ Perforating arteries (three).
  - ▹ Muscular branches.

SUPERFICIAL FEMORAL ARTERY

- This vessel typically starts at the level of the femoral head but, depending upon the level of CFA bifurcation, may begin higher or lower than this level.
- The normal luminal diameter in an adult male is 6 - 7 mm and 5 - 7 mm in a female. The vessel tapers proximal to distal, so that the distal end is nearly 1 mm smaller in luminal diameter than the proximal.
- It terminates at the hiatus in the adductor magnus, and its continuation becomes the popliteal artery. This is at the level of the junction of the mid and distal thirds of the thigh.
- Its branches include:
  - ▹ Muscular branches
  - ▹ Descending genicular branch (also known as anastomotica magna artery).

POPLITEAL ARTERY

- The popliteal artery is the continuation of the SFA.
- It starts at the hiatus in the adductor magnus, at the level of the junction of the mid and lower thirds of the thigh.
- It terminates by bifurcating into the anterior tibial artery and tibioperoneal trunk, 3 - 4 cm below the knee joint
- Branches of the popliteal artery include:
  - ▹ Medial and lateral superior genicular arteries.
  - ▹ Medial and lateral inferior genicular arteries.
  - ▹ Middle genicular artery.
  - ▹ Muscular branches.

DIAGNOSIS

- Although isolated calf claudication can sometimes be due to inflow or outflow disease, it is mostly due to femoro-popliteal occlusive disease.
- On examination, diminished popliteal and pedal pulses are clues to femoro-popliteal occlusive disease.
- Other signs of PAD such as diminished distal hair growth and dystrophic toe-nails are also often present.
- On non-invasive testing, in patients with isolated femoro-popliteal dis-

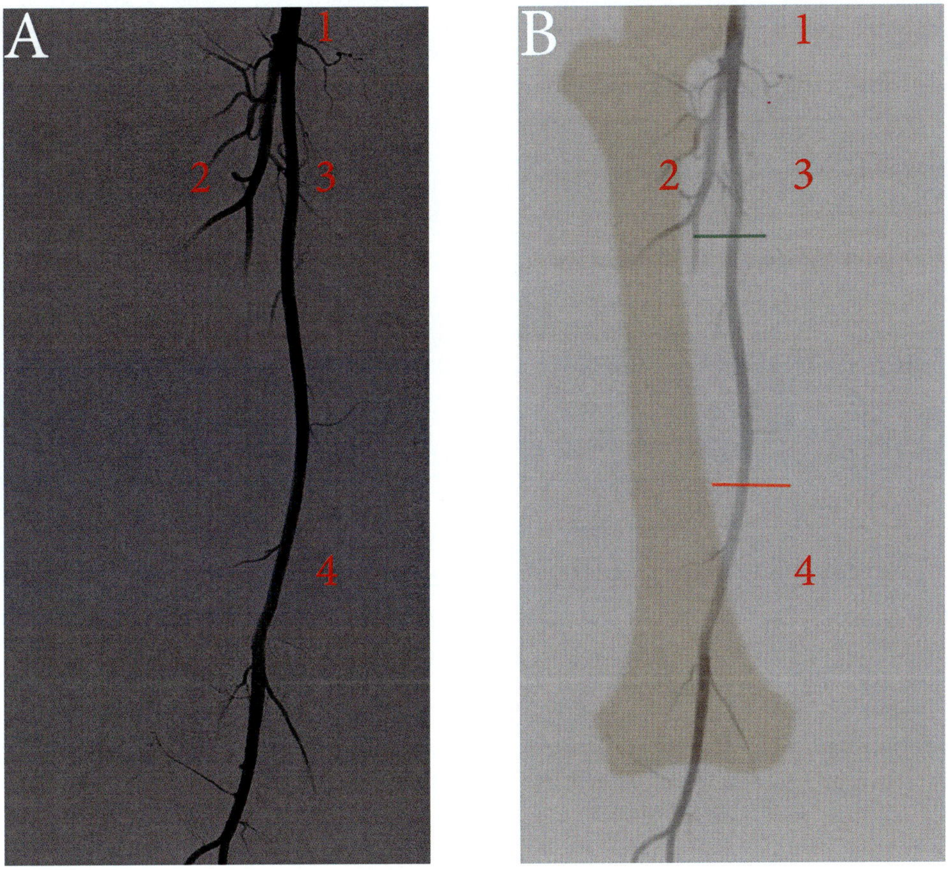

Figure 1.

Angiographic anatomy of femoro-popliteal arteries.

A. 1- Common femoral artery.
2- Profunda femoris artery.
3- Superficial femoral artery.
4- Popliteal artery.

B. This schematic shows the relationship of the arteries to the femur. The segment between the green and red lines represents the Hunter's canal. The hiatus in the adductor magnus is at the level of the red line and represents the beginning of the popliteal artery.

ease, the resting ABI is usually moderate-to-severely reduced (0.4 -0.8 range).

- Exercise testing is usually not necessary to make the diagnosis.
- Pulse volume recordings (PVRs) and Doppler measurements help to localize the disease. In the absence of concurrent aorto-iliac/CFA disease, proximal measurements at the CFA (high-thigh) level are usually normal. Abnormal popliteal and pedal (calf and ankle) measurements are the hallmark.
- The SFA and popliteal arteries lend themselves to relatively easy imaging with arterial duplex. This test should be considered in almost all patients (being evaluated for revascularization) in order to determine the burden of disease, the length of the diseased/occluded segment and the level of the proximal occlusion and distal reconstitution. This information can help in planning the revascularization strategy.

DIAGNOSTIC ANGIOGRAPHY

ACCESS

Site:

- For diagnostic angiography, contralateral common femoral artery access is used.
- Rarely, brachial artery access may be used for this purpose.

Sheath Size:

- We prefer to obtain initial access with a 21 gauge needle and 0.018" wire (e.g. 4 Fr. micropuncture kit), regardless of the site of access.
- This is then upsized to a 5Fr. sheath that is at least 20 cm long over a 0.035" wire.
- The tip of the this sheath is advanced to the distal CIA.

ABDOMINAL AORTOGRAPHY

Catheter:

- Any of the flush catheters such as a pigtail, tennis racquet (TR), or Omni flush catheter can be used.

Imaging:

- We start in an AP projection. Routine oblique projections are not necessary.
- However, these may be beneficial (see below) to delineate the disease in the distal common iliac (CIA), external iliac (EIA) and internal iliac arteries (IIA).
- Abdominal aortography should allow visualization of juxta/infra-renal aorta to the common femoral arteries bilaterally
  - ▹ Place the catheter at L1/L2 level and include bilateral femoral heads in the field of view.
- Use digital subtraction angiography (DSA) as this will reduce the volume of dye administration without compromising image quality.

Dye Injection:

- A power injector is used for aortography.
- The rate of power injection depends upon the patient's body habitus, renal function and ability to comply with instructions of breath-hold during aortography.
- In patients with normal renal function, we use 70% dye with 30% heparinized normal saline dilution (70:30). If the creatinine clearance is less than 30 mL/min, then we use 50% dilution (50:50). Alternatively, CO_2 may be used for this purpose in patients with renal insufficiency.
- Injection settings are 10-20 cc/sec for a total volume of 20-40 cc at 500-750 psi.

SELECTIVE ANGIOGRAPHY

Catheter:

- Via contralateral femoral artery access, a selective catheter such as an IM (our first choice), Omni flush, Rim or Simmons catheter is used and placed at the aortic bifurcation to engage the CIA ostium.
- Via brachial access a multi-purpose, angled hydrophilic catheter (e.g. angled Glidecath®), JR4 or Vert catheter can be used. These catheters may not reach beyond the aortic bifurcation. Hence, after engaging the appropriate CIA, they are exchanged out for a straight catheter such as the 150 cm long 0.035" QuickCross® catheter.

Imaging:

CFA/PFA/Proximal SFA:

- Perform angiography of the CFA and its bifurcation in an AP and ipsilateral oblique projection (around 30°) with or without a 20°- 30° cranial angulation. The latter view is particularly well-suited for visualizing the ostial/proximal SFA, the entire PFA as well as the CFA bifurcation.

Mid/Distal SFA:

- This is best imaged in an AP projection. Keep the catheter at CFA bifurcation for this.

Popliteal artery:

- This is also best imaged in an AP projection except in patients with knee joint prostheses. In these patients, an extreme ipsilateral oblique or a true lateral projection may be required.

Infra-popliteal arteries:

- These are ideally imaged in a view that allows separation of the tibia and fibula and minimizes the overlap among the tibial arteries.
- In a sedated patient the lower leg is externally rotated and the bony overlap is maximized in an AP projection. In this case, an ipsilateral oblique view separates the bones and removes much of the overlap of the infra-popliteal arteries. Since this latter maneuver may expose the operator(s) to excessive radiation, it may be desirable to secure the legs of the patient in two Velcro straps, just above the knee and just above the ankles, so that the optimal imaging window is available in an AP projection.
- In the presence of a significant SFA stenosis/occlusion, run-off angiography should be performed with the catheter still above this diseased segment. If feasible, a 50/50 contrast-saline mix is used for this purpose. This reduces the overall volume of dye used but also decreases patient calf discomfort precipitated by dye.

Dye Injection:

- DSA is used with manual injections of 5 - 8 cc of dye through the selective catheter.
- In the setting of occluded femoro-popliteal arteries, obtain prolonged cine runs to visualize collaterals to the distal reconstituted segment.

HEMODYNAMICS

- Hemodynamic assessment of femoro-popliteal disease is usually not as critical as for AIOD.

Hemodynamics:
If the ostium of the SFA is suspected to have a significant stenosis that is not evident angiographically, measure a gradient across it or do an IVUS.

- The measurement of a hemodynamic gradient across less obvious lesions that appear to be 'borderline-significant' on angiography is imperative prior to intervention.
- The significance of measuring a gradient across the entire length of a diffusely diseased SFA is somewhat controversial.
- If the peak gradient is barely > 20 mm Hg and no focal lesions are noted, intervening in an attempt to 'fix' this gradient is not advised.
- Pressure measurements should be made before and after hyperemia.
- Induce hyperemia by administering 200-300 mcg of intra-arterial (IA) NTG distal to the stenosis.
- Papaverine may also be used for inducing hyperemia.
- If a hemodynamically significant stenosis is identified but not elucidated on angiography, further angiographic views should be obtained to identify it.
- If none of these views are revealing, IVUS should be considered.

TABLE 1 - EQUIPMENT FOR DIAGNOSTIC FEMORO-POPLITEAL ANGIOGRAPHY	
Sheaths	• 5 Fr./20 cm long
Wires	• 0.035" 260 cm ▹ Angled stiff hydrophilic wire (e.g. Glidewire®) ▹ Angled non-stiff hydrophilic wire
Flush Catheters	• 4 Fr. Tennis racquet
Selective Catheters	• Contra • IM • Omni-flush • Rim • Simmons 1
Superselective Catheters	• 0.035", 0.038" ▹ Glidecath® (Terumo) ▹ QuickCross® (Spectranetics)

INTERVENTION

ACCESS

Contralateral CFA (Retrograde):

- In the vast majority of cases, retrograde access in the contralateral CFA is used for intervention.
- This obviously requires the use of a cross-over sheath, that is advanced over the aortic bifurcation and into the ipsilateral (on the side of the intervention) EIA.

Ipsilateral CFA (Antegrade):

- Antegrade access in the ipsilateral CFA may be necessary in cases when the aortic bifurcation cannot be crossed over e.g. in patients with:
  - ▹ Crisscrossing aorto-iliac stents
  - ▹ Excessive aorto-iliac tortuosity
  - ▹ Bifurcated abdominal aortic endografts
  - ▹ Aorto-femoral bypass grafts.

EQUIPMENT

GUIDING SHEATH

Size:

- Depending upon the choice of the final interventional modality, sheaths ranging from 5 - 7 Fr. may be required.

Balloon Angioplasty:

- ▹ A 5 Fr. sheath is adequate for balloon angioplasty.

Non-covered Stents:

- ▹ A 5 Fr. sheath may be adequate for certain 0.018" stents (e.g. Zilver®, Cook Medical).
- ▹ Most other non-covered stents, however, are 0.035" wire compatible systems that require a 6 Fr. sheath.

Covered Stents:

- ▹ Covered stents in 5.0 and 6.0 mm diameter are now available in a 6 Fr. system over a 0.018" wire. (e.g. Viabahn®).
- ▹ Covered stents in diameter > 6.0 mm require a 7 Fr. sheath.

Atherectomy:

- ▹ Orbital atherectomy requires a 5 Fr. sheath.
- ▹ Most other atherectomy devices in sizes appropriate for femoro-popliteal arteries require a 7 Fr. sheath.

Thrombectomy & thrombolysis:

- ▹ Rheolytic thrombectomy is performed through a 6 Fr. sheath.
- ▹ Ultrasound assisted thrombolysis catheters (EKOS® Corp.) are also 6 Fr. compatible.
- ▹ Simple infusion catheters (e.g. Fountain® catheter, Merit Medical and Unifuse®, Angiodynamics) are 4 Fr. compatible.

Shape:

Straight:

- In the case of antegrade CFA or brachial artery access for intervention, a straight sheath should be used.
  - ▹ Brachial: 90 - 110 cm length.
  - ▹ Ipsilateral femoral: >20 cm length.

Curved:

- When using contralateral CFA access for intervention, a curved or pre-shaped 'cross-over' sheath is used (e.g. Ansel ®& Balkin®). However, certain straight and kink-resistant sheaths (e.g. Raabe®) may also be used for this purpose.

WIRES

- Guidewires used for femoro-popliteal artery interventions are commonly 0.035" in diameter.
- Polymer-coated hydrophilic wires (e.g. Glidewire®) are the preferred workhorse wires for crossing these lesions. Once across the lesion(s), these wires can be exchanged out for an appropriate caliber wire (0.014"-0.035") non-hydrophilic, 260-300 cm long wires.
- The shape of the 'crossing' hydrophilic wire is chosen based on the lesion severity and morphology.
- Sometimes 0.014" wires may be needed to cross femoro-popliteal lesions that consist of long serpiginous micro-channels.

Total-occlusion:

- A straight, non-stiff or stiff, hydrophilic wire (the latter is preferred) is the wire of choice, supported by a 4 Fr. hydrophilic catheter (straight or angled).

- After initial penetration of the proximal cap, if this wire does not successfully cross the length of the occlusion, it may be switched out for an angled stiff hydrophilic wire to allow for better steerability.
- The two wires supported by a catheter may have to be used interchangeably until the occlusion is successfully crossed.

Non-total Occlusion:

- An angled ***non***-stiff hydrophilic wire is the preferred wire to traverse a severe/critical stenosis. Its stiff counterpart is not as torqueable and can dissect the artery while crossing a tight stenosis.

BALLOONS

Predilatation:

- Predilatation with a balloon is recommended as it creates a channel for smooth delivery of most other interventional devices such as stents, atherectomy devices etc. Additionally, this allows for a more accurate size estimation for the final device size.
    - ▹ Choose a balloon that is at least 2 - 3 mm smaller in diameter than the luminal diameter of the reference vessel.

Postdilatation:

- This is desirable as it allows for appropriate vessel expansion.
    - ▹ Choose a balloon with a diameter no greater than the luminal diameter of the reference vessel.
    - ▹ Gentle balloon inflation at 2 - 8 ATM is performed for up to 3 minutes
        - ♦ PTA alone < nominal pressure
        - ♦ Stent postdilatation ~ nominal pressure
        - ♦ Post-atherectomy 'dilatation' - 2 to 4 ATM, while the nominal pressure for most of these balloons is 7 - 8 ATM.

STENTS

Balloon-Expandable Stents (BES):

- These stents are not used for femoro-popliteal artery treatment.

Self Expandable Stents (SES):

- Only self-expanding stents, typically made of nitinol, should be used in femoro-popliteal arteries.
    - ▹ These stents should be selected at least 1 mm greater in diameter compared to the reference vessel ('up-sizing').
    - ▹ These stents require postdilatation in almost all cases.

ANTICOAGULATION

- IV unfractionated heparin is the preferred anticoagulant.
- Goal ACT > 250 sec.
- Bivalirudin or GP IIb/IIIa inhibitors are reserved for those patients who have visible thrombus or a predilection for intra-operative thrombosis.

TECHNIQUE

ACCESS & GUIDING SHEATH

As noted above, diagnostic angiography is usually performed via the contralateral CFA. For intervention this sheath is exchanged out for a 'guiding' sheath, as noted above.

The following steps are undertaken for this.

- In patients with left leg symptoms, right CFA access is used for diagnostic angiography (including aortography) and vice versa.
- An angled stiff hydrophlic guidewire is advanced through a selective/ cross-over catheter e.g. IM, Rim etc.
- This wire is advanced into a perforator branch of the PFA.
- The catheter and the CFA sheath are removed.
- A cross-over sheath (e.g. Ansel 1) is advanced over the aortic bifurcation under fluoroscopic guidance.
- The tip of the sheath is advanced into the distal EIA or the proximal CFA.
- The dilator is removed and the sheath flushed.
- The wire is then removed from the PFA.

Subsequent steps of crossing the SFA depend upon the anatomy of the diseased vessel and the morphology of the lesion (figure 2). The goal is to try to stay intraluminal. Once this process is successfully accomplished, the steps taken in the ensuing intervention remain more or less similar.

SFA OSTIAL TOTAL OCCLUSION

CATHETER & WIRE SELECTION

- We prefer a straight tip hydrophilic wire for crossing CTOs.
- After advancing the 'cross-over' sheath over the aortic bifurcation, the wire is withdrawn from the PFA into the CFA.
- Under a higher magnification a 'road map' is obtained in ipsilateral 30° projection.
- The wire is advanced to the tip of the proximal cap of the occlusion, without probing it.
- A 65 cm 0.035" compatible crossing catheter (e.g. Glidecath) is loaded on the wire and advanced to the occlusion.
    - ▹ At this stage, a shorter catheter provides better support and wire control necessary to cross the proximal cap. Once across the cap, this catheter may be exchanged out for a longer one.
- The shape of the catheter (straight or angled) is determined by the morphology of the proximal cap of the occlusion (figure 3).
    - ▹ In cases where the vessel tapers down to an eccentric 'beak' at the occlusion or there is a flush ostial occlusion, an angled catheter is preferred. In the former case it allows the wire to selectively cannulate the 'beak' and direct the wire towards the center of the lumen as the wire is advanced across the proximal cap.
    - ▹ If instead, the proximal cap is 'blunt' or has a central 'beak' a

straight catheter is preferred as it offers the best chance of advancing the wire while maintaining an intraluminal position. An angled catheter in such cases is more likely to push the wire off the 'trajectory' and end up in a subintimal plane.

TECHNIQUE

- The angled wire is exchanged out for a straight stiff 260 cm hydrophilic wire for crossing the occlusion.

Penetrating the Proximal Cap:

- Under 'road map' guidance, this wire is advanced to the proximal cap.
  - ▹ The support catheter is kept against the proximal cap during this maneuver to provide support and prevent buckling of the wire.
- Once the catheter is situated deeply in or against the proximal cap, the wire is slowly advanced to the tip of the occlusion (figures 2, 3 and 4), then withdrawn 2 - 3 inches to take a 'running start', building momentum and then quickly advanced it with a 'jabbing' motion to poke across the proximal cap.
- In a vast majority of cases, the straight stiff hydrophilic wire crosses an SFA occlusion without significant difficulty.
- Sometimes, however, this wire can be advanced only a partial length of the occlusion and either the catheter, the wire or both have to be switched out for their angled counterparts.

Advancing Through the Occlusion:

Leading with the wire tip:

- If after crossing the proximal cap the wire appears to be 'on course', continue advancing it.
- Every 2" - 3" follow the wire with the catheter to provide backup support.
- If the wire appears to have successfully crossed the distal cap, follow with the catheter.
- If, however, it appears to have hit a 'road block' or gone 'off course', the catheter and wire position is reassessed to determine if a switch needs to be made to either an angled stiff hydrophilic wire, an angled stiff hydrophilic catheter or both. This allows the wire and catheter to be redirected towards the distal cap.

Leading with a loop/knuckle:

- If while advancing the wire within the occluded segment, the wire goes sub-intimal and cannot be easily redirected back to the true lumen it may be better to let it form a loop/knuckle at the tip, which can then be advanced with better pushability.
- To form the loop, the wire is advanced to a point where the tip is engaged into plaque or subintimal tissue. Wire advancement is continued so that shaft takes a U-shaped configuration at the leading edge.
- This loop is easier to form with an angled wire than with a straight wire. The wire should be supported by a catheter.
- It is important to keep the loop narrow. The loop at its widest should be no wider than the diameter of the artery.
- A wide loop is less likely to penetrate the distal cap.
- A loop that has a spiral configuration suggests that the wire is in a dissection plane.

Looping an Angled Wire:

- An angled non-stiff wire is easier to loop than its stiff counterpart.
- However, the latter has more pushability.

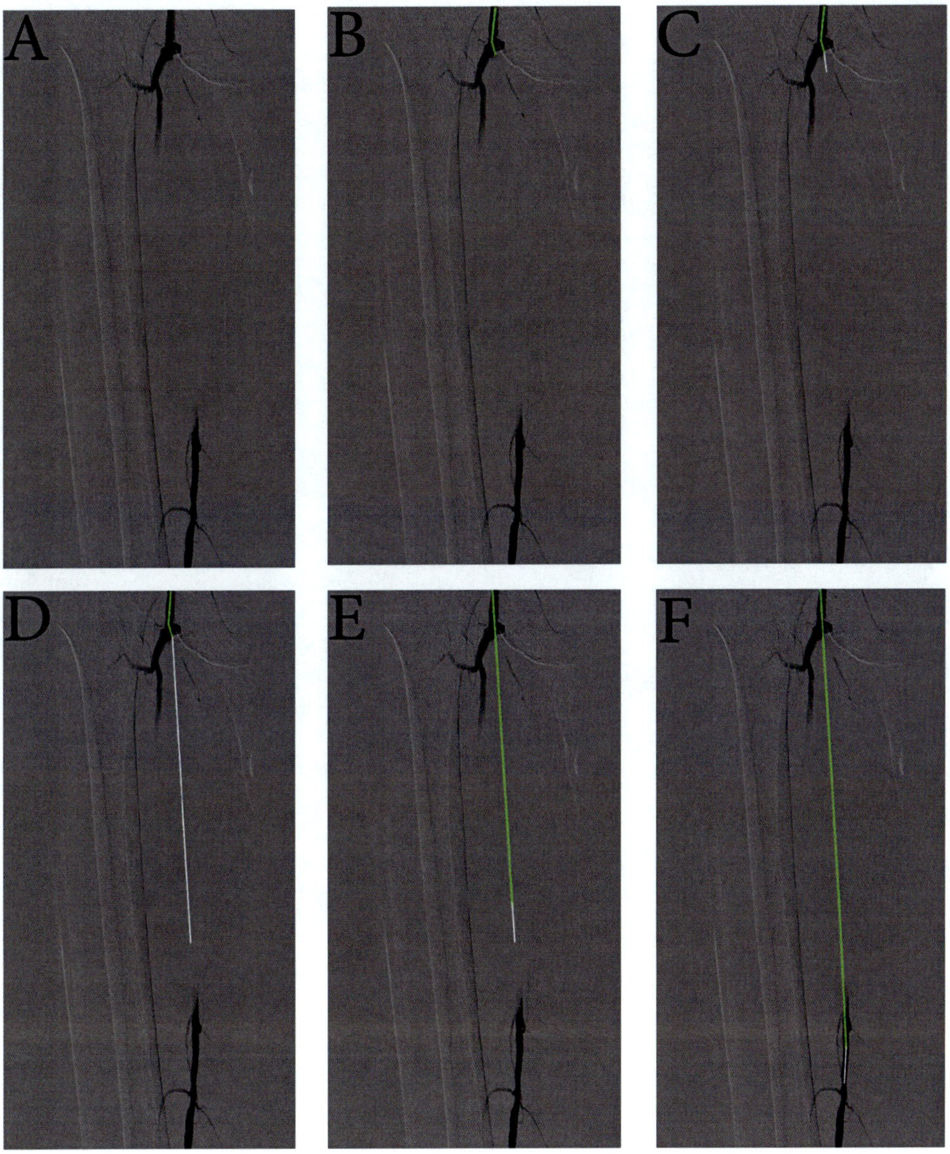

Figure 2.
Steps in crossing an SFA CTO.

A. An angiogram shows ostial SFA occlusion with reconstitution in the Hunter's canal. There is no obvious micro-channel at the ostium.
B. An angled 4 Fr. hydrophilic catheter (e.g. Glidecath) is used to 'engage' the ostium of the occluded SFA.
C. A 0.035" straight stiff hydrophilic wire is advanced to the tip of the catheter.
D. The wire is advanced across the proximal cap of the occlusion.
E. The catheter is then advanced across the proximal cap, following the wire to support its further advance.
F. The wire finally crosses the distal cap and the catheter is advanced over it.

Confirmatory Angiogram:
- Advance the catheter across the stenosis.
- Remove the wire
- Aspirate and flush.
- Inject 2-3 cc of diluted contrast to confirm intraluminal position.

- As the loop 'pops' across the distal cap the pressure at the leading end of the wire is felt and seen to 'give'.
- Usually the loop passes through a narrower passage as it crosses the distal cap and narrows transiently at this point before it becomes as wide as the distal luminal diameter.
- If the loop bunches up or becomes wider than the vessel diameter it is an indication that it is moving away from the vessel lumen.
- If above-the-knee or trans-genicular popliteal artery is patent it is not advisable to advance this loop beyond the patella. A continuing dissection may disrupt collateral flow and worsen the limb ischemia. Attempts should be made to re-enter the reconstituted vessel above the knee.

SFA Atherectomy:
- We prefer atherectomy as it debulks the lesion.
- We predilate a severely narrowed lesion with an undersized balloon (3.0 - 4.0 mm diameter) to make a channel for device delivery.
- We postdilate almost routinely.
- This is done with a 5.0 mm - 6.0 mm balloon at low pressures (2-4 ATM) to minimize the risk of a balloon induced dissection that may necessitate stenting.

Crossing the Distal Cap:
- Once the wire/catheter appears to be 'on course' again, depending on the lay of the wire, a straight or an angled catheter/wire may be necessary to cross the distal cap (figure 4 A-C).
    - ▹ If the wire appears 'online' with the distal cap a 'straight combination' wire will usually work.
    - ▹ If instead the wire lies lateral or medial to the reconstituted vessel, an angled catheter with the straight stiff hydrophilic wire will work better.
- If the wire seems to be across the distal cap of the occlusion but the catheter cannot be advanced due to resistance, it should be exchanged for a stiffer catheter (e.g. QuickCross) or a low profile, small (e.g. 3.0 - 4.0 mm x 40 - 60 mm) balloon.
- Often the stiffer catheter or balloon will cross without much resistance due to a stiffer shaft.
- Sometimes, however, angioplasty of the proximal portion of the vessel has to be performed with this balloon as it is advanced across the occlusion. It is acceptable to perform PTA in a subintimal plane as long as distally the wire has re-entered the true lumen.
- Brief inflations (< 30 sec.) should be performed (so that the balloon does not become 'winged' or deformed) until the entire length of the occlusion is crossed and the balloon re-enters the distal true lumen.
- The wire is removed, the catheter flushed and a confirmatory angiogram is performed through the wire lumen of the catheter.

SFA Stenting:
- Only self-expanding stents are used in SFA.
- Always predilate to ensure that the stent will expand well.
- Deploying a self-expanding stent in a severely stenotic segment without predilatation may lead to entrapment of the stent catheter within a poorly expanded stent.

Definitive Intervention:
- Once confirmed that the catheter is in the true lumen beyond the distal cap, a wire of appropriate caliber (0.014"-0.035") is reintroduced through the catheter and the catheter is removed.
- Definitive intervention can now be performed as planned.
    - ▹ Our preference is to treat severe SFA stenoses/occlusions primarily with PTA and atherectomy.
    - ▹ In the case of an ostial/very proximal SFA CTO, we prefer final treatment with a covered, self-expanding stent-graft. Extreme caution is to be exercised in this maneuver to not cover the PFA. It is strongly recommended that stent deployment at this location be performed under a road map obtained on a high magnification.

Final Angiogram:
- At the conclusion of the intervention, angiography should be repeated,

from the CFA to the foot, in order to rule out distal embolization, wire related perforations and other complications.

SFA MID/DISTAL TOTAL OCCLUSION

TECHNIQUE

- The general principles of crossing the occlusion remain the same.
- The sheath is still parked in the distal EIA. The wire and catheter are advanced to the proximal cap of the occlusion. The steps outlined in the above section are followed.
- The essential difference lies in the choice of the final intervention.
  - ▹ We do not treat all mid/distal occlusions with stents unless the patient has CLI. In claudicants, we prefer to perform a combination of angioplasty and atherectomy. Restenosis is much easier to deal with if there is no stent in the vessel!
  - ▹ In claudicants, we reserve the use of stents in the SFA for cases of flow-limiting dissections, perforations, or lesions with aggressive recoil.
  - ▹ It may not be prudent to use a stent in a vessel that is heavily calcified, not allowing the device to completely expand and hence risk thrombosis.

Removing A Cross-over Sheath:

Always remove this sheath over a 0.035" wire.

- The wire is advanced to the sheath tip (distal EIA)
- The sheath is then withdrawn under fluoroscopy
- This prevents a potential whip-lash from the sheath as it is withdrawn into the CIA and aorta, that can result in aorto-iliac dissection.
- Inserting the dilator in the sheath for this step is optional and not necessary.

SFA STENOSIS

CATHETER & WIRE

- An angled non-stiff hydrophilic wire is the best choice for crossing an SFA stenosis.
- We prefer the use of a 0.035" angled non-stiff hydrophilic wire. However, if the channel is critically narrowed, long and serpiginous, a 0.014" hydrophilic wire may be preferable (e.g. Choice PT Floppy).

TECHNIQUE

- Obtain a 'road map' of the vessel at an appropriate magnification.
- Cross the diseased segment with a suitable wire.
- Treat the lesion as desired. The same principles outlined above apply:
  - ▹ Our preference is to perform PTA and atherectomy without planned stenting unless the patient has CLI.
  - ▹ Stenting is reserved for flow-limiting dissections, perforations, or lesions with aggressive recoil.
  - ▹ Stents should be placed cautiously in a vessel that is heavily calcified, as poor expansion may lead to stent thrombosis.

POPLITEAL ARTERY TOTAL OCCLUSION

- In general, popliteal artery occlusions are more formidable than SFA occlusions as any additional compromise of flow due to an iatrogenic

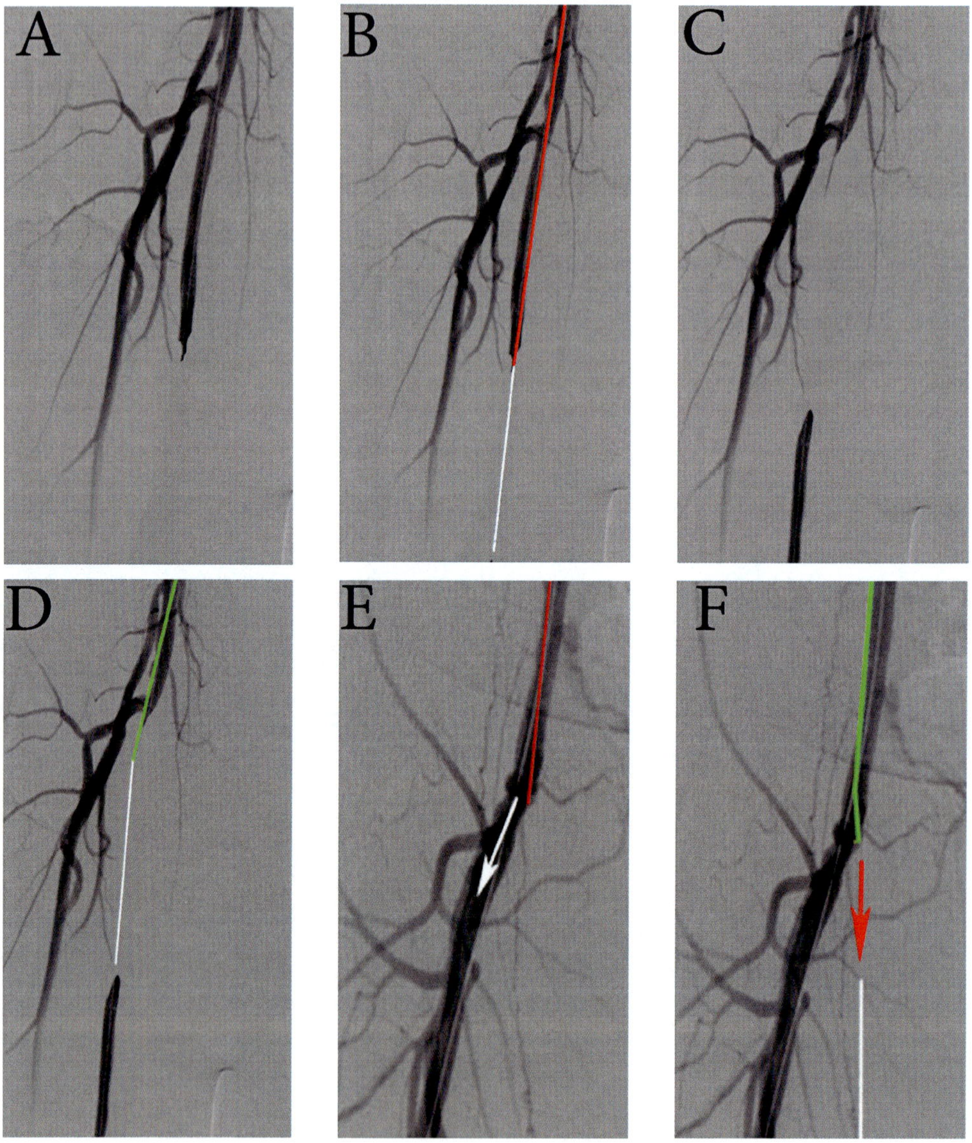

Figure 3.

Crossing the proximal cap of proximal/ostial SFA CTO.

A. Proximal CTO with a central 'beak'.
B. A 4 Fr. straight Glidecath (red line) and a 0.035" straight stiff Glidewire (white line) is used to cross the proximal cap.
C. Proximal CTO with an eccentric 'beak'.
D. A 4 Fr. angled Glidecath (green line) and a 0.035" straight stiff Glidewire (white line) is used to cross the proximal cap.
E. A 'flush' ostial occlusion of the SFA. A Straight catheter (red line) would not be able to direct the wire in the correct direction. It will likely enter the PFA (white arrow).
F. Instead an angled catheter directs the wire (white) towards the SFA (red arrow).

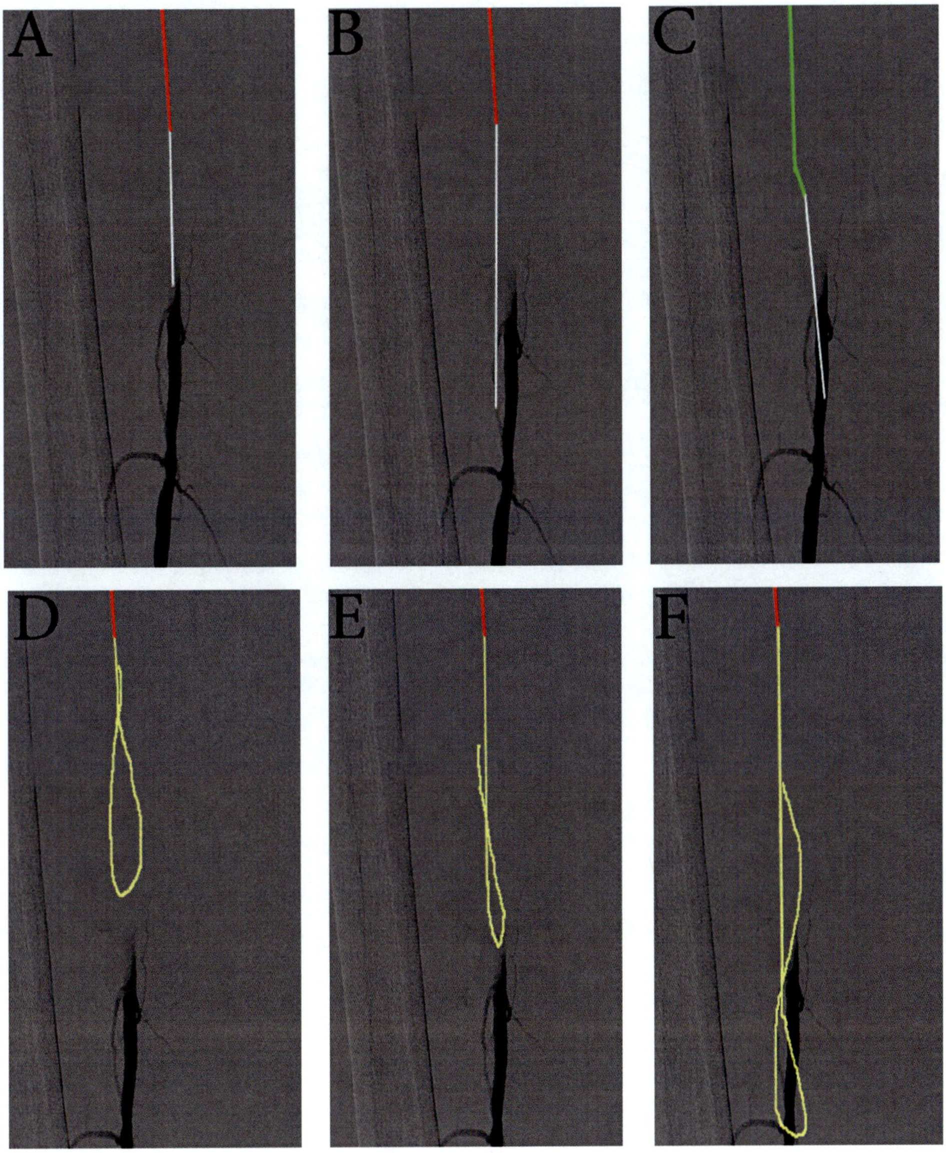

Figure 4.

Advancing through an SFA CTO and crossing the distal cap.

Leading with the wire tip:

A. A 4 Fr. Glidecath is advanced over a SSGW through the SFA CTO.

B. The wire is advanced further and the catheter is followed over it. However, the SSGW fails to cross the distal cap and enters subintimal space. It cannot be steered back into the true lumen.

C. The straight Glidecath is switched out for an angled one which points the straight wire towards the true lumen. The distal cap is then crossed successfully.

Leading with a loop:

D. An ANSGW is used to form a loop.

E. The looped wire is advanced to the distal cap and may actually cross it successfully.

F. However, the loop may continue in the dissection plane. An indication of this is that the diameter of the leading edge of the loop is wider than the vessel diameter.

complication may result in acute limb ischemia.

- Most patients with a popliteal artery total occlusion have CLI rather than claudication.
- As these patients have a limited perfusion reserve compared to claudicants, their ability to withstand any further reduction in flow (e.g. due to dissections leading to loss of collateral flow, etc) is low. For example, due to the short length of the vessel, if re-entry into the true lumen does not occur promptly and an extensive dissection plane is created instead, it is more likely to disrupt collateral flow, jeopardizing limb perfusion and converting CLI into acute limb ischemia (ALI).

ACCESS

- Ipsilateral CFA access in antegrade direction or contralateral CFA access in retrograde direction may be used as necessary. In addition, pedal access or antegrade SFA access may be necessary for recanalization in difficult cases.

CATHETER & WIRE

- For an occluded popliteal artery, we prefer the use of a 0.035" straight stiff hydrophilic wire and a 4 Fr. hydrophilic crossing catheter (at least 120 cm in length if using contralateral CFA access).
- However, we keep the threshold for switching to a 0.014" CTO wire fairly low. If the 0.035" wire is unsuccessful in re-entering or stays 'off-course', we select a Confianza Pro wire supported by either a coronary OTW balloon or a 0.014" support catheter (e.g. QuickCross).
- Caution should be exercised with the use of Confianza and other similar CTO wires. They provide minimal tactile feedback when exiting the vessel and going extra-vascular.
- If the occlusion is subtotal and a channel is present it is advisable to use a more steerable, hydrophilic 0.014" wire such as the Choice PT Floppy or Fielder XT to cross.

TECHNIQUE

- The general principles and technique of crossing a popliteal artery occlusion remain the same as those for SFA occlusions.
- However, there are important differences that should be recognized.
  - ▹ The occlusion is usually shorter in length.
  - ▹ The occlusion is typically present in the segment that is across the knee.
  - ▹ The occlusion is distal to the segment that usually receives collaterals from the PFA.
  - ▹ Reconstitution is often in the terminal popliteal artery or the tibioperoneal trunk.
  - ▹ The popliteal artery tapers off more rapidly than the SFA.
  - ▹ A popliteal artery aneurysm should be ruled out (IVUS, bilateral popliteal artery duplex, CTA or MRA), especially in patients with aortic or femoral aneurysms.
- Thus keeping in mind the above facts, it is critical not to disrupt any further flow in the popliteal artery or its collaterals.
- A brief initial trial of a 0.035" wire/catheter system should be performed if allowed by the anatomy.

- ▹ A straight stiff hydrophilic wire is advanced, supported by a 4 Fr. catheter to the proximal cap of the occlusion.
- ▹ A road map is obtained.
- ▹ The wire is advanced across the proximal cap into the occlusion with a rapid but gentle 'jabbing' motion.
- ▹ Once the wire is in the proximal cap, it is followed with the support catheter.
- ▹ This is continued until the wire goes across the distal cap and into the true lumen.
- ▹ The wire is removed, the catheter flushed and intra-luminal position confirmed with an angiogram.
- ▹ The wire system is then switched out to the system of choice and definitive intervention (PTA, atherectomy, stenting, etc) is performed.

- If unsuccessful, a prompt switch should be made to a 0.014" system and crossing should be continued from above. The same steps as outlined above are followed for the 0.035" system.
- If antegrade crossing remains unsuccessful, retrograde pedal artery access should be considered. Retrograde attempts are more likely to be successful.

BTK Popliteal Artery Stenting:

- PTA and atherectomy are very reasonable first-line choices for popliteal artery intervention.
- However, the result is seldom perfect from a hemodynamic perspective.
- While such a result may be acceptable for claudicants, in CLI patients dependence on popliteal artery inflow is paramount.
- Thus stenting of BTK popliteal artery in CLI patients is a justified first-line option (particularly where drug-eluting balloons are not available).
- We have used drug-eluting (coronary) stents with considerable success in this location.

HOW TO CROSS A CHALLENGING CTO

A challenging CTO is defined as an occlusion that cannot be crossed successfully with the use of a single access point employing traditional catheter and wire system(s).

When such failure is encountered despite multiple 'traditional' attempts (during the same procedure), an analysis of the causes of failure and alternate measures need to be considered.

CAUSES & POSSIBLE SOLUTIONS

The major causes of failure with suggested solutions are listed below.

Lack of Proximal Support:

- Failure to advance a wire/catheter across the occlusion may be due to lack of support at the level of the CFA.
  - ▹ Advancing the sheath over the support catheter and wire to the ostium of the SFA may provide the necessary support.
- The lack of proximal support may also result from tortuous and calcified aorto-iliac arteries leading to an inability to advance the wire/catheter.
  - ▹ This problem may be solved, in certain cases, by obtaining an antegrade access in the ipsilateral CFA, thus obviating the need to traverse the tortuous iliac arteries.
  - ▹ If antegrade access is not feasible, retrograde pedal/popliteal artery access should be considered.
  - ▹ Alternatively, the cross-over sheath may have to be up-sized by 1-2 Fr. sizes. This may not always be feasible or safe.

Inability to Cross the Proximal Cap:

- It is rare that this is the sole cause of failure in the absence of support issues.

- If, however, the tip of the straight stiff hydrophilic wire cannot be advanced across the proximal cap the following might succeed:
  - ▹ Use of the Frontrunner® device in lieu of the wire.
  - ▹ Careful use of the 'back end' of a stiff hydrophilic wire as the 'probing' end has worked on some occasions. After crossing the proximal cap, a catheter is advanced over the wire and the wire is re-oriented to its normal configuration, tip forward, to avoid the risk of perforation.

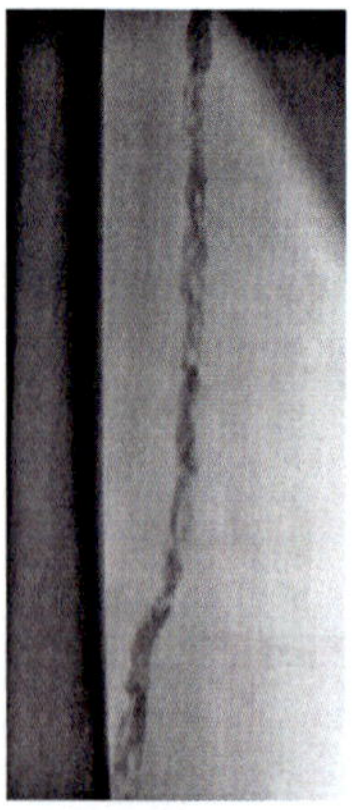

Figure 5. Heavy calcification in the femoro-popliteal arteries.

Heavily Calcified Arteries:

- Intrinsic femoro-popliteal arterial calcification may make certain occlusions 'untreatable' with current percutaneous therapies due to the sheer inability to cross the occlusive burden.
  - ▹ Surgical revascularization may be the only reasonable option in these patients with extreme calcification (figure 5).

Inability to Cross the Distal Cap:

- In an overwhelmingly large number of cases of challenging CTOs, the inability to cross the distal cap of the occlusion and to re-enter the true lumen remains the rate limiting step.
  - ▹ Retrograde access in the popliteal/pedal arteries, depending upon operator experience and comfort level, may be the most straightforward solution in many cases.
    - ♦ Once the wire has been 'externalized' intervention proceeds as usual via the CFA access (contralateral or ipsilateral).
  - ▹ CTO crossing devices, if available and familiar to the operator, may also offer the solution. The devices, listed below, are discussed separately in the following section:
    - ♦ Crosser™
    - ♦ Frontrunner™
    - ♦ Outback™ LTD
    - ♦ Pioneer™
    - ♦ Turbo Elite™ Laser
    - ♦ TruePath™

CTO ASSIST DEVICES

Dedicated devices designed for treating CTOs are listed below with their salient characteristics.

CROSSER™

- This is a crossing catheter for distal entry into the true lumen.
- It is 6 Fr. sheath compatible.
- The device has no integrated IVUS.
- It is compatible with a 0.014" guidewire and requires a single wire.
- Shaft length of the device is 120 cm.
- There is a hydrophilic coating along the entire length.

THE FRONTRUNNER™

- This device is also 6 Fr. sheath compatible.
- Its crossing profile is 0.039"; with actuating jaws that open to 2.3 mm.
- It has a hydrophilic coating along the entire length.
- The distal tip is shapeable to a certain extent.
- The device has to be supported by a 4 - 4.5 Fr. catheter. The latter is supplied separately by the vendor. However, can be used with a 4 Fr. hydrophilic catheter (e.g. Glidecath®).
- Frontrunner® catheter comes in two lengths: 90 cm and 140 cm.
- The respective support catheters by Cordis are 76 cm and 126 cm long.

THE OUTBACK™

- This is a re-entry catheter for distal entry into the true lumen.
- It is 6 Fr. sheath compatible.
- The device has no integrated IVUS.
- It is compatible with a 0.014" guidewire and requires a single wire.
- Shaft length of the device is 120 cm.
- There is a hydrophilic coating along the entire length.

THE PIONEER™

- This is also a re-entry catheter for distal entry into the true lumen.
- The device is 7 Fr. sheath compatible.
- It has an integrated IVUS.
- The Pioneer is compatible with a 0.014" guidewire. It requires two wires - one over which the catheter is advanced and the other for advancing through the needle.
- The device is compatible with Volcano® IVUS only.

TRUEPATH™

- This crossing device is intended for intraluminal crossing of a CTO.
- It works with a 0.018" compatible catheter < 135 cm in length.
- There is no integrated IVUS.
- The device is compatible with a 0.018" guidewire and requires a single wire.
- The shaft length is 165 cm.

TURBO ELITE LASER ABLATION CATHETER

- The device utilizes UV light to vaporize plaque into micro-particles. It is designed for crossing a CTO via ablation of obstructive tissue in the true lumen.
- It is available in 0.9 mm - 2.5 mm diameter.
- The device is 4 - 8 Fr. sheath compatible.
- There is no integrated IVUS.
- It is available in Rx and OTW configurations, compatible with a 0.014" guidewire (Rx) and 0.018" (larger OTW sizes). It requires a single wire.
- The shaft length of this catheter is 110 - 150 cm.
- There is a hydrophilic coating along the entire length.

TABLE 2 - EQUIPMENT FOR FEMORO-POPLITEAL INTERVENTION:	
Sheaths	• 5 - 7 Fr./45 cm Ansel™ 1
Wires	• 0.035" (260 cm) ▹ Angled stiff hydrophilic wire (e.g. Glidewire®). ▹ Angled non-stiff hydrophilic wire. ▹ Straight stiff hydrophilic wire. ▹ Straight non-stiff hydrophilic wire . • 0.018" (260 cm) ▹ Spartacore®. ▹ Steelcore®. ▹ V-18®. ▹ Platinum Plus (also available in 0.014").
Catheters	• 0.035" support/crossing catheters ▹ Glidecath™ (Terumo). ▹ QuickCross™ (Spectranetics). ▹ CXI™ (Cook Medical).
Balloons	• 0.014" ▹ Coyote (BSC) ▹ Viatrac® (Abbott Vascular). ▹ Armada 14® (Abbott Vascular). ▹ Advance 14 LP® (Cook). • 0.018" ▹ Armada® (Abbott Vascular). ▹ Fox SV® (Abbott Vascular). ▹ Savvy® (Cordis), etc. ▹ Sterling (BSC) • 0.035" ▹ Armada® (Abbott). ▹ Mustang® (BSC). ▹ OptaPro® (Cordis). ▹ Powerflex® (Cordis).
Stents	• 0.018" ▹ Zilver® (Cook). ▹ Supera® (IDEV). • 0.035" ▹ SmartControl® (Cordis). ▹ Absolute® (Abbott Vascular). ▹ Protege® (eV3) etc.
Stent-grafts	• 0.018" ▹ Viabahn® (Gore). • 0.035" ▹ Viabahn® (Gore).

MANAGEMENT OF FEMORO-POPLITEAL PERFORATION

Perforation of the femoro-popliteal arteries is less common than an iliac perforation. These perforations must be treated promptly as well, but are less catastrophic by virtue of their location (thigh vs. retroperitoneum).

CAUSES

This complication is mostly caused by iatrogenic trauma. The following serve as risk factors:

- Female gender.
- Heavy calcification.
- Chronic total occlusion.

Perforation occurs due to vessel trauma caused by interventional devices such as:

- Angioplasty balloons.
- Atherectomy devices.
- CTO re-entry devices.
- Wire tip.

Iatrogenic Trauma:
If during atherectomy the patient develops sudden, sharp pain in the thigh a perforation should be ruled out.

DIAGNOSIS

If a perforation is suspected, prompt angiography at the site of the most recent intervention should be performed. DSA should be used if possible.

Illustration

- A 63-year old male with claudication was undergoing left SFA excisional atherectomy with a SilverHawk® LS-M device for severe occlusive disease (through a 7 Fr. sheath) (figure 6A-C).
- The patient experienced severe pain in the thigh during a routine 'cut'.
- Angiography revealed perforation of the proximal SFA at the site of the most recent 'cut' (figure 6D).
- A 6.0 mm x 80 mm balloon was quickly introduced and inflated across the perforation site.
- A 6.0 mm x 150 mm Viabahn was then deployed over the existing 0.035" wire through the existing 7 Fr. sheath.
- The distal SFA which was also severely diseased was then treated with PTA and stenting. The stents were postdilated with the same 6.0 mm balloon.
- Subsequent angiography revealed complete sealing of the perforation site.

Viabahn® sizes for SFA:
6 Fr. sheath compatible:
- 5 & 6 mm, 0.018" wire platforms

7 Fr. sheath compatible:
- 5 & 6 mm, 0.035" wire platform
- 7 mm, 0.018" wire platform

8 Fr. sheath compatible:
- 7 mm, 0.035" wire compatible.

TREATMENT

Once identified the treatment should be swift and methodical:

Balloon Tamponade:

- Immediate balloon tamponade should be obtained at the site of perforation with an appropriate size balloon (5 - 7 mm diameter x 40 - 100 mm length. Inflation should be at a sub-nominal pressure (4 - 8 ATM) (figure 6E).

Sheath Selection:

- The perforation will have to be treated by placing a covered self-expanding stent.

6 Fr. sheath:

- ♦ For a 5 - 6 mm diameter stent, a 0.018" wire compatible Viabahn can be used. Lengths are 25 mm, 50 mm, 100 mm and 150 mm.

7 Fr. sheath:

- ♦ For a 7 mm diameter stent, a 0.018" wire compatible Viabahn can be used. Alternatively, if only 0.035" wire compatible Viabahn stents are available then the 5 - 6 mm diameter stent will also require a 7 Fr. sheath. A 7 mm 0.035" system stent will require an 8 Fr. sheath in this case.

- An appropriate size sheath and wire should be prepared if not already in place.

Stent Selection:

- For most perforations in the femoro-popliteal arteries, a 5.0 - 7.0 mm diameter self-expanding covered stent (e.g. Viabahn®(Gore Medical) is adequate.

Balloon Removal:

- If an appropriate caliber wire (0.018" or 0.035" based on the stent platform as noted above) is not already in use, it should be placed through the balloon catheter prior to the removal of the balloon.
- Viabahn stents track well over a 0.018" wire but less optimally over a 0.014" wire.

Sheath Up Sizing:

- A 6 Fr. sheath is required at a minimum (for a 0.018" wire compatible system).
- A 7 Fr. sheath may be necessary for other stent sizes (for a 0.035" wire compatible system or for a 7.0 mm diameter 0.018" wire compatible stent).

Obtaining A 'Road Map':

- For precise stent placement a 'road map' should be obtained prior to the insertion of the stent in the sheath.

Stent Delivery:

- The stent is delivered to the desired location, positioning is confirmed and then it is deployed.

Repeat Angiogram:

- A repeat angiogram should be obtained to assess if postdilatation or placement of another stent is necessary (figure 6F).
- DSA should be used to evaluate continued extravasation.
- Appropriate imaging at this juncture is critical to exclude ongoing bleeding.

POST-PROCEDURE FOLLOW-UP

- After successful treatment with a covered stent, continued bleeding from an SFA perforation is highly unlikely.
- However, a new post-op baseline for Hb/Hct should be established by checking once at the conclusion of the intervention.
- The need for transfusion of pRBCs is extremely rare.

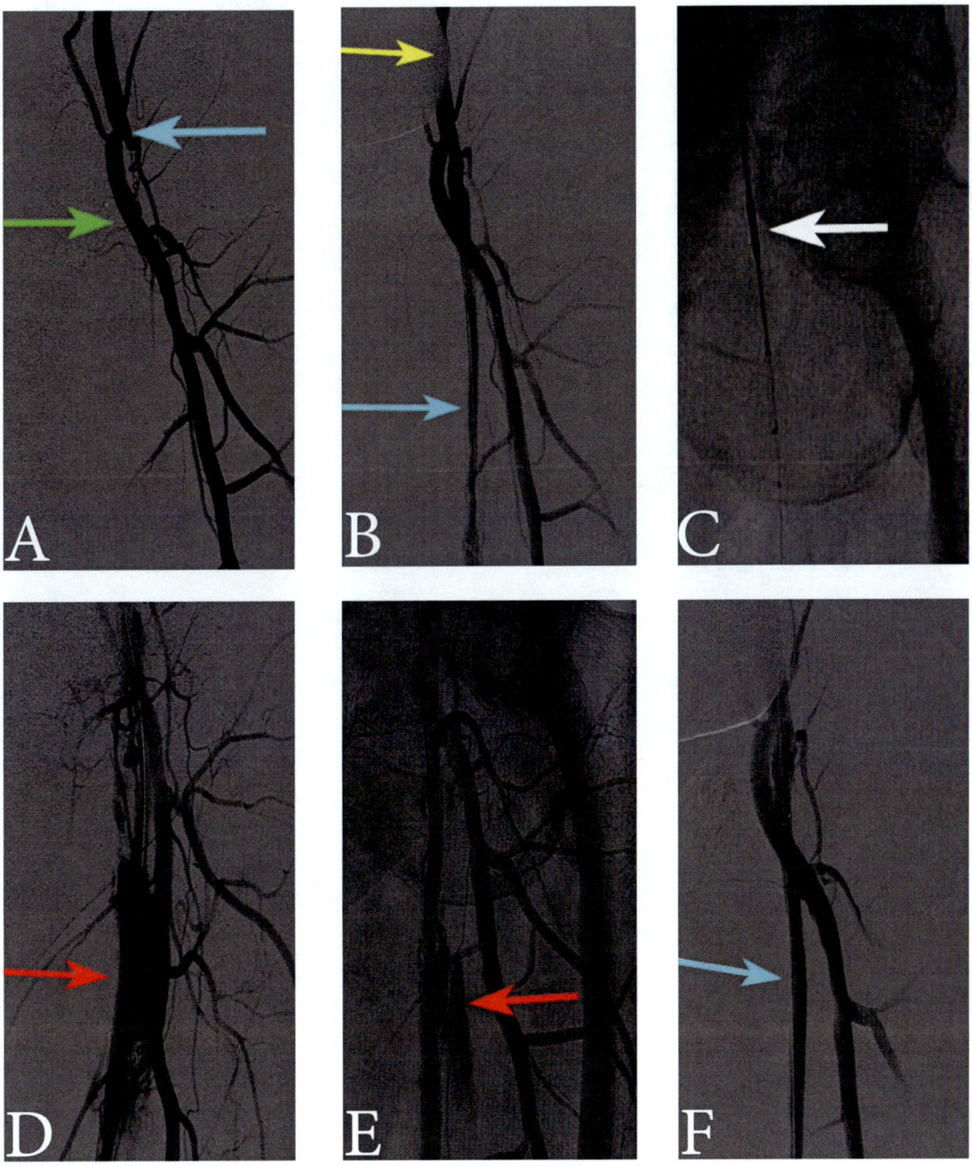

Figure 6.
Management of an SFA perforation.

A. Proximal occlusion of left SFA (blue arrow). The PFA (green arrow) originates from the postero-medial aspect of the CFA.

B. After successfully crossing the CTO, the SFA was treated with gentle PTA in anticipation of atherectomy. The yellow arrow points out an artifact caused by the bladder overshadowing the CFA in this oblique projection.

C. Atherectomy with an LS-M SilverHawk® device is performed. The arrow points to the 'cutter' of the device. The patient experienced sudden sharp pain in the upper thigh.

D. An angiogram is performed after removing the device. A major perforation of the proximal SFA is noted (red arrow).

E. Balloon angioplasty was performed and the wire was switched out through the balloon lumen to a 0.035" wire. A subsequent angiogram revealed continued extravasation (red arrow).

F. A 6.0 mm x 150 mm Viabahn® was then deployed which successfully sealed the perforation (blue arrow).

CASE 1

Mr. B is a 46 year old diabetic male with severe intermittent claudication affecting his right calf. His physical exam showed normally palpable femoral pulses and only weakly Dopplerable right pedal pulses.

Noninvasive vascular studies revealed severe femoro-popliteal PAD in the right leg.

Access and Diagnostic Angiography:

- For initial angiography, a 5 Fr./23 cm Cordis Britetip sheath was used in the left CFA.
- 2000 units IV UFH were given.
- Abdominal aortography was performed with a 4 Fr. TR catheter using 40 cc of Visipaque at 20 cc/sec at 500 psi.
- The TR catheter was then exchanged over an ASGW for a 5 Fr. IM catheter.
- The Right CIA was engaged with the IM catheter and the ASGW advanced into the PFA.
- The IM catheter was exchanged out for a 4 Fr./100 cm straight Glidecath®.
- The Glidecath was positioned in the CFA and images of the right SFA were obtained.
- The SFA was patent in the proximal two thirds with a focal area of severe stenosis in the distal third (figure 7A).

Interventional Technique:

- The ASGW was parked in the distal PFA and the sheath was exchanged out for a 5 Fr./45 cm Ansel® 1 sheath. This was advanced over the aortic bifurcation under fluoroscopic guidance.
- The patient was given an additional 3000 units of heparin.
- A road map was obtained.
- An ANSGW was used to cross the distal stenosis using the Glidecath for support. After crossing, this wire was switched out for an ASGW.
- The catheter was removed and angioplasty performed with a 5.0 mm x 60 mm balloon for 3 minutes (figure 7B).
- The balloon was removed over the wire and the catheter reinserted.

Final Angiography:

- Final images were obtained using the Glidecath revealing a small non flow limiting dissection which was left to heal on its own (figures 7C & 7D).

Take Home Message:

- Non-flow limiting dissections in the SFA do not always require stenting.

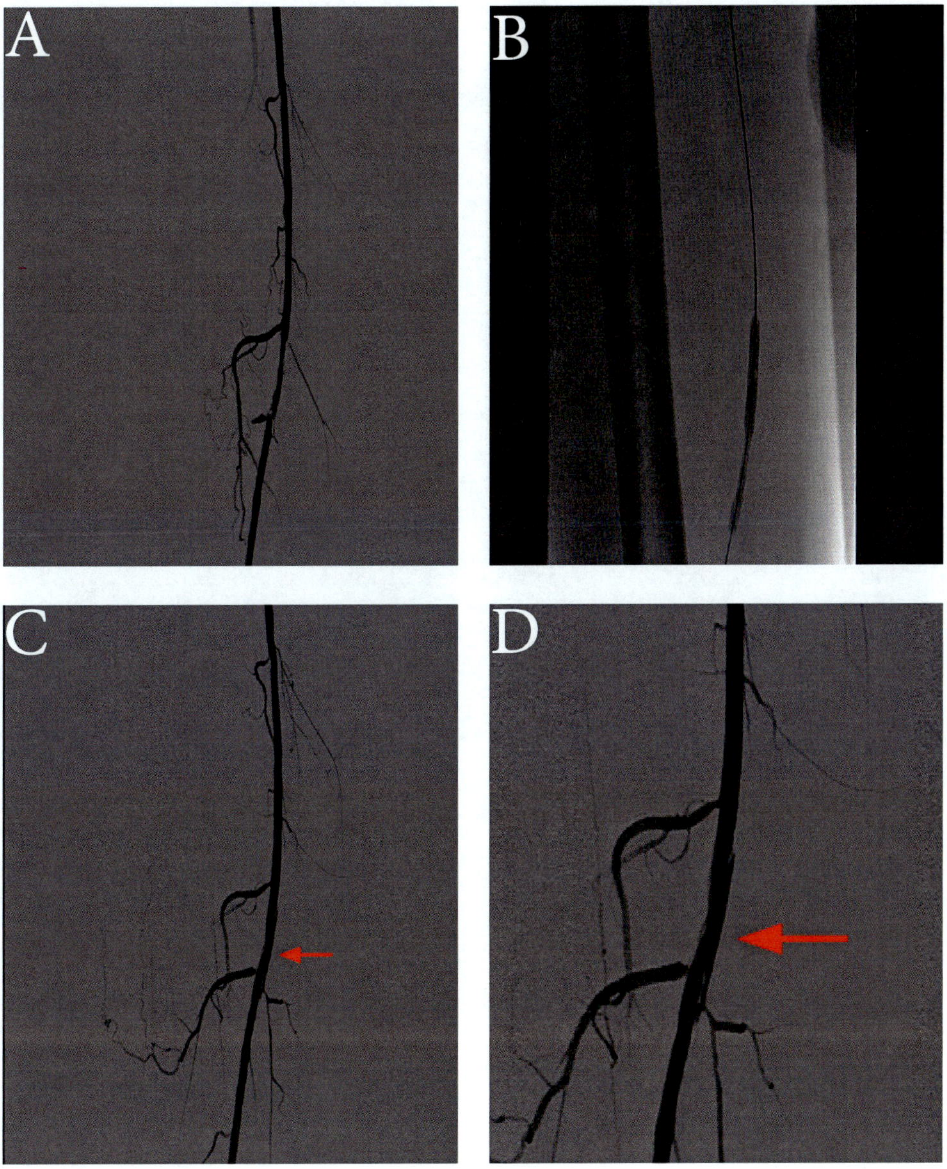

Figure 7.

A. Selective angiogram of the right SFA showing severe focal disease in the distal segment.
B. PTA of the focal lesion being performed with a 5.0 mm x 60 mm balloon.
C. Post-PTA angiogram shows significant improvement in the caliber of the vessel at the treatment site.
D. A focal dissection (red arrow) is seen in this magnified view. However, the dissection is non-flow-limiting and hence was not treated with repeat PTA or stenting. Dissections like this heal spontaneously in 2-6 weeks and if necessary, can be managed at that time with gentle PTA or atherectomy to avoid putting in a stent.

CASE 2

Mr. C, a 58 year old male with a history of CAD, prior CABG and left subclavian artery stenting, presented with symptoms of severe claudication in his both calves, left worse than right.

Resting non-invasive arterial testing did not suggest any significant PAD in either leg. However, given his classic and disabling symptoms as well as his risk factors, it was decided to proceed with angiography (of his left leg first due to worse symptoms).

Access and Diagnostic Angiography:

- For initial angiography a 5 Fr./23 cm Cordis Britetip sheath was used in the right CFA.
- 2000 units IV heparin were given.
- Abdominal aortography was performed with a 4 Fr. TR using 40 cc of Visipaque at 20 cc/sec at 500 psi.
- The TR was then exchanged over an ASGW for a 5 Fr. IM catheter.
- The left CIA was engaged with the IM catheter and the ASGW advanced into the PFA.
- The IM catheter was exchanged out for a 4 Fr./100 cm straight Glidecath.
- The Glidecath was positioned in the CFA and images of the left SFA were obtained.
- There was an area of diffuse disease in the distal left SFA (figure 8A). The ASGW was used to cross the distal stenosis.
- The Glidecath was advanced beyond the stenosis, the wire removed and hemodynamic gradients obtained.
- There was more than a 20 mm Hg resting gradient between the right CFA and the distal left SFA. This increased to 40 mm Hg after 200 mcg of IA nitroglycerin was given into the popliteal artery.
- A pullback gradient was also performed which revealed complete resolution of this gradient at the level of the mid SFA.

Interventional Technique:

- The patient was given an additional 3000 units of heparin.
- The ASGW was reinserted through the Glidecath and parked in the popliteal artery. The Glidecath was removed and a 7 Fr./45 cm Ansel 1 sheath was inserted over the wire.
- A Choice PT floppy wire was advanced across the lesion and into the popliteal artery.
- A 6 mm Spider filter device was then inserted over the floppy wire and then the wire removed.
- A road map was performed.
- Excisional atherectomy was performed using a SilverHawk® LS-M device.
- An angiogram was obtained (figure 8B).
- Angioplasty was then done with a 5.0 mm x 100 mm balloon at 2 ATM x 3 minutes (figure 8C).
- The balloon was removed over the wire.

Final Angiography:

- Final images were obtained through the sheath revealing a widely patent left SFA and above the knee popliteal artery with TIMI III flow and no residual stenosis, dissection or thrombus (figure 8D).
- Final hemodynamics revealed a provoked gradient of less than 20 mm Hg across the treated segment.

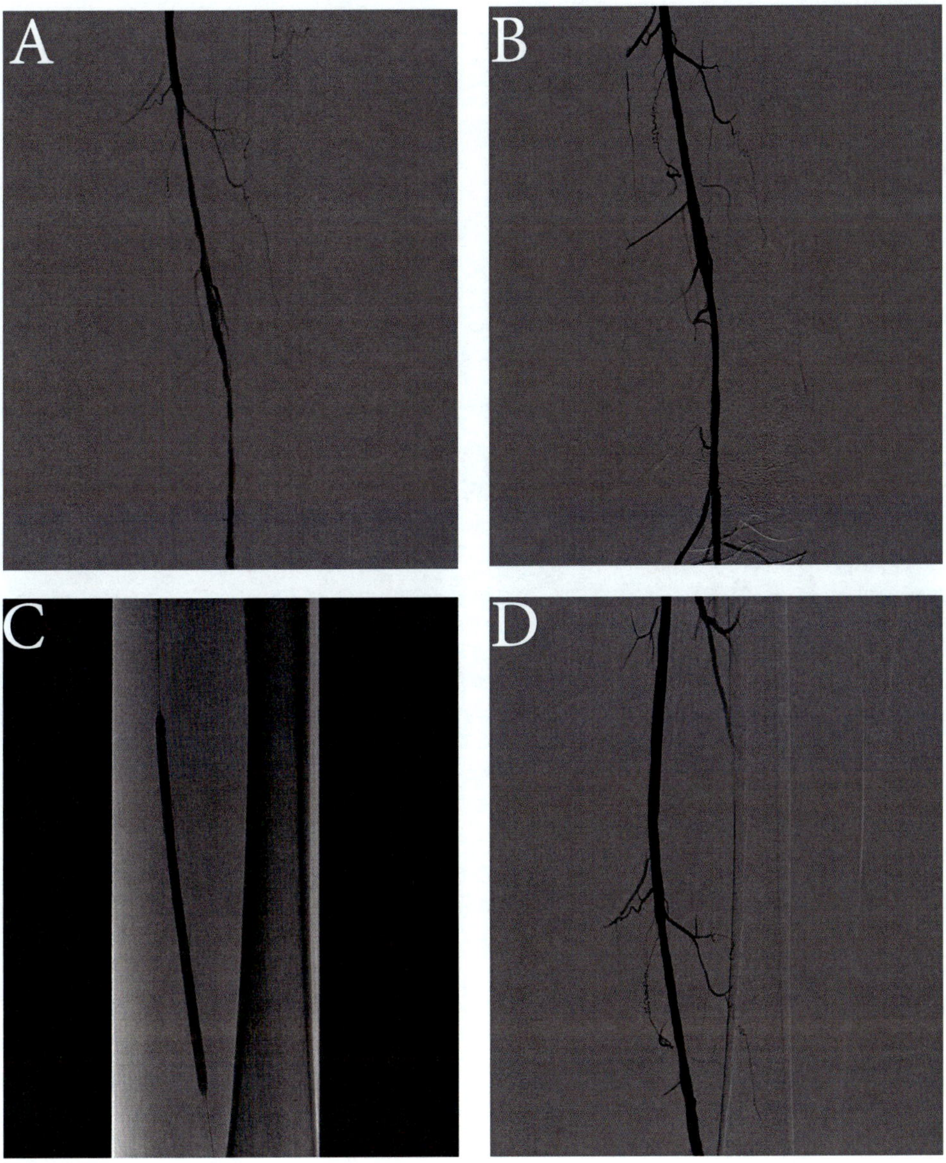

Figure 8.

A. Pre-treatment mid SFA diffuse moderate stenosis.
B. Angiogram obtained after atherectomy of mid SFA with an LS-M SilverHawk®.
C. Postdilatation with a 5.0 mm x 100 mm balloon at 2 ATM.
D. Final selective angiogram demonstrating a widely patent SFA with excellent flow.

CASE 3

Mrs. G is a 76 year old female with a history of hypertension and dyslipidemia. She had been experiencing disabling progressively worsening intermittent claudication in her right lower extremity for past several months.

Noninvasive testing suggested significant right femoral popliteal occlusive disease.

Access and Diagnostic Angiography:

- For initial angiography a 5 Fr./23 cm Cordis Britetip sheath was used in the left CFA.
- 2000 units IV heparin were given.
- Abdominal aortography was performed with a 4 Fr. TR catheter using 40 cc of Visipaque at 20 cc/sec at 500 psi.
- The TR catheter was then exchanged over an ASGW for a 5 Fr. IM catheter.
- The right CIA was engaged with the IM catheter and the ASGW advanced into the PFA.
- The IM catheter was exchanged out for a 4 Fr./100 cm straight Glidecath.
- The catheter was positioned in the CFA and images of the right SFA were obtained.
- The right SFA was occluded in the mid segment (figure 9A) and there was distal reconstitution at the level of the above-knee popliteal artery (figure 9B).

Interventional Technique:

- The patient was given an additional 3000 units of IV heparin.
- ASGW was reinserted through the Glidecath and parked in the right PFA. Glidecath was removed and a 6 Fr./45 cm Ansel 1 sheath was inserted over the wire. The tip of the Ansel sheath was positioned at the level of the distal right EIA.
- The wire was then removed and exchanged for a straight stiff glide wire (SSGW).
- A 4 Fr. Glidecath was then inserted and unsuccessful attempts were made to cross the chronic total occlusion with the straight stiff wire using the Glidecath for support. Attempts to cross with the ASGW were also unsuccessful.
- The ASGW was then exchanged for an ANSGW which was looped and used to crosse the CTO without difficulty (figure 9C).
- The Glidecath was advanced over this wire into the popliteal artery and a confirmatory angiogram was done.
- The wire was exchanged for the ASGW using the catheter. The stiff wire was parked in the peroneal artery.
- Over this wire, a 4.0 mm x 100 mm balloon was advanced into the SFA and angioplasty was performed in the previously occluded segment (figure 9D).
- This was followed by a 5.0 mm x 100 mm balloon x 3 minutes (figure 9E).

Final Angiography:

- Final images were obtained through the sheath revealing a widely patent SFA and popliteal artery and no residual stenosis, dissection or thrombus (figure 9E).

Take Home Message:

- Sometimes multiple variations of wires will be needed to cross chronic total occlusions using a catheter for support.
- Formation of a loop with a non-stiff wire is beneficial at times and can be used to cross CTOs.
- It is not mandatory to treat an SFA CTO with a stent. If post-PTA or post-atherectomy angiography is satisfactory stenting should be avoided.
- Non-flow-limiting dissections do not warrant treatment with stents.

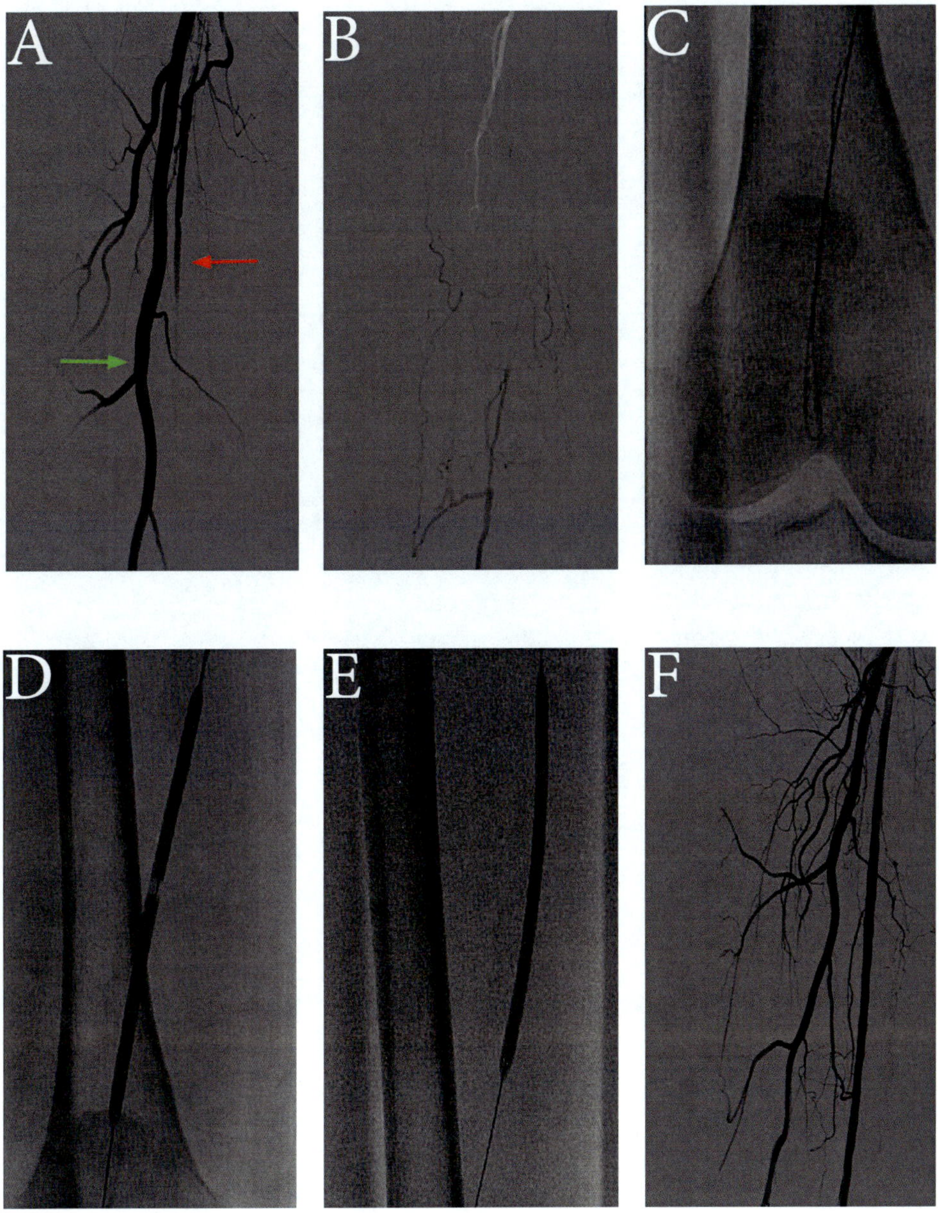

Figure 9.

A. Pre-treatment angiogram showing a widely patent PFA (green arrow) with diffuse severe disease in proximal SFA which occludes totally in the mid segment.
B. Angiogram showing distal reconstitution above the knee joint.
C. Occlusion crossed successfully with a looped ANSGW.
D & E. PTA of the diseased segments of the SFA.
F. Final angiography revealed widely patent SFA.

CASE 4

Mr. F is an 84 year old male with nonhealing ulcers in his right lower extremity. He has had the ulcers for nearly 6 months and has failed conservative medical therapy and wound care treatment.

Noninvasive studies suggest severe right femoro-popliteal disease.

Access and Diagnostic Angiography:

- For initial angiography a 5 Fr. /23 cm Cordis Britetip® sheath was used in the left CFA.
- 2000 units IV heparin were given.
- Abdominal aortography was performed with a 4 Fr. TR catheter using 40 cc of Visipaque at 20 cc/sec at 500 psi.
- The TR catheter was then exchanged over a ASGW for a 5 Fr. IM catheter.
- The right CIA was engaged with the IM catheter and the ASGW advance into the right PFA.
- The IM catheter was exchanged out for a 4 Fr./100 cm straight Glidecath.
- The catheter was positioned in the CFA and images of the right SFA were obtained.
- The right CFA was patent and there was 100% occlusion of the right SFA at the ostium with distal reconstitution in the above-knee popliteal artery, at the level of the patella.

Interventional Technique:

- The patient was given an additional 3000 units of IV heparin.
- The angled stiff wire was reinserted through the Glidecath and parked in the right PFA. The Glidecath was removed and a 7 Fr./45 cm Ansel 1 sheath was inserted into the left CFA over the wire, and advanced under fluoroscopic guidance over the aortic bifurcation.
- The tip of the Ansel sheath was positioned at the level of the distal right EIA.
- A straight stiff wire was then used along with a 4Fr./65cm angled Glidecath for support to cross the occlusion. The wire was advanced into the mid segment where it appeared to enter a dissection plane. Multiple unsuccessful attempts were made to redirect the wire distally into the true lumen using various combinations of angled and straight stiff glidewires along with straight and angled Glidecaths.
- An Outback® catheter was then used in a standard fashion. A 0.014" Platinum Plus wire was advanced into the dissection plane parallel to the distal reconstituted popliteal artery at the level of the patella. The Outback® catheter was advanced over this wire and then the wire withdrawn from the device.
- Appropriate position of the Outback® catheter and its needle orientation were confirmed in 2 orthogonal views.
- The needle was then advanced into the true lumen followed by advancement of the 0.014" Platinum Plus® wire through the needle into the true lumen. The wire was parked distally in the below-knee popliteal artery.
- The Outback® catheter was then exchanged out for a coronary 4.0 mm x 20 mm balloon. The balloon was advanced beyond the site of reentry and a confirmatory angiogram was performed through its lumen after removing the wire.
- The wire was then re-introduced and angioplasty was performed at the site of re-entry.
- The balloon catheter was exchanged for a QuickCross Extreme® catheter though which the angled stiff wire was inserted into the popliteal artery. The catheter was exchanged for a 6.0 mm x 100 mm Powerflex® balloon. Angioplasty of the SFA was then performed .
- Due to significant recoil, dissection and a suboptimal result adjunctive stenting was performed. A 5.0 mm x 50 mm Viabahn was deployed distally in the above-knee popliteal artery and at the site of re-entry. This was overlapped proximally with a 6.0 mm x 80 mm Supera stent. The Supera stent was overlapped proximally up to the ostium of the SFA with a 6.0 mm x 150 mm Viabahn. The stents were all postdilated with a 6.0 mm balloon.

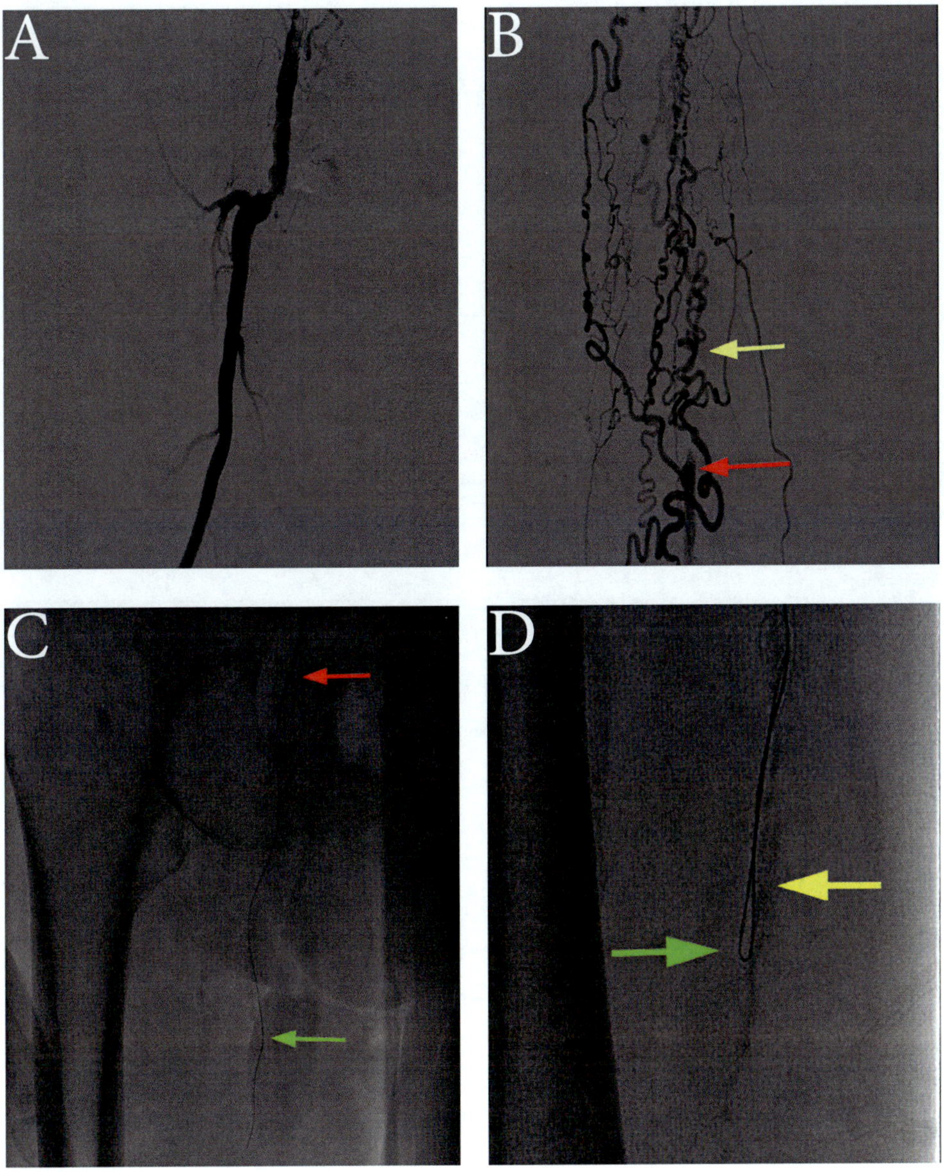

Figure 10.

A. Pre-treatment proximal SFA showing ostial total occlusion.
B. Angiogram showing distal reconstitution above the knee joint (red arrow) via PFA collaterals (yellow arrow).
C. An ANSGW (green arrow) supported by a Glidecath (red arrow) is noted in a dissection plane.
D The wire is looped (green arrow) and still remains in the subintimal space. A subtle dye stain (yellow arrow) is also seen in the dissection plane.

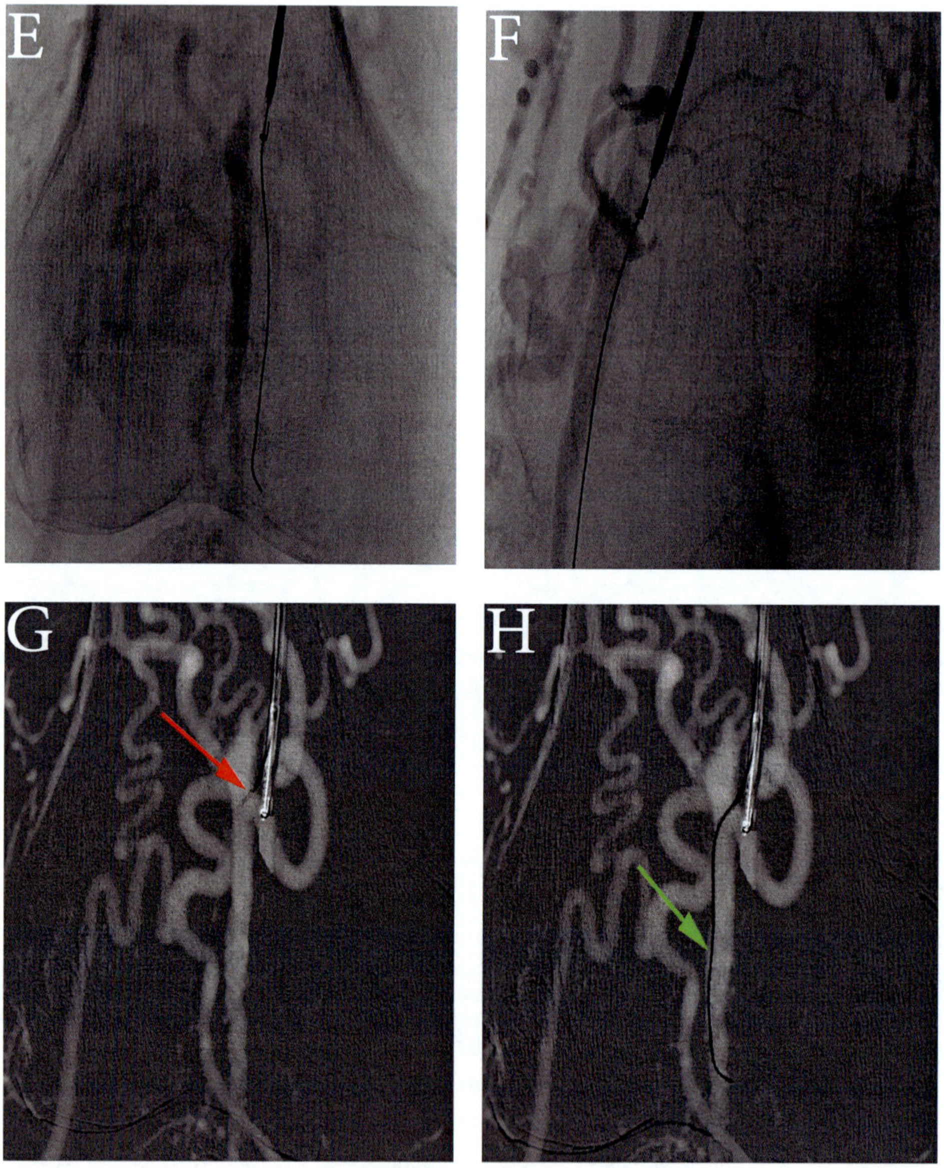

Figure 10 (continued):

E. The Outback® LTD catheter is being repositioned adjacent to the true lumen beyond the total occlusion. The needle is in the retracted position at this time. The marker on the tip of the catheter, as in this case, should look like an 'L' or a laterally inverted 'L' pointing towards the true lumen.

F. The Outback® catheter over a 0.014" wire is seen overlying the true lumen of the SFA beyond the total occlusion. The needle is still in the retracted position. The marker on the tip of the catheter assumes an upside down 'T' position.

G. Under 'road map' guidance , after withdrawing the 0.014" wire into the catheter, the needle (red arrow) of the Outback® catheter is advanced ('D'eployed) to enter the true lumen. This sequence of steps 'L', 'T' and 'D' is summarized in the name of the device.

H. The wire (green arrow) is then advanced through the needle into the true lumen.

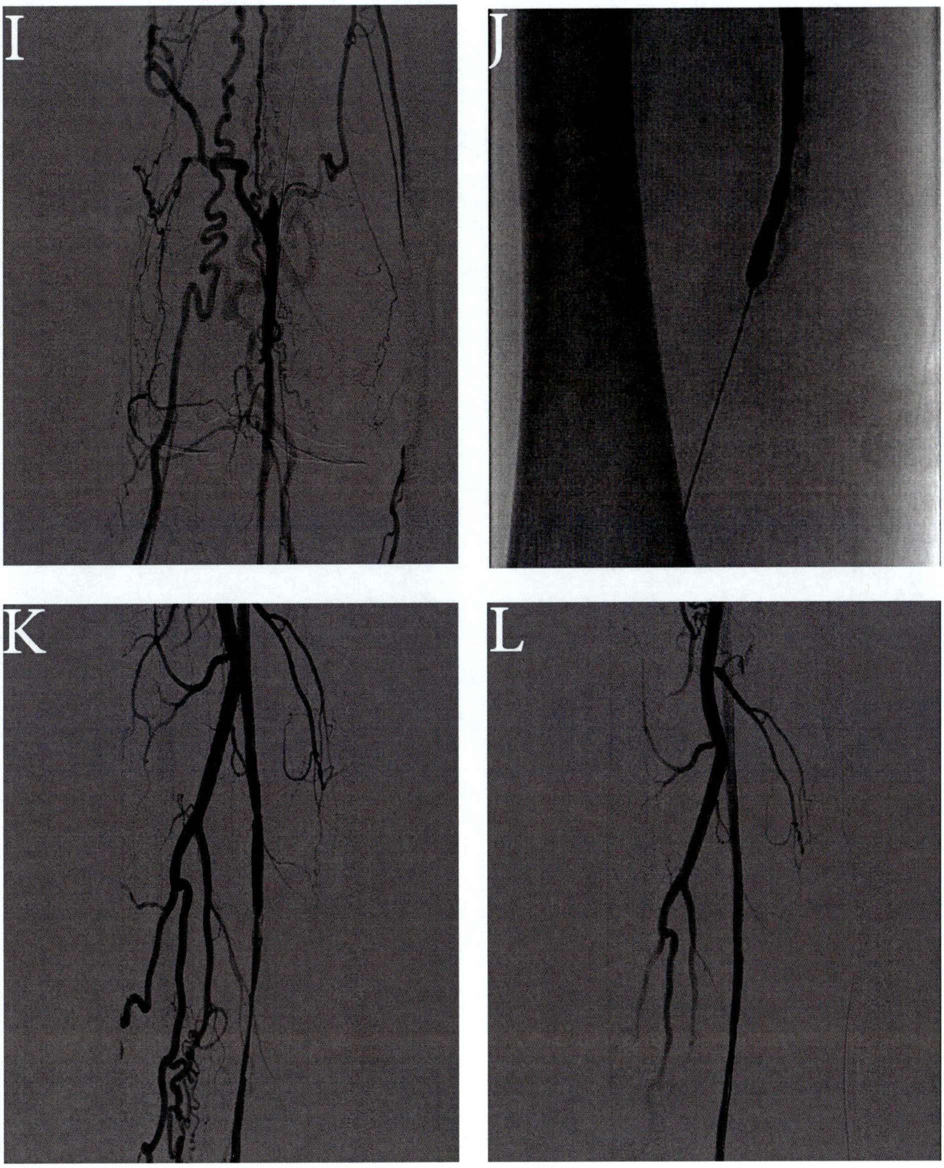

Figure 10 (continued).

I. The intra-luminal position of the 0.014" wire via a femoral injection from above.

J. After the wire is exchanged out for a 0.035" wire, to get more support, PTA is performed with a 6.0 mm x 100 mm balloon.

K. Angiogram post-PTA.

L. Angiogram post-stenting with covered Viabahn stents.

Final Angiography:

- Final images were obtained through the sheath revealing a widely patent SFA, PFA and popliteal arteries.

Take Home Message:

- The operator should be familiar with the use of CTO crossing devices (e.g. the distal re-entry device Outback). These come in handy on the rare occasion when traditional antegrade and/or retrograde approaches fail to cross the CTO.
- Treatment of an SFA CTO is acceptable in patients with CLI, particularly in the elderly patients.
- Stenting with Viabahn through the entire length of a long SFA occlusion may be unnecessary. However, the proximal and distal diseased segments and perhaps the profunda collaterals should be covered to diminish competetive flow that may later risk stent thrombosis.

7

INFRA-POPLITEAL ARTERY INTERVENTION

Infra-popliteal disease is the hallmark of critical limb ischemia (CLI). Typically, severe infra-popliteal disease is present in diabetics, patients with chronic kidney disease and the elderly. When present, it is usually diffuse and involves all three infra-popliteal arteries to some degree. These arteries tend to be calcified and when occluded have long segment CTOs.

Rarely, solitary infra-popliteal disease may also cause disabling intermittent claudication of the calf and/or the foot. When this happens it is primarily due to ostial/proximal disease involving the anterior tibial (AT) artery and/or tibio-peroneal (TP) trunk. This functionally serves as inflow disease for the calf and may mimic severe popliteal occlusive disease clinically.

ANATOMY

The infra-popliteal arteries consist of three/four named vessels, their branches and continuations:

ANTERIOR TIBIAL ARTERY

- The AT (figures 1A & 1B) is one of the two terminal continuations of the popliteal artery and usually arises from its lateral surface. Uncommonly, it may arise higher than the terminal popliteal artery and sometimes from its medial surface, coursing laterally.
- It is usually 3-4 mm in luminal diameter and tapers very gradually in size as it descends through the leg.
- It courses above the interosseous membrane (that connects the fibula and tibia) and enters the anterior compartment of the leg.
- Here it courses inferiorly, between the fibula and tibia, on the anterior surface of this membrane.

- As it descends in the leg it courses lateral to medial and in the lower fourth it lies anterior to the tibia, crossing the ankle joint.
- Beyond this point it continues as the dorsalis pedis (DP) artery.
- The AT is accompanied by paired veins, one on its each side, throughout its course. These are the AT veins.

Variants:

- It may arise higher than usual.
- It may arise from the medial aspect of the popliteal artery.
- The AT may be diminutive in size or completely absent.
- In the latter case, the peroneal or posterior tibial (PT) artery usually supply the terminal part of the vessel.

DORSALIS PEDIS ARTERY

Branches of the DP:
The branches of the DP are:
- Lateral Tarsal.
- Arcuate.
- Medial Tarsal.
- First Dorsal Metatarsal.
- Deep Plantar.

- The dorsalis pedis artery (figure 1B) is the continuation of the AT.
- It is usually 2-3 mm in diameter.
- It starts at the level of the ankle-joint and courses along the dorsum of the foot, along its medial third.
- It gives off lateral and medial tarsal arteries.
- The arcuate artery arises beyond the lateral tarsal artery and supplies the second, third and fourth dorsal metatarsal arteries.
- It terminates by dividing into the first dorsal metatarsal and the deep plantar arteries in the first inter-metatarsal space.
- The latter anastomoses with the lateral plantar artery (see below) and forms the plantar arch.

TIBIO-PERONEAL TRUNK

- This is the second terminal continuation of the popliteal artery.
- The TP trunk is usually 3-5 mm in luminal diameter and 4-6 cm in length.
- It terminates by bifurcating into the peroneal and posterior tibial arteries.
- Occasionally, this vessel may be absent altogether, with the PT and peroneal arising from the popliteal artery either together (a true 'trifurcation' of the popliteal artery into the AT, peroneal and PT) or at separate levels.

POSTERIOR TIBIAL ARTERY

Branches of the PT:
The branches of the PT are:
- Posterior medial malleolar.
- Nutrient.
- Communicating.
- Muscular.
- Medial calcaneal.

- The PT is the medial branch at the bifurcation of the TP trunk.
- Its luminal diameter is 3-4 mm.
- It descends along the medial aspect of the leg, lying posterior to the tibia.
- In the proximal part of the foot it lies midway between the medial malleolus and the medial process of the calcaneal tuberosity.
- Beyond this point it bifurcates into its terminal branches, the medial and lateral plantar arteries.
- The lateral plantar artery is usually the larger of the two terminal branches and anastomoses with the deep plantar artery, a branch of the DP, to form the plantar arch.
- The PT is accompanied by paired veins, one on its each side, throughout its course. These are the PT veins.

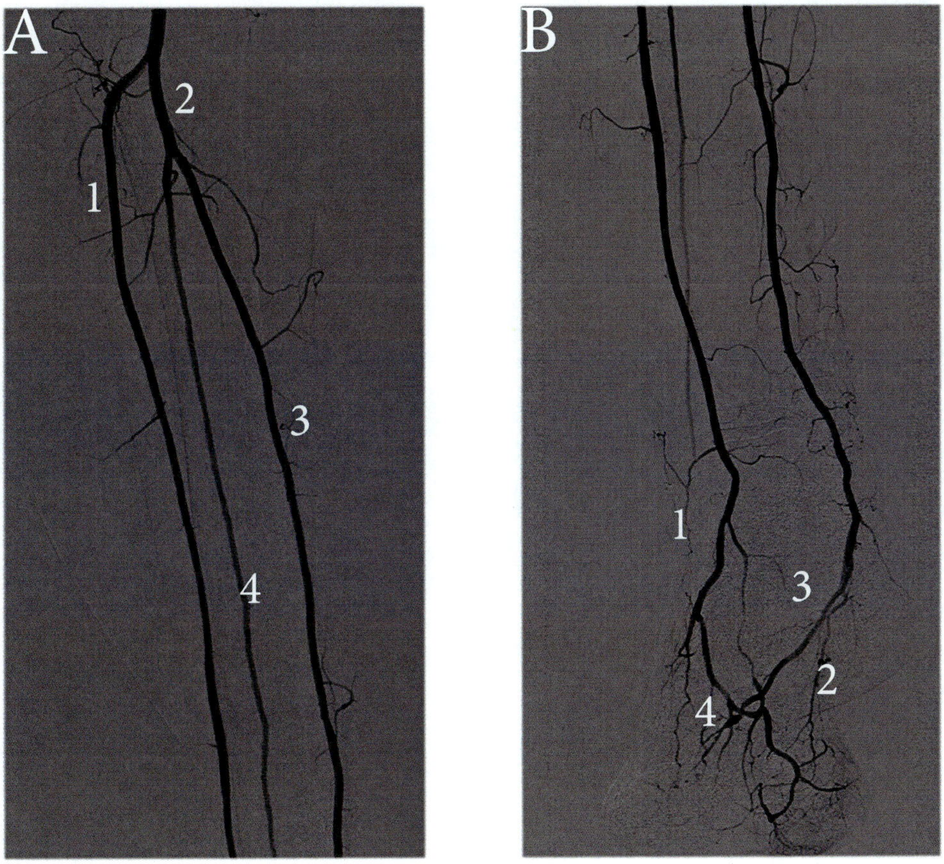

Figure 1.

Angiographic anatomy of the lower leg and foot.

A.
1- Anterior tibial artery
2- Tibio-peroneal trunk
3- Posterior tibial artery
4- Peroneal artery

B.
1- Dorsalis pedis artery
2- Medial plantar artery
3- Lateral plantar artery
4- Plantar arch

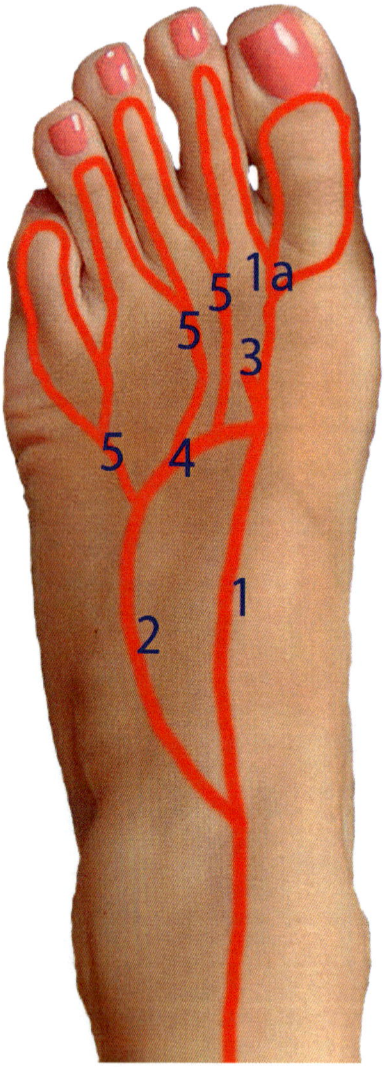

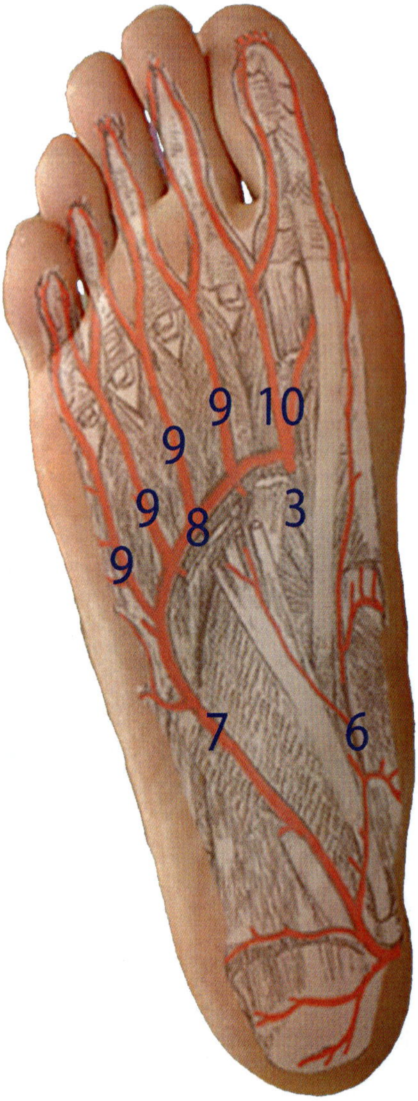

Figure 2.

Anatomy of dorsal arterial supply to the foot.

1. Dorsalis pedis artery.

1a. First dorsal metatarsal artery.

2. Lateral tarsal artery.

3. The deep plantar artery.

4. The arcuate artery.

5. Second, third and fourth dorsal metatarsal arteries.

Figure 3.

Anatomy of plantar arterial supply to the foot.

3. The deep plantar artery.

6. The medial plantar artery.

7. The lateral plantar artery.

8. The plantar arch.

9. The plantar metatarsal arteries.

10. The first plantar metatarsal artery.

Variants:

- The PT may be absent or smaller than usual, with the peroneal artery continuing distally into its course and supply the medial and lateral plantar arteries.

THE PERONEAL ARTERY

- The peroneal artery is the lateral branch at the bifurcation of the TP trunk and in essence forms its continuation.
- It is usually 3-4 mm in diameter.
- It descends somewhat laterally, between the tibia and fibula.
- It then runs behind the tibiofibular syndesmosis and divides into lateral calcaneal branches which supply the calcaneus.
- It often collateralizes the terminal DP and/or PT when these vessels have severe occlusive disease proximally.
- The peroneal artery is accompanied by paired veins, one on its each side, throughout its course. These are the peroneal veins.

Branches of the Peroneal:
The branches of the peroneal are:
- Perforating.
- Nutrient.
- Communicating.
- Muscular.
- Lateral calcaneal.

Variants:

- The peroneal artery may arise higher or lower than usual - directly from the popliteal artery or lower in the leg from the PT.
- It may be larger than usual or, rarely, diminutive/absent.

LEG AND FOOT ANGIOSOMES

- An angiosome, analogous to a dermatome, is a composite vascular territory (skin, soft tissue, muscle, nerve and bone) supplied by a particular artery (e.g. AT, PT and peroneal arteries).
- Tissue loss in an angiosome should be treated with revascularization of the source artery. This is called direct revascularization.
- When, instead of the source artery, a collateralizing artery is revascularized to improve perfusion to the affected tissue, this method of revascularization is termed 'indirect' revascularization.
- It has been shown that patients with tissue loss benefit more from direct rather than indirect revascularization.
- Thus, whenever possible, the source artery of an angiosome should be targeted for treatement of CLI patients.

CRITICAL LIMB ISCHEMIA

Critical limb ischemia (CLI) is characerized by acute/subacute worsening of chronic severe reduction in perfusion to the (lower) limbs.
The duration of symptoms is typically > 2 weeks.

In CLI, chronic ischemia is typically worsened by an acute increase in the metabolic demands of the distal limb tissue e.g. due to the development of a skin lesion/wound or progressive reduction in tissue perfusion (in case of rest pain).

CLI differs from acute limb ischemia (ALI) in that in the latter there is an acute thrombo-embolic insult that results in an abrupt acute/sub-acute occlusion of a large vessel (e.g. aorta, iliac, femoral or popliteal arteries or bypass grafts).

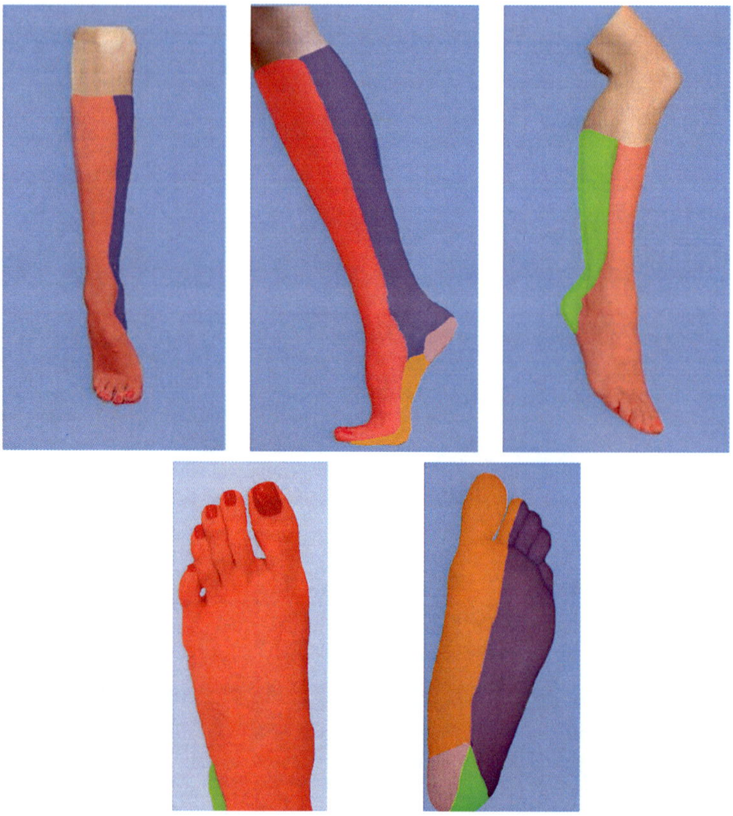

Figure 4.

Angiosomes of the lower leg and foot.

Red: Anterior tibial artery.

Purple: Posterior tibial artery.

Green: Peroneal artery.

Orange: Medial plantar branch of the PT.

Pink: Calcaneal branches of the PT.

Patients with CLI typically present with rest foot pain, ulcers or gangrene. The rest pain is classically present at night time when the patient lies supine and the augmentation in perfusion from gravity is decreased. Hence, these patients tend to wake up 2-4 hours into sleep with severe pain in the foot (e.g. 'toothache') and dangle the affected leg on the edge of the bed, sit up or even 'walk the pain out' to get temporary relief from the pain. Many patients 'learn' to sleep in a recliner to get comfortable at night.

Prompt diagnosis is the key to limb preservation. If the affected limb is not revascularized in a timely fashion limb loss is certain. These patients are also at an increased risk of systemic cardiovascular complications including MI and death. If revascularization is not performed, up to half of the patients die in the following 24 months.

DIAGNOSIS

- When non-invasive vascular assessment is performed in patients with CLI, typically multi-level PAD or isolated, critical infra-popliteal disease is noted.
- However, some automated PVR systems may not reflect the true severity of PAD in this vascular bed.
- If the history and physical examination is suggestive, strong consideration should be given to either conventional or non-invasive angiography (CTA or MRA) prior to ruling out disease in this territory.
- The ankle pressure for patients with rest pain is typically < 50 mm Hg (higher in patients with diabetes and/or CKD due to non-compressible arteries) with a toe pressure < 30 mm

Hg.
- In patients with tissue loss (ulcer, gangrene) the ankle pressure is < 70 mm Hg with a toe pressue of < 50 mm Hg.

DIAGNOSTIC ANGIOGRAPHY

ACCESS

Site:
- For diagnostic angiography, contralateral common femoral artery access is used.

Sheath Size:
- Regardless of the chosen site of access, we prefer to obtain initial access with a 21 gauge needle and 0.018" wire (e.g. 4 Fr. micropuncture kit).
- This is then up sized to a 5 Fr. sheath that is at least 20 cm long over a 0.035" wire (e.g. angled stiff hydrophilic wire, Wholey wire etc).

ABDOMINAL AORTOGRAPHY

Unless previous angiography has been performed or severe renal dysfunction is present, we perform aortography as usual.

Catheter:
- A flush catheter (e.g. TR) is used.

Imaging:
- As discussed in the previous chapter we start in an AP projection.

Dye Injection:
- A power injector is used for aortography.
- 70:30 or 50:50 diluted dye can be used.
- Injection settings are 10 - 15 cc/sec. for a total volume of 20 - 40 cc at 500-750 psi.

SELECTIVE ANGIOGRAPHY

Catheter:
- After removal of the flush catheter over a wire, a selective catheter such as an Omni flush, IM, Rim or Simmons catheter is advanced to the aortic bifurcation to engage the CIA ostium.

Imaging:

Femoro-popliteal artery:
- Other than the CFA, angiography of the femoro-popliteal arteries is performed in the AP projection.

Infra-popliteal arteries:
- If the femoro-popliteal arteries are patent, the catheter should be brought to the level of the knee joint for infra-popliteal angiography.
- Use IA NTG (150-300 mcg) prior to selective angiography to obtain better runoff.
- Ideally, these arteries are imaged in a view that allows separation of the tibia and fibula and minimizes the overlap among the tibial arteries.
  - ▹ In a sedated patient, the legs are usually externally rotated at the hip and this may mean a steep ipsilateral oblique view. This, however, exposes the operator to more radiation.
  - ▹ One way around this is to have the legs internally rotated at the hips and a velcro strap applied to the ankles to hold them in this position.

- In the presence of a significant SFA stenosis/occlusion perform run-off angiography with the catheter still above this diseased segment. If feasible, use a 50:50 contrast-saline mix for this purpose. This not only minimizes the overall volume of dye used but also reduces the likelihood of patient calf discomfort precipitated by dye.

Dye Injection:

- DSA is used with manual injections of 5-8 cc of 50:50 dye through the selective catheter.
- In the setting of occluded femoro-popliteal arteries, inject dye at least 5-10 sec prior to performing cine to give it enough time to reach the infra-popliteal arteries.
- Obtain prolonged cine runs to visualize collaterals to the distal reconstituted segment.

HEMODYNAMICS

- Infra-popliteal intervention is mostly performed for CLI patients.
- Benefit of this intervention in claudicants is minimal except in the case of critical proximal vessel stenoses that may compromise the 'inflow' to the calf.
- Thus, assessment of border-line lesions is seldom necessary in the infra-popliteal arteries.
- Measuring gradients across diseased segments of infra-popliteal arteries is neither practical nor feasible.
- These arteries are small and even a 4 Fr. catheter may be mostly occlusive thus making any pressure measurements unreliable.
- A pressure-wire could be used for this purpose but would be unpractical given the long length of these vessels.
- Typically, angiography is definitive.

INTERVENTION

GOAL

- To establish at least a single vessel continuous pulsatile flow to the foot.
- For patients with tissue loss, it is recommended that revascularization of the culprit artery for the affected angiosome be performed.

ACCESS

- CFA access is the primary access for these cases. However, antegrade popliteal artery or retrograde tibial artery access may also be used to at least cross occlusive disease.

ANTICOAGULATION

- IV unfractionated heparin is used.
- Goal ACT > 250 sec.
- Bivalirudin or GP IIb/IIIa inhibitors are not necessary.
- Vasodilators such as intra-arterial NTG or nitroprusside should be used abundantly, pre- and post intervention. These arteries are prone to severe vasospasm that may mimic severe occlusive disease prompting unnecessary intervention.

EQUIPMENT

WIRES

- 0.014" coronary guide wires are ideal. Always use 300 cm length rather than 190 cm as most devices are OTW.
- Choice PT™ floppy is our workhorse wire for infra-popliteal arteries. Others include: Fielder

XT™, Runthrough™, Miracle Brother® wires particularly Confianza Pro™ (for CTOs).
- 0.018” wires seldom have adequate steerability to be used as crossing wires. We rarely utilize the 0.018” angled Glidewire™ for crossing tough CTOs.
- 0.035” hydrophilic wires (straight tip) may rarely be used when other wires fail to cross.

SHEATHS

Contralateral (retrograde) CFA Access:

- In case of contralateral CFA access, a cross-over sheath is used.
  - ▹ If non-obstructive femoro-popliteal disease is present a 5 Fr./90 cm sheath (e.g. Shuttle sheath from Cook) is ideal to be brought to the level of the knee joint. This allows better visualization of distal arteries due to local dye injections and reduces the amount of contrast used.
  - ▹ If significant femoro-popliteal disease is present, a 5 Fr. /45 cm cross-over sheath can be used (e.g. the Ansel 1 from Cook). The tip of this sheath is ‘parked’ in the distal EIA/CFA.

Ipsilateral (antegrade) CFA Access:

- For antegrade access, a 5 Fr./20-40 cm long straight sheath can be used (e.g. Brite tip, Ansel 0 etc).
- The tip of this sheath can be parked in the distal SFA or popliteal artery.

BALLOONS

The diameter of normal infra-popliteal arteries is between 2.0 mm - 4.0 mm. The tibio-peroneal trunk is 3.0 - 5.0 mm in diameter.

At the time of this writing no drug-coated or drug-eluting balloons were available in the U.S.

0.014” Wire Compatible Balloons:

- Coronary balloons (1.5 mm - 4.0 mm diameter)
  - ▹ Longest is 40 mm.
  - ▹ Longest shaft length is typically 135 cm.
  - ▹ Monorail as well as over-the-wire (OTW).
- Non-coronary (peripheral) balloons
  - ▹ Longer balloon lengths (40 mm - 220 mm).
  - ▹ Longer shaft lengths (up to 200 cm).
  - ▹ Mostly OTW.

Coronary balloons can be used due to their lower profiles at least initially when crossing and treating diseased segments. However, due to their short lengths, coronary balloons (the longest is ~ 40 mm) are time consuming for performing angioplasty on long lengths of infra-popliteal arteries. Longer balloons designed specifically for infra-popliteal arteries are available. These offer lengths of up to 220 mm.

0.018” Wire Compatible Balloons:

- More pushable.
- Diameters range from 2.0 mm to > 5.0 mm.

TECHNIQUE

SHEATH EXCHANGE

- A 5 Fr. sheath is adequate to treat infra-popliteal disease unless atherectomy is being considered. The latter requires a 6 Fr. sheath.

- If no significant femoro-popliteal disease is noticed on diagnostic angiography, a sheath long enough to reach the popliteal artery should be used.
  - ▹ From the contralateral CFA, we use a 5 Fr./90 cm Cook Shuttle sheath for this purpose.
  - ▹ From ipsilateral antegrade access, a 5 Fr./20-40 cm long straight sheath is used.
- Prior to exchanging the initial sheath, a 0.035" angled stiff hydrophilic wire is advanced over the aortic bifurcation into the popliteal artery using a cross-over catheter.
- Sheath exchange is then carried out and under fluoroscopic guidance the Shuttle sheath is advanced over the aortic bifurcation down to the level of the knee joint. The dilator is kept in the sheath until it reaches its destination.
- Since the dilator sticks out several centimeters beyond the tip of the sheath it is important to make sure that enough wire length is present distal to the tip so that the dilator is never advanced beyond the wire tip.
- It should also be ensured that the wire is in the true popliteal artery or distal infra-popliteal branches rather than in a geniculate branch as inadvertent trauma to the vessel from the dilator can result in dissection or perforation.
- Once the dilator is removed and the sheath is flushed the patient should be fully anticoagulated to achieve an ACT of > 250 seconds.
- Having a long sheath through the length of the SFA can cause enough flow reduction and possible thrombus formation which could be catastrophic.

VESSEL CANNULATION

- Once the sheath is in the desired location, a road-map is obtained.
- A 0.014 " wire is then advanced into the target artery.
- In case of chronically occluded arteries, the tip of the wire should ideally have a very short bend. This allows better engagement of the proximal cap, often in a slow 'burrowing' motion. This short bend at the tip may make it difficult to engage the AT or the PT and may require the use of a 'guiding' catheter.
  - ▹ In these cases a 4 Fr./120 cm hydrophilic catheter may be used as a 'guiding' catheter through the sheath. This catheter should be > 120 in length. A 100 cm long catheter is not long enough as the sheath alone is 90 cm.
    - ♦ For anterior or posterior tibial arteries an angled catheter is used to selectively 'engage' the artery. This catheter does not have to physically engage the ostium of the artery of interest, rather it can be advanced close enough to perform sub-selective cannulation ('no touch technique'). The wire can then be advanced into the artery.
    - ♦ For the peroneal artery, a straight catheter is usually adequate.

CROSSING THE OCCLUSION

- Once the wire is in the desired artery, the 4 Fr. 'guiding' catheter may be exchanged out for a 0.014" support catheter (e.g. QuickCross or FineCross) or for a 0.014" OTW coronary balloon catheter.
  - ▹ OTW balloon is preferred to a monorail/rapid-exchange balloon as the former allows the ability to exchange the wire out without losing position.
- 'Road map', image reference, vascular calcification and bony landmarks should all be utilized in getting guidance as to the direction of wire advancement.
- Advance and lead with the wire gradually. Every 2-3 inches the supporting catheter should be advanced to catch up with the wire. If the wire appears to be 'on course' but resistance is met when advancing the catheter, consider performing an angioplasty at the site with a low profile coronary balloon.
- A 2.0 mm balloon is a good initial choice, perhaps 20-30 mm in length.
- Longer balloons are difficult to advance at first pass through long segments of diseased/chronically occluded arteries.

CONFIRMATORY ANGIOGRAPHY

- Once the wire has been successfully advanced across the occluded segment and re-entered the reconstituted segment, a confirmatory angiogram through the catheter is desirable.
- Following this, angioplasty of the entire occluded segment is advised. Start with a smaller diameter balloon (2.0 mm). Inflations do not have to be more than 6 ATM as long as no significant 'waist' is present, and often a lower pressure inflation such as 2-4 ATM may be adequate to create an initial channel. These low pressure inflations are the key to avoiding dissections.
- To avoid significant recoil, prolonged inflations up to 3 minutes are recommended.
- Initial angioplasty should be followed by definitive therapy.

DEFINITIVE INTERVENTION

ANGIOPLASTY

- PTA alone may be adequate as the definitive therapy in infra-popliteal arteries.
    - ▹ Prolonged (up to 3 minutes at a time) gentle inflations, as noted above, should be performed.
- This modality:
    - ▹ Allows treatment of long diseased segments in a relatively short time.
    - ▹ Usually requires 1-2 balloons of incremental diameters/lengths and is thus relatively low cost.
    - ▹ Can be, and often is, combined with treatment modalities e.g. proximal stenting or atherectomy.
    - ▹ No additional set up is required.

STENTING

- Bare-metal stents have a poor patency rate in the infra-popliteal arteries compared to drug-eluting stents (DES).
- Since non-coronary DES are not available in the U.S., coronary stents have been used in this vascular territory.
- These stents are exclusively balloon-expandable.
- DES have shown excellent results in the ostial/proximal tibial arteries. In the more distal segments, they tend to be less useful due to the diffuse nature of disease and tend to get affected by aggressive restenosis.
- The incidence of restenosis even with DES, due to poorly understood reasons, is higher than in the coronary bed.
- DESTINY (Drug Eluting Stents in the Critically Ischemic Lower Leg), a randomized, controlled multi-center trial from Europe showed that compared to bare-metal stents (Vision Multi-link), everolimus eluting coronary stents (Xience Prime) had significantly improved patency at 6 and 12 months (85.2% vs 54.5%; $p=0.0001$) in patients with CLI, in lesions < 40 mm.
- Self-expanding non-drug eluting stents have a higher restenosis rate than coronary DES and may be used only if appropriate size coronary DES are not available.

ATHERECTOMY

- Excisional and orbital atherectomy are both very useful in the treatment of infra-popliteal arterial disease. The former is better suited in the management of diffuse disease whereas the latter for calcified, focal lesions. Both, however, can lead to significant distal embolization of atheromatous debris leading to limb-threatening ischemia.

Excisional Atherectomy (SilverHawk®):

- ▹ Does not require additional setup.
- ▹ 0.014" wire and 6 Fr. sheath compatible.
- ▹ It is not supplied with its own wire. The operator can use the interventional wire that is across the lesion.
- ▹ In tibial arteries, embolic protection may not be feasible due to lesion location and vessel size.

Orbital Atherectomy (CSI®):

- ▹ Requires additional setup.
- ▹ Is supplied with its proprietary wire which must be used.
- ▹ Ideal for diffuse as well as non-diffuse calcific lesions.
- ▹ Adjunctive PTA is always used after atherectomy.

HOW TO CROSS A CHALLENGING TIBIAL CTO

A challenging CTO is defined as an occlusion that cannot be crossed successfully with the use of a single access point employing traditional catheter and wire system(s).

When such failure is encountered despite multiple 'traditional' attempts (during the same procedure), an analysis of the causes of failure and alternate measures needs to be considered.

CAUSES & POSSIBLE SOLUTIONS

The major causes of failure include:

Lack of Proximal Support:

- Failure to advance a wire/catheter across the occlusion may be due to lack of support at the level of the proximal cap. This is particularly relevant for an ostial occlusion of the anterior tibial artery.
  - ▹ Using an angled tip catheter with a relatively short tip may provide the necessary support.
- The lack of proximal support may also result from tortuous and calcified aorto-iliac arteries and leads to an inability to advance wire/catheter.
  - ▹ This problem may be solved, in certain cases, by obtaining an antegrade access in the ipsilateral CFA, thus obviating the need to traverse the tortuous iliac arteries.
  - ▹ If antegrade access is not feasible, retrograde pedal artery access should be considered.
  - ▹ Alternatively, the cross-over sheath may have to be up-sized by 1-2 Fr. sizes. This may not always be feasible or safe.

Inability to Cross the Proximal Cap:

- This may be the sole cause of failure in the absence of support issues.
- If, however, the tip of the initial wire (Choice PT floppy) cannot be advanced across the proximal cap, the use of a dedicated CTO wire (e.g. Asahi wires) instead might succeed.

Heavily Calcified Arteries:

- Intrinsic tibial arterial calcification may make certain occlusions 'untreatable' with current percutaneous therapies due to the sheer inability to cross the occlusive burden.

BTK Popliteal Artery Stenting:

- PTA and atherectomy are very reasonable first-line choices for popliteal artery intervention.
- However, the result is seldom perfect from a hemodynamic perspective.
- While such a result may be acceptable for claudicants, in CLI patients dependence on popliteal artery inflow is paramount.
- Thus stenting of BTK popliteal artery in CLI patients is a justified first-line option (particularly where drug-eluting balloons are not available).
- We have used drug-eluting (coronary) stents with considerable success in this location.

- ▹ Surgical revascularization may be the only reasonable option in these patients with extreme calcification.

Inability to Cross the Distal Cap:

- In an overwhelmingly large number of cases of challenging CTOs, the inability to cross the distal cap of the occlusion and to re-enter the true lumen remains the rate limiting step.
  - ▹ Retrograde access in the pedal arteries, depending upon operator experience and comfort level, may be the most straightforward solution in many cases.
    - ♦ Once the wire has been 'externalized', intervention proceeds as normal via the CFA access (contralateral or ipsilateral).
  - ▹ CTO crossing devices, if available and familiar to the operator, may also offer the solution. These devices are discussed separately in the following section.

TIBIAL CTO ASSIST DEVICES

CROSSER™

- Crossing catheter for distal entry into the true lumen.
- 6 Fr. sheath compatible.
- No integrated IVUS.
- Compatible with a 0.014" guidewire. Requires a single wire.
- Shaft length 120 cm.
- Hydrophilic coating along the entire length.

TRUEPATH™

- Intended for intraluminal crossing of a CTO.
- 0.018" compatible catheters < 135 cm in length.
- No integrated IVUS.
- Compatible with a 0.018" guidewire. Requires a single wire.
- Shaft length 165 cm.

VIANCE™ CROSSING CATHETER

- 0.014" wire compatible, OTW, CTO-crossing device for:
  - ▹ Intra-luminal crossing
  - ▹ Sub-intimal path crossing.
- The total working length of this catheter is 150 cm.
- 0.038" crossing profile.
- 5 Fr. sheath compatible.
- The atraumatic distal rounded tip is 0.040" in diameter as well as length.
- The catheter is capable of 'drilling' through the proximal cap of an occlusion by virtue of its high-fidelity torque delivery.
- This torque is transmitted using bi-directional rotation of the proximal torque device with the FastSpin technique.
- Increased torque delivery with gradual forward push enhances 'drilling' capability and the likelihood of crossing.

ENTEER™ RE-ENTRY SYSTEM

- 0.014" - 0.018" wire compatible CTO crossing device for sub-intimal path crossing.
- It has two components:

Re-entry Catheter:

- An OTW balloon catheter.
- 5 Fr. sheath compatible.
- Crossing profile is 0.066".
- Elliptical (flattened) balloon with offset wire exit ports (one on each surface), 180 degrees apart.
- Balloon dimensions are 3.75 mm x 1.5 mm x 20 mm (width x height x length) for ATK use, and 2.75 mm x 1.0 mm x 20 mm (width x height x length) for BTK use.
- Working length of the catheter is 135 cm for the ATK catheter and 150 cm for the BTK catheter.

Guide Wire:

- 0.014"/300 cm.
- Tip angle is 28°.
- The tip of the wire has a 0.007" long, sharp, central 'spike' that is 0.003" in diameter. This creates a puncture site for re-entry into the true lumen.
- The wire tip is very sharp and should be handled very carefully once in the distal lumen. Manipulation should be kept to a minimum and it should not be advanced beyond the first 1 cm after re-entry.
- The wire comes in three levels of stiffness - flexible, standard and stiff.

Technique:

- This device is used when while attempting to cross a CTO, the 'workhorse' 0.014" wire enters the subintimal plane and is unable to enter the true lumen.
- The Enteer catheter is advanced over the wire into this subintimal space.
- It is important to keep this dissection (subintimal) plane as small as possible.
- Once the Enteer catheter is in the sub-intimal space, the balloon is advanced adjacent to the vessel lumen.
- It is critical to work in an angiographic view that shows the distal reconstituted segment in long axis, with the balloon either medial or lateral to it, not superimposed on the vessel. This allows the operator to determine if the exiting Enteer wire is coming towards or away from the lumen. In a view where the balloon is superimposed on the vessel this judgment is difficult, if not impossible.
- Position adjacent to the vessel lumen is confirmed under fluoroscopy and the balloon is inflated at 4 ATM.
- The initial ('workhorse') wire is withdrawn and the Enteer wire is introduced.
- This wire will exit through one of the two ports on the balloon's surface.
- Once the Enteer wire is seen to enter the true lumen (the operator may feel a 'pop') it is advanced 5-10 mm.
- At this point it is assumed that the wire is distally in the true lumen. This wire will be removed and exchanged for a Pilot 200 wire.
- At this point, it is critical to not move the balloon, the table, the patient or the camera.
- The wire (Pilot 200) is advanced gradually so it enters the distal true lumen through the port of the balloon that is lined up with the re-entry puncture site.
- The wire is advanced distally into the vessel under fluoroscopic guidance.
- The Enteer balloon is exchanged out and intervention proceeds as usual.
- Atherectomy should not be performed along the length of the CTO as it may disrupt the subintimal plane and lead to perforation.

CASE 1

- Mr. W. is a 45-year-old male smoker with known PAD. He presented with symptoms of severe left calf and foot claudication. No tissue loss or ulceration was reported in either lower extremity.
- Physical exam in the office was unremarkable with the exception of non-palpable pedal pulses in the left foot with monophasic Doppler signal in both the DP and PT.
- Noninvasive arterial testing revealed a resting ABI of 0.87 in the right leg and 0.45 in the left leg. PVR suggested severe left infra-popliteal occlusive disease.
- Due to the patient's disabling symptoms a decision was made to proceed to angiography.

Access & Diagnostic Angiography:

- For initial angiography, we used a 5 Fr./23 cm Cordis Brite-tip sheath in the right CFA.
- 2000 units IV heparin was given.
- Over an 0.035" 260 cm angled stiff hydrophilic wire, a 4 Fr./65 cm TR catheter was advanced into the distal abdominal aorta.
- Hemodynamic gradients were measured between the distal abdominal aorta and the right CIA and subsequently EIA. As always, a resting and provoked gradient was measured at each of these levels.
- For a provoked gradient, 200 mcg of intra-arterial nitroglycerin was administered.
- In AP projection, distal abdominal aortography is performed using the following settings: 10 cc per second for a total of 20 cc at 500 PSI under DSA.
- After the initial aortogram, the catheter was exchanged out for a 5 Fr./65 cm IM catheter which was used to engage the ostium of the left CIA.
- The wire was advanced into the distal left PFA.
- Over the wire, the IM catheter was exchanged out for 4 Fr./100 cm straight hydrophilic catheter.
- The wire was removed, the catheter flushed and pulled back into the left common femoral artery.
- Hemodynamic gradient was recorded between the right CIA and left CFA both at rest and after intra-arterial nitroglycerin.
- Left CFA and proximal SFA angiography was performed in LAO 30° projection.
- The rest of the SFA and the proximal popliteal artery were imaged in the AP projection with the catheter still in the left CFA.
- For runoff angiography, the catheter was advanced into the left popliteal artery and manual angiography was performed. Slight LAO projection was used.
- Runoff angiography revealed severe ostial stenosis of the AT, subtotally occluded TP trunk, patent PT and occluded peroneal arteries (figure 5A).

Interventional Technique:

- The 0.035" hydrophilic wire was reinserted via the catheter into the left popliteal artery.
- The right CFA sheath was exchanged out for a 5 Fr./90 cm Shuttle sheath.
- This was advanced under fluoroscopic guidance over the aortic birufcation and the tip was parked in the left popliteal artery just below the knee joint.
- Additional 4000 units of IV UFH was given to achieve a therapeutic ACT of >250 sec.

Crossing the Diseased TP Trunk

- A road-map was obtained.
- A 0.014" Choice PT Floppy wire was used to cross the tibio-peroneal trunk without difficulty. This wire was advanced in the PT (figure 5B).
- A separate 0.014" Choice PT Floppy wire was used to cross the ostial AT stenosis and advanced in the distal vessel (figure 5C).

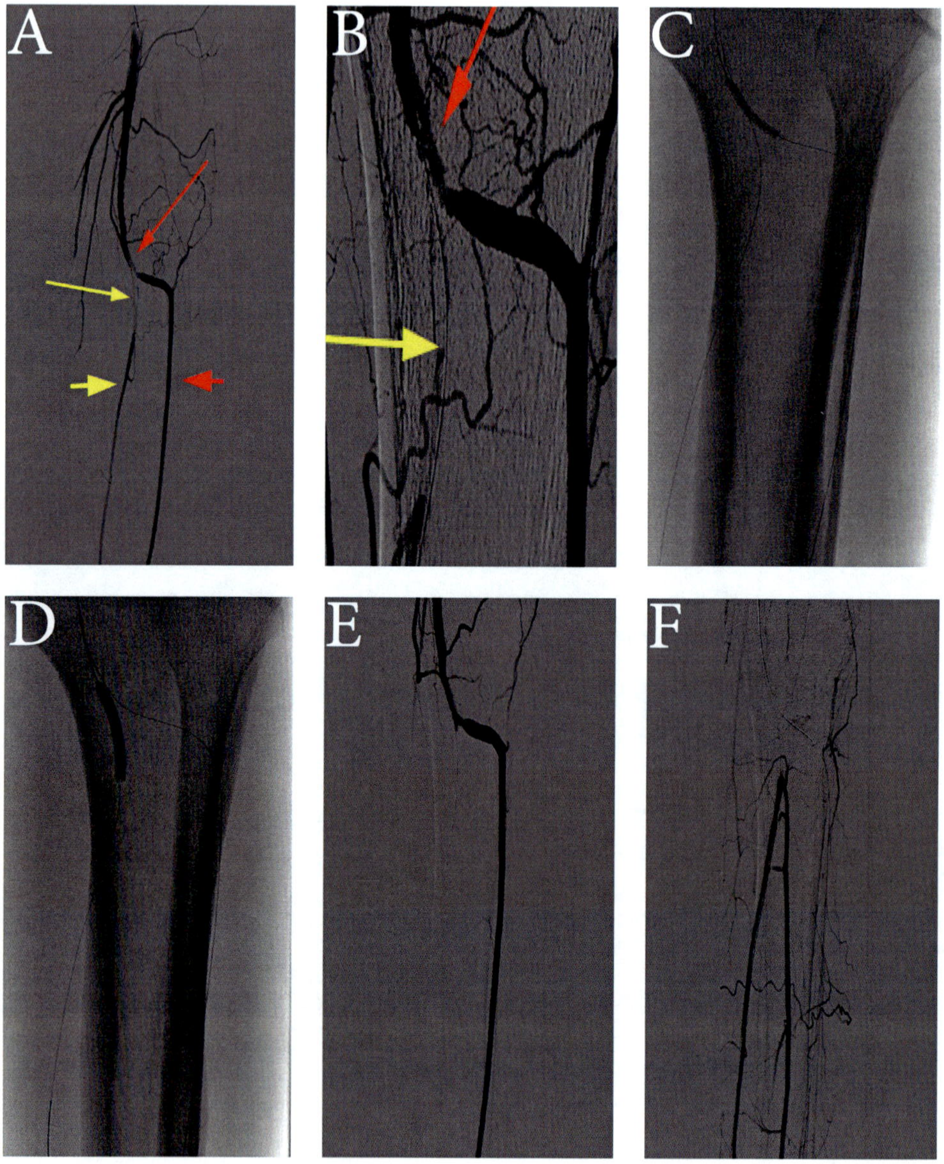

Figure 5.

A. Selective infra-popliteal angiogram of the left leg shows a severe stenosis involving the origin of the AT (long red arrow) with a subtotal occlusion of the TP trunk (long yellow arrow). The PT (short yellow arrow) fills distally and so does the AT (short red arrow). The peroneal artery does not fill.
B. A zoomed in view of the above angiogram suggesting a thrombotic lesion in the distal popliteal/AT ostium with a subtotal occlusion of the TP trunk.
C. A wire is seen in the PT and one in the AT. The ostium of the AT is being treated with PTA.
D. The TP trunk is being treated with PTA with a wire protecting the AT.
E. Improved flow in the AT with no flow in the TP trunk.
F. Delayed flow is noted in the PT and peroneal arteries via collaterals.

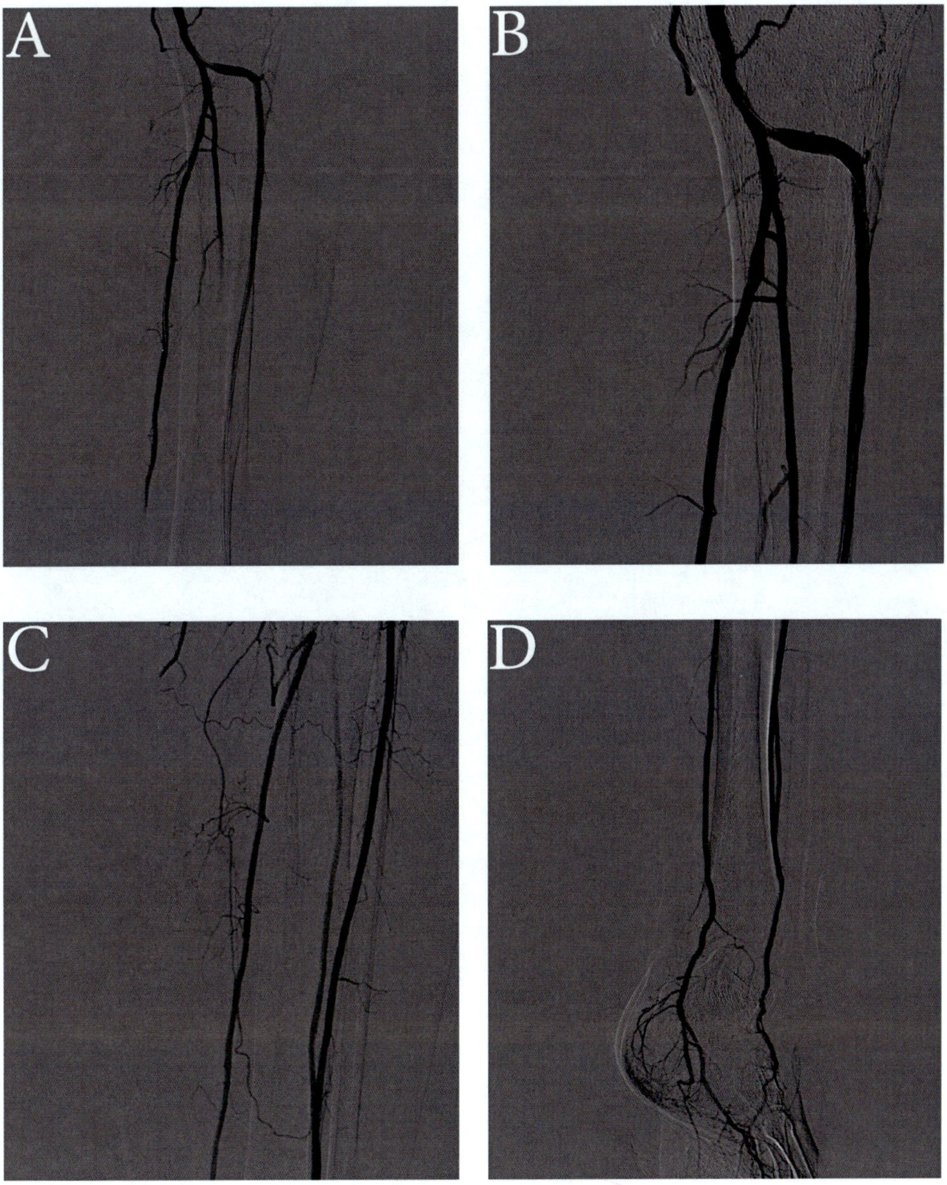

Figure 6.

A. Angiography after repeat angioplasty of the AT and TP trunk.
B. Angiography post-IA NTG administration. A 'stent-like' result is noted at the site of the original lesions.
C. Angiography post-IA NTG administration with improved flow in all three runoff arteries.
D. Excellent flow is noted in the runoff arteries all the way into the foot.

Aspiration Thrombectomy & Balloon Angioplasty

- With both wires in place, sequential aspiration thrombectomy was performed on both diseased segments.
- PTA of the diseased segment of the tibio-peroneal trunk with a 2.5 mm x 20 mm balloon (figure 5D).
- Each inflation was 3 minutes long.
- This was followed by angioplasty of the ostial/proximal anterior tibial artery (AT).
- Subsequently, the balloons were up-sized to 3.5 mm x 20 mm for the tibio-peroneal trunk and 3.0 mm x 20 mm for the AT.

Final Angiography

- Intra-arterial NTG was given (300 mcg).
- PT, peroneal and AT remained patent into the foot and had brisk flow (figures 6A & 6B).
- No significant residual stenosis, dissection or vessel cutoff was noted (figures 6C & 6D).
- The Shuttle was then removed over the stiff wire and a Perclose was deployed in the RIGHT CFA for closure.

Follow-up:

- At 30-day and one year follow-up, the patient remained asymptomatic. His claudication was completely resolved.
- He had palpable pedal pulses in the left foot.
- His resting ABI and PVRs improved considerably.
- He was maintained on ASA 325 mg/d indefinitely and Plavix 75 mg/d for 30 days.

Take Home Points:

- Solitary infra-popliteal disease may cause disabling intermittent claudication.
- Percutaneous intervention on proximal focal infra-popliteal disease is beneficial for relief of claudication symptoms.
- Infra-popliteal angioplasty has reasonable success rate particularly for focal disease.
- Stent-like result may be achieved with PTA.
- Consider performing aspiration thrombectomy first in lesions with a likely thrombo-embolic etiology even if somewhat chronic.
- PTA may be adequate for focal lesions despite their severity.
- Surgery should not be considered the first line of therapy in these patients.

CASE 2

- Mr. Y is a 72-year-old male diabetic with known PAD. He presented with severe rest pain in the left foot x 6 weeks with a non-healing ulcer on the tip of the first toe. He denied intermittent claudication in either lower extremity, even prior to this presentation.
- Physical exam was remarkable for non-palpable pedal pulses in both feet. On Doppler a very faint monophasic signal was noted in both the left DP and PT. The right pedal signals were also monophasic but stronger.
- Noninvasive arterial testing revealed a resting ABI of 0.48 in the right leg and 0.35 in the left leg. PVR suggested severe bilateral infra-popliteal disease, left worse than right with a non-pulsatile tracing on the left big toe.
- A diagnosis of CLI was made and the patient underwent angiography.

Access & Diagnostic Angiography:

- For initial angiography, we used a 5 Fr./23 cm Cordis Brite-tip sheath.
- 2000 units IV heparin was given.
- Abdominal aortography was performed.
- This was followed by selective left iliac and infra-inguinal angiography with runoff.

Findings:

- No obstructive disease was noted in the aorto-iliac arteries.
- The femoro-popliteal arteries were noted to have mild to moderate diffuse disease. However, there was no gradient between the left CFA and the BTK popliteal artery.
- The AT was noted to be occluded shortly after its origin, and remained occluded throughout. A severely diseased DP with distal occlusion was noted.
- The TP trunk was patent with mild disease and the peroneal artery was noted to have diffuse severe disease. It provided perfusion to the proximal DP.
- The PT was occluded proximally as well and remained occluded throughout.

Interventional Technique:

Antegrade access in left CFA

- Due to severe aorto-iliac tortuosity it was decided to obtain antegrade access in the left CFA.
- This was done under fluoroscopic guidance, with the Glidecath still in the left CFA to serve as a target marker.
- A 4 Fr. micropuncture kit was used in the standard fashion.
- An ASGW was then advanced through the microcatheter into the left popliteal artery just below the knee joint.
- A 5 Fr./45 cm Cordis Brite tip sheath was inserted over this wire into the left CFA.
- Additional 4000 units of IV UFH was given to achieve a therapeutic ACT of >250 sec.

Treatment of the PT CTO

- A road-map was obtained.
- A 0.014" Choice PT Floppy wire was used to cross the PT without difficulty. This wire was advanced into the foot without difficulty.
- This wire entered the lateral plantar artery.
- PTA of this vessel was performed along its entire length using a 3.0 mm x 150 mm balloon.

Treatment of the AT CTO

- The Choice PT was then directed to the AT.
- This CTO was also crossed without much difficulty.
- PTA was then performed using the same balloon as in the PT.
- Repeat angiography revealed satisfactory flow in the AT and PT.

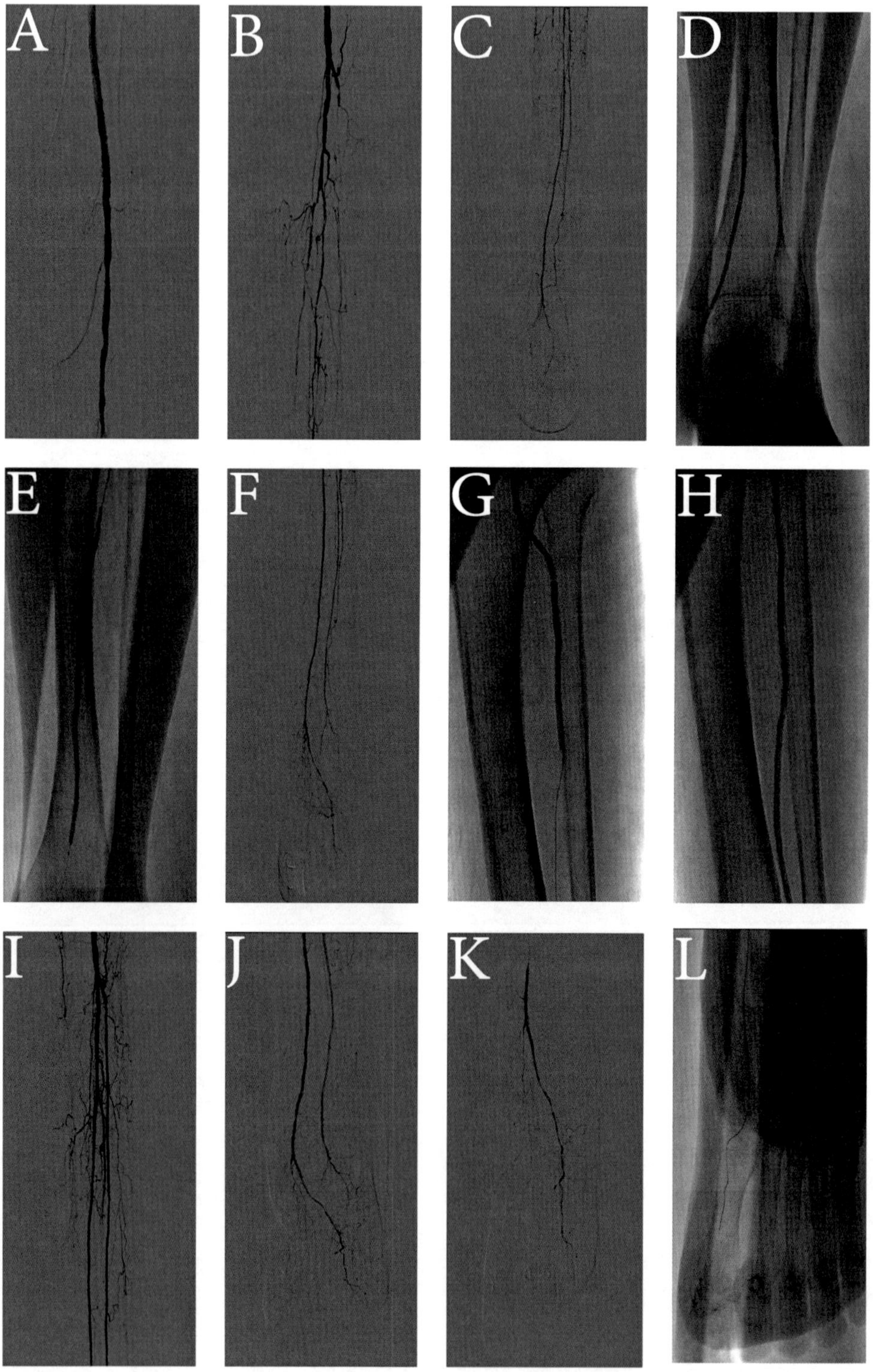
A
B
C
D
E
F
G
H
I
J
K
L

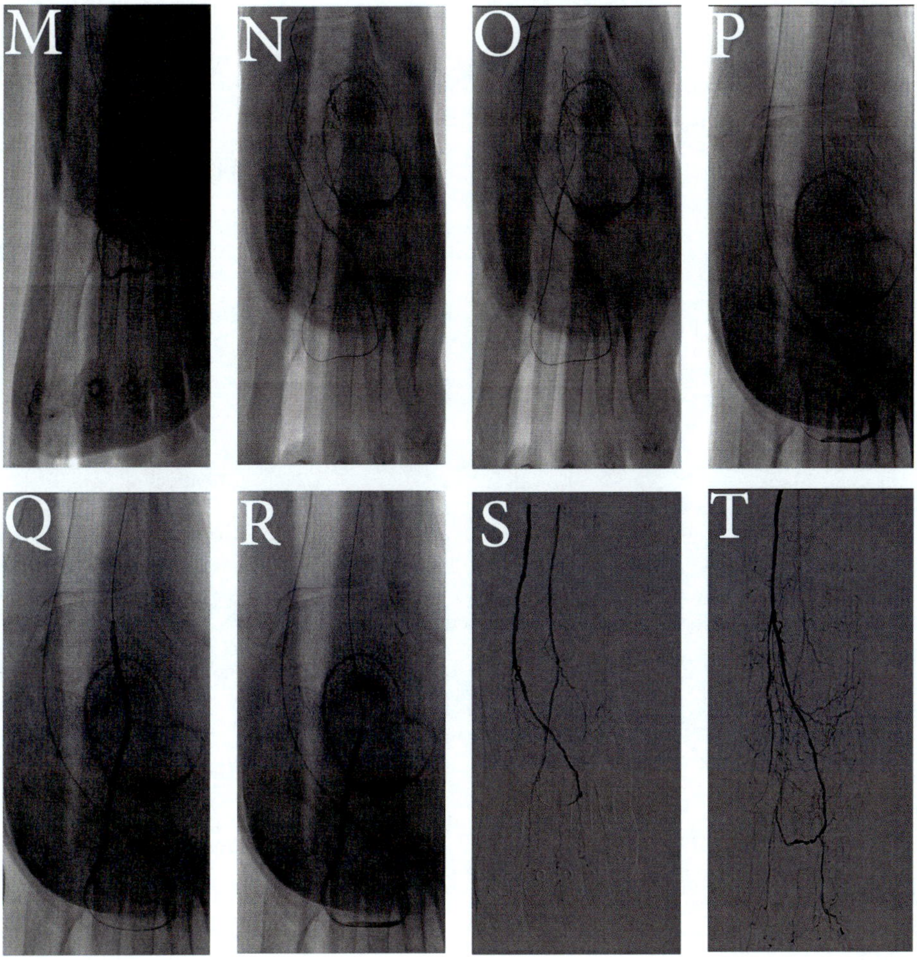

Figure 7.

A. SFA & popliteal artery.
B. BTK popliteal and runoff.
C. Distal runoff vessels.
D. PTA of distal PT.
E. PTA of mid PT.
F. Angio after PTA of PT.
G. PTA of AT.
H. PTA of mid/distal AT.
I. Angio after PTA of AT.
J. Angiogram of the foot after PTA of AT and PT. The DP is still occluded distally.
K. Selective angiogram of the PT showing the diseased lateral plantar artery without any flow in the plantar arch.
L. PTA of the medial plantar artery with a 1.5 mm balloon.
M. PTA of the lateral plantar artery and the plantar arch.
N. PTA of the deep plantar artery.
O. PTA of the DP with the 1.5 mm balloon.
P. PTA of the lateral plantar and the arch with a 2.0 mm balloon.
Q. PTA of the DP with the 2.0 mm balloon.
R. PTA of the arch and distal DP with the 2.0 mm balloon.
S. Angiogram with improved flow in the lateral plantar and the distal DP.
T. Late phase of the same angiogram show flow in the arch and the distal DP as well as the digital arteries.

- The DP remained occluded distally.
- The lateral plantar artery was severely diseased and the plantar arch did not fill.
- The medial plantar artery also was severely diseased.

Pedal loop intervention

- The Choice PT could not be advanced in an antegrade fashion across the DP occlusion.
- The wire was the removed from the AT and redirected to the PT.
- In the foot it was first directed into the medial plantar artery which was treated with PTA using a 1.5 mm x 15 mm coronary balloon.
- The wire was then directed into the lateral plantar artery and the plantar arch.
- PTA of this segment was performed using the same balloon.
- The wire was then maneuvered acrossed the deep plantar artery and into the distal DP in a retrograde fashion via the PT.
- The DP CTO was crossed successfully.
- This was then treated with the 1.5 mm balloon. Repeat angiography showed this balloon to be undersized and severe residual disease.
- The balloon was then switched out for a 2.0 mm x 80 mm Savvy balloon.

Final angiography

- Intra-arterial NTG was given (300 mcg).
- PT and AT remained patent into the foot and had brisk flow.
- No significant residual stenosis, dissection or vessel cutoff was noted.
- The medial and lateral plantar arteries as well as the plantar arch were noted to have satisfactory flow.
- The DP was also noted to be patent with reasonable flow in it.

Follow-up:

- At 30-day follow-up the patient remained asymptomatic. His ulcer completely resolved.
- He had 1+ palpable pedal pulses in the left foot.
- His resting ABI and PVRs improved considerably.
- He was maintained on ASA 325 mg/d indefinitely and Plavix 75 mg/d for 30 days.

Take Home Points:

- Isolated, critical infra-popliteal disease may cause disabling intermittent claudication.
- Percutaneous intervention should be first line therpay in most patients with CLI.
- PTA alone may provide satisfactory results.
- More than a single vessel runoff should be obtained, particularly the source artery of the affected angiosome should be targeted.
- Surgery should not be considered the first line of therapy in these patients.

8

RENAL ARTERY STENTING

Renal artery intervention is usually an elective procedure and is most commonly performed for obstructive disease. Rarely, intervention may be necessary for renal artery aneurysms and/or embolization. However, the most common indications are:

NON-CARDIAC RELATED

- Onset of HTN before age 30 years or after age 55 years.
- *Accelerated hypertension* - progressive worsening of previously well-controlled BP.
- *Refractory hypertension* - poorly controlled BP despite optimal doses of 3 anti-hypertensive agents including at least one diuretic.
- *Malignant hypertension* - poorly controlled BP with acute end-organ damage (e.g. acute renal failure, acute CHF).
- Worsening renal function upon initiation of ACEI or ARBs.
- Size discrepancy (height) between the two kidneys of > 1.5 cm or an atrophic kidney with no reasonable alternative explanation.
- Unexplained renal dysfunction even in those patients being initiated on renal replacement therapy.

CARDIAC RELATED

- Otherwise unexplained CHF.
- Refractory angina.
- Multi-vessel CAD.

ANATOMY

It is relevant to briefly discuss the anatomy of the renal arteries.

RENAL ARTERY

- Approximately 70% of the time there is a single main renal artery on each side (figure 1).
- The main renal arteries typically have a luminal diameter of 5-7 mm.
- The renal arteries arise off the aorta most commonly at L1 - L2 level. The right is slightly higher than the left.
- The right renal artery is also somewhat longer than the left, and usually runs behind the IVC.

VARIANTS

Number:

- The commonest variation, approximately 30% of the time, is the presence of an 'additional' artery which usually arises from the aorta.
- These 'additional' arteries are often called 'accessory' arteries, a misnomer as they are not extra and are in fact end-arteries that typically supply the upper or lower pole of the kidney.
- These 'additional' arteries usually arise proximal rather than distal to the main renal artery origin off the aorta.
- Each kidney may receive 1-6 arteries.
- There are no gender differences in the multiplicity of renal arteries.

Origin Location:

- The origin of the right and left renal arteries may be from a common trunk that arises anteriorly off the aorta.
- The origin can be as low as the aortic bifurcation or even from a pelvic vessel such as the CIA, EIA, IIA, or middle sacral artery.

Branches:

- The renal arteries, main and 'additional', sometimes give branches that usually arise from other arteries such as the hepatic, supra-renal, pancreatic, colic, gonadal and lumbar arteries.

DIAGNOSIS

A renal artery stenosis (RAS) that causes 60% or more narrowing of the artery is considered significant. This diagnosis requires an imaging modality that allows assessment of:

- The burden of renal artery disease.
- Location and severity of stenoses.
- Presence, location and patency of accessory renal arteries.
- Anatomy of peri-renal aorta (aneurysmal vs. obstructive etc).
- Size of the kidney.

While conventional angiography remains the 'gold standard' for the diagnosis of RAS, non-invasive methods such as duplex ultrasound (US), computed tomography angiography (CTA) or magnetic resonance angiography (MRA) are preferred first-line modalities for obvious reasons.

Duplex US is a cheap, non-invasive and risk-free test. However, like other US procedures, it is largely dependent on the technical skills of the operator and the patient's body habitus. In a non-obese patient, this should be the first test ordered to rule out renovascular disease (RVD), particularly in an institution where this test is easily available and sonographers are well experienced. An adequate US study can give useful information regarding the presence or absence of a significant stenosis and can also predict improvement in BP and renal function after an intervention (using renal resistive index or RRI assessment). It is also an ideal test for follow-up of patients who have undergone prior renal artery revascularization.

CTA and MRA are more expensive imaging modalities as compared to US but provide more valuable information regarding the origin, orientation and configuration of the renal arteries, presence of accessory renal arteries, assessment of the peri-renal aorta as well as the status of ileo-femoral arteries. This anatomical information is very useful in terms of planning invasive workup and therapy. For both these tests, however, the presence of pre-existing renal dysfunction is an issue. CTA requires the administration of 100-150 cc of contrast, which may cause further compromise in renal function in those with chronic kidney disease (CKD). MRA with gadolinium in these patients is limited by the risk of nephrogenic systemic fibrosis. MRA is also more likely to cause false positive findings in pa-

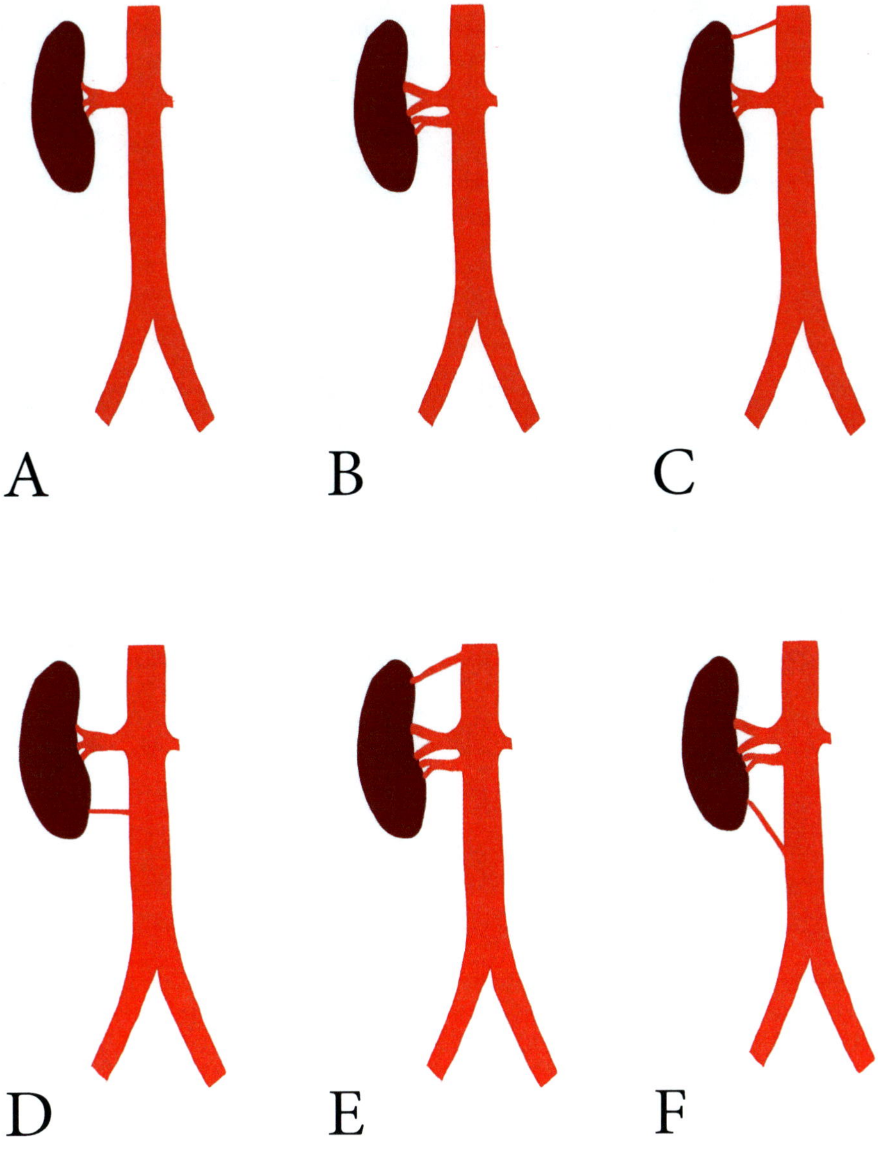

Figure 1.
Variations in renal artery origin.

A. Single renal artery.
B. Duplicated renal arteries.
C. Single main renal artery with an upper pole artery.
D. Single main renal artery with a lower pole artery.
E. Duplicated main renal arteries with an upper pole artery.
F. Duplicated main renal arteries with a lower pole artery.

tients with atherosclerotic disease and false negative in patients with fibromuscular dysplasia (FMD). Additionally, due to metal artifact from a previous stent, an MRA is not able to discern the presence of restenosis. CTA and US are better suited for detection of in-stent restenosis.

If the US study is suggestive of significant RAS, further investigation should follow with a conventional angiogram. If a satisfactory US study cannot be obtained and the patient has normal renal function, then a CTA should be pursued as a first line test. In patients in whom neither an US nor a CTA is feasible, conventional angiography is the next best step in the diagnosis of RAS.

PRE-OPERATIVE CONSIDERATIONS

We recommend discontinuing the use of diuretics the night before. The patients should also be adequately hydrated prior to and during the procedure to keep them euvolemic. Not only does this facilitate the management of post-operative hypotension but also reduces the likelihood of renal dysfunction from contrast-induced nephropathy (CIN).

ACCESS

SITE

Femoral:

For renal angiography, common femoral artery (CFA) access is primarily used (>90%). When using femoral access some operators choose to go to the contralateral CFA in order to get better 'support' from the aortic bifurcation/wall during engagement of the renal artery as well as for intervention. Although a reasonable approach for some cases, it is not applicable to majority of the cases. Therefore, we routinely start with right CFA access regardless of the laterality of the renal artery to be intervened upon.

Brachial/Radial:

Occasionally, femoral access may not be feasible and brachial/radial artery access may be necessary. Reasons for brachial access include:

- Inferior takeoff (caudal angulation) of the renal arteries.
- Occluded, severely tortuous or 'hostile' infra-renal aorta (with severe atheromatous or thrombotic burden), iliac or common femoral arteries.

Sheath Type & Size:

- For intervention, a 6 Fr. sheath is required. However, if the likelihood of performing an intervention is low (due to severe renal dysfunction, known aorto-iliac tortuosity), consider starting with a 4 Fr. or 5 Fr. sheath.
- Choosing a sheath that is at least 23-24 cm in length reduces the impact of aorto-iliac tortuosity on the ability of the diagnostic/guiding catheter to engage the renal arteries.
- A guiding sheath (pre-shaped sheath) may be used. An example is the Pinnacle Destination sheaths from Terumo.
- These come in shapes such as JR, HS, RDC etc.

DIAGNOSTIC ANGIOGRAPHY

AORTOGRAPHY

Catheter:

- Any flush catheter can be used for this purpose such as a pigtail, tennis racquet, Omni flush, etc.

- Use the tennis racquet catheter as it leads to less cephalad dye reflux and hence, a lesser chance of SMA opacification.
- The catheter should be positioned at the level of L1-L2 or just across the origin of the highest renal artery.

Imaging:

- For aortography, use an AP- LAO 45° projection (figure 2).
- Digital subtraction angiography (DSA) should be used as this may reduce the volume of dye administration without compromising image quality.
- Remember to use 'breath hold' with DSA.

Dye Injection:

- A power injector is used for aortography.
- The rate of power injection depends upon the patient's body habitus, renal function and ability to comply with instructions of breath-hold during aortography.
- Injection settings are 10 cc/sec for a total volume of 30-40 cc at 500-750 psi with no rate rise.
- In patients with normal renal function, use 70% dye with 30% heparinized normal saline dilution (70:30). If the creatinine clearance is less than 30 mL/min, then use 50% dilution (50:50). CO_2 may be used for this purpose.

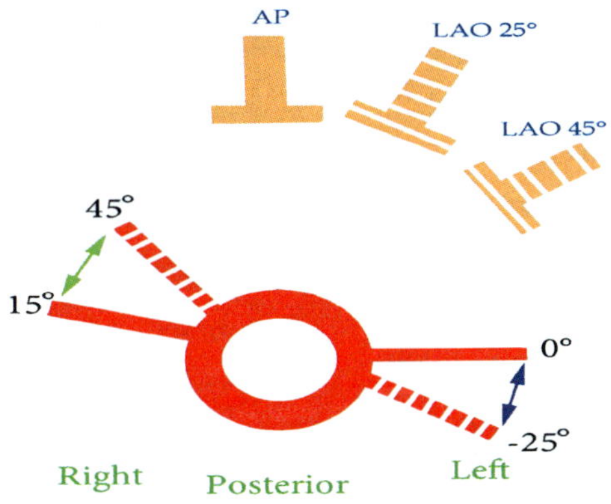

Figure 2.
Orientation of renal artery origin.
This schematic shows a cross-sectional view of the aorta (red circle) with the origin of the main renal arteries (red lines). The green arrows reflect the suggested gantry angle for optimal imaging of the left renal artery and black arrows those of the right.

- A test injection should be performed to make sure that the catheter is just below the level of the SMA. Otherwise, contrast opacification of the SMA and its branches may obscure the origins of the renal arteries and confound interpretation of the true burden of disease.

SELECTIVE ANGIOGRAPHY

Catheter:

- For selective angiography, a 4 Fr. or 5 Fr. catheter is used. Preference is for the latter, as with 4 Fr. catheters a streaming artifact is common and an adequate flow rate is difficult to achieve without causing the catheter to 'blow out' of the renal artery ostium.
- From a femoral approach, use a 5 Fr./65 cm IM catheter. From a brachial/radial artery approach, a 5 Fr. multipurpose (MPA) catheter is an ideal selection. Of note, for radial approach a 125 cm length is required.
- Ideally, a 'no-touch' technique should be used for engaging the ostium of the renal artery. In this technique, one wire (0.018") is used to keep the diagnostic catheter off the aortic wall while the second wire (0.014") is used to engage the renal artery once the catheter is in the appropriate position.

Dye Injection:

- DSA is used with manual injections of 3-5 mL through the selective catheter.

HEMODYNAMICS

It is not necessary to measure a gradient across a renal artery stenosis if it appears severe (>60% narrowed) angiographically.
However, we recommend assessing the hemodynamic significance of a proximal stenosis if its severity is unclear angiographically.

Historically, this has been done by advancing an end-hole catheter (e.g. a 4 Fr. Glidecath) across the stenosis to measure the pressure gradient by 'pullback' or by simultaneous recording.

- A peak-to-peak systolic gradient of > 20 mm Hg had been defined as the cutoff for significance.
- Alternatively, a mean gradient of 5-10 mm Hg has been used.

However, there is no proven physiologic correlation between these values and significant renal hypoperfusion. Furthermore, due to the small size of the renal arteries, such an exercise may not be very informative and may actually be harmful.

For the above reasons, we do not utilize this technique in the evaluation of renal artery stenoses. Instead, we recommend the use of a 0.014" pressure-wire to measure the translesional gradient.

- Unlike the coronary circulation, no vasoactive stimulus is necessary.
- A resting pressure gradient is obtained by measuring the pressure distal to the lesion (Pd) and in the aorta (Pa).
- The ratio Pd/Pa of less than 0.90 has been shown to be hemodynamically significant and if feasible should be treated with percutaneous transluminal renal angioplasty (PTRA) and stenting.
- A Pd/Pa greater than or equal to 0.90 is considered insignificant.

INTERVENTION

ACCESS

Once diagnostic angiography confirms the presence of severe renal artery stenosis, intervention is planned. Femoral artery access is used for diagnostic angiography.

ANTICOAGULATION

- IV UFH is preferred.
- Bivalirudin or GP IIb/IIIa inhibitors should be avoided because of the inability to reverse the effect.
- Goal ACT > 250 sec.

GUIDING CATHETER /SHEATH

Femoral Approach:

- From the femoral artery approach, a 6 Fr./55 cm guiding catheter is used for most renal artery interventions.
- The exception is when a (PTFE) covered balloon expandable stent (e.g. iCast) is being considered (e.g. in cases of in-stent restenosis, renal artery aneurysm exclusion, etc). For these cases, a 6 Fr. pre-shaped 'guiding sheath' is used. The bulky profile of these stents requires delivery through at least a 6 Fr. guiding sheath or an 8 Fr. guiding catheter. Terumo makes

'guiding sheaths' in 5 Fr.- 8 Fr. sizes with renal configurations (IM, hockey-stick, RDC and multi-purpose) in 45 cm, 65 cm and 90 cm lengths.
- An IM configuration is the most commonly used guide. Other commonly used catheters are hockey-stick, JR4 and renal double-curve. (RDC).

Brachial Approach:

- From the brachial artery approach, a 6 Fr./90 cm or longer guiding catheter is used for most renal artery interventions.
- The configuration used most often is a 'multipurpose' (MPA) guiding catheter.
- If a guiding sheath has to be used, the same configuration is available in a 90 cm length.

Radial Approach:

- Radial artery access requires longer catheters (125 cm). Thus a 6 Fr./125 cm long MPA guiding catheter is used for most renal artery interventions.
- There are no guiding sheaths available in this length at this time for renal interventions using a radial artery access. Brachial artery access should be used instead in these cases.

WIRES

- Guidewires used for renal artery interventions are mostly 0.014" system and 180-190 cm long.
- Medium stiffness, non-hydrophilic tip wires are preferred. Hydrophilic wires should be avoided as then can lead to distal perforation. However, hydrophilic wires may be needed to cross a critical ostial stenosis. Once the lesion is crossed, exchange the wire for a non-hydrophilic tip.
- When feasible, attempt to use a distal embolic protection device such as the Spider (eV3), Angioguard etc. However, it is mandatory to have enough clearance for the landing zone of these devices to avoid distal trauma as well as the system getting 'trapped' behind a stent.

BALLOONS

- Luminal diameter of normal renal arteries is 5 - 7 mm. Balloons used for renal artery intervention are 0.014" wire compatible.
- For a femoral access case, monorail balloons in 70 - 80 cm length.

Predilatation Balloons:

- Predilatation balloon diameter is typically 3.0 mm - 5.0 mm with a length of 8-15 mm.
- Conservative predilatation of the lesion for sizing purposes.

Postdilatation Balloons:

- Postdilatation balloons are sized 1:1 to the luminal diameter of the normal vessel. Typical diameters are 5.0 - 7.0 mm.

STENTS

- Balloon-expandable are used exclusively.
- 1:1 sizing to the reference vessel luminal diameter is used to avoid vessel perforation.
- Use IVUS if unsure of the vessel luminal diameter, although predilatation might help determine the appropriate size.
- For ostial disease, position the stent so that it protrudes 1-2 mm into the aorta.
- Covered (PTFE) stents are used for in-stent restenosis, exclusion of renal artery aneurysms, vascular perforation, and cases of endovascular aortic repair that require the creation of a renal 'snorkel' or 'chimney'.

TECHNIQUE

The two techniques that are utilized for renal artery stenting are 'telescoping' and 'no-touch' techniques. These are discussed separately below.

'TELESCOPING' TECHNIQUE

In this technique, a 4 Fr. diagnostic catheter is pre-loaded into a 6 Fr. guiding catheter prior to introduction into the patient's vasculature. Once introduced, the 'telescoping assembly' allows the diagnostic catheter to lead and engage the renal artery of choice, and then advance and 'guide' the guiding catheter over the diagnostic catheter into that renal artery. This technique is particularly useful in engaging the visceral arteries in patients with challenging aortic anatomy, as it provides a solid rail (i.e. the diagnostic catheter) for guiding catheter delivery. It also serves to minimize trauma to the aortic wall from the guiding catheter tip.

Vessel Cannulation:

- After obtaining an aortogram, the flush catheter is removed.
- A diagnostic catheter, e.g. a 4 Fr./100 cm IM catheter is flushed and introduced into a 6 Fr./55 cm guiding catheter through a Tuohy-Borst connector attached to the guiding catheter. This assembly is best created outside the body, on the back table.
- The diagnostic/guiding catheter assembly is advanced over a 0.035" wire into the abdominal aorta (figure 4A).
- The wire is removed and the diagnostic catheter is used to engage the renal artery to be intervened upon, in the usual fashion.
- A 0.014" wire is advanced through the diagnostic catheter into the renal artery to secure a 'foot hold' in the vessel (figure 4B).

Telescoping the Guiding Catheter:

- The guiding catheter is then slowly advanced over the diagnostic catheter, under fluoroscopic guidance (figure 4C).
- It is gradually advanced into the ostium.

Removal of the Diagnostic Catheter:

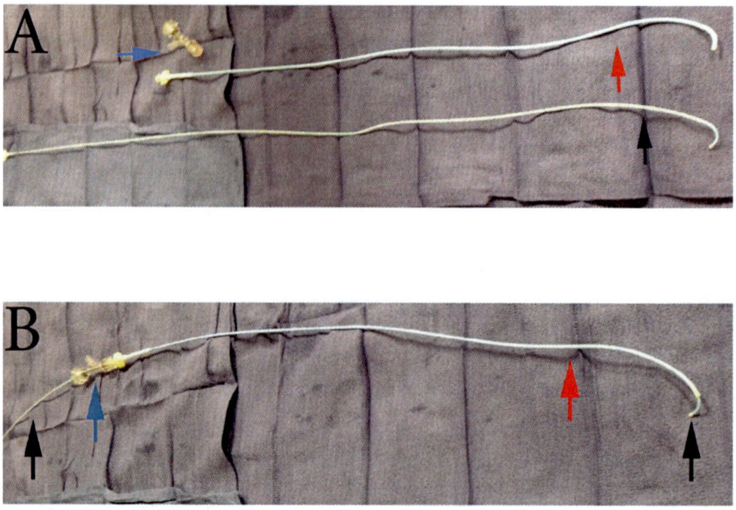

Figure 3.

Telescoping IM guiding catheter/diagnostic catheter.

A.
A Tuohy-Borst adapter (blue arrow), a 6 Fr./55 cm IM guiding catheter (red arrow) and a 4 Fr./100 cm diagnostic IM catheter (black arrows) are used to create the telescoping assembly.

B.
After connecting the Tuohy-Borst adapter to the guiding catheter, the diagnostic IM is loaded into the guide. The diagnostic catheter is advanced until its tip exits that of the guide's. After flushing, the assembly is ready to be advanced over a wire into the distal abdominal aorta.

- The diagnostic catheter is then slowly withdrawn, taking care not to dislodge the guiding catheter in this process (figures 4D & 4E).
- The intervention on the renal artery then proceeds in the following usual fashion.

Embolic Protection:

- In some cases, a 0.014" distal embolic protection device (i.e. a filter) may be appropriate (baseline renal dysfunction, hostile aorta, feasible landing zone).
- A filter (e.g. Spider®) of appropriate size can then be advanced over the existing wire. This particular filter has the advantage of using a bare wire of choice for crossing the lesion and then removing it after the delivery catheter is in place, then deploying the Spider.

Predilatation:

- The lesion is predilated with a slightly undersized balloon (figure 5B).
- This step is recommended as it facilitates in appropriately sizing the stent.
- During this step, the guiding catheter may back out slightly in the aorta. As the balloon is being pulled out of the vessel the catheter should be 'shepherded' back over the balloon into the vessel.
- Care should also be taken to not allow the filter to migrate distally as it may cause spasm or dissection in the vessel.

Stenting:

- Only balloon-expandable, 0.014" system stents are used.
- These stents should be sized 1:1 to the normal reference vessel. Oversizing may lead to vessel trauma and perforation.
- The stent length is chosen keeping in consideration the fact that the stent will shorten somewhat with expansion.
- Additionally, for ostial stenoses, the stent should be positioned such that the proximal end protrudes into the aorta about 1-2 mm (figure 5C).
- The guiding catheter is backed out slightly by pulling it back, to expose the stent.
- During deployment a slight but constant forward pressure should be applied to the guide so that the stent does not migrate back too much into the aorta.
- The stent should be deployed at the nominal pressure (figure 5D).
- However, if the patient develops flank pain, further inflation should be curtailed beyond that point. After prompt deflation, a quick angiogram should be obtained to rule out vessel dissection/perforation.
- The stent balloon is pulled out halfway and the guiding catheter is 'shepherded' back over it into the stent. Caution should be used to prevent damaging the stent with the guide tip.
- Once the guide is comfortably inside the stent, backward traction should be applied as the balloon is pulled out of the vessel.
- The balloon is removed and an angiogram is obtained to determine the need to post-dilate the stent.

Postdilatation:

- Postdilatation is recommended, even if the stent appears to be perfectly expanded.
- A compliant balloon, shorter than the length of the stent is used.
- After post-dilating the distal portion of the stent, the balloon is withdrawn into the proximal half.
- This step allows an opportunity to flare the ostial/aortic segment of the stent.
- The proximal half is left 'hanging" in the aorta. The guide is pulled back just enough to expose the balloon.
- The balloon is inflated with a slight forward pressure on the guide.

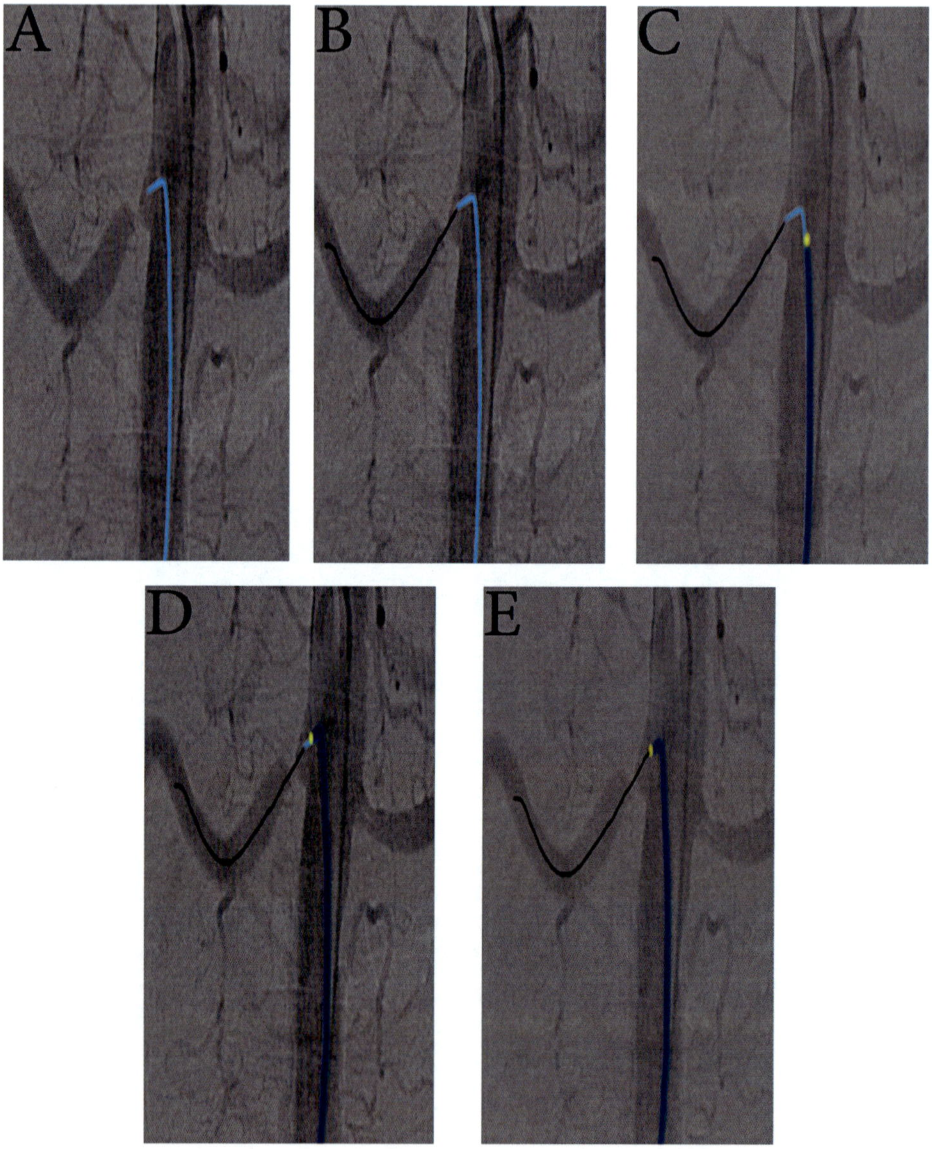

Figure 4.

'Telescoping' technique.

A. A 4 Fr. diagnostic catheter is used to engage the right renal artery.
B. A 0.014" non-hydrophilic wire is advanced into the renal artery.
C. A 6 Fr. IM renal guide is advanced over the diagnostic catheter.
D. The guide is gradually advanced into the ostium of the renal artery.
E. The diagnostic catheter is withdrawn without dislodging the guide.

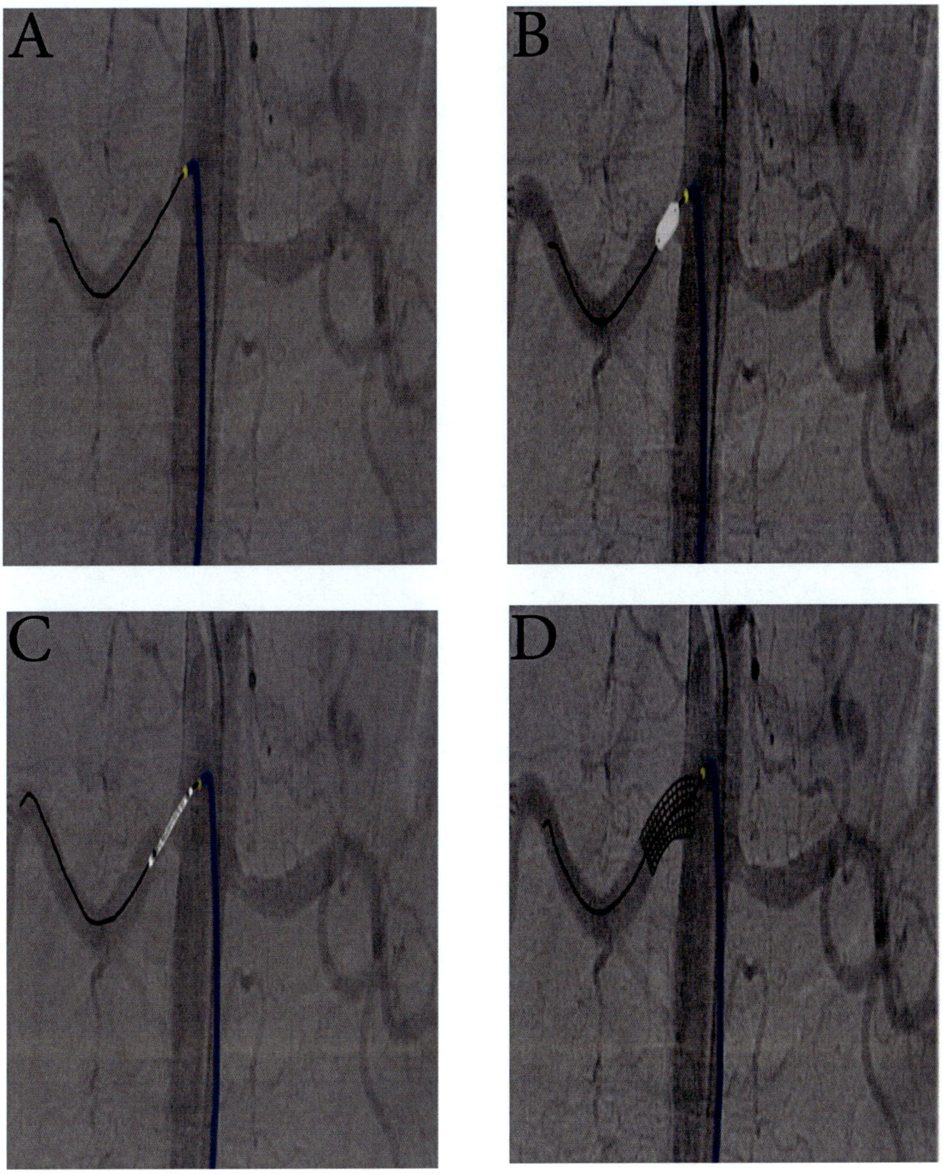

Figure 5.

Renal Intervention.

A. After placement of a 6 Fr. guiding catheter and a 0.014" wire in the right renal artery.
B. A predilatation balloon is advanced into the renal artery and the vessel is predilated.
C. After removal of the balloon, a balloon-expandable stent is loaded on the wire and positioned across the lesion.
D. The stent is deployed at nominal pressure and the balloon is removed. Angiography reveals a widely patent renal artery. Note the stent protrudes 1-2 mm into the aorta.

Final Angiogram:

- As the postdilatation balloon is being removed, the guiding catheter is slowly 'shepherded' over it into the stent, avoiding any trauma to the latter.
- Caution should be used to bring in the guide slowly so that the stent is not damaged by it.
- Repeat angiography to review the final result.
- Post-stenting images should be carefully analyzed for perforation, vessel dissection, and distal embolization.
- Pull the wire out back into the stented segment. If an embolic protection device was used instead, that is retrieved.
- Remove the guide slowly by pulling back and turning the tip in the contralateral direction, while maintaining some traction on the wire.

'NO TOUCH' TECHNIQUE

In this technique the guiding catheter is advanced into the renal artery while maintaining a wire 'rail' in the supra-renal aorta to avoid dragging the catheter tip in aortic plaque/thrombus. The following steps are used:

- The femoral sheath should be around 20-23 cm long so that its tip is in the distal aorta.
- After aortography, the flush catheter is removed over a 0.035" wire.
- A 0.018" stiff wire is advanced into the supra-renal aorta (figure 6A).
- A 6 Fr. guiding catheter connected to a Tuohy-Borst adapter is advanced over this wire into the abdominal aorta.
- Adjust the length of the protruding segment of the wire, so that the catheter tip does not come in contact with the aortic wall.
- The catheter is adequately flushed clear of debris and air in the infra-renal segment of the aorta.
- It is advanced to the level of T12 - L1 vertebrae.
- A puff is obtained to ascertain positioning as the catheter is slowly withdrawn to get to the level of the renal artery of interest (figure 6B).
- A separate 0.014" interventional wire is introduced along with the pre-existing 0.018" wire into the guide.
- The 0.014" wire is advanced to the tip of the guide.
- When the catheter is right across from the renal artery of interest, the 0.014" wire is advanced into the renal artery (figure 6C & 6D).
- The 0.018" wire is then withdrawn slowly to allow the guide to 'drop' into the renal artery (figure 6E & 6F).
- Renal intervention is performed as outlined in the previous section.

Re-engaging the guide after stenting:
If the guiding catheter drops below the stent after removal of the balloon, the following maneuver should be used to re-engage:

- Partially loosen the Tuohy-Borst adapter.
- Turn the guide slightly contralateral, so the tip is beyond the midline.
- Push up to the level of the stent opening.
- Rotate back slowly in the ipsilateral direction to engage the stent.

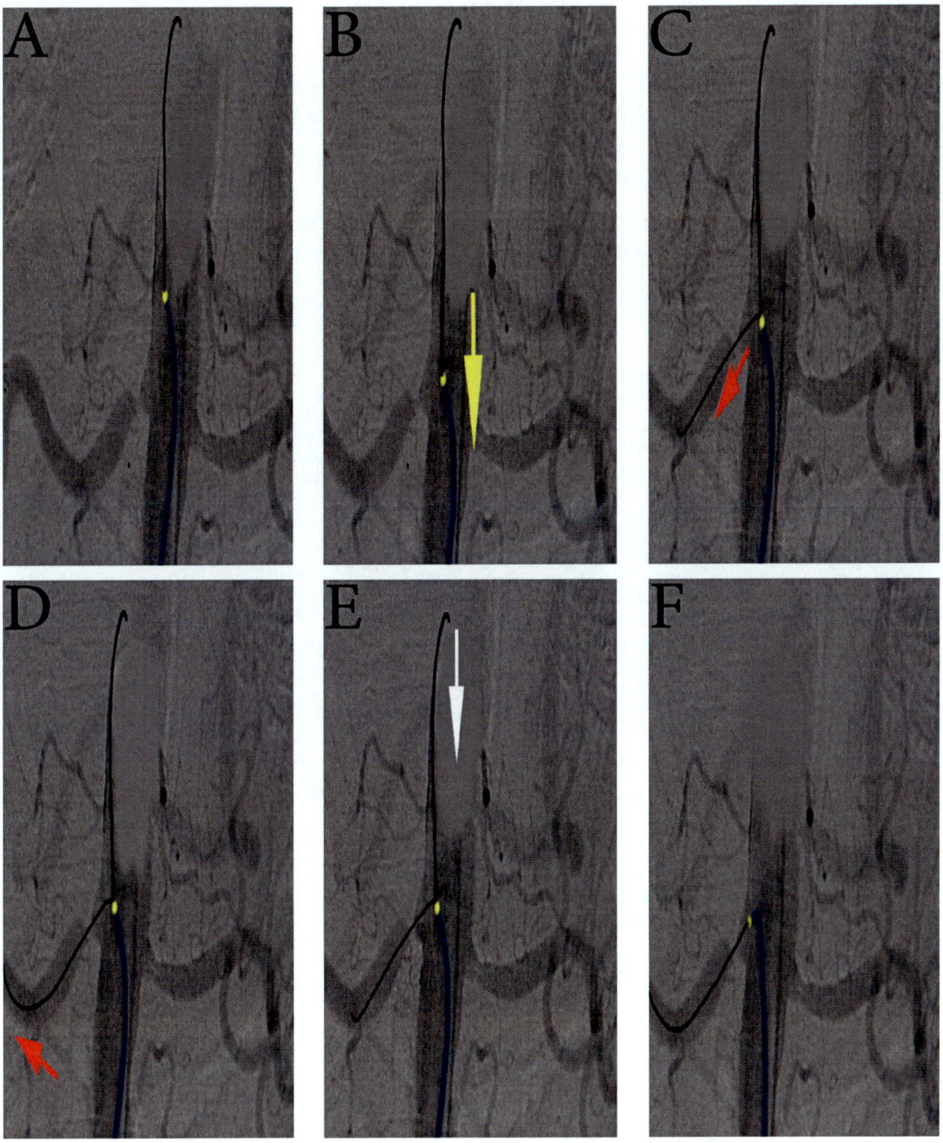

Figure 6.

'No-touch' technique.

A. A 0.018" stiff wire is advanced into the supra-renal aorta. A 6 Fr. guide is advanced over this wire to a level above the origin of the renal artery of interest.
B. The catheter is withdrawn to the level of the renal artery.
C. A 0.014" wire is advanced along side the 0.018" wire in the guide and is used to cannulate the renal artery.
D. This wire is advanced into the distal renal artery.
E. The 0.018" wire is withdrawn.
F. The guide 'drops' into the renal artery.

CASE

Mr. H is a 76 year old male with known PAD. He had severe HTN and despite being on maximal doses of four anti-hypertensive agents he continued to have poorly controlled blood pressure.

After undergoing a renal artery duplex that suggested hemodynamicallly signficant stenoses in bilateral renal artery ostia he underwent renal angiography. His renal function was preserved with a normal creatinine clearance.

Access and Diagnostic Angiography:

- For initial angiography, a 5 Fr./23 cm Cordis Britetip® sheath was used in the right CFA.
- 2000 units IV heparin were given.
- Abdominal aortography was performed with a 4 Fr. TR catheter using 40 cc of Visipaque at 20 cc/sec at 750 psi (figure 7A).
- The TR catheter was then exchanged over a J-tip wire for a 5 Fr. IM catheter.
- Selective angiography of bilateral renal arteries was performed.

Interventional Technique:

- The patient was given an additional 3000 units of IV heparin.
- Over a 0.035" J-wire, the 5 Fr. sheath was exchanged out for a short 6 Fr. sheath.
- A 6 Fr./55 cm IM guiding catheter was used.
- This guide was connected to a Tuohy-Borst adapter.
- Through this adapter, a 4 Fr./100 cm IM diagnostic catheter was introduced over the 0.035" J-tip wire.
- The tip of the diagnostic catheter was pushed beyond the tip of the guiding catheter.
- This assembly of 'telescoping' diagnostic/guiding catheter was then introduced via the 6 Fr. sheath into the abdominal aorta.
- The wire was advanced above the diaphragm.
- The diagnostic catheter was advanced over the wire to a level above the left renal artery origin.
- The wire was removed and the diagnostic catheter was used to engage the ostium of the left renal artery.
- Once the artery was engaged, a 0.014" ATW wire was used to cross the lesion and advanced in the distal renal artery.
- Over this wire/diagnostic catheter 'rail,' the guiding catheter was advanced into the left renal artery.
- Once the guide was engaged in the ositum of the renal artery, the diagnostic catheter was removed, maintaining the wire position (figure 7B).
- The lesion was then predilated with a 5.0 mm x 20 mm balloon, one third of it hanging in the aorta, at 6 ATM.
- The balloon was removed, the guide was re-engaged.
- A 6.0 mm x 18 mm stent was the positioned in the lesion, with 2 mm of the stent in the aorta, proximal to the ostium of the left renal artery.
- The stent was deployed at nominal pressure.
- The balloon was partially withdrawn, the guiding catheter pulled proximal to the balloon while maintaining wire position, and the ositum was 'flared'.
- The guide was then shepherded back over the balloon into the renal artery ostium.
- The balloon was then removed.
- An angiogram was performed to confirm a satisfactory result (figure 7C).The wire and guide were then withdrawn.
- The ostium of the right renal artery was then engaged and the same steps repeated (figure 7D).

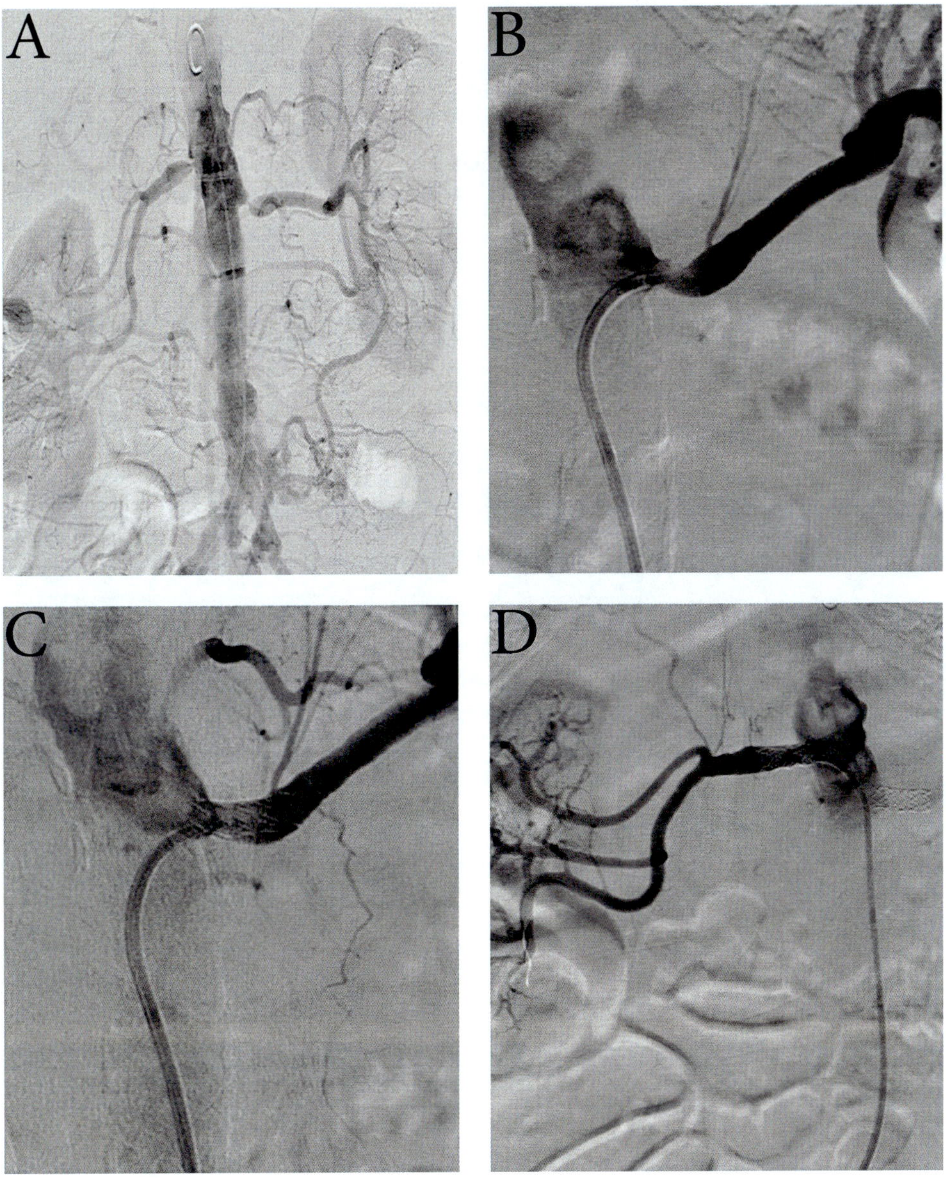

Figure 7.

A. Pre-treatment abdominal aortogram showing bilateral renal artery stenosis.
B. Selective angiogram of the left renal artery pre-intervention.
C. Left renal artery post-stenting.
D. Right renal artery post-stenting.

- The right renal artery was also stented in a similar fashion.

Final Angiography:

- Final angiography revealed widely patent bilateral renal arteries.

Follow-up:

- At 30-day follow-up, the patient's blood pressure was noted to be normal on his pre-intervention drug regimen.
- A follow-up renal artery duplex maintained normal flow velocities in bilateral renal arteries with widely patent stents by sonographic criteria.

9

MESENTERIC ARTERY STENTING

The three 'mesenteric' arteries that perfuse the gastrointestinal tract are:

- The celiac trunk/axis
- The superior mesenteric artery
- The inferior mesenteric artery

Indications for angiography and intervention on these vessels generally include:

- Mesenteric ischemia.
- Coronary ischemia in patients with a gastro-epiploic artery coronary bypass.
- Pre-operative evaluation in patients with pancreatic and hepatic malignancy.
- Acute/sub-acute gastrointestinal bleeding.

Acute mesenteric ischemia rarely requires percutaneous intervention. This life-threatening emergency generally requires emergent laparotomy with resection of any infarcted gut and is usually a result of either acute embolic phenomenon or dissection involving the mesenteric arteries. A select group of these patients with proximal thrombosis or dissection of mesenteric arteries may benefit from angiography followed by PTA/stenting or thrombolytic therapy administration.

Rarely, acute mesenteric ischemia may occur due to diminished visceral perfusion as a result of sustained systemic hypotension (shock). This condition, often referred to as non-occlusive mesenteric ischemia (NOMI), is characterized by hypoperfusion and severe spasm of the splanchnic arteries. Patients suffering from NOMI may benefit from early angiography as it allows for administration of papaverine (60 mg bolus and selective transcatheter infusion at 30-60 mg/hour) or other vasodilators to relieve vasospasm.

Chronic mesenteric ischemia is rare. In general, for severe chronic mesenteric ischemia to occur and cause significant symptoms, two out of the three mesenteric arteries have to be severely diseased or occluded. Presentation includes abdominal pain and ongoing weight loss. However, the classic triad of 'post-prandial angina', food aversion and weight loss is seldom present.

DIAGNOSIS

Initial work-up of patients with chronic mesenteric ischemia is often focused on alternate diagnoses and results in a delay in suspecting this condition. The diagnosis should be considered in all patients with chronic abdominal pain, progressive weight loss and known PAD.

Abdominal US with duplex scanning of the visceral arteries is a cheap, non-invasive and risk-free test. In experienced hands, this test can confirm the presence of significant disease with relative ease. However, the yield of this test can be affected by subjective factors such as the skill and experience of the technician, patient body habitus, presence of bowel gas and respiratory motion.

In patients with suspicion of mesenteric ischemia and evidence of significant disease on duplex scanning, conventional angiography should be considered. If, however, the US is non-diagnostic and the index of suspicion remains high, a CTA or MRA should be considered.

Conventional angiography remains the gold-standard and should be considered in all patients who are candidates for percutaneous or surgical revascularization.

ANATOMY

CELIAC TRUNK

- The celiac trunk is the first visceral branch of the abdominal aorta.
- It arises at the level of T12-L1 vertebrae from the anterior aspect of the aorta.
- It is usually 6 - 9 mm in diameter and 1 - 2 cm long.
- In a vast majority of cases, it gives rise to three major branches: first the left gastric and then bifurcating into the common hepatic and splenic arteries. As their names indicate, branches of the celiac trunk supply the stomach, liver and spleen. They also supply the pancreas and duodenum.
- The three major branches of this vessel include:
  1. Left gastric artery
  2. Common hepatic artery
     - ♦ Gastro-duodenal artery
       - ▫ Anterior and posterior pancreatico-duodenal arcades
       - ▫ Right gastro-epiploic artery (main artery to the stomach and sometimes used as a conduit for coronary bypass usually to the RCA or distal circumflex artery).
     - ♦ Right gastric artery
     - ♦ Right hepatic artery
     - ♦ Left hepatic artery
  3. Splenic artery (supplies the pancreas, stomach and the spleen)
     - ♦ Dorsal pancreatic artery
     - ♦ Transverse pancreatic artery
     - ♦ Arteria pancreatica magna
     - ♦ Left gastro-epiploic artery
     - ♦ Superior and inferior segmental splenic arteries

SUPERIOR MESENTERIC ARTERY

- The superior mesenteric artery (SMA) arises from the anterior aspect of the aorta at the level of L1. It comes off nearly 1.5 cm below the celiac trunk and just above the renal arteries.
- It too arises from the anterior surface of the aorta.

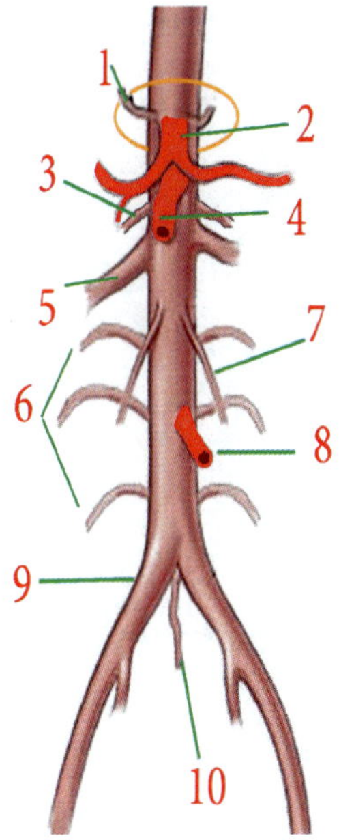

Figure 1.

Anatomy of the abdominal aorta

The orange circle represents the level of the aortic hiatus in the diaphragm.

1. Inferior phrenic artery.
2. Celiac axis.
3. Middle supra renal arteries.
4. SMA.
5. Renal artery.
6. Lumbar arteries.
7. Gonadal artery.
8. Inferior mesenteric artery.
9. Common iliac arteries.
10. Median sacral artery.

- It is 5-8 mm in diameter and supplies the small intestine (including the duodenum, jejunum, ileum, cecum), ascending colon and a large part of the transverse colon.
- Its main branches include:
    1. Jejunal and ileal arteries
    2. Ileocolic artery
    3. Right colic artery
    4. Middle colic artery

INFERIOR MESENTERIC ARTERY

- The inferior mesenteric artery (IMA) arises about 3 - 4 cm above the aortic bifurcation, at the level of L3 - L4.
- It is smaller than the SMA, measuring 3 - 7 mm in diameter.
- It supplies the transverse colon, descending colon, sigmoid, and part of the rectum.
- The IMA travels downward into the pelvis crossing the left common iliac artery to become the superior rectal artery.
- Main branches of the IMA include:
    1. Left colic artery
    2. Sigmoidal branches
    3. Superior rectal artery (terminal branch, descending into the pelvis)

Variations:

- Most variations in the origin of the mesenteric arteries pertain to the celiac trunk and SMA.
- Couinaud described 8 variants of the celiac trunk:
    1. Classic celiac trunk (65-75%) - left gastric, common hepatic and splenic
    2. Hepato-lineal trunk
    3. Hepato-gastric trunk

Figure 2.

Mesenteric collaterals.

A. Collaterals i.e. prominent pancreatico-duodenal arcades (red arrows) from the SMA (green arrowhead) to celiac artery (yellow arrowhead) in a patient with an occluded celiac axis.

B. Collaterals from the SMA (yellow arrowhead) to the IMA in a patient with occluded IMA. The red arrows show the arc of Riolan. The green arrows show the artery of Drummond.

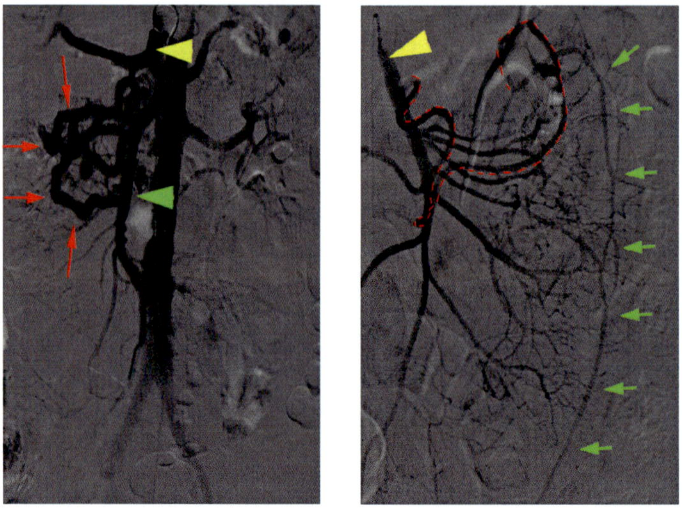

A B

4. Hepato-spleno-mesenteric trunk
5. Gastro-lineal trunk
6. Celiacomesenteric trunk
7. Celiac colic trunk
8. No celiac trunk – separate origin of all three arteries (left gastric, hepatic and splenic) from the aorta.

Collaterals:

- Multiple interconnecting branches of the celiac, SMA and IMA (figure 1A) are present normally and protect the bowel against ischemia or infarction when perfusion through one or two of the mesenteric arteries is compromised. It is important to be able to recognize these collateral pathways angiographically, as they are commonly seen in patients with chronic mesenteric ischemia.

Communications between the celiac and SMA:

- Collaterals that connect the celiac and SMA include the pancreatico-duodenal arcades (anterior and posterior).

Communications between the SMA and IMA:

- The SMA and IMA are connected through an arterial arcade called the 'marginal artery of Drummond' (named after Sir David Drummond, an English physician:1852-1932). This 'vessel' runs in the mesentery, parallel to the colon. It receives contributions from the ileocolic, right colic, middle colic, left colic and sigmoidal arteries.
- The 'arc of Riolan' (named after Jean Riolan, a French anatomist:1580-1657) or the 'meandering' (windy, tortuous) artery, as it is sometimes called, is an inconstant vessel which forms a redundant 'windy' loop to connect the middle colic branch of the SMA to the left colic branch of the IMA. Depending upon whether the SMA or the IMA is more severely diseased, blood flows through the arc towards the former or the latter.

ACCESS

SITE

Femoral:

- Right CFA is the primary access site used for mesenteric angiography.

Brachial/Radial:

- Occasionally, brachial/radial artery access may be preferred due to:
  - An inferiorly directed takeoff of mesenteric arteries.
  - Occluded, severely tortuous or 'hostile' (with severe atheromatous burden) infra-renal aorta, iliac or common femoral arteries.
  - Planned or rescue mesenteric 'snorkeling' or 'chimney' creation (stenting of aortic branch vessel during aortic endograft procedures for aneurysmal disease).

SHEATH SIZE AND TYPE

- Regardless of the site, initial arterial access is obtained with a micropuncture kit (21 guage needle and a 0.018" wire).
- Over a 0.035" wire, this is then up-sized to a 5 or 6 Fr. sheath.
- For intervention, a 6 Fr. sheath is required. However, if the likelihood of performing an intervention is low (due to severe renal dysfunction, known aorto-iliac tortuosity, etc.), consider starting with a 5 Fr. sheath.
- Choosing a sheath that is at least 23-25 cm in length reduces the impact of aorto-iliac tortuosity on the ability of the diagnostic/guiding catheter to engage the mesenteric arteries.

DIAGNOSTIC ANGIOGRAPHY

ABDOMINAL AORTOGRAPHY

Catheter:

- Any flush catheter can be used for this purpose such as the pigtail, tennis racquet, or Omni flush.
- We prefer to use a 5 Fr. TR to improve flow rates and obtain better imaging.
- The catheter should be positioned at the level of T12-L1 or just across the origin of the celiac trunk. Make sure not to place the catheter more than 0.5 cm above this position, as motion of the diaphragm during respiration may render the images uninterpretable.

Imaging:

- Since the celiac trunk and SMA arise off the aorta anteriorly, a lateral aortogram is essential to detect ostial/proximal segment disease.
- Both AP and lateral abdominal aortography should be performed.
- Digital subtraction angiography (DSA) should be used as this may reduce the volume of dye administration without compromising image quality.
- Remember to use 'breath hold' with DSA.

Dye Injection:

- A power injector is used for aortography.
- The rate of power injection depends upon the patient's body habitus, renal function and ability to comply with instructions of breath-hold during aortography.
- Injection settings are 20 cc/sec for a total volume of 40 cc at 500-750 psi with no rate rise.
- In patients with normal renal function, use 100% dye without any dilution. If the creatinine

Radial Access:
- Use standard radial technique.
- Give a 'cocktail' of IA NTG (200-300 mcg), verapamil (2-5mg) and UFH (3000-6000 U).
- Do not use a J-wire initially while traversing the radial artery as it may induce severe spasm.
- Advance your first catheter over an angled stiff hydrophilic wire.

clearance is less than 30 mL/min., then use 30% dilution (70:30). CO2 may be used for this purpose.

- A test injection should be performed to make sure that the catheter is at the level of the celiac trunk or just above it.

SELECTIVE ANGIOGRAPHY

Catheter Selection:

- For selective angiography, a 5 Fr. catheter is used.

Femoral access:

- Via CFA access, the following catheters are most commonly used:
  - ▹ JR4
  - ▹ IM
  - ▹ Omni-Sos
  - ▹ Simmons 1
  - ▹ Cobra 2 or 3
  - ▹ Shepherd's crook

Brachial/radial access:

- Via brachial/radial artery access the following catheters are used:
  - ▹ Multipurpose (MPA)
  - ▹ Amplatz left 1 or 2 (AL1 or 2)
- Of note, for radial approach, a 125 cm length catheter is required which may still not reach the IMA in a tall patient.

Dye Injection & Imaging:

- DSA is used for optimal imaging.
- Manual injections of 3 - 5 mL are performed through the selective catheter.
- Since the celiac and SMA takeoff anteriorly from the aorta, they are foreshortened in an AP projection and one can easily miss ostial/proximal lesions. So both AP and lateral projections must be obtained.
- Prolong the cine run to capture collaterals (e.g. the marginal artery of Drummond or the arc of Riolan) which could indicate the presence of significant obstructive disease.

HEMODYNAMICS

The presence of compelling symptoms and non-invasive testing suggestive of severe disease usually leads to angiography. Moderate lesions on angiography should not be dismissed without measurement of a translesional gradient using a pressure-wire in a similar fashion as in the assessment of renal artery stenoses. Although robust data do not exist on this, the same parameters used for the renovascular bed can be used here to define a 'severe' lesion using this modality. A baseline pressure index (meseneteric artery systolic pressure: aortic systolic pressure) of < 0.90 is considered significant. We believe that in order to mimic the physiology of splanchnic vasodilatation in the postprandial state, a vasodilating agent such as IA NTG (300 mcg) should be used. In our opinion, this hemodynamic assessment (baseline and provoked) is mandatory. Additionally, IVUS can be used to assess the severity of a stenosis.

INTERVENTION

ACCESS

- Once diagnostic angiography confirms the presence of severe stenosis, intervention is planned. Depending on the diagnostic findings femoral, brachial or radial artery access is used for the intervention.

ANTICOAGULATION

- IV UFH is preferred.
- Bivalirudin or GP IIb/IIIa inbibitors should be avoided because of the inability to reverse the effect.
- Goal ACT > 250 sec.

EQUIPMENT

GUIDES

Guiding catheters and guiding sheaths both can be used for mesenteric interventions (table 1). Typically, a 6 Fr. guiding catheter or guiding sheath is adequate. The former are used more commonly as they allow more freedom in terms of positioning and placement (its easier to adjust the position of a catheter inserted through a sheath than to adjust the position of the guiding sheath over a 0.014" wire during intervention). Additionally, a 6 Fr. guiding catheter has adequate internal diameter to allow delivery of most 0.014" balloon expandable stents. We use guiding sheaths only when considering the use of a bulkier stent such as a covered stent (stent-graft). These require delivery through an 8 Fr. guiding catheter or a 6 Fr. guiding sheath.

Femoral Access:

From femoral access we use the following guide shapes:

- ▹ IM
- ▹ JR4
- ▹ RDC
- ▹ Hockey-stick

Brachial/Radial Access:

For upper extremity access the following guide shapes are used:

- Multipurpose (90 cm long)
- Amplatz left 1 or 2 (AL1 or 2)

The guiding sheaths (Terumo International Systems, Inc.) are available in the following shapes:

- ▹ IM (45 cm)
- ▹ RDC (45 cm)
- ▹ Hockey-stick (45 cm)
- ▹ Multi-purpose (90 cm; as it is usually used via upper extremity access only)

WIRES

- 0.014" coronary guidewires are ideal.
- 0.018" guidewires are used for bulkier balloons or stents.
- 0.035" guidewires are used to allow better guide and sheath positioning and are exchanged out for a 0.014" system for the actual intervention.

BALLOONS

- The diameter of normal mesenteric arteries ranges between 3 - 9 mm.
- However, a 9.0 mm diameter balloon is almost never used for mesenteric intervention.
- Balloons should be appropriately sized to avoid dissection and perforation.
- Although, 0.014" wire platform balloons are preferred due to their lower profile and superior deliverability, 0.018" and 0.035" wire platform balloons may also be used, if necessary.
- For a femoral access case, monorail balloons in 70 - 80 cm shaft length are preferred.

Predilatation Balloons:

- Predilatation balloon diameter is typically 3.0 - 5.0 mm with a length of 15 - 20 mm.
- Conservative predilatation of the lesion should be performed. This allows creation of a better channel for delivering a stent and also facilitates in sizing the vessels (estimating vessel diameter compared to the balloon diameter).

Postdilatation Balloons:

- Postdilatation balloons are sized 1:1 to the luminal diameter of the normal vessel. Typical diameters are 5.0 - 7.0 mm.

TABLE 1 - EQUIPMENT FOR MESENTERIC INTERVENTION	
Wires	• 0.014" ▹ ATW® ▹ BMW® • 0.018" ▹ Steelecore®
Guiding Catheters/ Sheaths	• IM • JR4 • HS • RDC • MPA (brachial/radial access) • AL 1 or 2 (brachial/radial access)
Balloons	• 0.014" ▹ Viatrac (Abbott), etc • 0.018" ▹ Armada (Abbott) ▹ Savvy (Cordis), etc
Stents	• 0.014" ▹ Palmaz Blue (Cordis) ▹ Herculink, etc (Abbott) • 0.018" ▹ Express SD (BSC), etc
Stent-grafts	• 0.018" ▹ Viabahn (Gore) • 0.035" ▹ iCast (Atrium)

STENTS

- For the vast majority of mesenteric intervention cases, only balloon-expandable stents are used.
- 1:1 sizing to the reference vessel diameter is used to avoid vessel perforation.
- Use of IVUS is recommended if unsure of the vessel diameter, although predilatation might help determine the appropriate size.
- For ostial disease, position the stent so that it protrudes 1 - 2 mm into the aorta.

- Covered (PTFE) stents are used for in-stent restenosis, vascular perforation, and cases of endovascular aortic repair that require the creation of a fenestrated endograft or a mesenteric 'snorkel'/'chimney'/'periscope'. In the latter case, self-expanding PTFE covered stents such as the Viabahn (Gore Medical, Flagstaff AZ) may be used.

TECHNIQUE

The technique of intervention on all three mesenteric arteries is nearly identical. For the purpose of illustration, technique of celiac artery intervention will be described here. The same technique should be used for SMA or IMA intervention.

'TELESCOPING' TECHNIQUE

In this technique, a 4 Fr. catheter (e.g. 100 cm IM catheter) is advanced through a 6 Fr. guiding catheter (e.g. 55 cm IM guide) in a coaxial 'telescoping' fashion to minimize trauma to the aortic wall from the catheter tip. The assembly (the diagnostic catheter through the guide) is made prior to introducing the diagnostic catheter in the body.

- This technique is illustrated by the following scenario in which aortography via femoral access demonstrates a severe ostial stenosis of the celiac trunk (figure 3).

Access:

- Right CFA access for the diagnostic angiogram and intervention.

Vessel Cannulation:

- After obtaining an aortogram, the flush catheter is removed.
- A selective catheter, e.g. a 4 Fr./100 cm IM catheter is flushed and introduced into a 6 Fr./55 cm guiding catheter of choice (e.g. IM) through a Tuohy-Borst connector attached to the guiding catheter.
- The diagnostic catheter is advanced over a 0.035" wire (J-tip, Wholey, etc) into the abdominal aorta.
- The wire is removed and the diagnostic catheter is used to engage the celiac trunk, in the lateral projection (LAO 90 degrees).
- The lateral projection is the working view for the intervention.
- A 0.014" non-hydrophilic wire is advanced through the diagnostic catheter into the celiac trunk and then into the splenic or common hepatic artery, to secure a 'foot hold' in the vessel. This provides support for advancing the guide over the diagnostic catheter.

Telescoping the Guiding Catheter:

- The guiding catheter is then slowly advanced over the diagnostic catheter and wire combination, under fluoroscopic guidance, into the ostium.

Removal of the Diagnostic Catheter:

- Once the guiding catheter is adequately seated in the vessel, the diagnostic catheter is slowly withdrawn cautiously over the wire.
- The intervention then proceeds in the following usual fashion.

Predilatation:

- The lesion is predilated with a slightly undersized balloon.
- This step is recommended as it facilitates in appropriately sizing the stent.
- During this step, the guiding catheter may back out slightly in the aorta. As the balloon is

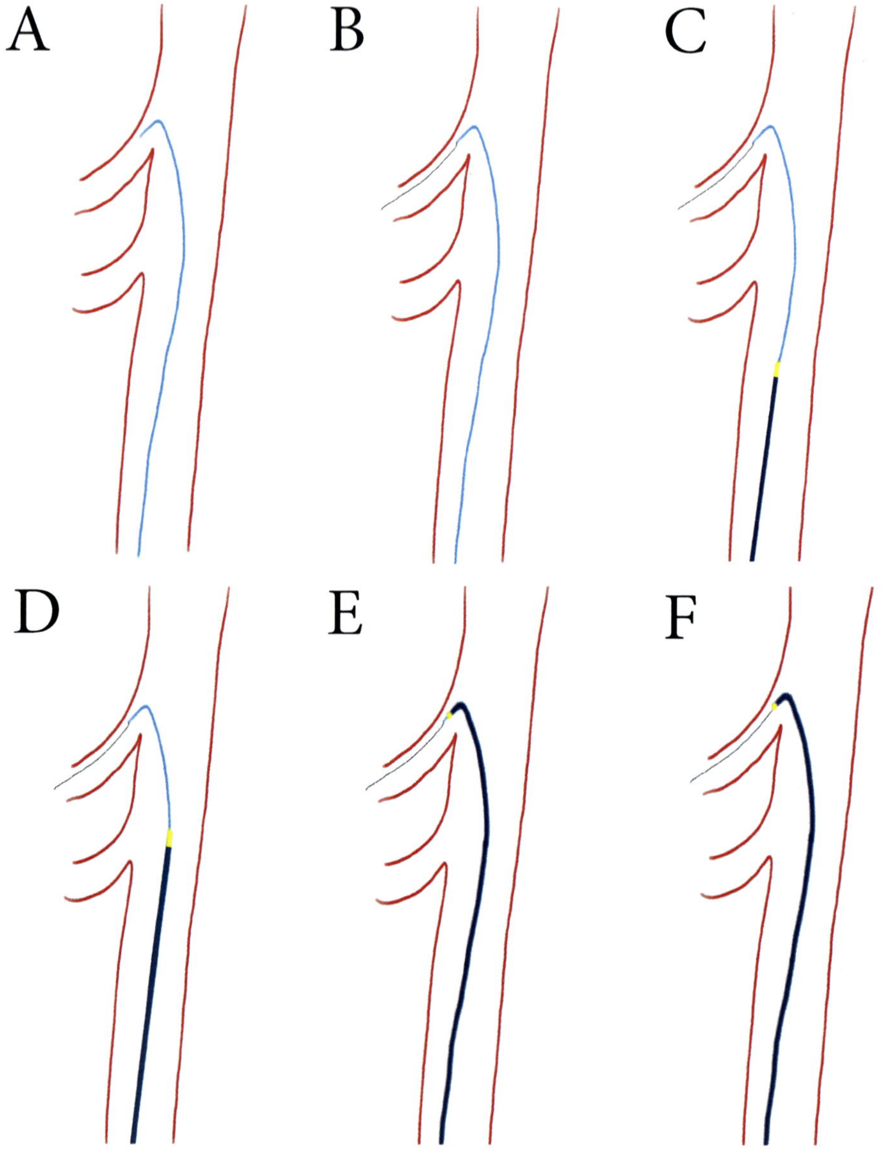

Figure 3.

'Telescoping' technique of mesenteric intervention.

A. A 4 Fr. diagnostic catheter (pre-loaded through the guide) is advanced over a 0.018" SteelCore® wire and used
to engage the celiac trunk.
B. A 0.014" non-hydrophilic wire is advanced into the vessel.
C. A 6 Fr./55 cm pre-loaded IM guide is advanced over the diagnostic catheter.
D. The guide is gradually advanced into the ostium of the artery.
E. The diagnostic catheter is withdrawn without dislodging the guide.
F. The guide is then used to intervene in the usual manner.

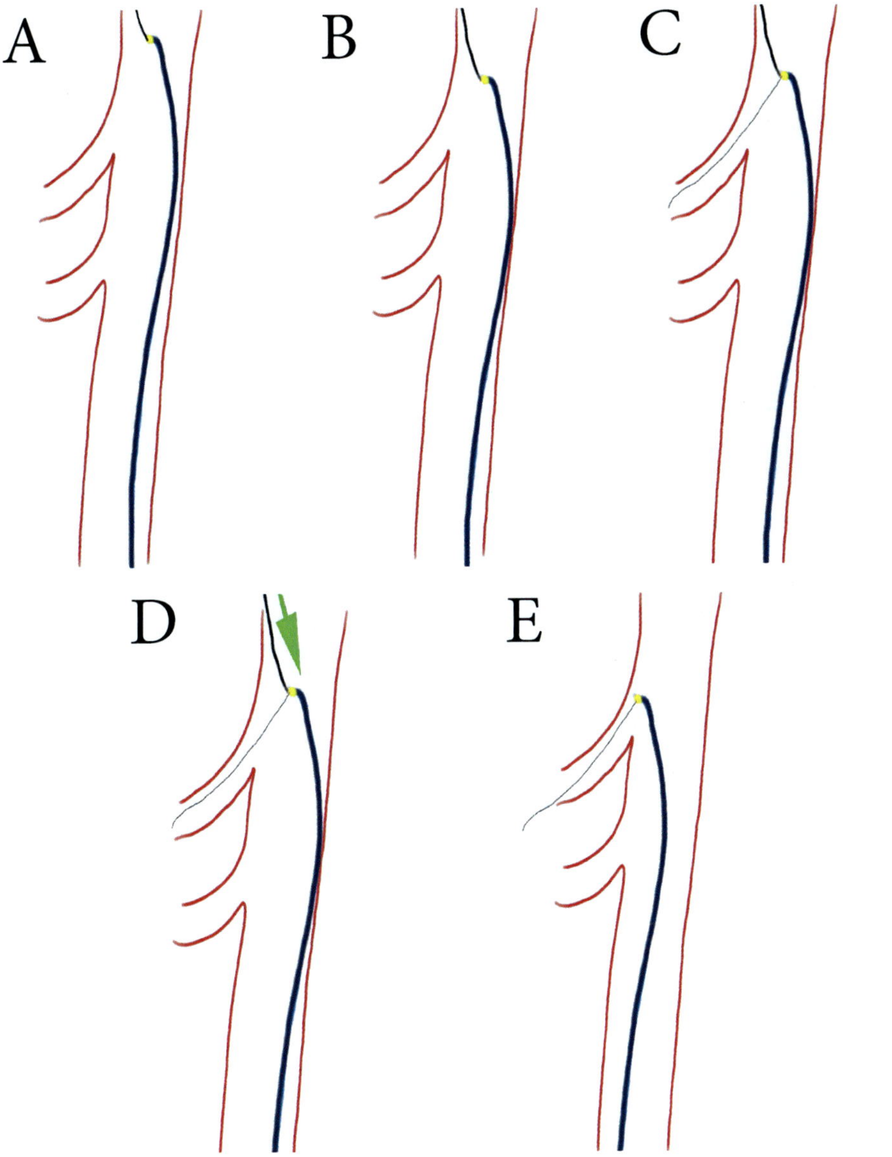

Figure 4.

'No-touch' technique of mesenteric intervention.

A. A 0.018" stiff wire is advanced into the distal thoracic aorta. A 6 Fr. guide is advanced over this wire to a level above the origin of the celiac trunk.
B. The catheter is withdrawn to the level of the celiac trunk.
C. A 0.014" wire is advanced along side the 0.018" wire in the guide and is used to cannulate the celiac trunk.
D. The 0.018" wire is withdrawn.
E. The guide 'drops' into the celiac trunk.

being pulled out of the vessel, the catheter should be 'shepherded' back over the balloon into the vessel.

- Care should also be taken to not allow the wire to migrate distally as it may cause spasm or dissection in the vessel.

Stenting:

- Only balloon-expandable, 0.014" system stents are used.
- These stents should be sized 1:1 in diameter to the normal reference vessel. Oversizing may lead to vessel trauma and perforation.
- The stent length is chosen keeping in consideration the fact that the stent will shorten somewhat with expansion.
- Additionally, for ostial stenoses, the stent should be positioned such that the proximal end protrudes into the aorta about 1 - 2 mm.
- The guiding catheter is backed out slightly by pulling it back, to expose the stent.
- During deployment, a slight but constant forward pressure should be applied to the guide so that the stent does not migrate back too much into the aorta.
- The stent should be deployed at the nominal pressure.
- However, if the patient develops abdominal/back pain further inflation should be curtailed beyond that point. After prompt deflation a quick angiogram should be obtained to rule out vessel dissection/perforation.
- The stent balloon is pulled out halfway and the guiding catheter is 'shepherded' back over it into the stent. Caution should be used to prevent damaging the stent with the guide tip.
- Once the guide is comfortably inside the stent, the balloon is pulled out of the vessel.
- An angiogram is obtained to determine the need to post-dilate the stent.

Postdilatation:

- Postdilatation is recommended, even if the stent appears to be perfectly expanded.
- A compliant balloon, shorter than the length of the stent is used.
- After post-dilating the distal portion of the stent, the balloon is withdrawn into the proximal half in order to flare the ostial/aortic segment of the stent.
- The proximal half is left "hanging" in the aorta. The guide is pulled back just enough to expose the balloon.
- The balloon is inflated with a slight forward pressure on the guide.

Final Angiogram:

- As the postdilatation balloon is being removed, the guiding catheter is slowly 'shepherded' over it into the stent, avoiding any trauma to the latter.
- Repeat angiography is performed to review the final result.
- Post-stenting images should be carefully analyzed for perforation, vessel dissection, and distal embolization.
- Pull the wire out back into the stented segment.
- Remove the guide slowly by pulling back and turning the tip backwards toward the spine, while maintaining some traction on the wire.

'NO TOUCH' TECHNIQUE

In this technique, the guiding catheter is advanced into the celiac artery while maintaining a wire 'rail' in the descending thoracic aorta to avoid dragging the catheter tip in aortic plaque/thrombus. The following steps (figure 4) are used:

- The femoral sheath should be around 23 - 25 cm long so that its tip is in the distal aorta.
- After aortography, the flush catheter is removed over a 0.035" wire. This wire is removed.
- A 0.018" stiff wire is advanced into the descending thoracic aorta through the sheath.
- A 6 Fr. guiding catheter connected to a Tuohy-Borst adapter is advanced over this wire into

the abdominal aorta.

- Adjust the length of the protruding segment of the wire, so that the catheter tip does not come in contact with the aortic wall.
- The catheter is adequately flushed clear of debris and air in infra-renal aorta.
- It is advanced to the level of T10-T11 vertebrae.
- A puff is obtained to ascertain positioning as the catheter is slowly withdrawn to get to the level of the celiac trunk (or other mesenteric artery of interest).
- A separate 0.014" interventional wire is introduced along with the pre-existing 0.018" wire into the guide.
- The 0.014" wire is advanced to the tip of the guide.
- When the catheter is right across from the celiac trunk, the 0.014" wire is advanced into this artery.
- The 0.018" wire is then withdrawn slowly to allow the guide to 'drop' into the celiac trunk.
- Intervention is performed as outlined in the previous section.

BRACHIAL/RADIAL ACCESS INTERVENTION

- Brachial or radial artery access is used for mesenteric intervention for a variety of reasons, as was noted previously. Left arm access is preferred to right.
- Usually, either conventional or CT/MR angiography images have already been obtained and have guided this decision.
- If unsure about the anatomy and the likelihood of intervention, we start with a 5 Fr. sheath. Otherwise, we use a 6 Fr. sheath from the start.
- For radial access cases, a cocktail of intra-arterial NTG, verapamil (HR permitting) and unfractionated heparin is routinely used for access site patency.
- An angled stiff Glidewire is advanced into the abdominal aorta. A flush catheter is advanced over this for aortography.
- After aortography, a guiding catheter or sheath can be advanced and used for selective angiography if intervention is likely.
- A 6 Fr. MPA or Amplatz left (AL1 or 2) guiding catheter are usually adequate.
- Perform a selective angiogram of the celiac artery in the AP and lateral projections.
- The remaining steps of the intervention are similar to the steps performed with femoral access.

MEDIAN ARCUATE LIGAMENT SYNDROME (MALS)

- This syndrome is characterized by an abnormally low median arcuate ligament (a fibrous band that connects the crura of the diaphragm across the aortic hiatus).
- This low arcuate ligament compresses the proximal celiac artery and results in chronic abdominal pain and weight loss.
- It is found typically in thin young women, age 20-40 years.
- While this condition can be diagnosed on arterial duplex, it is best detected on conventional angiography and/or CTA (seen on sagittal and 3D reconstructions- figure 5). These studies usually show a severe narrowing of the proximal celiac artery, with a superior indentation (rather than a concentric stenosis typically seen with atherosclerosis) with post-stenotic dilatation and prominent collateral flow (via the SMA, resulting in a dilated gastroduodenal artery).
- This entity is a diagnosis of exclusion.
- Since the anatomic compression is not uncommon (up to 25%), treatment is considered only for symptomatic patients.
- Treatment consists of surgical decompression of the ligament. This can be done laproscopically.

- On angiography, the best view is a true lateral projection. The compression is most prominent during expiration. During inspiration, the celiac artery descends into the abdominal cavity and the compression is thus reduced.
- Endovascular therapy should NOT be considered, as delayed stent fracture from repetitive trauma is common.
- In severe cases, the compression remains obvious even during inspiration.
- This condition is sometimes referred to as 'celiac artery (or axis) compression syndrome' (CACS).

Figure 5.

Median arcuate ligament compression of celiac axis (sagittal view on CTA).

The red arrow represents the level of the compression. The blue oval represents the left renal vein in cross-section

1. Celiac axis.

2. SMA.

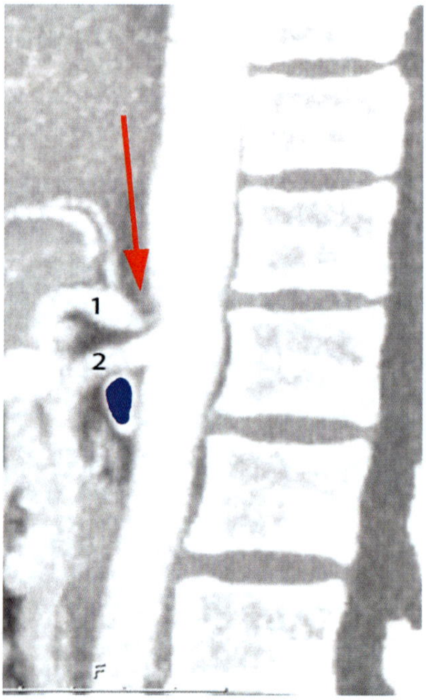

CASE

- Mrs. S is a 65 year old female ex-smoker with known PAD. She had undergone bilateral renal artery stents in the past for poorly controlled HTN. She presented with symptoms of severe abdominal pain and food aversion for 18 months. Already a petite lady, she had involuntary weight loss of > 20 lbs in this time. Her discomfort was post-prandial in nature and was associated with mild nausea. Her family physician had performed an extensive work-up for an underlying malignancy which was unrevealing.
- Physical exam in the office revealed a thin and frail middle-aged female. Her extremity pulses were normal. She had an abdominal bruit on either side of the umbilicus.
- An MRA had been done which suggested a patent celiac and IMA with an occluded SMA.

Access & Diagnostic Angiography:

- For initial angiography, a 5 Fr./23 cm Cordis Brite-tip sheath was used in the right CFA.
- 2000 units IV heparin was given.
- Abdominal aortography was performed with a 5 Fr. TR, using 40 cc of Visipaque at 20 cc/sec at 750 psi.
- Selective angiography was then performed with a 5 Fr. IM catheter.
- Both the celiac trunk and the SMA were occluded 100% shortly after take off.
- To engage the IMA an Omni-Sos catheter was used.
- Significant dampening of invasive pressure was noted on ostial engagement of this vessel.
- The IMA was noted to be the sole supply to the gut.
- A prominent arc of Riolan ('meandering' artery) and a less prominent marginal artery of Drummond were clearly visualized. Additionally, collaterals from the SMA were seen to the celiac trunk.
- A small infra-renal AAA was also noted. The MRA had shown this to be < 3 cm in diameter.
- Bilateral renal arteries had widely patent stents.

Interventional Technique:

- Since the take of the IMA was directed acutely inferiorly, it was decided to use left radial access.
- A 6 Fr./11 cm Terumo Glide sheath was inserted into the left brachial artery after initial access with a micropuncture technique.
- A 6 Fr. MPA guiding catheter was then advanced over an ASGW under fluoroscopic guidance through the subclavian artery and into the abdominal aorta.
- The IMA was engaged with it. A 300 cm ATW was used to cannulate the IMA and the wire was advanced into the left colic artery.
- IVUS of the ostium showed a critical stenosis. PTA was performed with a 3.0 mm x 15 mm coronary balloon.
- Based on IVUS measurements a 3.5 mm x 13 mm DES was used for the lesion and post-dilated to a 4.5 mm diameter.

Final Angiography:

- This revealed a widely patent IMA with TIMI-3 flow.
- Excellent flow into the SMA and celiac, via the arc of Riolan and the artery of Drummond, was also seen. Manual compression was applied to both arteriotomy sites.
- Aspirin and Clopidogrel were continued.

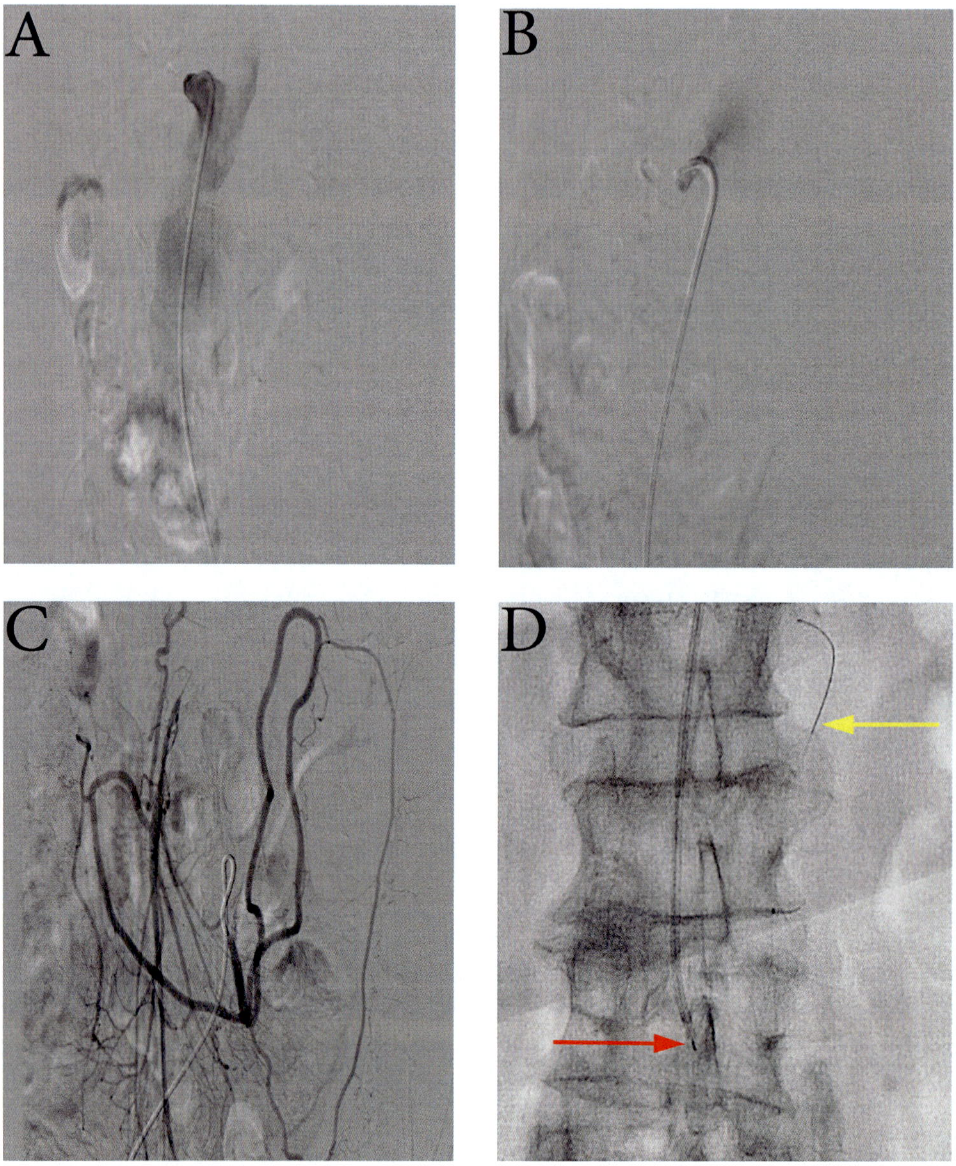

Figure 6.

A. Selective angiogram of the celiac artery showing 100% proximal occlusion.
B. Selective angiogram of the SMA which is also 100% occluded proximally.
C. Selective angiogram of the IMA demonstrating a severe ostial stenosis. A very prominent arc of Riolan is present.
The marginal artery of Drummond is also seen. This shows robust collateralization of the SMA and the celiac trunk via the IMA.
D. A 6 Fr. MP guiding catheter introduced via brachial artery access is seen introducing an IVUS catheter into the IMA (red arrow). A 0.014" ATW wire (yellow arrow) is seen in the arc of Riolan with an IVUS catheter at the ostium of the IMA.

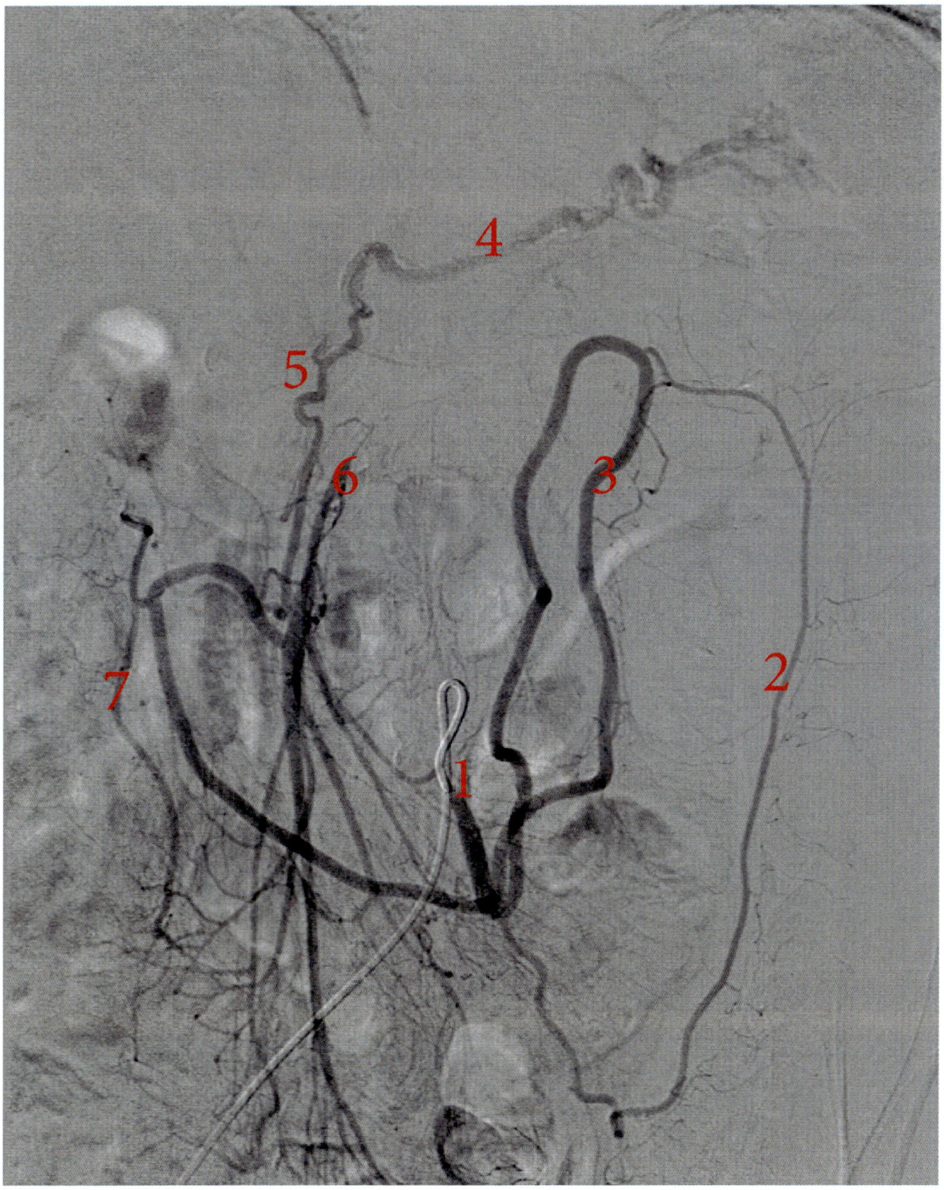

Figure 7.

Selective IMA angiogram.

1. IMA.
2. Marginal artery of Drummond.
3. Arc of Riolan.
4. Splenic artery.
5. Gastro-duodenal artery.
6. SMA.
7. Right colic artery.

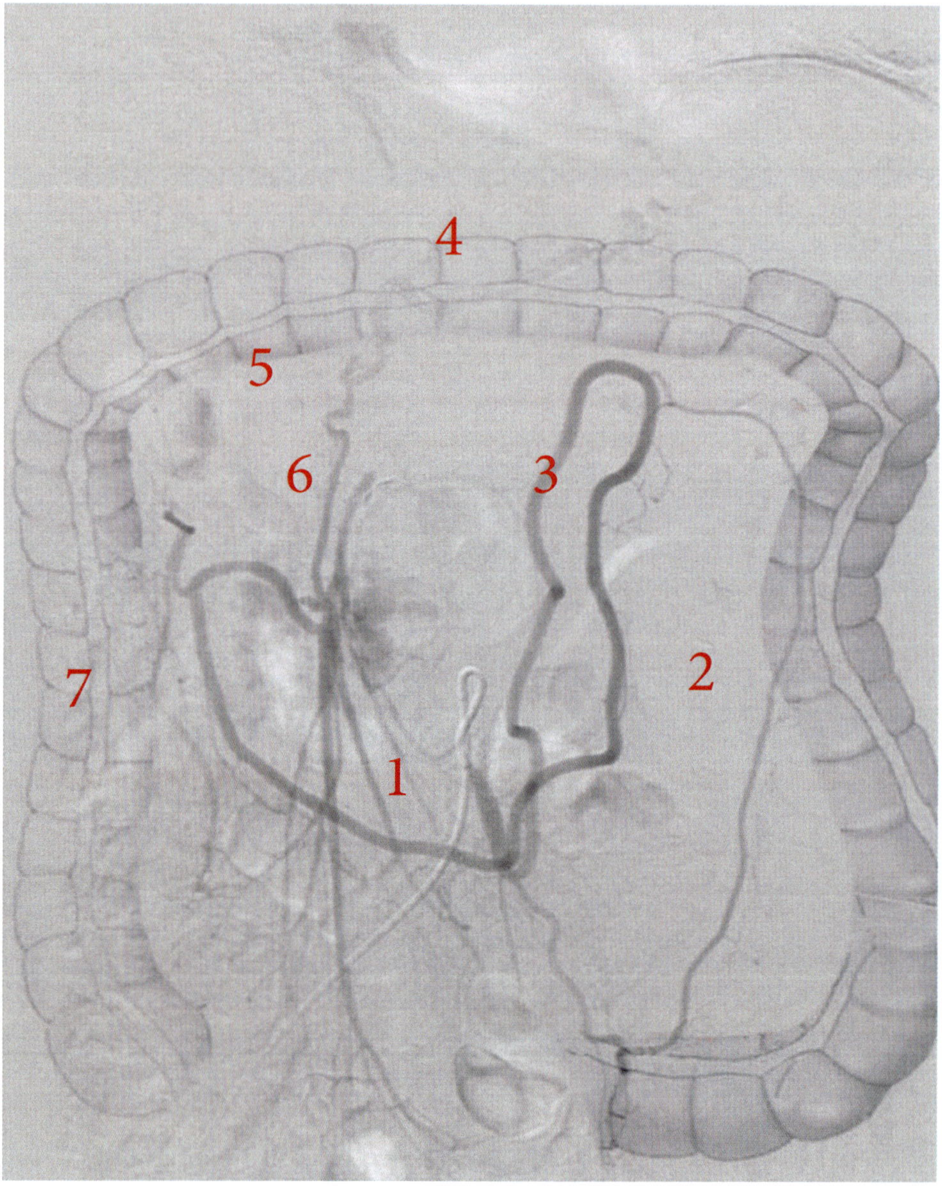

Figure 8.

Selective IMA angiogram demonstrating relationship to the surrounding colon.

1. IMA.
2. Marginal artery of Drummond.
3. Arc of Riolan.
4. Splenic artery.
5. Gastro-duodenal artery.
6. SMA.
7. Right colic artery.

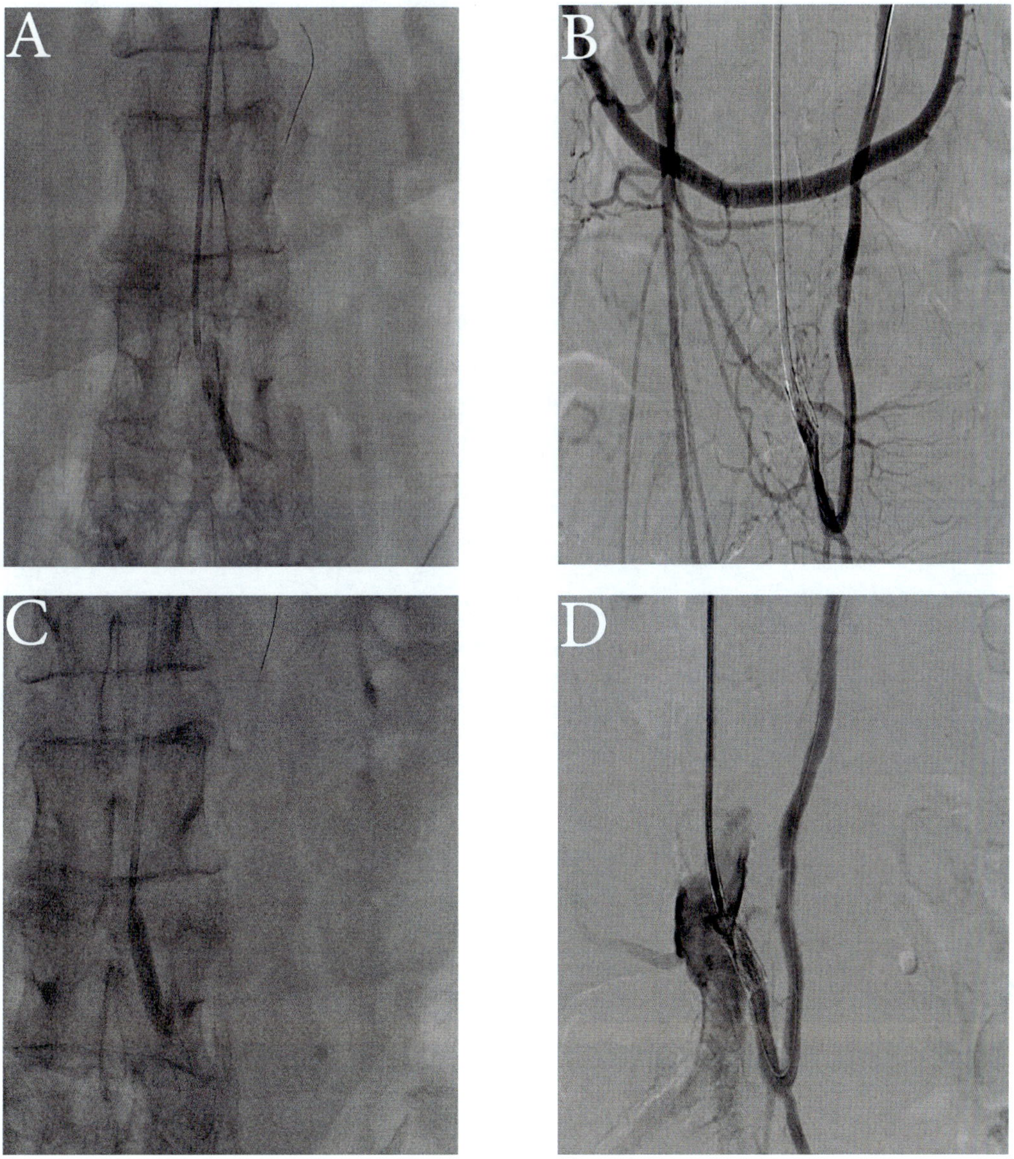

Figure 9.

Stenting of inferior mesenteric artery.

A. Predilatation of the ostial IMA stenosis.
B. Angiogram obtained after PTA of IMA.
C. Stent deployment followed by ostial postdilatation.
D. Final selective angiogram demonstrating a widely patent IMA with excellent flow. Note the reflux in the aorta
compared to the pre-intervention selective angiogram.

Follow-up:

- At 30-day follow-up the patient reported complete resolution of her symptoms. She had actually gained back 5 lbs of weight. Being a gourmet chef, she was enjoying her cooking once again!
- One year post-op she continued to gain weight in a desirable fashion.
- Her dual anti-platelet therapy was maintained, and if tolerated was to be continued indefinitely.

10

SUBCLAVIAN ARTERY STENTING

Subclavian artery occlusive disease is not uncommon. However, unlike lower extremity arterial disease, the symptoms of claudication are seldom manifest with upper extremity occlusive disease and is often diagnosed incidentally. Despite severe stenoses, patients usually have minimal impairment of function or lifestyle alteration.

Symptoms may include upper extremity claudication (with activities such as combing/shampooing the hair or increased use of the hand) and rarely ataxia and other vertebro-basilar insufficiency symptoms.

Additionally, when the internal mammary artery is used as an in-situ coronary artery bypass conduit, the patient may experience classic anginal and/or heart failure symptoms (coronary-subclavian steal) due to significant coronary ischemia in the presence of significant proximal subclavian artery stenosis.

ANATOMY

The subclavian arteries are major arteries that supply blood to the upper extremities. Branches of the subclavian artery supply the head and thorax.

- The right subclavian artery arises from the brachiocephalic (innominate) artery behind the right sternoclavicular joint while the left subclavian artery arises from the aortic arch.
- Each subclavian artery is divided into three portions:
  - ▹ First portion extends to the medial border of the scalenus anterior muscle.
  - ▹ Second portion lies behind the scalenus anterior muscle.
  - ▹ Third portion extends from the lateral margin to the outer border of the 1st rib.
- Rarely, the origin of the right or left subclavian artery is aberrant.
  - ▹ The aberrant artery usually arises just distal to the left subclavian artery and crosses in the posterior mediastinum usually behind the esophagus toward the right upper extremity.
  - ▹ This may create a vascular ring around the trachea and esophagus which leads to dys-

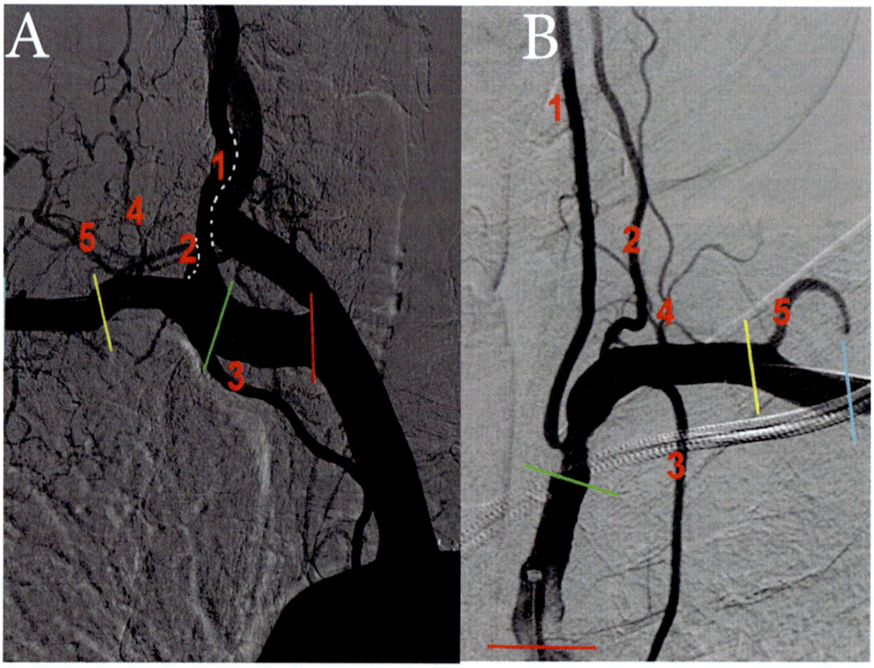

Figure 1.

Anatomy of Subclavian Artery.

A, Right subclavian artery and **B,** left subclavian artery **1,** Vertebral artery. **2,** Thyrocervical trunk. **3,** Internal mammary artery. **4,** Costocervical trunk. **5,** Dorsal scapular artery.

The red line indicates the origin of the subclavian artery. Proximal subclavian artery is enclosed by the red and green lines. Mid subclavian artery is enclosed by the green and yellow lines. Distal subclavian artery is between the yellow and blue lines. Axillary artery starts at the blue line.

phagia.

- ▹ Dysphagia due to the right subclavian artery is called dysphagia lusoria.

- The luminal diameter of the proximal subclavian artery is 8-10 mm in an adult male and 7-9 mm in an adult female.

Branches of subclavian artery:
Important branches include:
- Vertebral artery
- Internal mammary artery
- Thyrocervical trunk
- Costocervical trunk
- Dorsal scapular artery

DIAGNOSIS

Clinically, the diagnosis of significant unilateral disease can be made simply by palpating bilateral brachial artery pulses and checking non-invasive blood pressure in both upper extremities. A difference of greater than 20 mm Hg in systolic pressure suggests significant underlying subclavian artery disease on the side with the lower pressure. This suspicion should be further confirmed by obtaining a carotid duplex U/S (CDU). Obviously, the lack of a difference in systolic BP between the two arms does not exclude significant subclavian artery stenosis e.g. in patients with bilateral high-grade subclavian artery disease.

When the history, exam and a CDU suggest severe subclavian artery disease and the patient stands to benefit from revascularization further imaging should be considered. Depending upon the patient's age, underlying symptoms, and the risk of complications, conventional angiography is a consideration. This is particularly true for those patients with coronary-subclavian steal symptoms and evidence of inducible coronary ischemia.

ACCESS

SITE

Femoral:

For subclavian diagnostic angiography, mostly CFA access is used. For diagnostic angiography, a 5 Fr. sheath is adequate.

Brachial/Radial:

Diagnostic angiography can also be performed via ipsilateral or contralateral brachial/radial artery access if necessary (e.g. aorto-iliac occlusive disease).

SHEATH

Diagnostic angiography (aortography and selective subclavian angiography) should provide information regarding the use of either an 80-90 cm long sheath vs. a 'guiding' catheter for the intervention. If the former is used, then a 6 Fr. sheath is adequate as the inner diameter of a 6 Fr. sheath would easily allow delivery of most balloons and stents. On the contrary, if due to anatomic reasons a 'guiding catheter' is required, then at least an 8 Fr. system is necessary.

- For diagnostic angiography, a 5 Fr. sheath is adequate.
- For the intervention, upsize the sheath to a 6 Fr./90 cm shuttle sheath.
- If the decision is to use a guiding catheter, an 8 Fr. sheath is required.

DIAGNOSTIC ANGIOGRAPHY

ARCH AORTOGRAPHY

Catheter:

Use a pigtail catheter (non-angled), position it in the ascending aorta and perform a power injection.

Imaging:

- For aortography, an LAO 40° or greater projection is used – ideally the thoracic aorta should be imaged in a projection that shows the best separation of ascending, arch and descending aorta. In older patients and those with tortuous descending aorta, this may require more extreme LAO projection.
- In patients with suspected right subclavian artery disease, an RAO projection is also performed to visualize the course of the proximal subclavian artery.
- DSA should be used if possible as this may reduce the volume of dye administration without compromising image quality.

Dye Injection:

- The rate of power injection depends upon the patient's body habitus, renal function and ability to comply with instructions of breath-hold during aortography.
- We use 20 cc/sec for a total of 40 cc at 500 psi. In patients with normal renal function, we use 70% dye with 30% heparinized normal saline dilution (70:30). If the creatinine clearance is

less than 30 mL/min, then we use 50% dilution (50:50). Rarely, gadolinium may be used for this purpose. CO2 should NOT be used for subclavian angiography.
- The origins of all brachiocephalic arteries should be clearly visualized.

SELECTIVE ANGIOGRAPHY

Catheter Selection for Femoral Access:

- Several catheters are available (e.g. JR4, vertebral, IM, head-hunter, HS, Simmons,Vitek, etc.) to perform selective left subclavian and innominate angiography. In patients with a normal aortic arch (type I), we use a JR4 catheter.
- In patients with complex aortic arch anatomy (types II and III), Simmons or Vitek catheters are preferred. The curve of the former is 'formed' in the ascending aorta whereas the latter is best 'formed' in the proximal descending aorta.

Catheter Selection for Brachial/Radial Access:

- We prefer to use the Simmons catheter for selective angiography of the contralateral subclavian artery. For example, if access was obtained using the right brachial/radial artery, use the Simmons catheter to engage the left subclavian artery.

Dye Injection/Imaging:

- 5-10 cc manual injections of 70:30 (or 50:50 in those with renal dysfunction) contrast are adequate.
- LAO and AP projections are usually adequate for the left subclavian artery. However, an RAO projection may be very helpful in identifying the origin of the left vertebral artery. This latter projection is also very useful during ostial subclavian stent placement to appropriately visualize how much of the stent is protruding in the arch.
- For the right subclavian artery, straight RAO and RAO (30° - 40°) caudal (30°) projection is very helpful in showing the ostial segment.
- DSA and breath-hold should be utilized during imaging.
- In patients with subtotal or total occlusion of the proximal vessel, a prolonged cine run should be performed to allow the distal vessel to fill via collaterals.
- Identify the origins of the vertebral and mammary arteries. Ideally, these should not be covered with stents in order to avoid plaque shift and associated complications.

HEMODYNAMICS

Hemodynamic assessment is strongly recommended. In cases of angiographically obvious severe stenoses, the measurement of a hemodynamic gradient across these lesions may be less critical. However, for lesions that appear to be 'non-significant' on angiography, it is imperative to measure the trans-stenotic gradient.

- Ideally, the gradient is measured by using simultaneous pressure recordings proximal and distal to the stenosis.
- In order to obtain a reliable measurement, the caliber of the measuring catheters should be at least 1 French size apart. Otherwise, artifactual dampening of the pressure waveforms may be noted and misinterpreted as a significant gradient. For example, when using a 6 Fr. sheath or guiding catheter, a 4 Fr./120 cm catheter should be used to measure trans-lesional gradient. With an 8 Fr. guide even a 5 Fr. catheter e.g. Glidecath (Terumo, Somerset NJ) can be used.
- Measurements should be made before and after hyperemia. We induce hyperemia by administering 200-300 mcg of intra-arterial NTG distal to the stenosis. Some operators use Papaverine for the same purpose.

TABLE 1 - EQUIPMENT FOR DIAGNOSTIC SUBCLAVIAN ANGIOGRAPHY	
Sheaths	• 6 Fr. ; 11+ cm long
Wires	• 0.035" 260 cm ▹ Angled stiff hydrophilic wire (e.g. Glidewire®) ▹ Angled non-stiff hydrophilic wire
Flush Catheters	• Pigtail (non-angled)
Selective Catheters	• IM • JR4 • HS • Simmons • Vitek • Vert
Superselective Catheters	• 0.035", 0.038" ▹ QuickCross® (Spectranetics, Colorado Springs, CO) ▹ Glidecath® (Terumo, Japan)

INTERVENTION

ACCESS

Depending on the aortic configuration and lesion location/morphology, femoral, brachial or both accesses are used.

ANTICOAGULATION

- IV UFH is preferred.
- Goal ACT > 250 sec.
- Bivalirudin or GP IIb/IIIa inhibitors should be avoided because of the inability to reverse the effect.

EQUIPMENT

GUIDES

- Guiding sheaths or guiding catheters can be used for subclavian artery interventions.
- Typically, a 6 Fr./90 cm guiding sheath e.g. Shuttle® (Cook, Bloomington, IN) is adequate for subclavian artery intervention in patients with a normal arch (type I).
- In complex arch anatomy (type II and III), we prefer to use an 8 Fr. guiding catheter to provide more support to cross the lesion and allow for deliverability of appropriate sized stents.
- Most balloon-expandable stents required for this vessel size cannot be delivered through a smaller caliber guiding catheter.

Femoral Access:

We use the following guide shapes:

- JR4 (type I arch)
- HS (type II and III arch)
- AL1 (type II and III arch)

Brachial/Radial Access:

We use the following guide shapes:

- AL1-3(type I-III arch)

WIRES

- 0.035" hydrophilic guidewires are used to allow for better guide and sheath positioning as well as better deliverability of bulkier stents.

BALLOONS

- Normal luminal diameter of the subclavian arteries in males is 8-10 mm and 7-9 mm in females.

Pre-dilataion Balloons:

- Use an undersized balloon (4-6 mm x 20 mm) to predilate the lesion.
- In case of a CTO, start predilatation with a smaller balloon. Once a channel has been created, the balloon can be upsized to a slightly bigger balloon prior to final stent placement.

Post-dilatation Balloons:

- Post-dilatation balloon should be a little shorter in length than the stent and are sized 1:1 to the luminal diameter of the normal vessel.
- Position these balloons such that the distal marker is well within the stent because they have big 'shoulders' that may protrude outside the stent and traumatize the unstented adjacent arterial segment possibly leading to dissection or even plaque shift in the bordering vertebral or internal mammary artery.

STENTS

- Balloon-expandable stents are used for ostial/proximal vessel due to the need for precise placement and radial force in this location.
- For the mid and distal vessel, prior to intervention, thoracic outlet syndrome should be ruled out. If absent, angioplasty alone is usually preferred. However, self-expanding stents can be used for this location if necessary. A downside to putting stents in the mid/distal segment is that it may create a 'thoracic outlet syndrome' physiology there.
- For balloon-expandable stents, 1:1 sizing to the reference vessel diameter is used to avoid vessel perforation.

TECHNIQUE

LEFT SUBCLAVIAN ARTERY OSTIAL/PROXIMAL STENOSIS

Arch aortography via femoral access demonstrating a severe LEFT proximal subclavian stenosis.

Technique for Transfemoral Intervention:

- After selective angiography (figure 2A), advance a 0.035" angled non-stiff hydrophilic wire (ANSGW) through the diagnostic catheter, across the lesion and place it in the axillary or proximal brachial artery (figure 2B).
- Remove the diagnostic catheter and advance a 4 Fr./120 cm straight Glidecath over the 0.035" ANSGW and place it in the axillary or brachial artery (figure 2C).
- Exchange the ANSGW and advance a 0.035" 260 cm angled stiff hydrophilic wire (ASGW) through the catheter (figure 2D).
- Exchange out the 5 Fr. short sheath over the 0.035" ASGW for a 6 Fr./90 cm Shuttle® sheath (figure 2E).

- Advance this sheath along with its dilator proximal to the lesion in the subclavian artery if possible. Make sure to not advance the dilator into the lesion as it may dissect or perforate the vessel.
- Remove the dilator (figure 2F) and anticoagulate the patient with goal ACT >250 sec.
- Pre-dilate the lesion (figure 2G) with an undersized balloon (4 - 6 mm x 20 mm). In case of a CTO, start predilatation with a smaller balloon. Once a channel has been created, the balloon can be upsized to a slightly bigger balloon prior to final stent placement.
- This step prepares the lesion for stenting. It gives an idea about the compliance of the lesion and the likelihood of stent expansion but it also allows for better visualization during stent positioning. Furthermore, it also reduces the risk of stent dislodgement off the balloon while entering a severely stenotic lesion.
- Prior to removing the predilatation balloon, 'shepherd' the sheath over it and across the lesion (figure 2H).
- Exchange the balloon out for the stent of choice.
- Advance the stent across the lesion. Back the sheath out into the ostium. These steps also reduce the likelihood of stent dislodgement (figures 2I & 2J).
- Aim to have 1 - 2 mm of the stent protruding into the aorta outside the ostium. Small dye injections should be performed to determine if the stent placement is precise.
- When the stent is in the correct position, inflate the balloon slowly at 1 - 2 ATM/sec (figure 2K). The nominal pressure for most of these stent balloons is 7-8 ATM.
- Once the stent is deployed, deflate the balloon and pull it out into the aorta half-way. Advance the sheath over this balloon 'shepherding' it into the stent under careful fluoroscopic guidance. This maneuver allows the postdilatation balloon to be advanced inside the stent without any resistance from the ostial struts.
- Insert the postdilatation balloon, keep it within the stent and inflate at nominal pressure.
- If the original lesion was ostial and the stent covers it adequately, then this stented segment needs to be 'flared' with the postdilatation balloon (figure 2L). After an initial inflation within the stent, pull the balloon back half-way out of the stent and dilate again. Make sure the sheath has been pulled back enough and is not covering the balloon.
- Remove the postdilatation balloon and perform a selective angiogram to confirm adequate stent expansion and coverage of the lesion and to exclude stent edge dissection or plaque shift in the vertebral or internal mammary arteries.
- It is also advisable to repeat simultaneous pressure measurements in the aorta and beyond the stented segment.
- We also routinely measure a non-invasive brachial blood pressure against simultaneous central aortic pressure after the intervention to establish a new baseline.

Radial Access:

- Use standard radial technique.
- Give a 'cocktail' of IA NTG (200-300 mcg), verapamil (2-5 mg) and UFH (3000-6000 U) through the radial artery sheath.
- Do not use a J-wire initially while traversing the radial artery as it may induce severe spasm.
- Advance your first catheter over an angled stiff hydrophilic wire.

Technique for Intervention Via Brachial/Radial Access:

- Use a 6 Fr./65 cm straight sheath (e.g. Pinnacle Destination®; Terumo) for radial access and a 6 Fr./45 cm straight sheath (e.g. Ansel® 0) for brachial access on the ipsilateral side for the intervention.

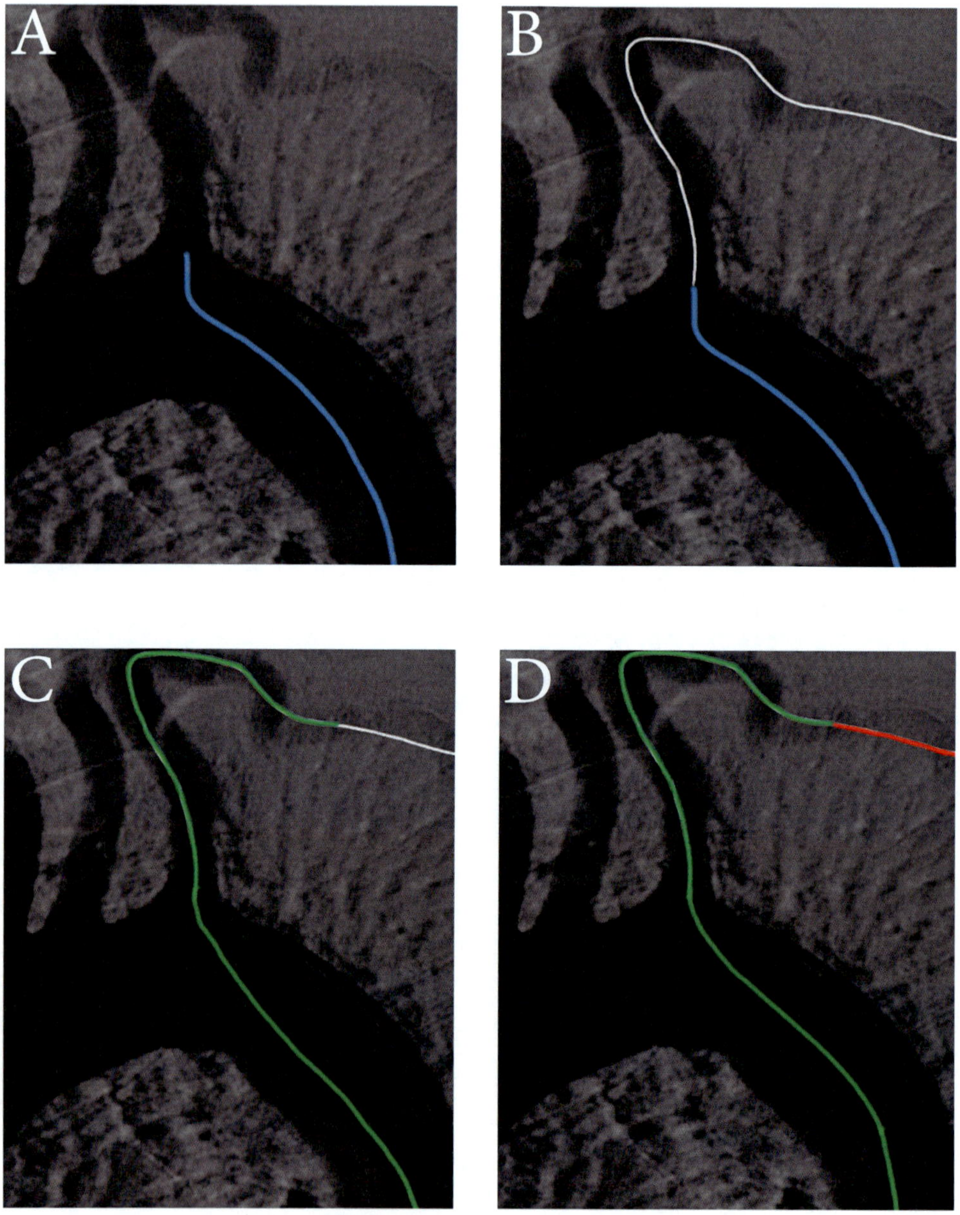

Figure 2.

Ostial/Proximal Left Subclavian Artery Intervention.

A. A 6 Fr. JR4 diagnostic catheter is used to engage the left SCA ostium.
B. An angled non-stiff hyrdophilic wire is advanced across the lesion and into the axillary artery.
C. The catheter is exchanged out for a straight hydrophilic catheter. This second catheter is advanced into the axil
lary artery.
D. Through this catheter the wire is exchanged out for a stiffer wire e.g. an angled stiff Glidewire®.

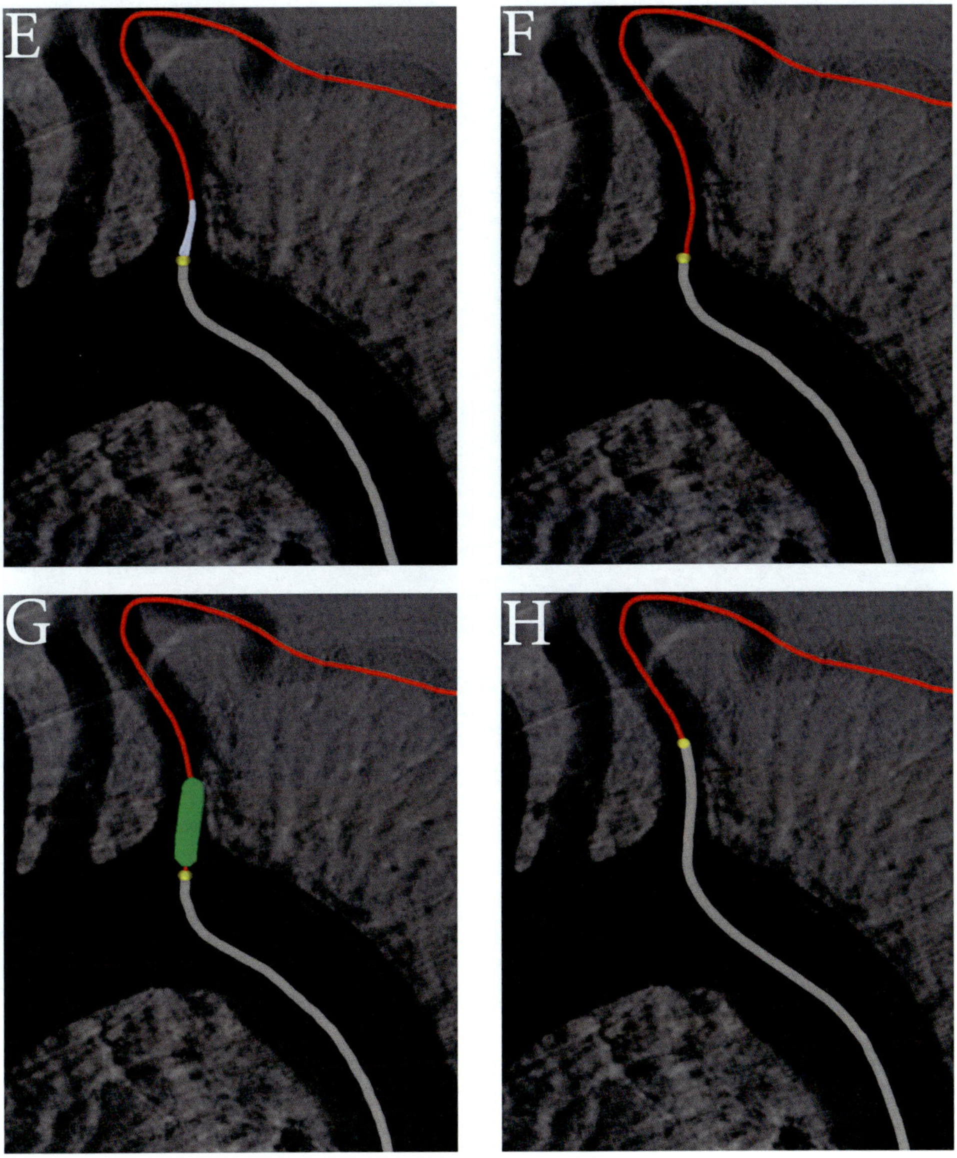

Figure 2.

Ostial/Proximal Left Subclavian Artery Intervention (continued).

E. The straight catheter and femoral sheath are exchanged for a 6 Fr./90 cm straight sheath e.g. Shuttle®. This
sheath is advanced into the descending aorta and ostial SCA.
F. The dilator of the sheath is removed.
G. The lesion in the ostial SCA is predilated with an under-sized balloon.
H. The sheath is 'shepherded' over the balloon, across the lesion.

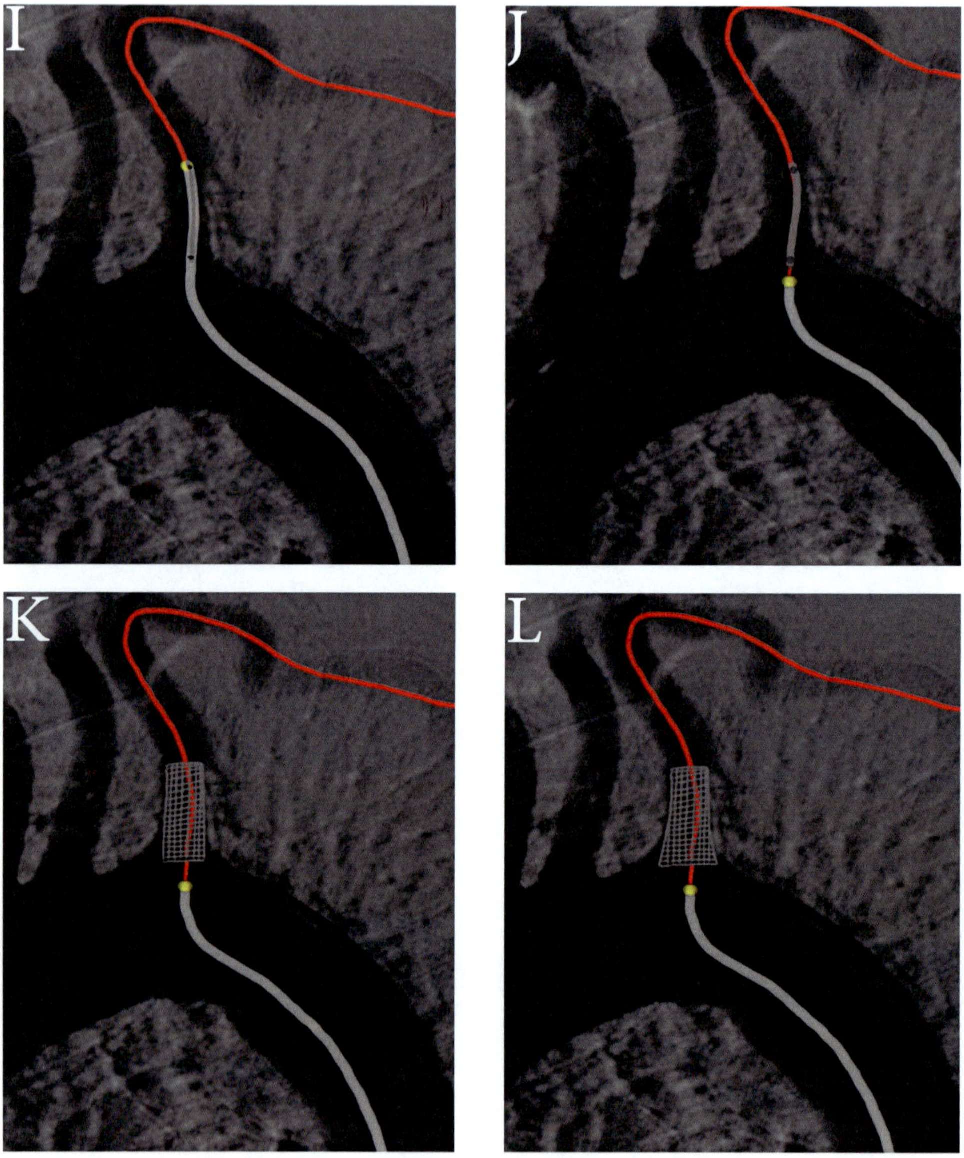

Figure 2.

Ostial/Proximal Left Subclavian Artery Intervention (continued).

I. The stent is positioned while the sheath is still across the lesion.
J. The sheath is withdrawn into the arch and the stent is exposed, positioned across the lesion.
K. The stent is deployed at nominal pressure if tolerated.
L. The ostium of the stent is flared with compliant balloon.

- A pigtail catheter is kept in the distal arch, via CFA or contralateral upper extremity access, to allow visualization of the aortic lumen.
- The lesion is crossed in a retrograde fashion.
- Imaging is performed via the brachial/radial access sheath and/or via the pigtail.
- After the lesion is crossed successfully, the steps for intervention are identical to those of the femoral approach.

LEFT SUBCLAVIAN ARTERY CHRONIC TOTAL OCCLUSION

Technique Via CFA Access:

- Right CFA access is used along with the ipsilateral (left) brachial artery access (affected subclavian).
- The technique is similar to the steps as mentioned above in the ostial/proximal left subclavian artery stenosis section.
- However, recanalization is preferably performed in a retrograde fashion via the left brachial artery.
- Do not fully anticoagulate the patient until a wire and catheter have successfully crossed the lesion.
- Use an under-sized balloon for predilatation to avoid the risk of perforation.
- Make sure the appropriate size covered stent (i.e. stent-graft) is available. For proximal segment disease, this should be a balloon-expandable stent (e.g. iCast®).

RIGHT SUBCLAVIAN ARTERY OSTIAL/PROXIMAL STENOSIS

Arch aortography via femoral access demonstrating a severe proximal right subclavian artery stenosis.

Technique for Intervention Via CFA Access:

The technique is similar to the steps as mentioned above in the ostial/proximal left subclavian artery stenosis section.
Key differences when intervening on the right subclavian:

Positioning of the stent:

- Ideally, the stent should only protrude 1 mm into the brachiocephalic artery in order to avoid covering the ostium of the common carotid artery.

Balloon postdilatation:

- Perform gentle postdilatation of the stent to prevent dissection into the common carotid artery.

RIGHT SUBCLAVIAN ARTERY CHRONIC TOTAL OCCLUSION

Technique for Intervention Via CFA Access:

- Right CFA access is used along with the right brachial artery access.
- The technique is similar to the steps mentioned in the above section.
- The recanalization is preferably performed in an antegrade fashion via transfemoral access. (figure 3)

TABLE 2 - EQUIPMENT FOR SUBCLAVIAN ARTERY INTERVENTION:	
Sheaths	• 6 Fr./90 cm Shuttle® • 8 Fr./11 cm
Guiding Catheters	• IM • JR4 • HS • AL1 or 2 (brachial/radial access for contralateral SCA)
Wires	• 0.035" (260 cm) ▹ Angled stiff hydrophilic wire ▹ Angled non-stiff hydrophilic wire ▹ Straight stiff hydrophilic wire ▹ Straight non-stiff hydrophilic wire • 0.018" (260cm) ▹ Spartacore® ▹ Steelcore® ▹ V-18®
Balloons	• 0.014" ▹ Viatrac® (Abbott Vascular, Santa Clara, CA) ▹ Armada® (Abbott Vascular) ▹ Advance 14LP® (Cook, Bloomington IN) • 0.018" ▹ Armada® ▹ Savvy® (Cordis, Bridgewater, NJ) • 0.035" ▹ Armada® ▹ Powerflex® (Cordis)
Stents	• 0.018" ▹ Omnilink® (Abbott Vascular) • 0.035" ▹ Genesis® (Cordis) ▹ Omnilink® (Abbott Vascular)
Stent-grafts	• 0.018" ▹ Viabahn® (Gore, Flagstaff, AZ) • 0.035" ▹ iCast® (Atrium Medical, Hudson, NH) ▹ Viabahn® (Gore, Flagstaff, AZ)

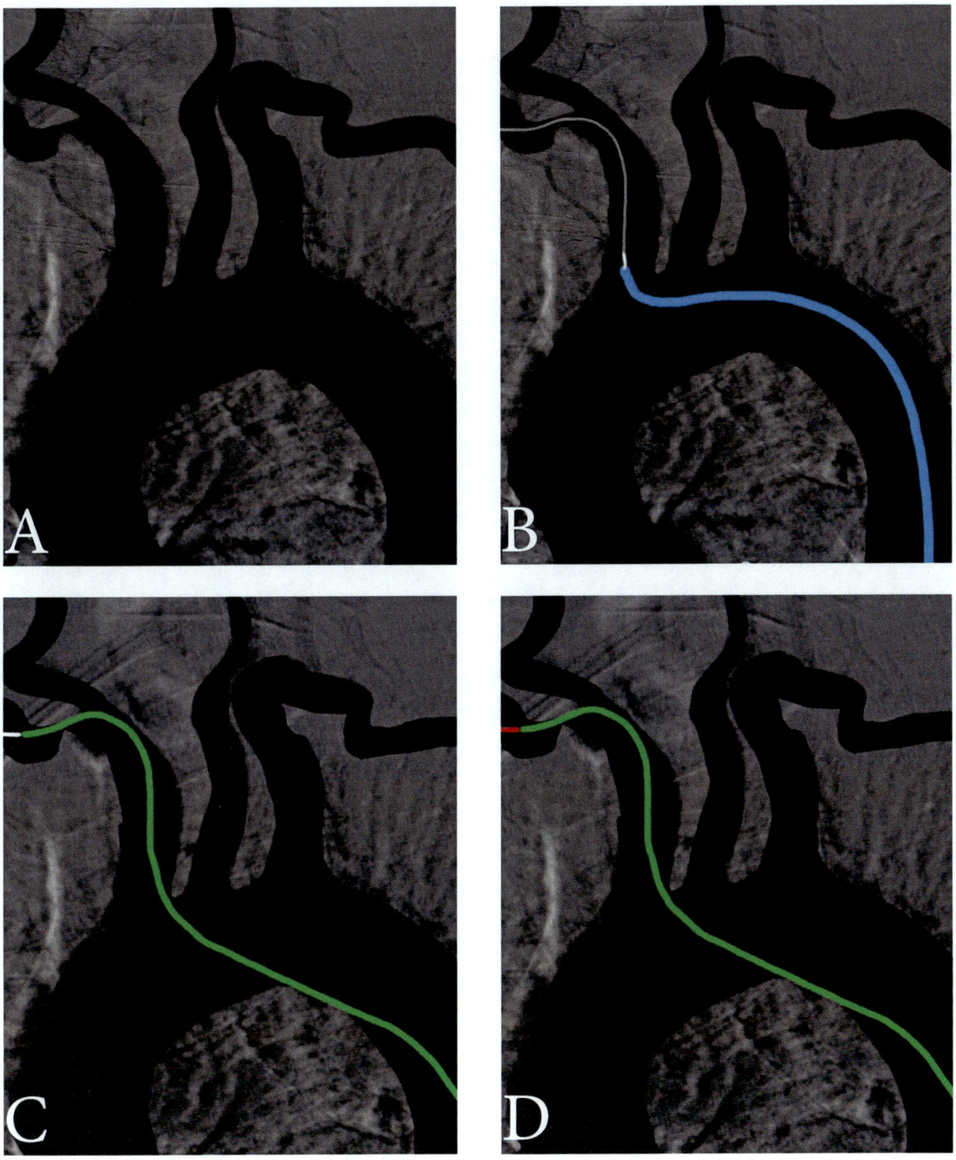

Figure 3.

Ostial/Proximal Right Subclavian Artery Intervention.

A. The aortogram shows a severe stenosis involving the proximal right SCA.
B. The innominate artery is engaged with a 6 Fr. JR4 (blue) diagnostic catheter. A non-stiff AGW is advanced into the axillary artery.
C. The JR4 is exchanged out for a straight Glidecath (green). This is advanced across the lesion into the distal SCA.
D. Through the Glidecath the ANSGW (white) is exchanged out for an ASGW (red). The Glidecath is then removed.

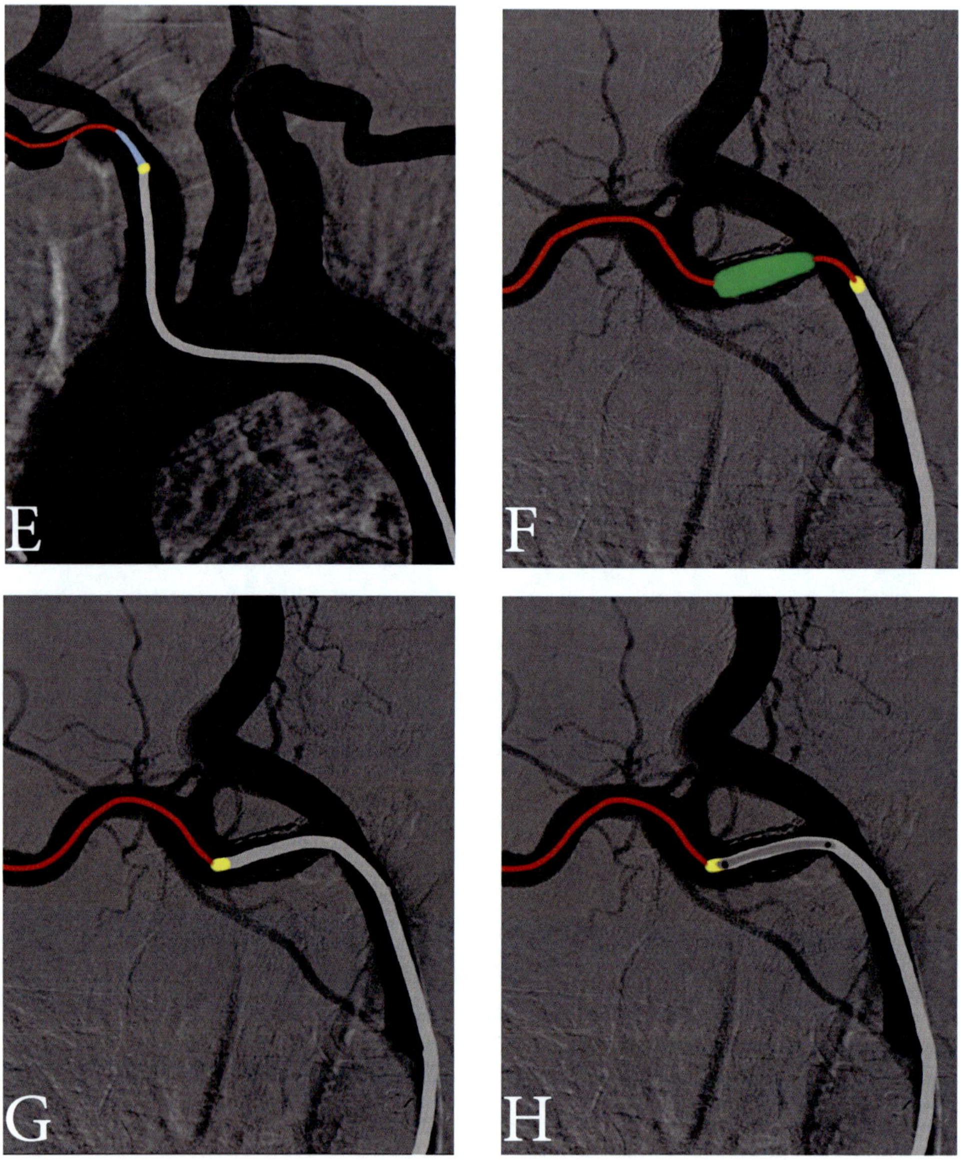

Figure 3.

Ostial/Proximal Right Subclavian Artery Intervention (continued).

E. Over the ASGW, a 6 Fr./90 cm Shuttle sheath (grey) is advanced into the innominate artery. The dilator (light
blue) is removed.
F. A predilatation balloon is advanced into the lesion and the lesion is predilated.
G. The sheath is carefully 'shepherded' over the balloon into the lesion.
H. A BES is then positioned within the sheath, across the lesion.

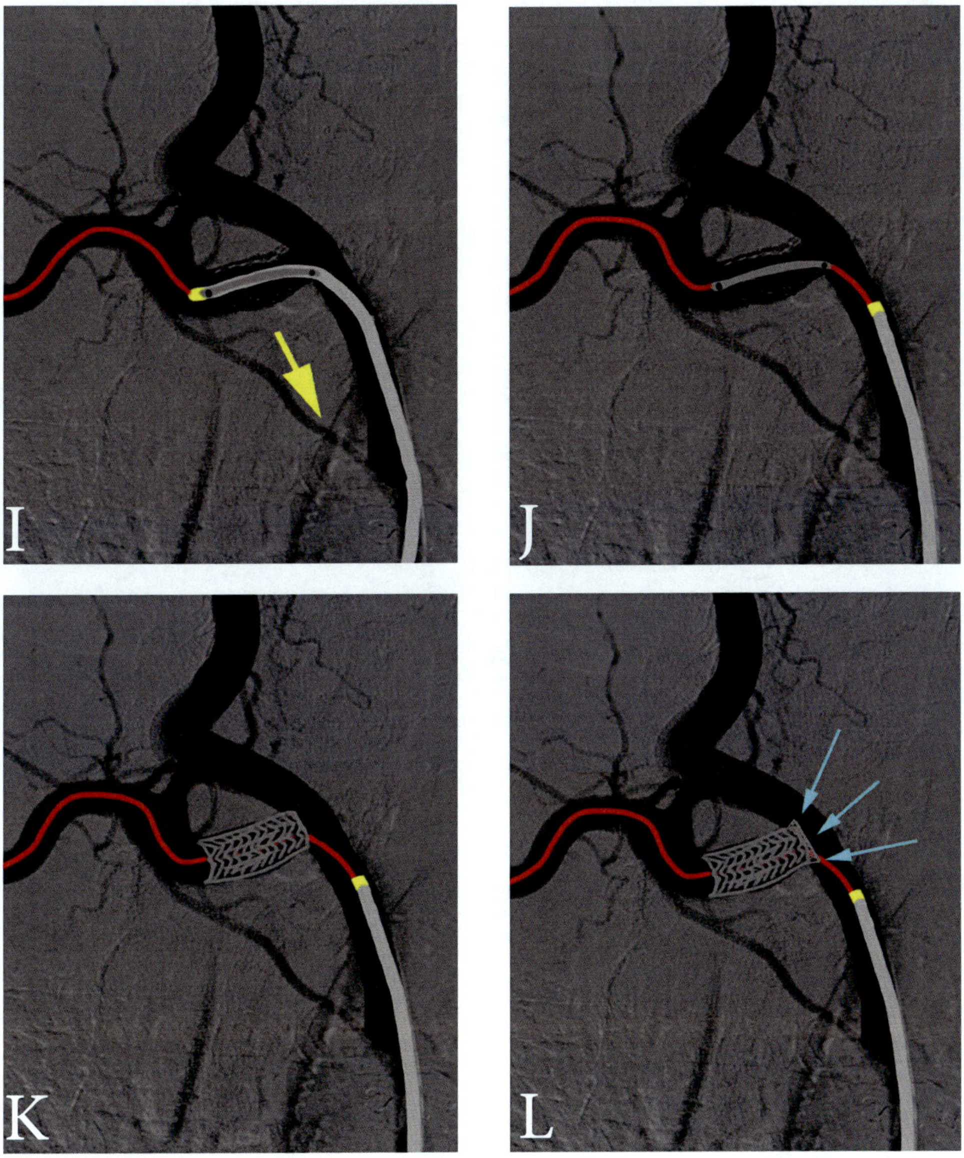

Figure 3.

Ostial/Proximal Right Subclavian Artery Intervention (continued).

I. The stent is positioned while the sheath is still across the lesion, and the sheath is then withdrawn slowly.
J. The sheath is withdrawn into the innominate artery and the stent is exposed, positioned across the lesion.
K. The stent is deployed at nominal pressure if tolerated.
L. The ostium of the stent is flared with a compliant balloon.

CASE

Mr. AL is a 55 year-old male with history of CAD. Two years prior to his presentation he underwent CABG with a left internal mammary artery (LIMA) bypass graft to the left anterior descending (LAD). Subsequently, he had recurrent crescendo angina and underwent a coronary angiogram. Physical exam prior to the angiogram was remarkable for a BP of 138/78 mm Hg on the right and 80/64 mm Hg on the left, as well as diminished left brachial and radial pulses.

Selective angiography of the left subclavian artery demonstrated an eccentric > 80% stenosis with a baseline trans-lesional gradient of 55 mm Hg in the proximal segment of the artery.

Diagnostic Angiography:

- Right CFA access was obtained and a 5 Fr. sheath was inserted.
- IV UFH 2000 units was given.
- Arch aortography was performed in LAO 40 °s with a 5 Fr. non-angled Pigtail using 40cc of Visipaque® at 20 cc/sec at 500 psi.
- Selective angiography was performed with a JR4 catheter (figures 4A & 4B).

Interventional Technique:

- An ANSGW was used to cross the lesion. The JR4 catheter was removed.
- A Glidecath® was advanced over the wire into the axillary artery.
- This wire was then exchanged out for ASGW. The catheter was removed.
- The 5 Fr. sheath was exchanged for a 6 Fr./90 cm Shuttle® sheath for the intervention. This sheath was advanced into the ostium of the left SCA.
- A road map was obtained to visualize the vertebral and internal mammary arteries well.
- A 5.0 mm x 20 mm balloon was advanced across the lesion and a gentle inflation was performed (figure 4C).
- The balloon was removed using the "shepherding" technique (figure 4D).
- An 8.0 mm x 29 mm balloon-expandable stent was advanced across the stenosis, positioned in the appropriate location, away from the vertebral and LIMA, and deployed (figure 4E).
- Again, the stent balloon was removed using the "shepherding" technique (figure 4F).
- An 8.0 mm x 30 mm balloon was used for postdilatation (figure 4G).

Final Angiography:

- This revealed a widely patent left subclavian artery, vertebral artery and the LIMA (figure 4H).
- Repeat pressure measurement in the SCA and the aortic arch revealed no residual gradient at rest or post-IA NTG.

Follow-up:

- He was discharged home on ASA 325 mg/d and Clopidogrel 75 mg/d.
- At 30-day follow-up, the patient was noted to be angina free and asymptomatic.
- There was no difference in BP between the two arms.
- He was maintained on ASA indefinitely with the intention to continue Clopidogrel for six months.

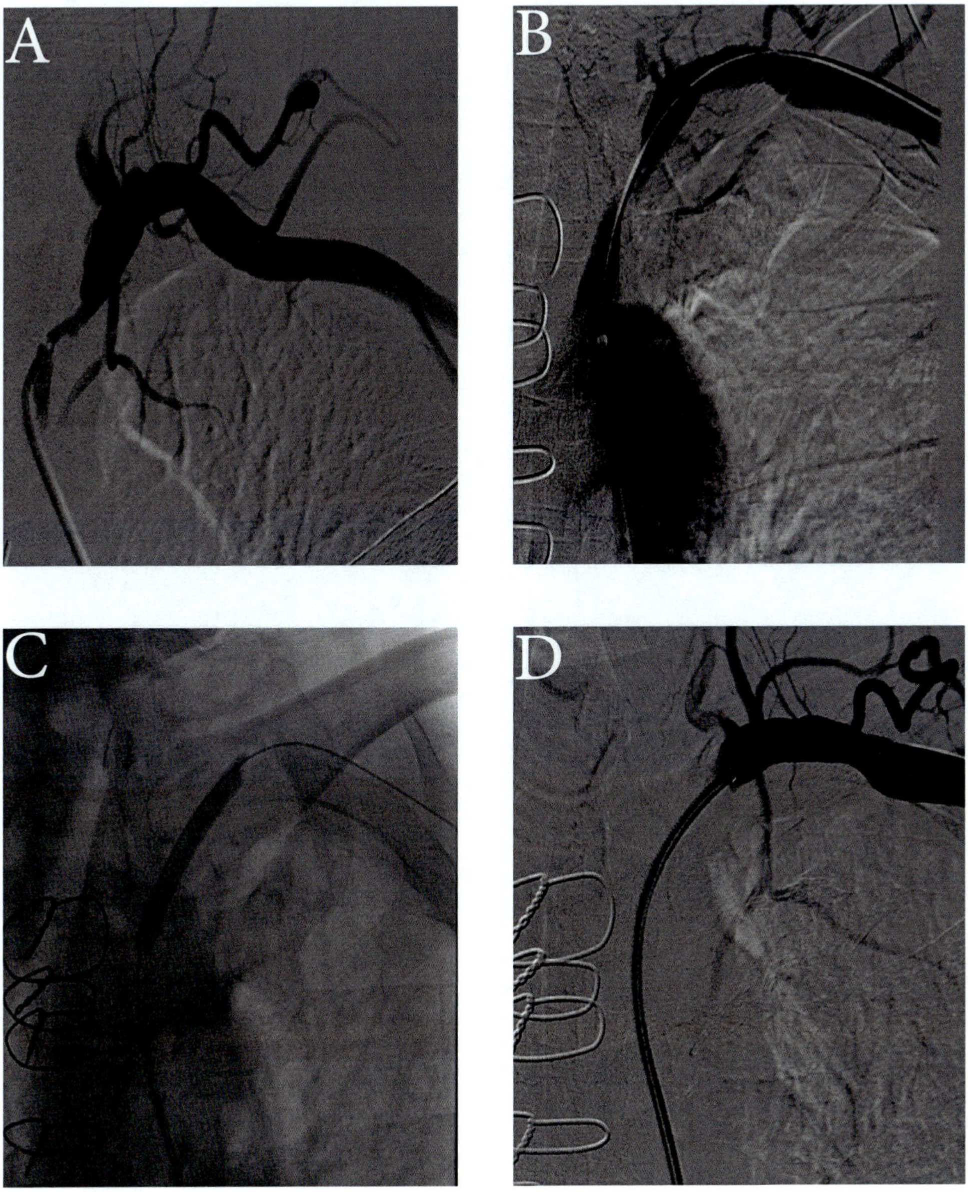

Figure 4.

Left subclavian artery stenting.

A. Selective angiogram of the left SCA showing a severe focal stenosis in the proximal segment. The origin of the
vertebral artery and LIMA is distal to the stenosis.
B. Sub-selective angiogram of the SCA showing the ostium of this vessel to be spared.
C. The lesion is gently predilated with an under-sized balloon through the Shuttle® sheath.
D. An angiogram is obtained after PTA. The Shuttle® has been 'shepherded' distal to the lesion to facilitate placement of a balloon-expandable stent.

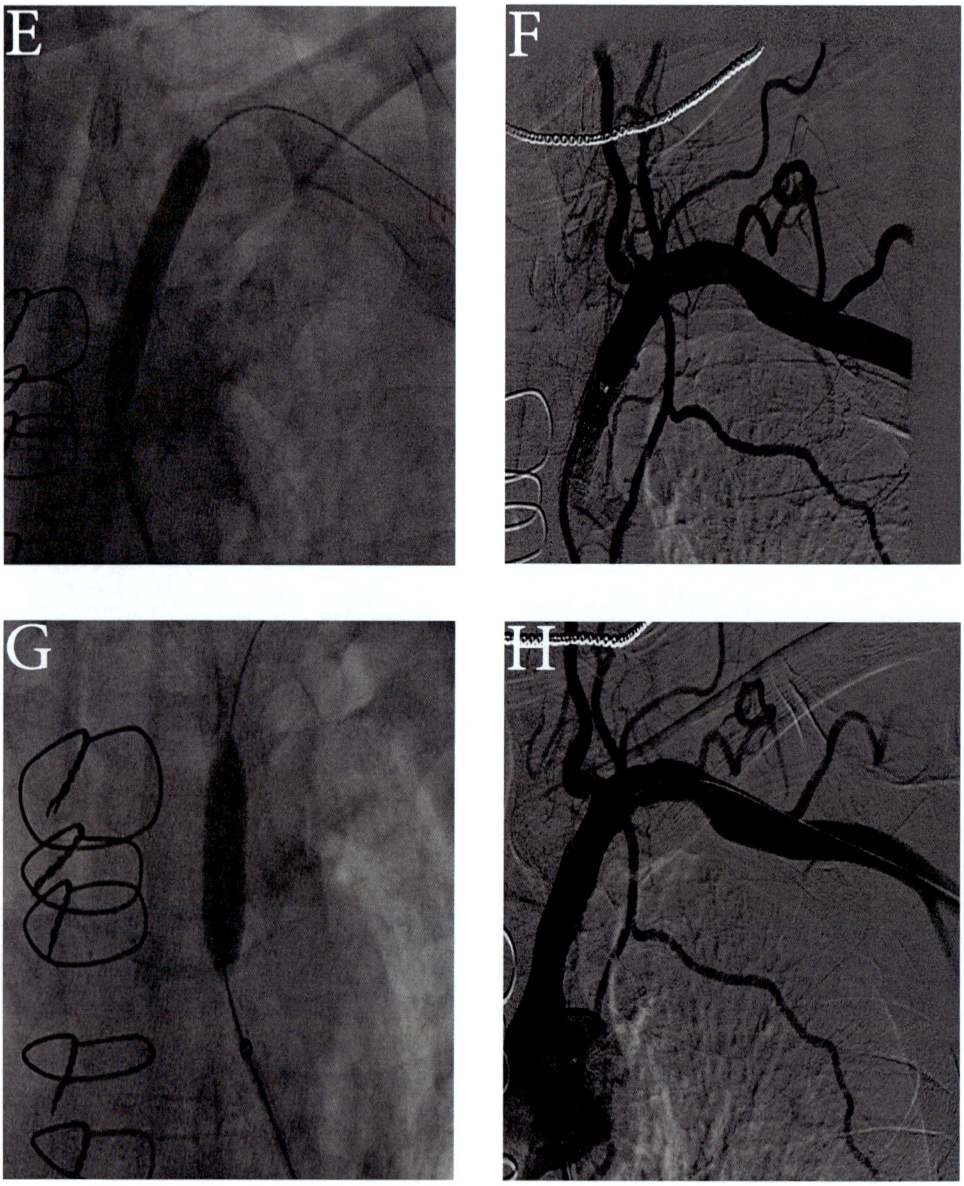

Figure 4.

Left subclavian artery stenting (continued).

E. Stent deployment in the stenosis.
F. Post-stent angiogram shows continued patency of the vertebral artery and LIMA.
G. The stent is postdilated.
H. Final angiogram is obtained through the Shuttle® in the arch.

11

CAROTID ARTERY STENTING

For most patients with carotid artery disease, stenting offers an excellent alternative to CEA. This modality has been studied in asymptomatic as well as symptomatic patients and has been shown to be at least equivalent to CEA in most patient subsets. However, octogenarians have been shown to derive less benefit from this therapy than younger patients due to a slightly higher risk of peri-operative TIA/CVA.

ANATOMY

It is relevant to briefly discuss the anatomy of the aortic arch along with the carotid arteries.

AORTIC ARCH

- The thoracic aorta has three distinct parts:
    - Ascending aorta
    - Aortic arch
    - Descending aorta
- The arch gives off branches to supply the head and upper extremities.
- The curvature of the arch and its relationship to the origins of the great vessels strongly influence the ability to perform carotid angiography and stenting.
- Aortography reveals the configuration of the aortic arch, its curvature as well as the origins of the great vessels.
- This information is used to classify the arch into one of three types. With respect to carotid angiography and stenting, this classification helps quantify the potential difficulty in selectively cannulating the common carotid arteries.
    - A straight horizontal line is drawn across the greater curvature of the arch and another

one parallel to it, off the lesser curvature. The relationship of the origin of the innominate artery relative to these lines/planes is used in classifying the type of arch (figure 1).

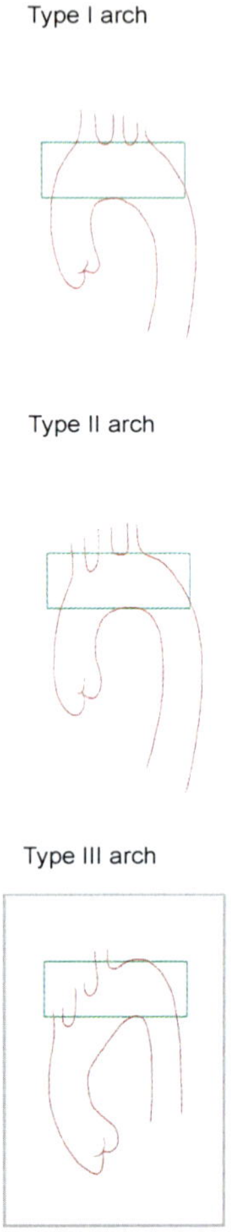

Figure 1.

Aortic arch classification.

Type I:

- Type I aortic arch is considered 'normal'. It is defined by the origin of the great vessels in one horizontal plane off the greater curvature of the arch. Thus the innominate, the left common carotid (CCA) and left subclavian artery (SCA) all arise at the same level (or alternatively, the innominate and left CCA are less than one CCA diameter apart in height).

Type II:

- A type II aortic arch is characterized by the origin of the innominate artery being between the two horizontal planes off the greater and lesser curvatures. Hence, the ostium of the innominate artery is lower than that of the left CCA (by more than the diameter of the CCA).

Type III:

- In type III aortic arch, the innominate artery appears to arise off the ascending aorta. The ostium of the innominate artery is below the horizontal plane off the lesser curve (the difference between the origins of the innominate and left CCA is more than twice the diameter of the CCA).

THE GREAT VESSELS

- Normally, the arch gives rise to the innominate, left common carotid (CCA) and left subclavian arteries (SCA), separately, in that order.

The Innominate Artery:

- It is the first branch of the aorta after the coronary arteries and arises at the level of the second right costal cartilage.
- The lumen of the innominate artery is usually 8-14 mm in diameter.
- Usually 4 - 5 cm in length, it divides into the right SCA and the right CCA at the level of the sterno-clavicular joint.
- Normally, it gives no branches.

Carotid Arteries:

Common carotid arteries:

- The left common carotid artery arises from the summit of the arch, between the innominate and left subclavian arteries, at the level of the second costal cartilage.
- The left CCA is somewhat longer than the right (which arises at the level of the sterno-clavicular joint) and has a thoracic portion in addition to the cervical segment. Otherwise, the right and the left CCA are identical in their caliber and course.
- Each is 8 - 12 mm in luminal diameter and normally gives no branches.
- Rarely, the CCA may give rise to the superior thyroidal, ascending pharyngeal or occipital arteries.
- The CCA terminates at the level of C4-C5 vertebrae (level of the thyroid cartilage) by bifurcating into the external (ECA) and internal carotid arteries (ICA). This bifurcation may occur as low as T2 or as high as C1.

Internal carotid artery:

- The ICA is the bigger of the two terminal divisions of the CCA.
- It is 5 - 7 mm in luminal diameter, normally gives no branches in the neck and supplies most of the blood flow to the ipsilateral eye and cerebral hemisphere.
- In 1996 Bouthillier and colleagues proposed a classification of the ICA into seven segments numbered numerically (C1 - C7) in the direction of blood flow. This nomenclature is widely accepted and it is recommended that the operator be familiar with it.
- These seven segments are (figure 2):
    1. C1 - cervical
    2. C2 - petrous
    3. C3 - lacerum
    4. C4 - cavernous
    5. C5 - clinoid
    6. C6 - ophthalmic
    7. C7 - communicating
- The ICA terminates by dividing into the anterior and middle cerebral arteries.

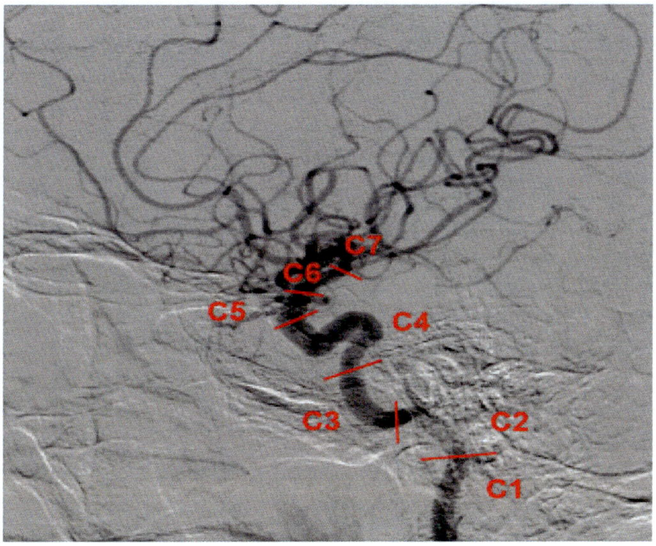

Figure 2.

Segments of ICA.

External carotid artery:

- This is the smaller of the two terminal branches of the CCA.
- It is usually 5 mm in luminal diameter in the proximal segment.
- The ECA gives rise to multiple branches, supplies the neck, the face and the scalp.

Left Subclavian Artery:

- The left SCA is the next branch off the arch. It is usually 8-10 mm in diameter. Its important branches include the vertebral artery, the internal mammary artery, the thyro-cervical and the costo-cervical trunks.

VARIATIONS IN GREAT VESSEL ORIGIN

- More than 25 variations in the origin of the great vessels off the aortic

ECA Branches:
- Superior thyroid artery
- Ascending pharyngeal
- Lingual artery
- Facial
- Occipital
- Posterior auricular
- Superficial temporal
- Internal maxillary

arch have been described. The arch may give rise to as few as one and as many as six vessels. Although a discussion of all these variants is beyond the scope of this chapter, it is important for the reader to know that the following vessels may arise directly from the arch in various combinations:

- ▹ The (right) innominate artery
- ▹ The right vertebral artery
- ▹ The right CCA
- ▹ The right SCA
- ▹ The right ICA
- ▹ The right ECA
- ▹ The 'left innominate' artery
- ▹ The left CCA
- ▹ The left SCA
- ▹ The left vertebral artery
- ▹ The thyroidea ima artery
- ▹ The left ICA
- ▹ The left ECA

The following four variants account for the most commonly found patterns of great vessel origin:

- **Normal arch (~80%):**
  - ▹ Innominate artery, left common carotid artery and left subclavian artery in that order (nearly 80%) (figure 3A).

- **'Bovine arch' (~20%):**
  - ▹ The most frequent variation in the origin of the left CCA is its common takeoff with the innominate artery from the arch (figure 3B). Infrequently, it may arise off the innominate artery - the 'bi-carotid trunk'. These two variations are sometimes referred to as a 'bovine' origin. The term 'bovine' arch is a misnomer. In cattle there is a single brachiocephalic trunk that first divides into right and left subclavian arteries and then a bi-carotid trunk, unlike the so-called 'bovine' arch. It has been suggested that 'bovine' may be a reference to the 'appearance' of the silhouette of the arch in these cases - 'head of the cow, 2 horns'.

Aberrant right subclavian artery:
When the right SCA occurs with a bi-carotid trunk, forms a vascular ring around the esophagus and trachea. This ring may cause symptoms of dysphagia or respiratory distress. This happens rarely, and the condition has been called dysphagia lusoria ('freak of nature').

- **Arch origin of the left vertebral artery (~5.8%)**
  - ▹ Origin of the left vertebral artery (LVA) off the arch, between the left CCA and left SCA is the next most common variant (figure 3C). In 80% of cases, the vertebral arteries originate from the SCA, 5% from the arch, and the rest of the time from the CCA or the innominate artery.

- **Arch origin of right SCA distal to LSCA (0.5%)**
  - ▹ The right SCA may arise directly from the aortic arch, most commonly, distal to the origin of the left SCA (figure 3D).

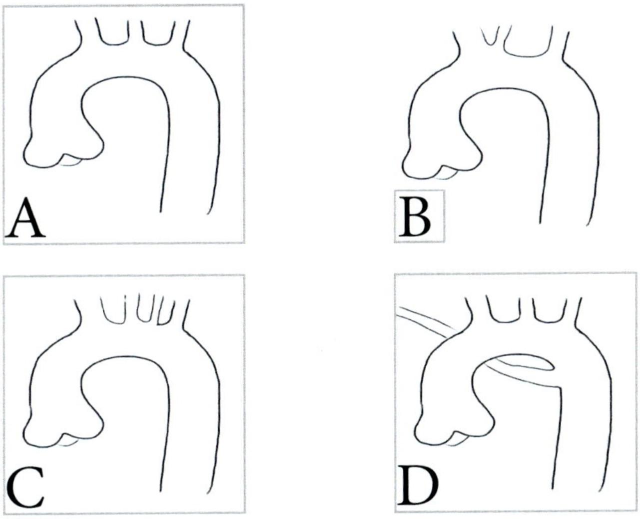

Figure 3.

Aortic arch variants.

A. Normal aortic arch.

B. 'Bovine' aortic arch.

C. LVA arising directly from the arch, between left CCA and SCA.

D. Right SCA arising from the arch, distal to the left SCA.

DIAGNOSIS

Carotid US is the most commonly used modality to make a diagnosis of carotid artery disease. CTA and MRA are used fairly commonly for this purpose as well and provide excellent anatomical details. Conventional angiography, although the gold standard, is used only when the other modalities fail to arrive at a consensus on disease severity.

PRE-OPERATIVE CONSIDERATIONS

We recommend that all patients being considered for elective carotid artery stenting (CAS) should undergo additional pre-operative imaging in the form of either a CTA or MRA of head and neck. This provides insight into the origin of the arch vessels, the burden of disease in the carotid arteries, and also detailed information about the cerebral circulation. Critical information such as the presence of an 'isolated hemisphere' may be picked up prior to the procedure allowing for appropriate planning.

CONTRAINDICATIONS TO CAROTID ARTERY STENTING

Carotid stenting should be avoided in the following subset of patients. CEA should be considered instead.

- Presence of thrombus in the ICA.
- Severe proximal CCA or distal ICA tortuosity.
- Subtotal occlusion ('string sign').

Neither CAS nor CEA should be considered for the following patients:

- Totally occluded ICA/CCA.
- Major residual neurologic deficit from a previous disabling stroke in the territory of the ICA

being considered for intervention.

PATIENT PREPARATION

- The patient is prepared in a standard fashion with both groins shaved.
- Arterial access is usually obtained in the right CFA with venous access in the left common femoral vein (if proceeding to stenting). The purpose of venous access is to be able to administer IV atropine and neo-synephrine/NTG for prompt delivery and action.
- We prefer to keep our patients awake and responsive. Sedation is not given. Small doses of Fentanyl may be used for those who are really anxious or have back pain etc.
- A metal washer or a coin (e.g. a dime) is taped to the side of the neck that is being imaged. This is placed just lateral to and at the level of the angle of the jaw. This serves as a 'rough' marker for the level of carotid bifurcation. Additionally, it can be used for performing QCA. A coin or radio-opaque tape can be used for this purpose (a U.S. dime measures 17.91 mm in diameter).
- IV medications that are prepared in anticipation of the procedure include IV neo-synephrine for BP support and IV NTG to control high BP. IV atropine is also routinely used for patients undergoing carotid stenting to prophylax against bradycardia and hypotension that occur after balloon angioplasty of the lesion. This response is elicited by balloon-induced baro-trauma of carotid baro-receptors.
- Insert a Foley catheter prior to the start of the procedure as the patient may have to lie flat in bed for 12-24 hours post-op.
- Staff will draw their own ACT through the venous sheath at appropriate intervals.
- Manifold with waste-bag; closed system.
- We prefer to use non-ionic, iso-osmolar contrast agents only (e.g. Visipaque).

EQUIPMENT FOR ANGIOGRAPHY

- A small rubber toy (e.g. a rubber duckie) that makes a sound when squeezed is placed in the patient's contralateral hand. During the procedure, the patient is asked to "squeeze the duck" in order to assess the motor function of the hand.
- 4 Fr. micropuncture kit.
- Sheaths: 6 Fr./11 cm and 6 Fr./90 cm Shuttle sheath; 5 Fr./11 cm Arrow sheath.
- Wires: 0.035" J-tip 260 cm wire; 0.038" angled non-stiff 260 cm Glidewire.
- Manifold with a waste-bag and a closed system for contrast injection and hemodynamic monitoring.
- 5 Fr. non-angled pigtail catheter.
- Power injector.
- 5 Fr./125 cm Vertebral (Vert.) catheter; substitutes 5 Fr. Vitek, Simmons (1 through 4).

OUTLINE OF THE PROCEDURE

- Patient preparation.
- Arterial access.
- Venous access (if proceeding to carotid stenting).
- Suture venous sheath in situ; connect extension tubing to venous sheath; meds ready.
- Anticoagulation (goal ACT~300 sec).
- Arch aortography.

- Exchange arterial sheath for 6 Fr./90 cm Shuttle sheath.
- Advance Shuttle to descending aorta.
- Selective catheter over non-stiff 0.038" angled Glidewire (260 cm).
- Cannulation of carotid artery with selective catheter and 0.038" wire.
- Check ACT.
- Selective angiography of the ipsilateral carotid artery.
- Advancement of selective catheter/ 0.038" Glidewire into the ECA under road-map.
- Telescopic advancement of Shuttle sheath into the respective CCA.
- Removal of catheter and wire; repeat road map.
- Introduction of embolic protection device into the ICA.
- Administration of atropine 1.0 mg IV prior to pre-dilatation.
- Predilatation.
- Stent deployment.
- Postdilatation.
- Repeat angiography.
- EPD retrieval.
- Final angiography.
- Removal of Shuttle.
- Arteriotomy management.

ACCESS

ARTERIAL ACCESS

- Carotid stenting can be performed via femoral, radial or brachial artery access.
- Femoral access is used in most cases.

Sheath Size & Type:

5 Fr. - Diagnostic angiography.
6 Fr. - Diagnostic angiography + carotid stenting.
8 Fr. - If using a guiding catheter for carotid stenting.

- For diagnostic angiography, a 5 Fr./11 cm sheath is adequate. Usually a 5 Fr. selective catheter is adequate to perform selective angiography.
- Most carotid stents are compatible with a 6 Fr. sheath or an 8 Fr. guiding catheter. In most cases, a long straight sheath is used as the 'guide'. This sheath is usually advanced into the distal CCA.
- We prefer to use a 6 Fr./90 cm Shuttle sheath (Cook, Bloomington IN). This sheath is hydrophilic and braided and comes with a Tuohy-Borst adapter.
- Guiding catheters are used in cases of complex aortic arch anatomy (type II-III) or CCA tortuosity due to the inherent difficulty in advancing a sheath into the CCA.
- During the case, the guiding catheter is kept in the aortic arch and not advanced into the CCA while engaging the ostium of the innominate or left CCA.

VENOUS ACCESS

Central venous access is obtained only if proceeding to carotid stenting. This is obtained preferably in the left common femoral vein. The main purpose of this access is to be able to administer medication promptly:

- Atropine – for prophylaxis against bradycardia.
- Neosynephrine – for BP support when SBP < 100 mm Hg.
- Nitroglycerin – for management of hypertension when SBP > 160 mm Hg.

Sheath Size & Type:

- A 5 Fr./11 cm flexible/kink-resistant Arrow sheath is used.
- Post-procedure this sheath is usually left in place overnight. Since, it is kink resistant the patient can sit up to 30 degrees upright in bed without damaging the vein.
- We connect a 60" connector tubing to this line and hand that off to the nursing staff. The tubing is flushed and connected to the IV pump that has neo-synephrine infusion on standby. Additionally, the staff has the ability to withdraw blood samples from the sheath through this tubing for checking the ACT. This prevents unnecessary operator distraction.

ANTICOAGULATION

- Fully anticoagulate the patient after arterial access has been obtained and prior to performing diagnostic angiography.
- Unfractionated heparin (UFH) is our anticoagulant of choice.
- However, in patients who are actively symptomatic, we prefer to use IV bivalirudin due to its more potent anti-thrombotic effect.
- Goal ACT is ~300 seconds.

DIAGNOSTIC ANGIOGRAPHY

ARCH AORTOGRAPHY

Catheter:

- A 5 Fr. non-angled pigtail catheter is used for arch aortography.

Imaging:

- This should be performed in a projection that opens up the aortic arch the most. This view provides optimal anatomic detail about the arch, the origin of the great vessels and their relationship to each other.
  - ▹ Typically, this means an LAO projection of 40 degrees or more. Optimal angulation is confirmed on fluoroscopy prior to injection, as the image intensifier is positioned above or below 40 degrees LAO. Place the catheter 1-2 cm proximal to the origin of the innominate artery (towards the ascending aorta).
  - ▹ Rarely, an AP or RAO projection may be necessary due to poor visualization of the great vessel origins in LAO. Alternatively, selective angiography could be used.
- Use digital subtraction angiography (DSA) as this will reduce the volume of dye administration without compromising image quality.
- Remember to use 'breath hold' with DSA.

Dye Injection:

- A power injector is used for aortography.
- 70:30 or 50:50 diluted dye can be used in patients with renal dysfunction.
- The rate of power injection depends upon the patient's body habitus, renal function and ability to comply with instructions of breath-hold during aortography.
- Injection settings are 20 cc/sec for a total volume of 30 - 40 cc at 500 -750 psi with no rate rise.
- CO2 MUST NOT be used for arch or cerebral angiography.

SELECTIVE ANGIOGRAPHY

Catheter:

Catheter choice for selective angiography depends upon the type of the arch. Selective catheters can be divided into two main groups:

Simple curve catheters:

- As the name implies these catheters have a single primary curve. Examples include (figure 4A):
  - ▹ Vertebral
  - ▹ JB
  - ▹ Headhunter
  - ▹ JR
- We prefer to use the vertebral catheter.
- These catheters are used for carotid angiography in patients with:
  - ▹ Type I arch without a levo-flexed left CCA (one that goes more acutely to the left than vertically up).
  - ▹ Type I arch without a "bovine" left CCA.

Complex-curve catheters:

- These catheters have a primary and a secondary curve, in the 'formed' state. Examples include (figure 4B):
  - ▹ Vitek
  - ▹ Simmons (1 through 4)
- We prefer to use the Vitek catheter. However, in patients with a type III arch with 'bovine' configuration , this catheter may not advance as easily into the left CCA due to its rigid secondary curve. In that case, it is preferred to use a Simmons 3 or 4 catheter, which is easier to advance due to its long tip and flexible curves.
- These catheters are used for carotid angiography in patients with:
  - ▹ Type II/III aortic arch
  - ▹ Type I arch with levo-flexed left CCA
  - ▹ Type I arch with "bovine" origin of left CCA

Technique:

Using a simple-curve catheter (e.g. Vertebral):

- The catheter is advanced over a 0.038" angled non-stiff hydrophilic wire into the aortic arch beyond the innominate artery.
- It is then rotated in a counterclockwise direction as it is gradually withdrawn.
- It should first engage the innominate artery. To engage the left CCA, counter clockwise torque is maintained and the catheter is withdrawn further. This maneuver pulls the catheter out of the innominate artery and into the left CCA.
- These steps are repeated to get out of the left CCA and into the left subclavian artery.
- Once the catheter is engaged in the artery of choice, the wire is advanced 2"-4" beyond the tip and the catheter is advanced over it. The wire is remove, the catheter flushed and selective angiography is performed as necessary.

Using a complex-curve catheter (Vitek):

- This catheter (figure 5) is advanced over a wire to the descending thoracic aorta.
- Just distal to the left subclavian artery the wire is removed and the catheter is 'formed' - it is rotated so that the convexity or 'butt' of the catheter is facing the patient's right and the tip towards the patient's left.
- The catheter is then advanced in this configuration into the arch, curve facing forward and tip backward.
- It will 'pop' into the left subclavian artery first. Once engaged, it will become deeply seated in the vessel if it is pulled back. If it is pushed forward the catheter will disengage from the

Figure 4.

Selective catheters for carotid angiography.

A. (left to right).
- Headhunter 1
- JB1
- JB2
- Vertebral

B. (left to right).
- Simmons 1
- Simmons 2
- Vitek

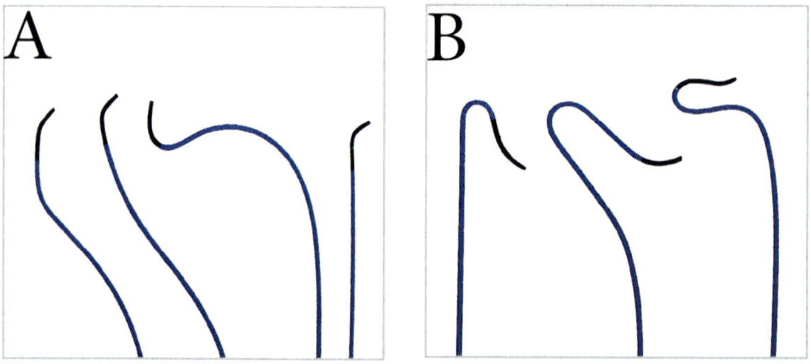

subclavian and will engage the left CCA.

- Follow the same steps to deeply seat in the left CCA vs. engagement of the innominate artery.

Using a complex-curve catheter (Simmons 1-4):

- The Simmons catheters (figures 6 & 7) have to be formed into their curve prior to engaging the carotid arteries. This is accomplished either in the descending thoracic/abdominal aorta or the ascending aorta. The former is preferred over the latter as the catheter then does not have to be withdrawn back in the arch in the formed configuration, a maneuver that in theory may increase athero-embolization from the catheter tip.
- In the descending thoracic/abdominal aorta, the catheter is advanced over a 0.038" wire into a branch vessel (e.g. the contralateral CIA, renal or the left subclavian artery) so that the tip is in the vessel. The wire is withdrawn proximal to the secondary curve of the catheter and the shaft is advanced cephalad. The tip acts as a hinge point and the catheter bends upon its secondary curve during this cephalad push assuming its formed shape.
- Once formed, the catheter is advanced like the Vitek catheter in the arch: 'curve facing forward and tip backward'. For the most part, the same maneuvers are used to engage or disengage an arch vessel as with the Vitek.

Shaping the Simmons Catheter:

- Our preferred approach is to form the Simmons in the in the abdominal aorta at the aortic bifurcation.

Imaging:

The view that profiles the bifurcation of the CCA optimally is the best view for carotid angiography and stenting.

CCA:

- The CCA is best visualized in an AP and lateral projection. However, an ipsilateral oblique projection (around 40 - 60 degrees) may also be helpful. Normally, the CCA bifurcates at C4-5 level but this may occur anywhere between C1 and T2. A non-bifurcating CCA has also been

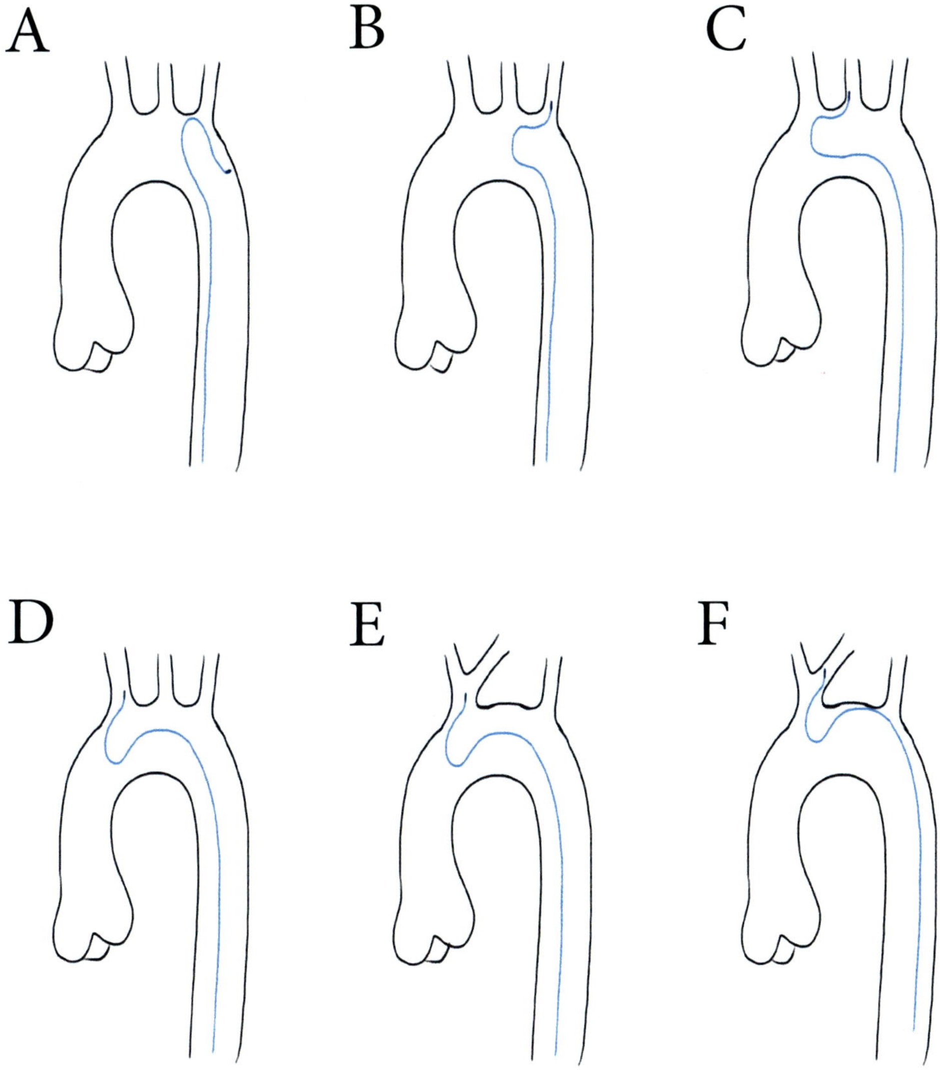

FIGURE 5.

Technique for Using the Vitek catheter.

A. The Vitek catheter is advanced to the level of the descending thoracic aorta, with its convexity ('butt') facing to the
the
patient's right and the tip to the left.
B. It is then advanced into the arch in this configuration, engaging the left SCA.
C. If pushed forward it disengages from the left SCA and 'pops' into the left CCA.
D. The same maneuver is repeated to advance it into the innominate artery.
E. To engage the ostium of the left CCA in a 'bovine' arch this catheter is advanced into the innominate artery.
F. Gentle backward tug on the catheter pulls it up into the ostium of the CCA.

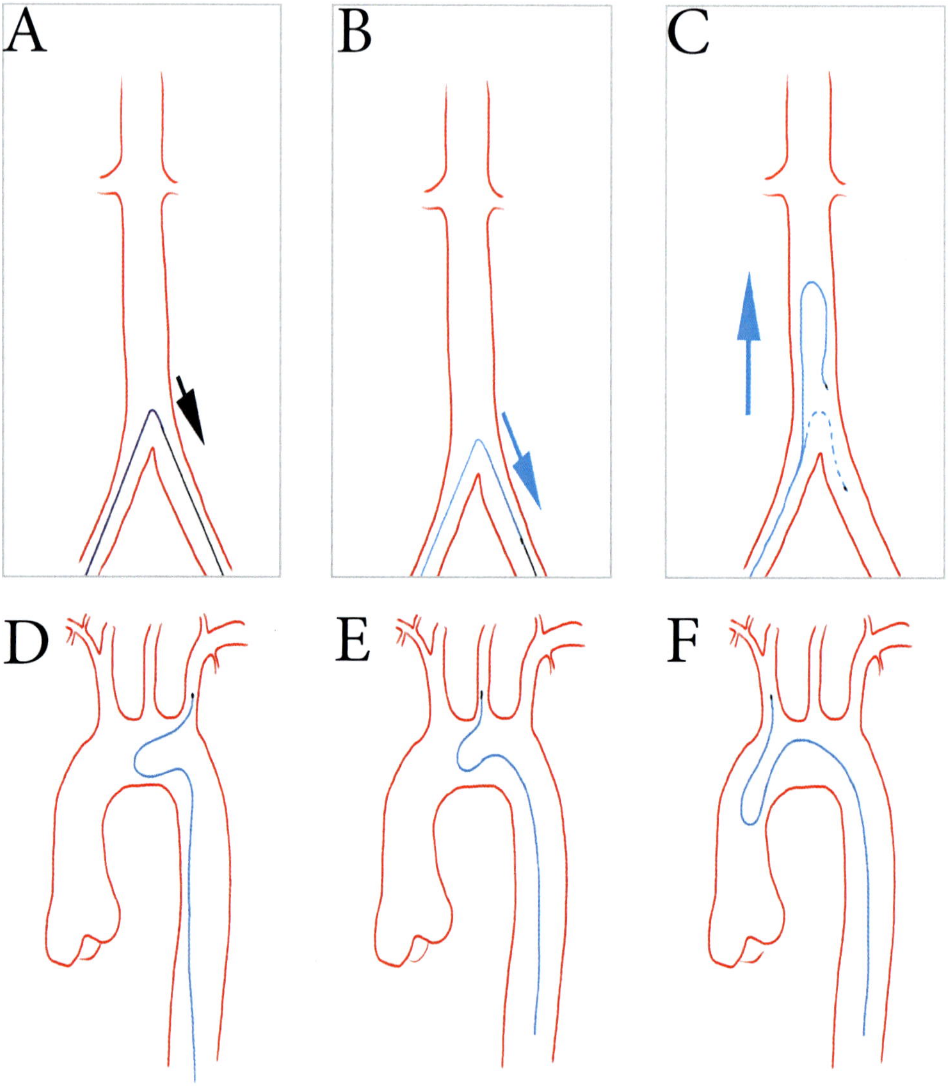

Figure 6.

Technique for using the Simmons catheter.

Our preferred technique is to form the Simmons catheter at the aortic bifurcation.

A. An IM catheter is used to engage the contralateral common iliac artery, a wire is advanced into the EIA.
B. The IM catheter is exchanged out for the Simmons catheter while maintaining wire position in the contralateral EIA/CFA.
C. The wire is withdrawn into the catheter, and the catheter is advanced. As the catheter moves up it forms its desired shape.
D. The catheter is then gradually advanced until it engages the left subcalvian artery.
E. Now, like the Vitek catheter, with a gentle push it can be disengaged from the SCA and advanced to engage the left CCA.
F. The same maneuver is used to engage the innominate artery.

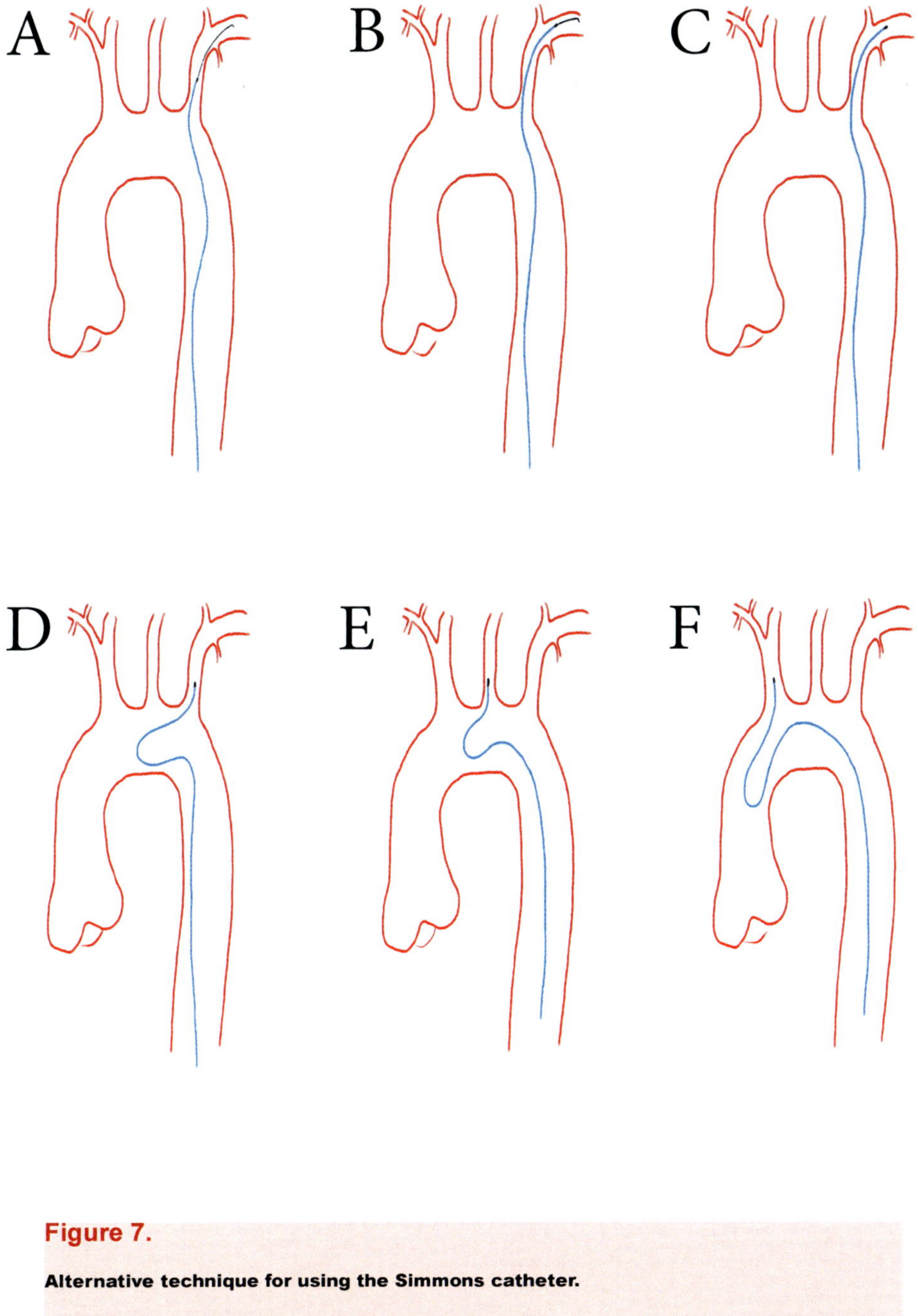

Figure 7.

Alternative technique for using the Simmons catheter.

A. The Simmons catheter is advanced into the left subclavian artery over a pre-existing wire (placed through a JR4 catheter).
B. A 0.038"hydrophilic stiff wire is advanced into the mid vessel and the catheter is pulled into mid SCA over the wire.
C. The wire is withdrawn into the catheter to a level below the arch.
D. The catheter is then gradually advanced until it forms its shape.
E. Now, like the Vitek catheter, with a gentle push it can be disengaged from the SCA and advanced to engage the left CCA.
F. The same maneuver is used to engage the innominate artery.

reported.

- To visualize the ostium and proximal segment of the right CCA, an RAO caudal projection is very helpful as it opens up the bifurcation of the innominate artery.

Cervical ICA:

- This is also best imaged in an AP or lateral projection. However, an oblique projection may be necessary to separate the ICA from the ECA.

Intracranial ICA and the circle of Willis:

- This segment of the ICA is best imaged in a true lateral and an AP cranial projection (Towne's view - AP cranial projection superimposes the petrous and superior orbital ridges) (figures 8A).

Vertebral arteries:

- A sub-selective injection from the SCA or a selective injection may be performed.
- Use IA NTG (150-300 mcg) prior to selective angiography to reduce ostial spasm.
- The origin of the vertebral arteries (V1 segment) is imaged in a Towne's view.
- The V2 segment is best visualized in the AP and lateral projections.
- The V3 and V4 segments are best imaged in the Water's (figure 8B) and the lateral projection.

Vertebral artery segments:
The vertebral artery is divided into four segments:
- V1 - origin to foramen transversarium (usually C6 vertebra).
- V2 - the nearly vertical segment running through the foramina transversaria of C6 to C2 vertebrae.
- V3 - from C2 to duramater.
- V4 - from the duramater to its confluence with the contralateral vertebral artery, forming the basilar artery.

Dye Injection:

- DSA is used with manual injections of 3 - 5 cc of 50:50 dye through the selective catheter.
- The patient is instructed as follows for every injection "do not breath, do not swallow, do not move".
- For intra-cranial imaging, obtain prolonged cine runs to visualize the venous sinus flow.

CAROTID INTERVENTION

GOAL

- To establish patency of ICA/CCA with balloon angioplasty and stenting using embolic protection.
- Residual stenosis of <30%.

ACCESS

- Femoral access is the primary access used for these cases. However, radial or brachial artery access may be used as necessary.

ANTICOAGULATION

As noted above:

- IV unfractionated heparin is used.
- Goal ACT~300 sec.
- Bivalirudin is used in symptomatic patients.
- GP IIb/IIIa inhibitors are unnecessary and not recommended.

INTERVENTIONAL EQUIPMENT

WIRES

0.038":

- To deliver the guiding sheath into the CCA, a 0.038" 260 cm non-stiff angled hydrophilic wire is used. This wire is supportive enough to allow sheath advancement but not too stiff to prevent advancing the system forward.

0.014":

- Carotid stents are 0.014" wire based systems, hence a 0.014" wire is used for the interventional part.
- Embolic protection filters range in diameter from 5.0 mm - 7.5 mm. An appropriate filter sized to cervical ICA should be selected.

SHEATHS

Guiding Sheath:

▹ In a vast majority of carotid stent cases (type I and most type II arch), a 6 Fr./90 cm straight sheath (e.g. Shuttle sheath) is used as the 'guiding' sheath. It is advanced into the distal CCA and the intervention is performed through it.

Guiding Catheter:

▹ For cases where a 'guiding sheath' cannot be used due to excessive carotid/aortic tortuosity and a complex arch configuration (some type II and all type III), a 'guiding' catheter has to be used instead. Due to the caliber of the devices (stents, balloons, etc.), the guiding catheter has to be at least 8 Fr. caliber. The following configurations are used most commonly:

- ♦ JR4
- ♦ Hockeystick
- ♦ Modified Amplatz left 1
- ♦ IMA

BALLOONS

Once again, 0.014" wire compatible, monorail balloon systems are used.

These balloons are prepared with a 30:70 dye dilution so that inflation and deflation are relatively quick.

Predilatation Balloons:

- ♦ Predilatation balloon diameter is typically 4.0 mm with a length of 20 mm.

Postdilatation Balloons:

- ♦ Postdilatation balloons are chosen in the following sizes 5.0 mm x 20 mm or 6.0 mm x 20 mm.

STENTS

- Nitinol self-expanding stents (diameter 7.0 – 10.0 mm and lengths 30 - 40 mm) for ICA and non-ostial CCA disease.
- These stents are available in tapered and straight configurations. Either can be used. There is no advantage of one type over the other.

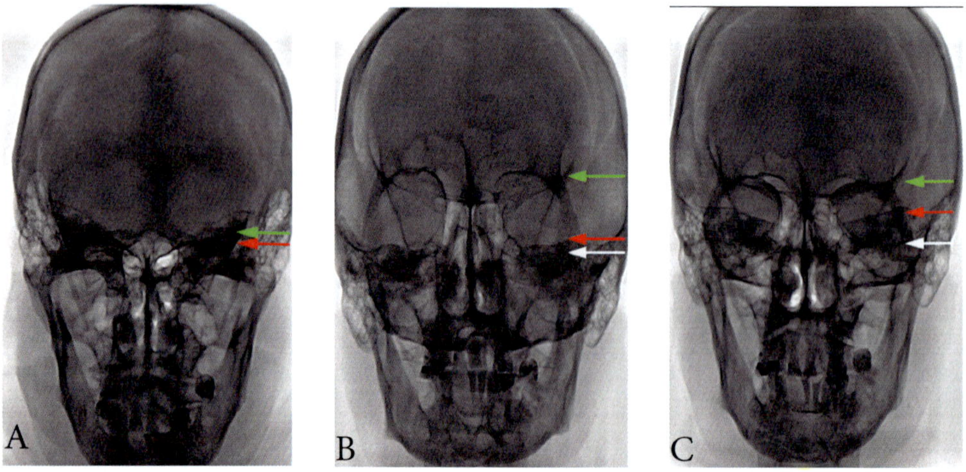

Figure 8.

Anterior views (of the skull) for cerebral angiography.

A. Towne's - The petrous bone (red arrow) is superimposed on the superior orbital ridge (green arrow).

B. Water's - The petrous bone (red arrow) is superimposed on the inferior orbital ridge (white arrow).

C. Caldwell - The pertrous bone (red arrow) is between the superior (green arrow) and inferior orbital ridges (white arrow).

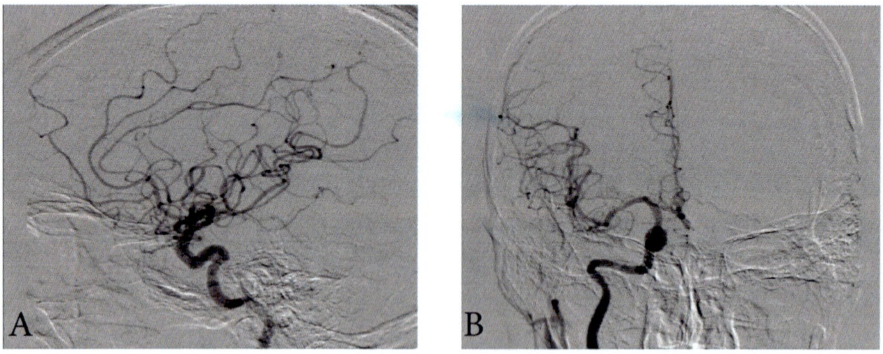

Figure 9.

Views for cerebral angiography.

A. Lateral angiogram of intra-cranial ICA.

B. Towne's view of intra-cranial ICA.

EMBOLIC PROTECTION DEVICES (EPD)

- The use of embolic protection devices during carotid stenting is mandatory.
- Multiple devices are available and depending on the type it may serve the purpose of the primary interventional wire as well.
- The original EPD in CAS was a distal occlusion balloon on a hypotube wire (Guardwire, Medtronic Medical). This caused flow interruption and in about 5% of the patients led to clinical intolerance during the procedure.
- Distal protection filters are the most commonly used EPDs.
- Filter based EPDs provide fairly reasonable protection against embolization during the procedure.
- However, proximal embolic protection (PEP) has been shown to be more protective. There are two available systems for PEP with and without flow reversal.

GUIDING SHEATH TECHNIQUE

VESSEL CANNULATION

- If pre-operative imaging is suggestive of a severe stenosis and carotid stenting appears to be feasible and safe, we proceed with this option as the defacto plan.
- After aortography (figure 10A), the pigtail is removed and over a 0.038" wire, the CFA sheath is exchanged out for a 6 Fr./90 cm straight Shuttle sheath (figure 10B).
- This sheath is advanced under fluoroscopic guidance into the descending thoracic aorta.
- A selective catheter is advanced through the sheath over the wire and the ipsilateral CCA is selectively cannulated. The wire is removed.
- If the plan is to perform bilateral carotid artery angiography, the side that is not being considered for intervention is imaged first.
- Selective angiography is performed to identify the ICA and ECA.

EQUIPMENT SELECTION & PREPARATION

- Next step is to select and prepare the pre-dilatation and post-dilatation balloons, the stent, the embolic protection device and the retrieval sheath.
- To expedite the procedure, we prepare and put these in a systematic pile on the foot end of the operating table: the retrieval sheath on the bottom, next the post-dilatation balloon, the stent and the pre-dilatation balloon on the top.

SHEATH ADVANCEMENT INTO THE CCA

- The ACT should be therapeutic prior to advancing the sheath into the CCA.
- A 0.038" angled non-stiff hydrophilic 260 cm wire is advanced through the selective catheter into the ECA under road map guidance (figure 10C).
- The selective catheter is then advanced over this wire into the distal CCA or the ECA (figure 10D).

Sheath Advancement:

If there is difficulty in advancing the sheath due to tortuosity in the CCA/innominate arteries or the arch is complex, a guiding catheter should be considered. However, if it is desired to use the sheath the following can be tried to advance the sheath over a stiffer wire:

- Over the 0.038" wire in the ECA, re-advance the selective catheter or 5 Fr. Vertebral catheter.
- Through this catheter switch the wire for a 0.035" straight-tip Superstiff Amplatz wire.
- Advance the sheath in a telescoping fashion over this stiff wire/catheter combination.
- If still unsuccessful use a guiding catheter instead or abandon the procedure and consider radial access or alternate therapy.

TABLE 1 - PROXIMAL EMBOLIC PROTECTION DEVICES	
Gore Flow Reversal	• Gore. • Proximal occlusion with active flow reversal via an external circuit and back into the femoral vein. • 9.5 Fr. Gore balloon sheath (with dilator)/Gore balloon wire for ECA and Gore External Filter. • Creates an external AV shunt via 6 Fr. femoral venous sheath. • ECA diameter < 6 mm and CCA diameter 6-12 mm. • Used with 9 Fr./10 cm Terumo Pinnacle® or Terumo RadioFocus®sheaths. • Non-Terumo sheaths > 10 Fr. required.
Mo.Ma Ultra	• Invatec Corp. • Proximal occlusion without active flow reversal. • 9 Fr. sheath/0.035" wire compatible/95 cm shaft. • ECA diameter < 6 mm and CCA diameter 6-12 mm. • Aspirate 60-80 cc of blood prior to balloon deflation.

TABLE 2 - DISTAL EMBOLIC PROTECTION DEVICES	
Accunet	• Abbott Vascular. • Fixed-wire filter. • 0.014" wire Rx- system 190 cm/300 cm lengths ; 4.5 mm - 7.5 mm diameter (for vessels 3.25 mm - 7.0 mm in diameter). • Maximum pore size of filter 150 um.
AngioGuard	• Cordis. • Fixed-wire filter. • 0.014" wire Rx- system 180 cm length; 4.0 mm - 8.0 mm diameter (for vessels 3.0 mm - 7.5 mm in diameter). • Maximum pore size of filter 100 um.
Emboshield Nav6	• Abbott Vascular. • BareWire technology; 0.014" PTFE-coated stainless steel wire with a 3 cm distal platinum/nickel tip; 190 cm/315 cm lengths. • Proximal end has Zebra-stripes; distal wire has a 0.019" stop. • Filter available in two sizes: Small/5.0 mm (2.5 mm - 4.8 mm vessels) and Large/7.2 mm (4.0 mm - 7.0 mm vessel diameter).
Filterwire® EZ	• Boston Scientific. • Fixed-wire filter. 0.014" wire. • 4 and 5 Fr./0.035" wire compatible; 45 cm/90 cm/ 135 cm length.
Gore Embolic Filter	• Gore. • Fixed-wire ePTFE filter. 0.014" PTFE coated wire 185 cm/300 cm length. • 0.014" wire/ 6 Fr. Guiding catheter compatible. • Two sizes 5 mm and 7 mm (for vessel diameter 2.5 mm - 5.5 mm). • Maximum pore size of filter 100 um.
Spider FX	• Covidien. • Fixed-wire Nitinol mesh filter. Cross lesion with guidewire of choice (0.014" or 0.018"). • 0.014" wire system; 190 cm/320 cm length. • 3 mm - 7 mm diameter (for coronary SVGs/carotids arteries 3.0 mm - 7.0 mm in diameter and lower extremity peripheral arteries 5.0 mm -7.0 mm in diameter).

- The Shuttle sheath is then advanced over this catheter/wire combination in a telescopic fashion, under fluoroscopic guidance into the distal CCA (figure 10E). It is assumed there is no severe ostial innominate or CCA disease prior to this maneuver. If such disease is suspected a guiding catheter should be used instead and perhaps ostial disease treated as well.
- Once in the distal CCA, the wire and catheter are removed while stabilizing the Shuttle. It is allowed to adequately back bleed through the hemostatic valve and it is flushed carefully to make sure no air or debris is present.
- A new road map is then obtained, making sure the tip of the sheath and the distal cervical ICA are visualized.
- If the patient is on IV nitroglycerin at this point for BP control, it is discontinued.

CROSSING THE LESION & EMBOLIC PROTECTION

- In all cases of carotid stenting use of embolic protection is mandatory. This may be proximal or distal (tables 1 & 2).
- In most cases, a 0.014" distal embolic protection (device) wire is used and serves as the interventional wire.
- An appropriate filter is selected (see table 3 & 4 below for examples of Accunet and Spider FX), sized to the distal ICA (usually 7.0-7.5 mm for adult males and 6.0-6.5 mm for adult females).
- Filter sizing is based on the C1 segment of the ICA where the filter will be 'parked'.
- This filter should be 'parked' in a straight segment of the distal ICA (figure 10F), making sure it will not cause severe spasm or dissection.
- The filter is deployed in a standard fashion and a timer is started. Ideally, the dwell time of this filter should be < 20 minutes. Longer dwell times have been associated with a higher rate of peri-operative TIA/CVA.
- Care should also be taken to not allow this filter to migrate into the petrous portion of the ICA where the likelihood of intense spasm and dissection is high.

TABLE 3 - SIZING OF ACCUNET FILTER

Size of fully expanded *Accunet* filter (mm)	Reference vessel diameter range (mm)
4.5	3.25-4.0
5.5	4.0-5.0
6.5	5.0-6.0
7.5	6.0-7.0

TABLE 4 - SIZING OF SPIDER FX FILTER

Size of fully expanded *Spider FX* filter (mm)	Reference vessel diameter range (mm)
3.0	3.0
4.0	3.1-4.0
5.0	4.1-5.0
6.0	4.5-6.0
7.0	5.5-7.0

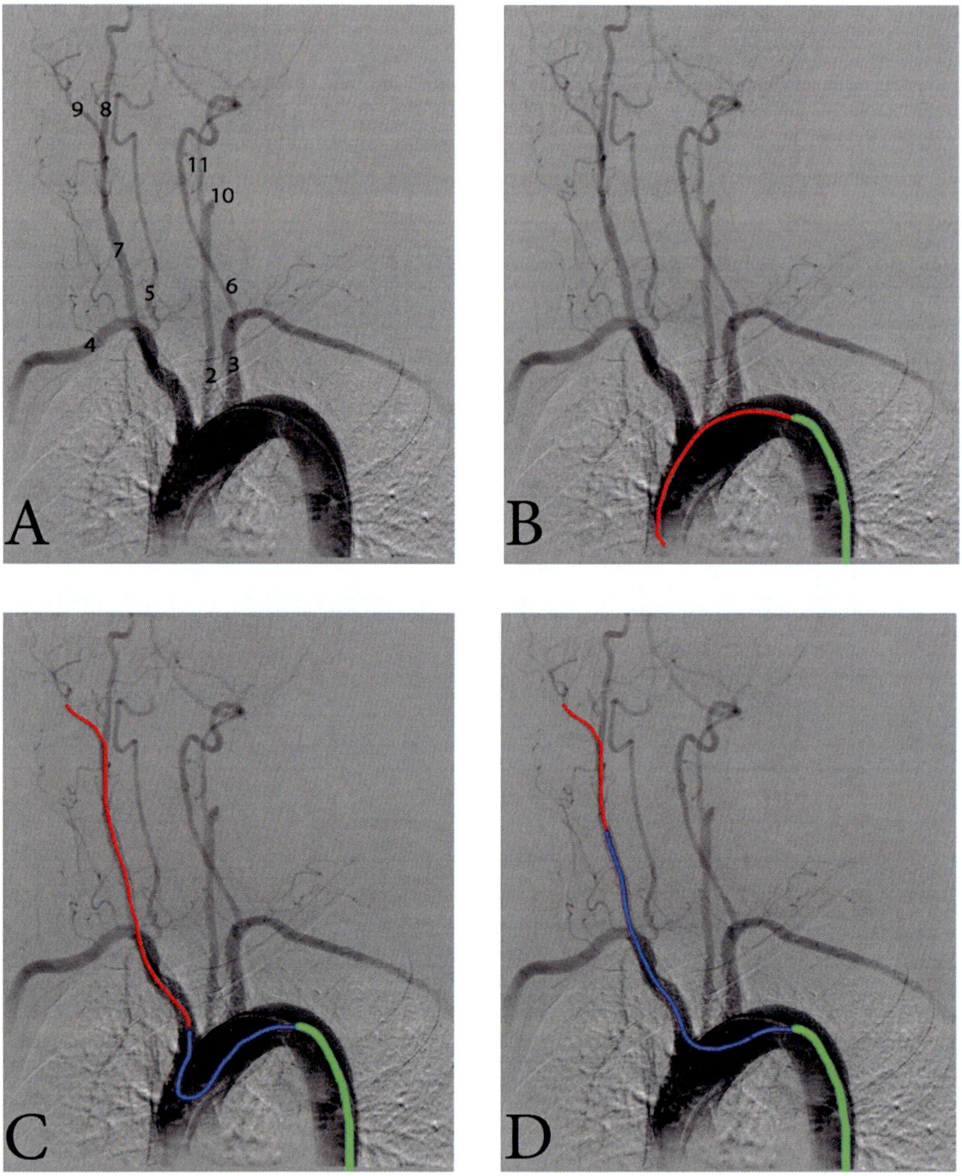

Figure 10.

Carotid stenting using a guiding sheath.

A. Arch aortogram: 1, Innominate artery; 2, left CCA; 3, left SCA; 4, right SCA; 5, right vertebral artery; 6, left vertebral artery;
7, right CCA; 8, right ICA; 9, right ECA; 10, left ICA (occluded 100%); 11, left ECA.
B. Shuttle sheath (green) in the descending aorta and a 0.038" ANGW (red) in the arch.
C. A Vitek catheter (blue) is used to engage the innominate artery and the 0.038" GW (red) is advanced in to the right ECA
under road map guidance.
D. The Vitek catheter is then advanced over the wire into the CCA. The Shuttle remains in the descending aorta at this time.

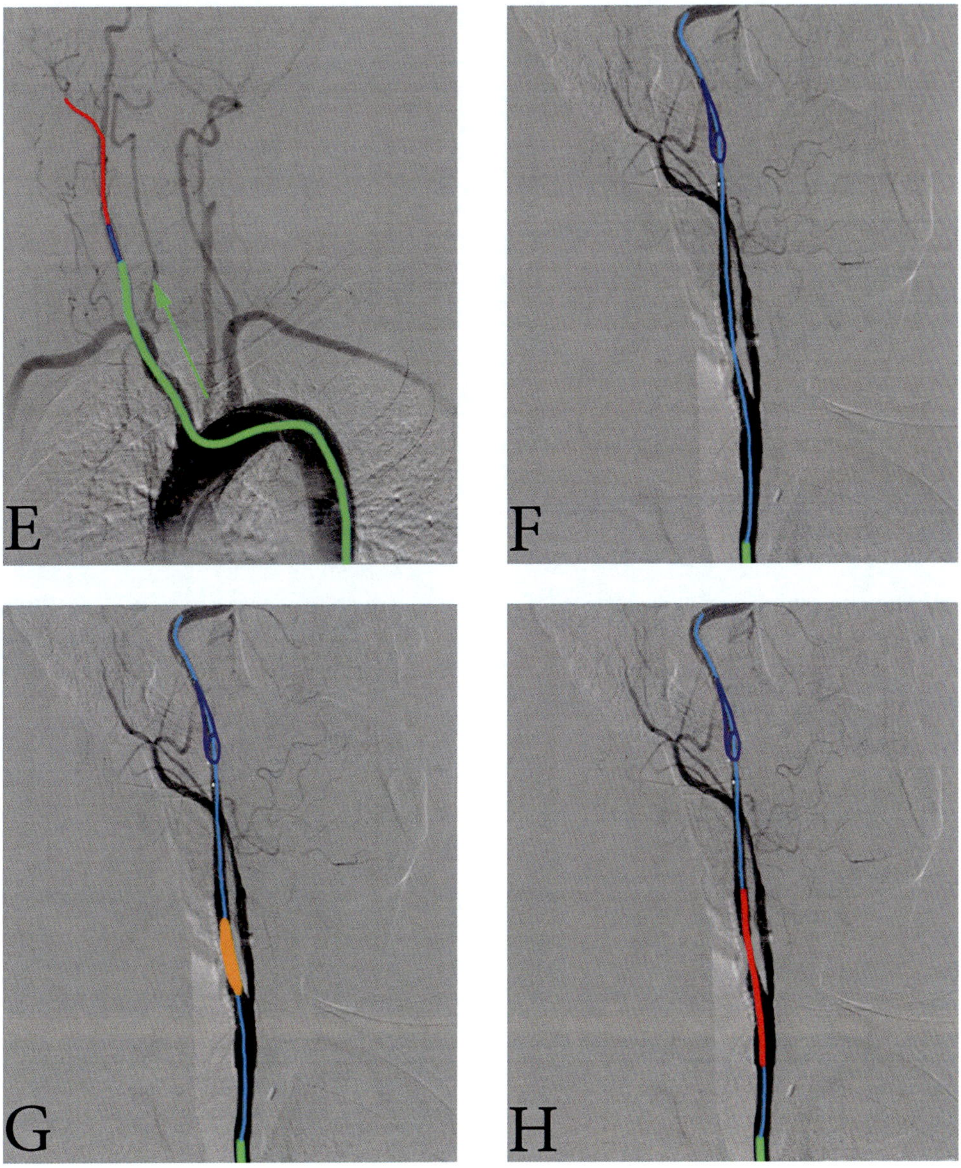

Figure 10.

Carotid stenting using a guiding sheath (continued).

E. The Shuttle sheath is advanced in a 'telescoping' fashion over the Vitek and 0.038" wire into the CCA.
F. After removal of the Vitek and the wire, the 0.014" embolic protection device (filter) is used to cross the lesion. The filter is
deployed in the distal C1 segment.
G. The lesion is pre-dilated with a 4.0 mmx 20 mm balloon.
H. An appropriate size stent is positioned across the lesion. The stent is often extended past the ECA into the distal CCA.

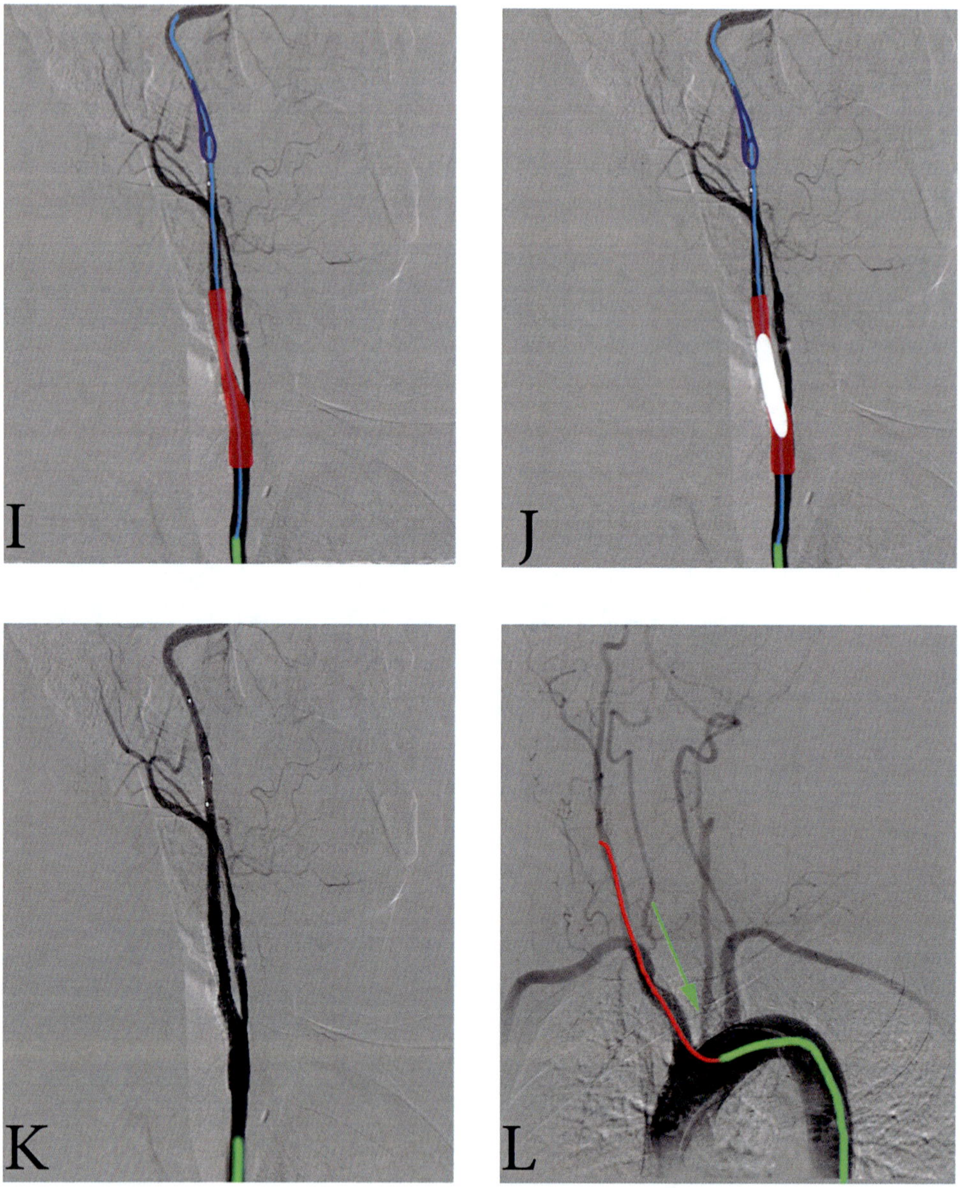

Figure 10.

Carotid stenting using a guiding sheath (continued).

I. The stent is deployed in a standard fashion.
J. The stent is then post-dilated with a 6.0 mm x 20 mm balloon. After repeat angiography, if normal flow is confirmed, the EPD is retrieved.
K. Final angiography is then performed. A widely patent right ICA is noted with a patent ECA.
L. The 0.038" wire is re-introduced into the mid CCA. The Shuttle is withdrawn into the aorta over this wire so that it does not 'scrape' the CCA causing a dissection.

PREDILATATION

- After filter deployment, the pre-dilatation balloon (4.0 mm x 20 mm monorail) is loaded on the wire.
- The patient is given 1 mg of IV atropine.
- Predilatation is performed (figure 10G) under road-map with a rapid deflation (inflation time < 5 seconds).
- The balloon is deflated and removed.
- The patient is asked to 'squeeze the toy' in the contralateral hand to grossly assess motor function as a surrogate for distal embolization.
- Heart rate and systolic BP should be closely monitored. If the BP dips to less than 100 mm Hg, a small bolus dose (5 mcg) of IV Neosynephrine is given.
- IV normal saline bolus is also given to augment intravascular volume.

STENT DEPLOYMENT

- Next, the stent is loaded on the wire and positioned across the lesion under road map guidance (figure 10H).
- It is deployed in a standard fashion (figure 10I).
- Make sure the Shuttle sheath has been pulled back to gain adequate room for the stent.
- The delivery platform is removed after stent deployment.
- Generally, there is no further drop in BP after stenting, prior to post-dilatation.

POSTDILATATION

- The post-dilatation balloon is loaded on the wire.
- Postdilatation is performed with a 5.0 mm or 6.0 mm x 20 mm balloon (figure 10J).
- Again, the inflation duration is < 5 seconds.
- The patient is asked to 'squeeze the toy' in the contralateral hand to grossly assess motor function as a surrogate for distal embolization.
- The patient may need additional doses of IV Neosynephrine for BP support at this point. If the systolic pressure continues to be less than 100 mm Hg, consider initiating a Neosynephrine drip.
- Patients with a contralateral occlusion are prone to have ischemic symptoms during this period of hypotension and hypoperfusion.
- Aim to keep SBP 100 - 160 mm Hg. Higher blood pressures can lead to intra-cranial HTN and hemorrhage due to cerebral dysautoregulation ("hyperperfusion syndrome").

REPEAT ANGIOGRAPHY

- Prior to retrieval of the embolic protection device, an angiogram should be performed to document patency and normal flow (figure 10K).
- Rarely, 'no re-flow' is seen due to filter 'overload' with debris from the lesion. There may be a combination of floating debris in the static column of blood plus debris lining the filter.
- This situation is best dealt with advancing the Shuttle sheath through the stent (under fluoroscopic guidance to not damage the stent) and close to the filter. A 60 cc luer-lock syringe is connected to the side-arm of the sheath and aspiration is performed. At least 20-30 mL of blood is aspirated and wasted.
- This should result in restoration of normal or near normal flow (TIMI 2-3). This is confirmed on repeat angiography.
- If the flow remains poor, quick repeat aspiration should be performed and another 30 mL blood should be removed.
- Additionally, a dissection at the stent edge or at the site of filter deployment should be ruled out.
- If no dissection is noted, and flow still remains sluggish the filter can be retrieved assuming that either the filter is 'full' or there is spasm at the site.

Field of view:
During the procedure, keep the following three areas in the field of view at ALL times:
- Filter/proximal marker on the filter.
- ICA lesion
- Tip of the guiding sheath.

Difficult Retrieval:
If the retrieval catheter cannot be advanced through the stent the following steps may be used:
- Having the patient turn their neck to one side or the other.
- Moving the sheath gradually either up or down.
- Advancing the sheath into the stent (making sure it does not damage the stent struts).
- Taking a shapeable retrievable catheter.
- Post-dilating the stent.
- Using a 5 Fr. 125 cm MPA catheter.
- Using a 'buddy wire'.

FILTER RETRIEVAL

- Make sure the retrieval catheter has been flushed.
- After removal of the post-dilatation balloon, the retrieval catheter is back loaded on the wire.
- It is advanced out of the sheath and into the stent under fluoroscopic guidance.
- Keep the deployed filter in field of view and well above the stent.
- Once through the distal end of the deployed stent, the retrieval catheter is advanced until the proximal loop of the filter (four markers on Accunet; elliptical hoop of the Spider FX) is just in the catheter.
- DO NOT attempt to contain the entire filter within the retrieval catheter. This can dislodge the trapped debris in addition to the risk of filter detachment.
- Make sure the Tuohy-Borst adapter is wide open.
- Gently pull the retrieval catheter and the filter as a unit into and through the stent.
- Once in the sheath they can be swiftly removed.
- Allow the sheath to back-bleed adequately and flush once again ensuring there is no air or debris.
- Repeat angiography should be performed to document normal flow after removal of the filter. If spasm is present and the patient's BP allows, local IA nitroglycerin can be given.
- The patient is asked to 'squeeze the toy' in the contralateral hand to grossly assess motor function as a surrogate for distal embolization.
- Cervical ICA and intra-cranial ICA, ACA and MCA should be imaged in lateral and Towne's views.

SHUTTLE WITHDRAWAL

- Prior to removal of the Shuttle sheath from the CCA, a 0.038" wire should be advanced into the vessel to the tip of the sheath, and the sheath withdrawn over it (figure 10L).
- This reduces the chance of the sheath 'dragging' along the vessel wall causing a dissection during withdrawal.
- The patient is asked to 'squeeze the toy' in the contralateral hand to grossly assess motor function as a surrogate for distal embolization.

ACCESS MANAGEMENT

- The arteriotomy is managed as appropriate.
- The venous sheath is left in situ overnight to infuse medication for BP management (Neosynephrine for SBP < 100 mm Hg and IV NTG for SBP > 160 mm Hg).

GUIDING CATHETER TECHNIQUE

- If the configuration of the aortic arch (type II/III) would not allow the procedure to be completed using a guiding sheath (e.g. Shuttle sheath), consideration is given to carotid stenting through an 8 Fr. guiding catheter.
- An 8 Fr. sheath is used in the CFA.
- The guide is advanced over a 0.035" wire into the arch and used to en-

gage the innominate (for right ICA) or left CCA (for left ICA).

- The guiding catheter is not advanced beyond the arch!
- Only distal embolic protection can be used with a guiding catheter. Proximal protection cannot be used.
- A roadmap of the CCA/ICA is obtained and the EPD is advanced through the guide across the lesion into the distal cervical ICA.
- The rest of the procedure is performed in the same fashion as with a guiding sheath.

CASE 1

A 59-year old male truck driver presented with an anterior STEMI. He successfully underwent primary PCI with DES to the proximal LAD. However, he gave a history of three episodes of right eye amaurosis fugax in the preceding two weeks leading up to his current presentation.

A carotid US on the day after his primary PCI revealed a critical stenosis in his right ICA (80-99%). Given his active symptoms of TIA, carotid revascularization was recommended. However, due to his recent STEMI he was considered to be a suboptimal candidate for CEA. Hence, carotid stenting was advised.

Initial Diagnostic Equipment:

- 4 Fr. micropuncture kit.
- 6 Fr./11 cm sheath.
- 5 Fr./11 cm Arrow flexible sheath.
- 0.038" angled non-stiff Glidewire.
- 5 Fr./100 cm pigtail catheter.
- Power injector.
- Squeezable toy in left hand.
- Washer on right neck.

Access & Diagnostic Angiography:

- Arterial access was obtained in the left CFA as coronary PCI had been performed via the right CFA 24 hours prior.
- Venous access was obtained in the left CFV and a venous extension tubing was connected and flushed.
- The venous sheath was sewn in place.
- No sedation was used.
- The patient was given IV Bivalirudin for anticoagulation.
- Arch aortography was performed using the 5 Fr. pigtail catheter.
- For selective angiography of the right CCA, a 5 Fr. Vert catheter was selected.

Interventional Technique:

Startup equipment

The following were prepared and setup on the table.

- 6 Fr./90 cm Shuttle sheath.
- 5 Fr./125 cm Vert catheter.
- Spider FX 7.0 mm filter.
- 4.0 mm x 20 mm Viatrac balloon for pre-dilatation.
- 6.0 mm x 20 mm Viatrac balloon for post-dilatation.
- 7 mm/10 mm x 30 mm Acculink stent.
- Spider retrieval catheter.

Engaging the right CCA

- The sheath in the left CFA was exchanged out for a 6 Fr./90 cm Shuttle sheath.
- The Shuttle was brought into the descending thoracic aorta, the dilator was removed and the sheath was flushed.
- Over the 0.038" 260 cm angled non-stiff Glidewire, the Vert catheter was advanced into the arch beyond the innominate.
- After selective cannulation, a road map of the right CCA was obtained and the wire was advanced into the ECA.

- The catheter was advanced into the distal CCA.
- The Shuttle was telescoped over the two into the distal CCA.
- The catheter and wire were removed and the Shuttle flushed.

Embolic protection

- A road map was obtained.
- Due to the critical stenosis (99%) in the proximal ICA (figure 11A), it was decided to cross the lesion with a 0.014" 300 cm Choice PT floppy wire.
- A 7.0 mm Spider FX filter was advanced over this wire into the straight segment of distal C1 (figure 11B).
- The filter was deployed in a standard fashion.

Predilatation

- The pre-dilatation balloon was loaded on the wire.
- The patient was given 0.6 mg of IV atropine.
- The lesion was pre-dilated for < 5 sec. at nominal pressure (figure 11C).
- The balloon was removed.
- The patient was asked to 'squeeze the toy'.

Stenting

- Next, the stent was loaded on the wire and advanced into the lesion.
- The sheath was withdrawn a couple of inches and the stent deployed in the standard manner.
- The patient was asked to 'squeeze the toy'.
- The stent catheter was removed.

Postdilatation

- The 6.0 mm x 20 mm balloon was loaded on the wire and post-dilatation was performed for < 5 sec (figure 11D).
- The patient was asked to 'squeeze the toy'.
- The balloon was removed.
- The patient was given 5 mcg of IV Neosynephrine due to hypotension (SBP 80 mm Hg).

Repeat angiography

- Repeat angiography revealed a widely patent ICA with < 20% residual stenosis.
- TIMI-3 flow was noted.

Filter retrieval

- The Spider was retrieved in the standard fashion with partial closure capture of the filter.
- Final angiograms were performed (figure 12) including intra-cranial views.

Shuttle removal

- The sheath was then removed from the CCA after the 0.038" angled Glidewire had been re-inserted into the CCA proximal to the stent.
- The sheath was withdrawn with the wire still being anchored in the CCA.

Arteriotomy management & patient disposition

- After removal of the Shuttle, a 6 Fr. Perclose Proglide device was used to achieve arterial closure.
- The venous sheath had been sutured in place.
- The patient was sent back to the CCU.
- He required IV Neosynephrine drip for the next 12 hours.
- He was continued on ASA 325 mg/d and Plavix for at least one year (due to his MI).

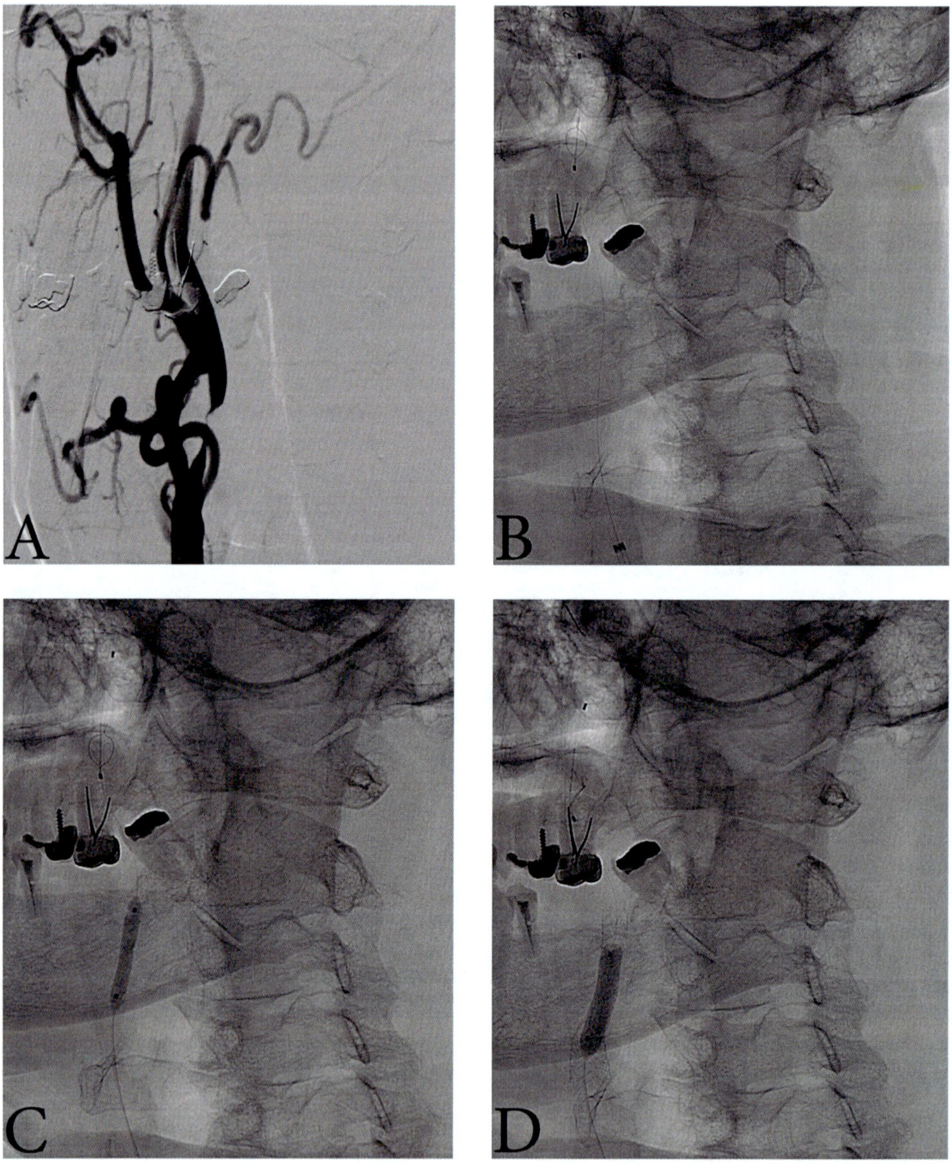

Figure 11.

Carotid stenting.

A. Diagnostic angiogram showing a critical right ICA stenosis.
B. Spider FX deployed in the distal C1.
C. Predilatation after embolic protection.
D. Postdilatation after stenting.

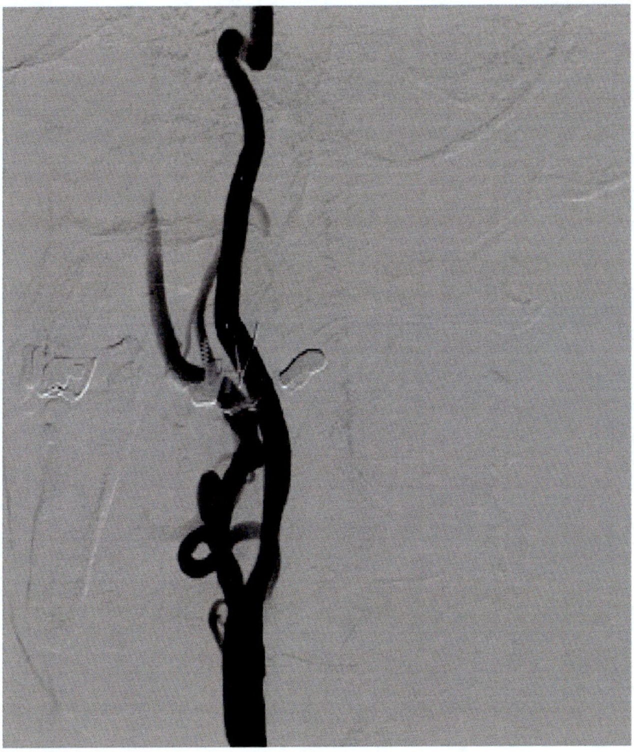

Figure 12.

Post-stent carotid angiography.

CASE 2

A 78-year old right-handed male with bilateral carotid bruits and increasing exertional angina underwent coronary angiography revealing severe triple vessel disease. A carotid US revealed severe (80-99%) stenosis in the ICA bilaterally. His left SCA was also noted to have an 80% proximal stenosis. These were confirmed angiographically at the time of cardiac catheterization. A type I aortic arch was noted (figure 13A). The left ICA was noted to have an ulcerated plaque in the proximal segment with a severe stenosis (figure 13B). He was advised to undergo carotid artery stenting followed by CABG.

It was decided to intervene on his left ICA first followed by the right ICA and left SCA in a staged fashion.

Initial Diagnostic Equipment:

- 4 Fr. micropuncture kit.
- 6 Fr./11 cm sheath.
- 9 Fr./23 cm sheath.
- 5 Fr./11 cm Arrow flexible sheath.
- 0.038" angled non-stiff Glidewire.
- 5 Fr./100 cm pigtail catheter.
- 5 Fr./125 cm Vert catheter.
- Power injector.
- Squeezable toy in right hand.
- Washer on left side of the neck.

Access & Diagnostic Angiography:

- Arterial access was obtained in the right CFA.
- Venous access was obtained in the left CFV and a venous extension tubing was connected and flushed.
- The venous sheath was sewn in place.
- No sedation was used.
- The patient was given IV UFH for anticoagulation. An ACT of > 250 sec was confirmed.
- Arch aortography was performed using the 5 Fr. pigtail catheter.
- For selective angiography of the left CCA, a 5 Fr. Vert catheter was selected.

Interventional Technique:

Startup equipment

A decision was made to use the MoMA Ultra proximal embolic protection system. This system is inserted through a 9 Fr. sheath that is at least 23 cm long.
The following were prepared and setup on the table.

- 9 Fr./23 cm Cordis Brite-tip sheath.
- 5 Fr./125 cm Vert catheter.
- 0.038" 260 cm ANSGW.
- 0.035" 260 cm SuperStiff Amplatz wire.
- 0.014" 180 cm Choice PT floppy wire.
- 4.0 mm x 20 mm Viatrac balloon for pre-dilatation.
- 6.0 mm x 20 mm Viatrac balloon for post-dilatation.
- 7 mm/10 mm x 30 mm Acculink stent.
- 20 cc luerlok syringe.

Engaging the left CCA

- The sheath in the right CFA was removed over a 0.035" J-tip wire and the arteriotomy was pre-closed with a single Perclose device in the standard fashion.

- A 9 Fr./23 cm CBT sheath was inserted in the right CFA and the tip advanced into the distal aorta. The j-tip wire was removed.
- Over the 0.038" 260 cm angled non-stiff Glidewire, the Vert catheter was advanced into the arch and the left CCA was engaged.

Wire advancement in the left ECA

- After selective cannulation, a road map of the left CCA was obtained and the wire was advanced into the distal ECA.
- The catheter was advanced over the wire into the mid ECA.
- The wire was then exchanged out for the 0.035" j-tip SuperStiff Amplatz wire. The tip of this wire was kept in the mid ECA.
- The Vert catheter was removed.

Introduction of the MoMa Ultra for embolic protection

- A MoMa Ultra proximal embolic protection system was prepared in the standard fashion.
- Under fluoroscopic guidance, the MoMa Ultra device was introduced through the 9 Fr. sheath and advanced into the left CCA (figure 13C & 13D) so that the tip was in the mid ECA, approximately 1.5 cm beyond the carotid bifurcation.
- The dilator/manderil was removed at this point.

Setup of proximal occlusion

- With the wire in place, the ECA balloon was inflated with a 20 cc syringe using a 50:50 contrast:saline mix (figure 13E).
- A confirmatory angiogram was obtained to ensure ECA occlusion (figure 13F).
- The SuperStiff Amplatz wire is then removed.
- The CCA balloon was then inflated in the same fashion (figure 13G).
- 10 cc blood was removed from the side-arm of the hemostatic valve.
- A 5 cc contrast injection was perfomred to confirm CCA occlusion (figure 13H).

Assesing patient tolerance to proximal occlusion

- With proximal occlusion in place the arterial (invasive) pressure measured through the MoMa was around 60 mm Hg which is adequate. The right arm cuff systolic pressure was 140 mm Hg.
- The patient was assessed neurologically and was noted to have no intolerance to the device.

Wiring of the ICA

- With proximal occlusion in place and tolerated well by the patient, a Choice PT floppy wire was used to cross the lesion in the ICA (figure 13I).
- The tip of the wire was advanced into the distal cervical ICA (C1) segment.

Predilatation

- The pre-dilatation balloon was loaded on the wire (figure 13J).
- The patient was given 1.0 mg of IV atropine.
- The lesion was pre-dilated for < 5 sec. at nominal pressure (figure 13J).
- The balloon was removed.
- The patient was asked to 'squeeze the toy'.

Stenting

- Next, the stent was loaded on the wire and advanced into the lesion.
- It was deployed in the standard manner.
- The patient was asked to 'squeeze the toy'.
- The stent catheter was removed.

Postdilatation

- The 6.0 mm x 20 mm balloon was loaded on the wire and post-dilatation was performed for < 5 sec (figure 13K).
- The patient was asked to 'squeeze the toy'.
- The balloon was removed.

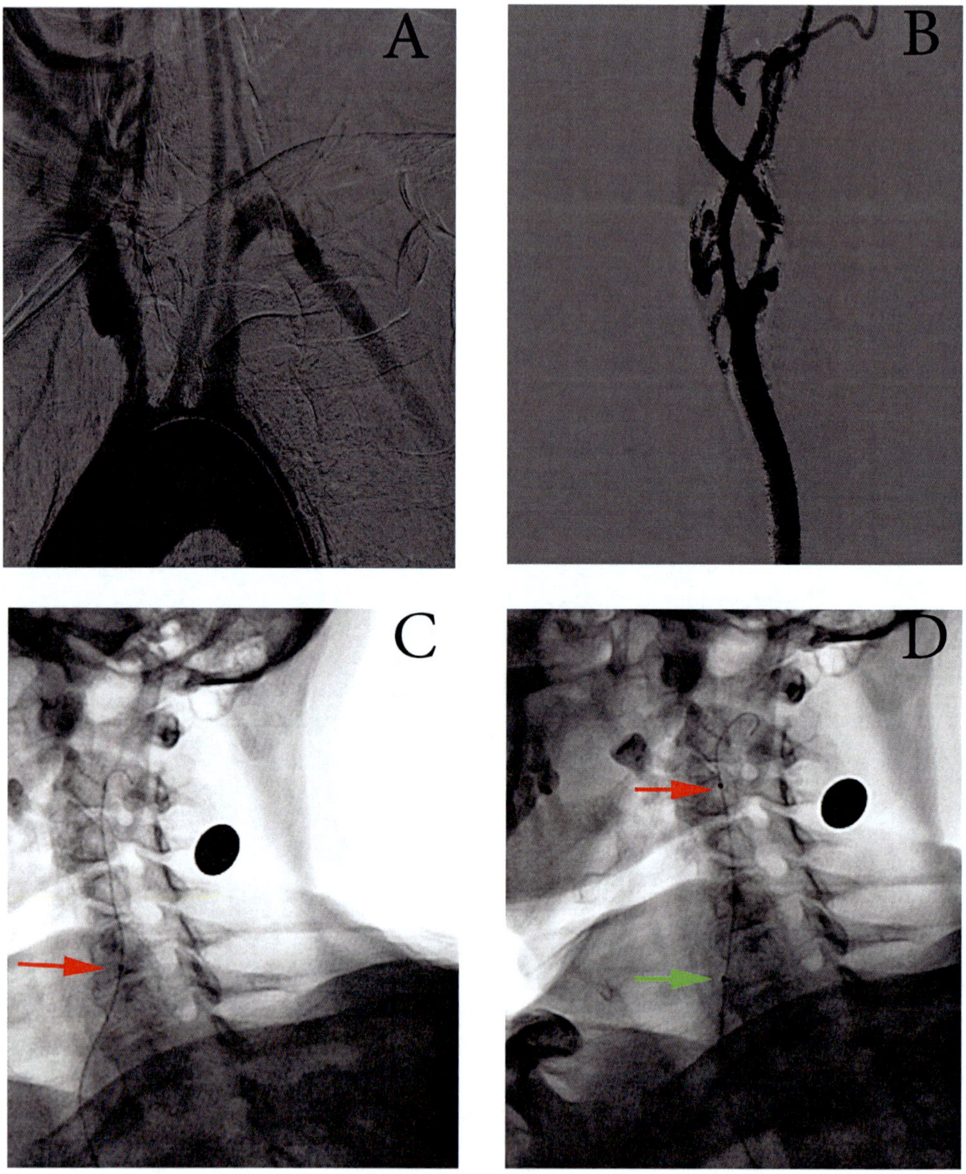

Figure 13.

Carotid stenting with proximal embolic protection.

A. Arch aortography reveals a type I arch.
B. Selective angiography of the left CCA shows a severe stenosis in the proximal ICA with an ulcerated plaque.
C. The MoMa device is being advanced into the CCA over a SuperStiff Amplatz wire.
D. The MoMa device is in the CCA with the ECA balloon (red arrow) in the ECA and the CCA balloon (green arrow) visible in the CCA.

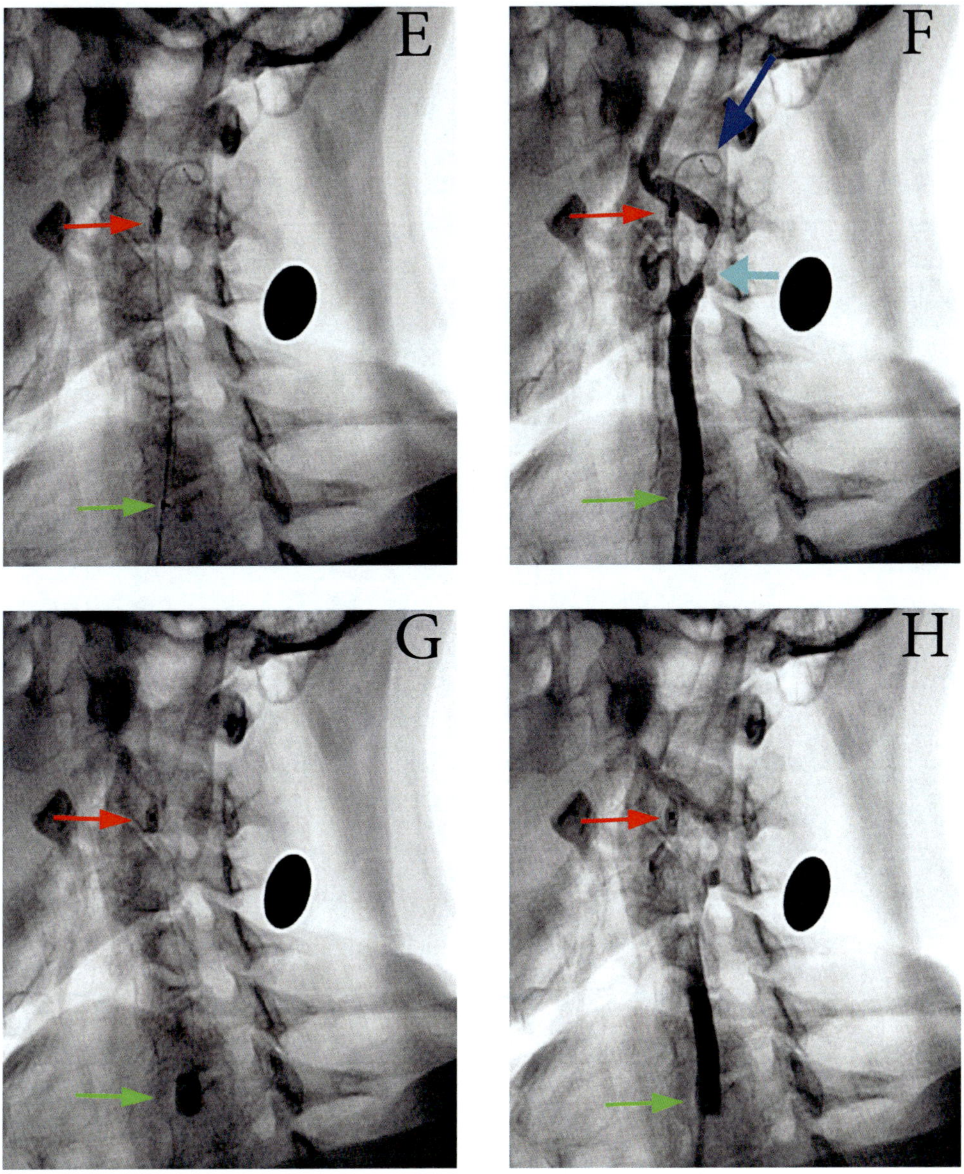

Figure 13.

Carotid stenting with proximal embolic protection (continued).

E. The ECA balloon (red arrow) has been inflated. The CCA balloon is still deflated.
F. Angiogram demonstrates successful ECA occlusion.
G. The CCA balloon (green arrow) has been inflated as well and the wire has been removed from the ECA.
H. Angiogram demonstrates cessation of flow in the CCA.

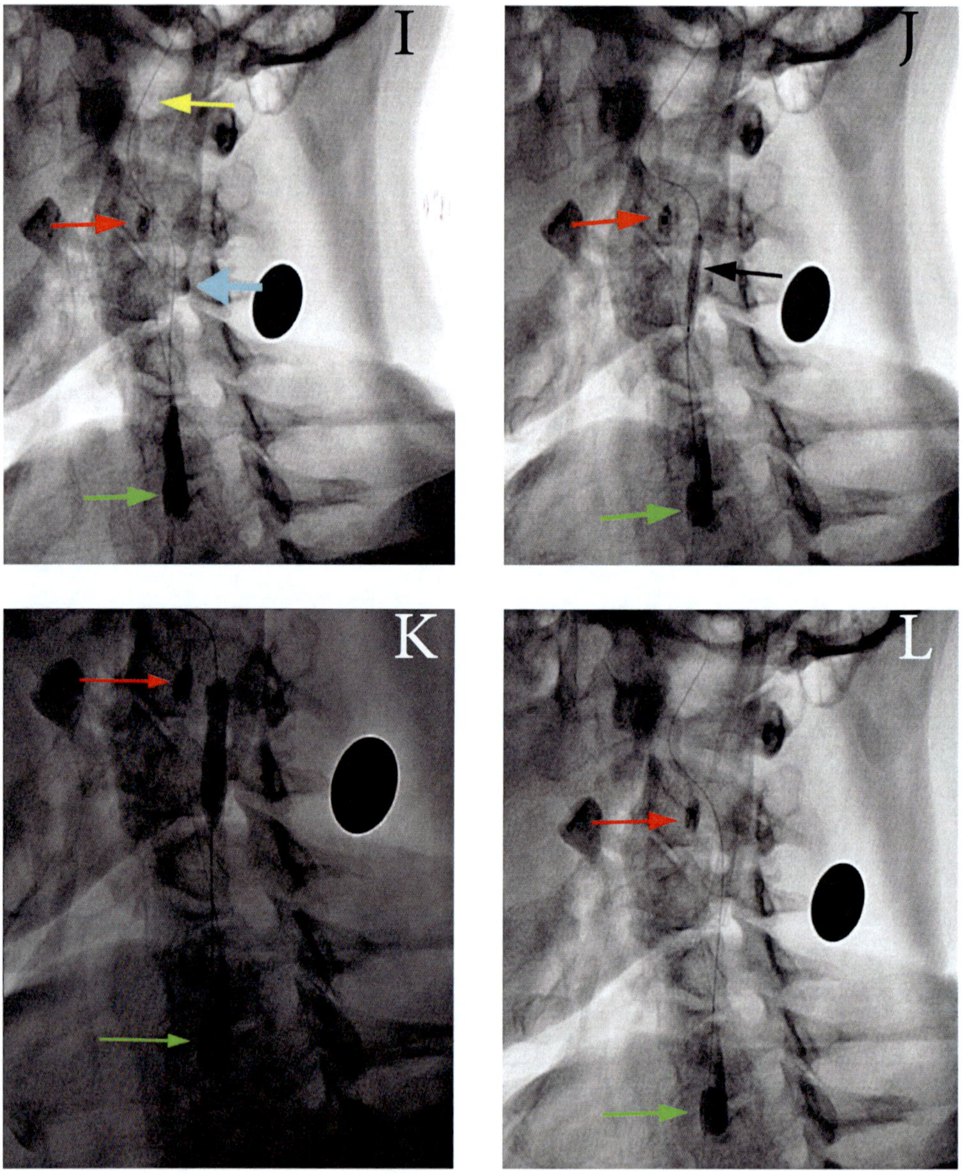

Figure 13.

Carotid stenting with proximal embolic protection (continued).

I. A Choice PT floppy wire (yellow arrow) is seen in the ICA.
J. The pre-dilatation balloon (black arrow) is inflated across the lesion.
K. The stent is being post-dilated.
L. After post-dilatation both ECA and CCA balloons are still up.

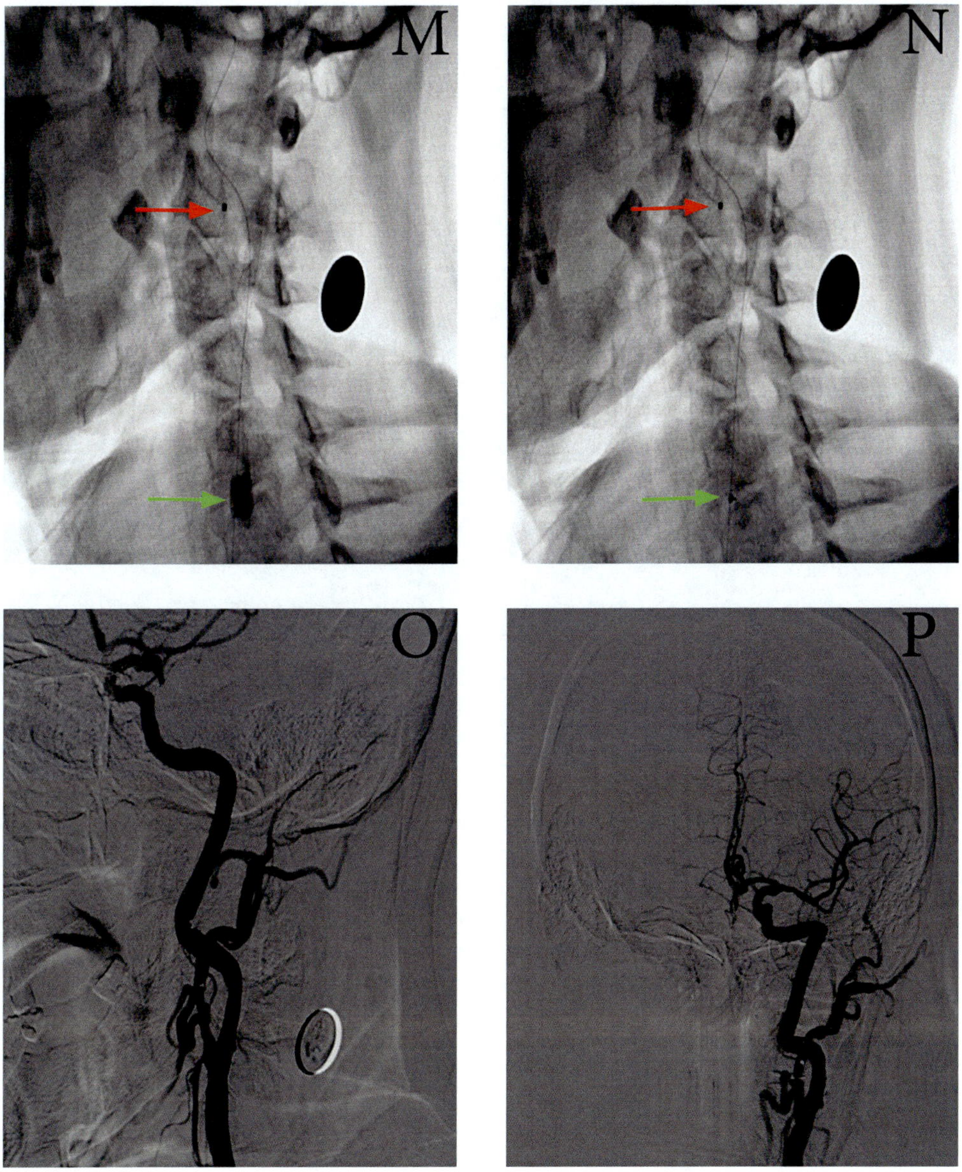

Figure 13.

Carotid stenting with proximal embolic protection (continued).

M. The ECA balloon has been deflated while the CCA balloon is still up.
N. The CCA balloon has been deflated as well.
O. Final angiography reveals a widely patent ICA. Mild spasm is seen in the ECA after the balloon deflation. The MoMa is still in place.
P. Intracranial angiogram shows a patent MCA and ACA. Filling of the right ACA is also seen via the anterior communicating artery.

- The patient did not require any Neosynephrine. Cuff pressure was maintained at 120 mm Hg.

Aspiration

- A 20 cc leurlok syringe was connected to the side-arm of the hemostatic valve and aspiration was performed with gradual removal of 20 mL of blood x 3 with both ECA and CCA balloons still inflated (figure 13L).
- The third syringe was 'filtered' through the provided basket to enure no visible debris was present.

Restoration of antegrade flow

- The ECA balloon was deflated first (figure 13M).
- This was followed by deflation of the CCA balloon (figure 13N).

Repeat angiography

- Repeat angiography including intra-cranial views revealed a widely patent ICA with < 20% residual stenosis (figures 13O & 13P). TIMI-3 flow was noted.

MoMa retrieval

- The MoMa device was removed under fluoroscopic guidance from behind the stent without any issues.
- The 0.038" ANSGW was then introduced into the device and positioned in the CCA below the stent.
- The MoMa was then withdrawn from the CCA into the arch and removed from the body.

Arteriotomy management & patient disposition

- After removal of the MoMa, the 9 Fr. sheath was removed from the right CFA and closure was performed using the previously deployed Perclose sutures.
- The venous sheath had been sutured in place.
- The patient was sent back to the CCU.
- He required IV Neosynephrine drip for the next 12 hours.
- ASA 325 mg/d and Plavix were continued and he was discharged home the following day.

12

INNOMINATE ARTERY STENTING

- Severe innominate artery stenosis causes impaired perfusion in the right carotid and subclavian artery distribution and also increases the risk of TIA/CVA in the anterior as well as the posterior circulation.
- Most of the lesions in this vessel are in the ostial location and result from disease of the aortic arch.
- Carotid US and CTA/MRA can easily make the diagnosis.
- Surgical revascularization entails a bigger procedure and PTA/stenting is considered first line therapy.
- However, if the burden of calcium in the innominate artery or the adjacent aortic arch is prohibitively high (as in CTO of the innominate artery), percutaneous revascularization may not be optimal or safe. In these cases, extra-anatomic bypasses such as carotid-carotid bypass or axillo-axillary bypass should be considered instead as safer alternatives.

ANATOMY

The anatomy of the aortic arch (see chapter 11) strongly influences the procedural success of innominate artery stenting.

THE INNOMINATE ARTERY

- It is the first branch of the aorta after the coronary arteries and arises at the level of the second right costal cartilage.
- The lumen of the innominate artery is usually 8 - 14 mm in diameter.
- Usually 4 - 5 cm in length, it divides into the right SCA and the right CCA at the level of the sterno-clavicular joint.
- Normally, it gives no branches.
- In case of a 'bovine arch' the left CCA may arise from the proximal innominate artery.

PRE-OPERATIVE IMAGING

Patients being considered for innominate artery stenting should undergo a CTA (preferably) or MRA of the aortic arch, descending thoracic aorta and the great vessels. This helps determine the feasibility of the percutaneous intervention and whether transfemoral or brachial approach should be used.

PATIENT PREPARATION

- The patient is prepared in a standard fashion with both groins shaved.
- The right arm should also be prepped and cleaned for possible radial/brachial artery access.
- Arterial access is usually obtained in the right CFA.
- Central venous access is not necessary, unlike CAS.
- We prefer to keep our patients awake and responsive. Sedation is not given. Small doses of Fentanyl may be used for those who are really anxious or have back pain etc.
- A small rubber toy (e.g. a rubber duckie) that makes a sound when squeezed is placed in the patient's left hand. During the procedure the patient is asked to "squeeze the duck" in order to assess the motor function of the hand.

ACCESS

ARTERIAL ACCESS

- Most cases are performed via femoral artery access as this allows for the use of embolic protection.
- In some cases, brachial/radial artery access has to be obtained in addition to femoral artery access - the wire is introduced via the brachial access and externalized via femoral access. This allows a solid rail over which the guiding sheath/catheter can be advanced via the CFA access, to the ostium of the innominate artery.
- Most stents used for innominate artery stenting are compatible with a 6 Fr. sheath or an 8 Fr. guiding catheter.
- In most cases, a long straight sheath is used as the 'guide'.
- We prefer to use a 6 Fr./90 cm Shuttle sheath (Cook, Bloomington IN). This sheath is hydrophilic and braided and comes with a Tuohy-Borst adapter.
- Guiding catheters are used in cases of complex aortic arch anatomy (type II-III).
- During the case, the guiding catheter is kept in the aortic arch and not advanced into the innominate artery.

VENOUS ACCESS

- Central venous access is not necessary for innominate artery stenting.

ANTICOAGULATION

- IV unfractionated heparin is used.
- Goal ACT > 250 sec.
- Bivalirudin or GP IIb/IIIa inhibitors should be avoided.

DIAGNOSTIC ANGIOGRAPHY

ARCH AORTOGRAPHY

Catheter:

- A 5 Fr. non-angled pigtail catheter is used for arch aortography.

Imaging:

- This should be performed in a projection that opens up the aortic arch the most. This view provides optimal anatomic detail about the arch, the origin of the great vessels and their relationship to each other.
  - ▹ Typically, this means an LAO projection of 45 degrees or more. Optimal angulation is confirmed on fluoroscopy prior to injection, as the image intensifier is positioned above or below 45 degrees LAO. Place the catheter 1-2 cm proximal to the origin of the innominate artery (towards the ascending aorta).
  - ▹ Rarely, an AP or RAO projection may be necessary due to poor visualization of the great vessel origins in LAO. Alternatively, selective angiography could be used.
- Use digital subtraction angiography (DSA) as this will reduce the volume of dye administration without compromising image quality.
- Remember to use 'breath hold' with DSA.

Dye Injection:

- A power injector is used for aortography.
- 70:30 or 50:50 diluted dye can be used in patients with renal dysfunction.
- The rate of power injection depends upon the patient's body habitus, renal function and ability to comply with instructions of breath-hold during aortography.
- Injection settings are 20 cc/sec for a total volume of 30 - 40 cc at 500 -750 psi with no rate rise.
- CO_2 MUST NOT be used for arch or cerebral angiography.

SELECTIVE ANGIOGRAPHY

- This is also performed in the same way as CCA angiography.
- However, the ostium of the diseased innominate artery should be engaged with caution as dissection or distal embolization may result from deeply seating a catheter.
- A gentle manual injection should be performed.

Catheter:

The same principles are used for selection of catheters for innominate artery angiography as for selective carotid angiography (chapter 11).

Imaging:

The view that profiles the arch optimally is the best view for innominate angiography and stenting.

- An LAO projection of 45-80° usually opens up the arch well.
- To visualize the ostium and proximal segment of the right CCA, an RAO caudal projection is very helpful as it opens up the bifurcation of the innominate artery.

Dye Injection:

- DSA is used with manual injections of 3 - 5 cc of 50:50 dye through the selective catheter.
- The patient is instructed as follows for every injection "do not breath, do not swallow, do not move".
- For intra-cranial imaging, obtain prolonged cine runs to visualize the venous flow.

INTERVENTIONAL EQUIPMENT

WIRES

The use of an embolic protection device (EPD) is strongly recommended for all cases of innominate artery stenting. When the procedure is performed entirely via a transfemoral approach, embolic protection use is accomplished by using a 'buddy wire' technique employing a 0.014" and a 0.018" wire (figure 1A). Together, these wires provide a 0.032" platform (0.014" + 0.018") for stent delivery. This improves balloon and stent trackability across the lesion.

However, if EPD use via transfemoral approach is not feasible and the right radial/brachial artery is used for delivering this instead, a 0.018"-0.035" wire may be used via the transfemoral route to deliver balloons and stents (figure 1B). Often, this 'transfemoral' wire is externalized via the upper extremity to provide more support to the guiding catheter/sheath.

0.014" Wire:

- The 0.014" wire chosen for innominate artery intervention is a carotid EPD.
- An EPD wire that can be extended to a 300 cm length should be considered as OTW platform stents are used.
- As in ICA stenting, the EPD (filter) is 'parked' in the distal cervial ICA (C1 segment).
- EPDs range in diameter from 5.0 mm - 7.5 mm. An appropriate filter sized to the cervical ICA should be selected. If the CCA is small (< 7 mm in diameter) the filter may be parked in this vessel instead. We, however, favor keeping the filter in the ICA to maximize embolic protection and to prevent the filter from being dragged down into the proximal CCA during intervention, risking prolapse of the guide into the aorta.

0.018" Wire:

- A 0.018" 300 cm wire such as the V-18 or Steelecore is used as the 'buddy wire' for anchoring purposes in the right axillary/brachial artery.
- If this wire cannot be advanced easily from the arch (via the transfemoral catheter) across the lesion and into the axillary artery, retrograde wiring via a brachial artery access should be considered.
- If this retrograde wiring technique is used, the 0.018" wire is externalized via the femoral approach and the transfemoral guiding sheath/guiding catheter is loaded over it and advanced into the arch ('body flossing' technique).

0.035" Wire:

- If the EPD is delivered via the radial/brachial access, a 0.035" wire may be used to deliver balloons/stents via the transfemoral route.
- A stiff, supportive wire such as an ASGW or an Amplatz SuperStiff should be used. In the case of the latter, this wire is delivered across the innominate artery lesion through an exchange catheter and not as the primary crossing wire.
- In cases where a higher degree of support is necessary for the guiding catheter/sheath (e.g. complex aortic arch), the 0.035" wire may be externalized via the radial/brachial sheath ('body flossing' technique).

SHEATHS

Guiding Sheath:

- In patients with a type I aortic arch, a 90 cm straight sheath (e.g. Shuttle sheath) can be used as the guiding sheath and the intervention is performed through it.
- In patients with a type II arch, the use of a guiding sheath may still be possible, with concur-

rent right radial/brachial artery access and wire externalization('body flossing' technique).

- For patients with a type III arch, the use of a transfemoral guiding sheath may not be feasible due to the inability to track the sheath to the innominate artery without the risk of dislodging the EPD from the ICA. A guiding catheter should be used instead. As noted above, if necessary the 'body flossing' technique maybe used to provide better support.
- If a bare-metal stent is being considered, a 6 Fr. sheath is adequate for sizes up to 10 mm diameter.
- If instead a covered BES (e.g. iCast®) is being considered, a 7 Fr. sheath is necessary.

Guiding Catheter:

- For most cases we prefer the use of a guiding catheter as it provides better support. This is particularly true in case of excessive aortic/iliac tortuosity and a complex arch configuration (type II/III).
- Due to the caliber of the devices (balloons, stents etc.) the guiding catheter has to be at least 8 Fr. in caliber.
- The following configurations are most commonly used:
  - ▹ Hockeystick (HS)
  - ▹ JR4
  - ▹ Modified Amplatz left 1
  - ▹ IMA

BALLOONS

- Balloons compatible with a 0.014" - 0.035" wires are used.
- These balloons are prepared with a 30:70 dye dilution so that inflation and deflation are relatively quick.

Predilatation Balloons:

- Predilatation balloon diameter is typically 4.0 - 8.0 mm with a length of 20 - 30 mm.
- 0.014" monorail balloons loaded on the EPD wire can be used for this purpose.

Postdilatation Balloons:

- Postdilatation balloons are chosen in the following sizes 8 - 12 mm in diameter and 20 - 40 mm in length.

STENTS

- Similar to ostial stenting elsewhere in aortic branches, innominate artery stenosis is treated with a balloon-expandable stent.
- Covered (with PTFE) BES are not necessary.
- However, if the lesion or the aortic arch is heavily calcified, then consideration should be given to placement of a covered BES. In theory, this lowers the risk of vessel perforation.
- If the burden of calcium is prohibitively high (as in CTO of the innominate artery or very heavy innominate/arch calcification), percutaneous revascularization may not be optimal or safe. In these cases, extra-anatomic bypasses such as carotid-carotid bypass or axillo-axillary bypass may be considered.

EMBOLIC PROTECTION DEVICES (EPD)

- The risk of intra-op TIA/CVA from innominate artery PTA/stenting is around 1.5%. Therefore, embolic protection should be used if possible.
- Due to the location of the lesion, proximal embolic protection is not possible and only distal EPD is used.
- The EPD is parked in the right ICA. If the CCA is < 7 mm in diameter the EPD may be parked in the CCA.
- In order to use a distal EPD, a 'buddy wire' technique is used. This consists of using a 0.014"

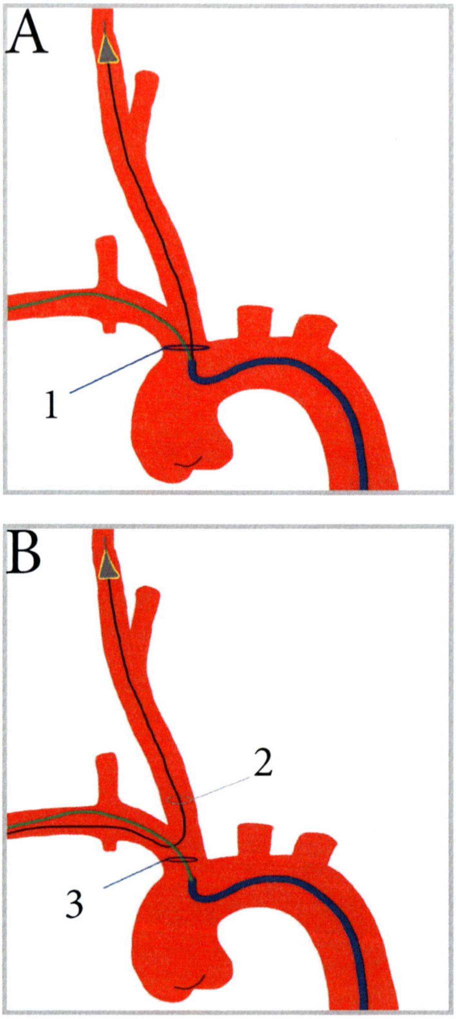

Figure 1.

Innominate artery stenting with embolic protection.

A. A transfemoral catheter is engaged in the innominate artery. The 'buddy wire' technique has been used with the EPD as well as the main wire (1) introduced through the transfemoral catheter.

B. This shows that the EPD (2) has been introduced via the radial/brachial artery access while the main wire (3) has been introduced via the transfemoral catheter.

EPD wire and a 0.018" 'buddy wire' that is parked in the right axillary/brachial artery.

- For a non-ostial stenosis (e.g. mid/distal innominate artery) the buddy wire may have to be parked in the ECA.
- A 0.035" wire compatible stent is used and is loaded on both the wires (through a transfemoral sheath).

TECHNIQUE

- We prefer to use an 8 Fr. guiding catheter via transfemoral access for innominate artery intervention as the guiding catheter provides better support than the straight sheath in most cases.
  - ▹ However, if aorto-iliac/CFA does not permit the use of an 8 Fr. system and the patient has a type I arch, a 6 Fr. Shuttle sheath can be used in a similar fashion.
  - ▹ For patients with a type II/III arch, where an 8 Fr. guiding catheter cannot be used via transfemoral access for some reason, radial/brachial artery access for intervention may be the only feasible route. Again, if feasible, EPD use is recommended. However, the use of an EPD via upper extremity access means that the balloons/stents must be loaded on a separate wire thus requiring a bigger sheath for the procedure. Therefore, in these cases an EPD is often not used. This should be an exception rather than the rule.

VESSEL CANNULATION

- After aortography (figure 2A), the pigtail is removed and over a 0.035" wire, the CFA sheath is exchanged out for an 8 Fr./11 cm CBT sheath.
- Through this an 8 Fr. coronary HS guiding catheter is advanced into the arch.
- The innominate artery is engaged.
- Selective angiography is performed to identify the SCA, CCA, ICA and ECA.

WIRING THE SUBCLAVIAN ARTERY

- A 0.018" V-18 wire is then advanced across the lesion, and navigated into the right axillary artery (figure 2B).
- This serves as the 'buddy wire' to anchor the guiding catheter in the innominate artery.
- A new road map is then obtained, making sure the tip of the guiding catheter and the distal cervical ICA are visualized.

CROSSING THE LESION & EMBOLIC PROTECTION

- In all cases, use of embolic protection is strongly recommended. However, transfemoral delivery of this device may not be feasible in all patients (e.g. aorto-iliac tortuosity and complex aortic arch configurations). As noted above, in these cases radial/brachial access may be used for EPD delivery. Despite this alternate access, the EPD may remain undeliverable to the ICA/CCA due to a sharp angulation between the right subclavian and CCA. In these cases, one may have to forego the use of an EPD and proceed directly to intervention.
- A 0.014" distal embolic protection (device) wire is used (figure 2C).
- An appropriate filter is selected (see table 3 & 4 for examples of the Accunet and Spider FX), sized to the distal ICA (usually 7.0-7.5 mm for adult males and 6.0-6.5 mm for adult females).
- Filter sizing is based on the C1 segment of the ICA where the filter will be 'parked'.
- This filter should be 'parked' in a straight segment of the distal ICA, making sure it will not cause severe spasm or dissection.
- The filter is deployed in a standard fashion and a timer is started. Ideally, the dwell time of this filter should be < 20 minutes. Longer dwell times have been associated with a higher rate

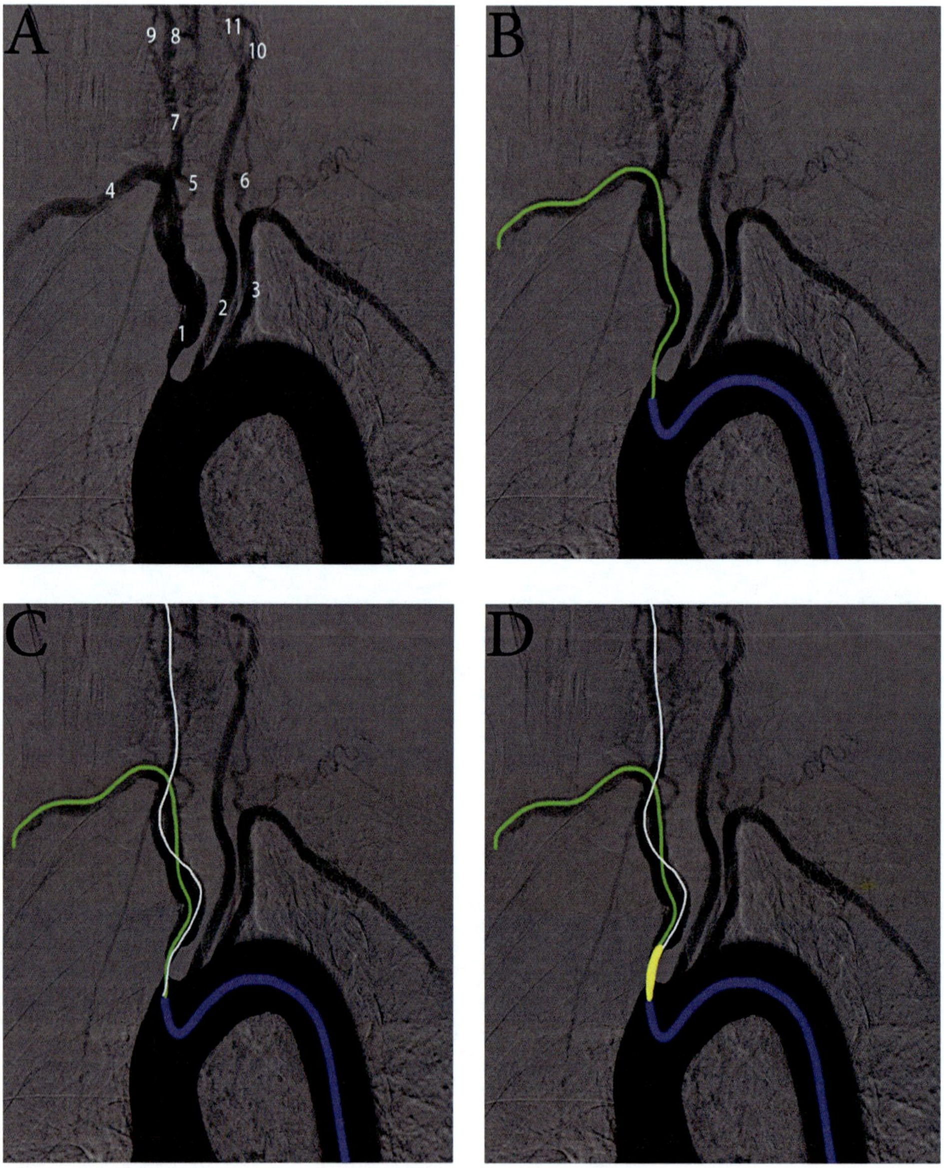

Figure 2.

Innominate artery stenting using a guiding catheter.

A. Arch aortogram showing a severe proximal stenosis of the innominate artery: 1, Innominate artery; 2, left CCA; 3, left SCA; 4, right SCA; 5, right vertebral artery; 6, left vertebral artery;7, right CCA; 8, right ICA; 9, right ECA; 10, left ICA; 11, left ECA.

B. An 8 Fr. Hockey stick guiding catheter (purple) is used to engage the innominate artery. A 0.018" "buddy" wire (green) has been advanced into the right axillary artery.

C. A 0.014" distal EPD (white) has been advanced into the C1 segment of the right ICA.

D. A pre-dilatation balloon is loaded on either wire and the lesion in the innominate artery is predilated.

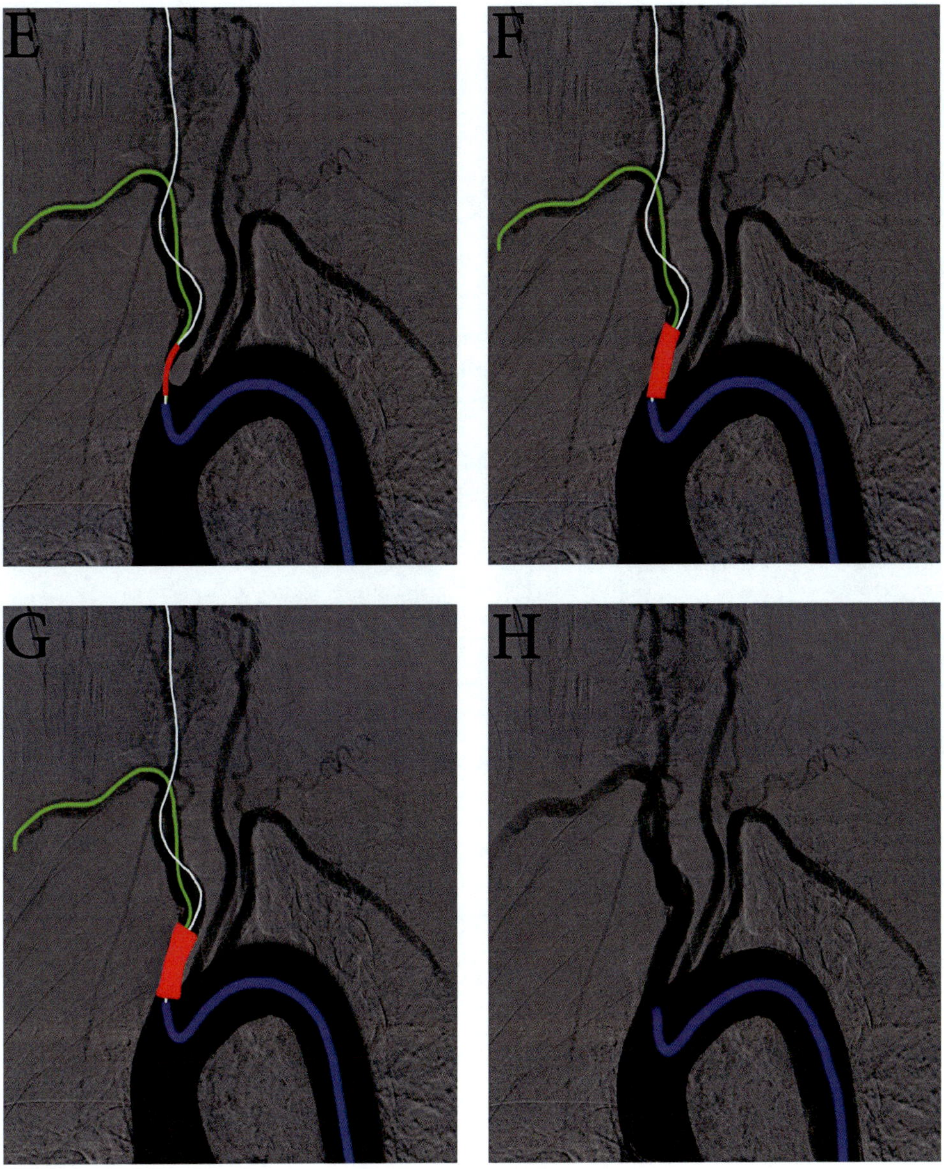

Figure 2.

Innominate artery stenting using a guiding catheter (continued).

E. A BES is loaded on BOTH wires and positioned across the lesion.
F. The stent is deployed in a standard fashion with a 1-2 mm protruding into the aortic arch.
G. The stent is then postdilated as necessary.
H. Final angiography after removal of the EPD and the 0.018" wire demonstrates a widely patent lumen.

of peri-operative TIA/CVA.

- Care should also be taken to not allow this filter to migrate into the petrous portion of the ICA where the likelihood of intense spasm and dissection is high.

PREDILATATION

- After filter deployment, the pre-dilatation balloon is loaded on the wire (figure 2D).
- Unlike ICA stenting, atropine is not indicated for innominate artery stenting as the carotid baroreceptor reflex is not activated by intervening on the innominate artery.
- Predilatation is performed under road-map.
- The balloon is deflated and removed.
- The patient is asked to 'squeeze the toy' in the left hand to grossly assess motor function as a surrogate for distal embolization.

Stent:
- It is critical to load the stent on BOTH the wires. Otherwise, the other wire will be 'jailed' and trapped behind the stent.
- This can be catastrophic if the EPD wire is trapped, necessitating surgery for removal of the device.

STENT DEPLOYMENT

- Next, the stent is loaded on BOTH the wires and positioned across the lesion under road map guidance (figure 2E).
- For ostial lesions, approximately 1 - 2 mm of the stent should be allowed to protrude into the arch (figure 2F).
- It is deployed in a standard fashion at the nominal pressure.
- Make sure the Shuttle sheath has been pulled back to gain adequate room for the stent.

POSTDILATATION

- The post-dilatation balloon is loaded on the wire.
- Postdilatation is performed with an appropriately sized balloon (figure 2G).
- Aggressive post-dilatation should be avoided to reduce the likelihood of vessel trauma and perforation/aortic dissection.
- The post-dilatation balloon is withdrawn partially into the aorta and the stent is 'flared.'
- The patient is asked to 'squeeze the toy' in the left hand to grossly assess motor function as a surrogate for distal embolization.
- Aim to keep SBP 100 - 160 mm Hg. Higher blood pressures can lead to intra-cranial HTN and hemorrhage due to cerebral dysautoregulation ("hyperperfusion syndrome").

REPEAT ANGIOGRAPHY

- Prior to retrieval of the embolic protection device, an angiogram should be performed to document patency and normal flow.
- Rarely, 'no re-flow' is seen in the ICA due to filter 'overload' with debris from the lesion. There may be a combination of floating debris in the static column of blood plus debris lining the filter.
- This situation is best dealt with advancing a 5 Fr./125 cm MP guiding catheter through the guiding catheter in a telescoping fashion.
- The MP catheter is advanced through the stent (under fluoroscopic guidance to not damage the stent) and as close to the filter as possible. A 60 cc luer-lock syringe is connected to the side-arm of the sheath and aspiration is performed. At least 20-30 mL of blood is aspirated and

wasted.

- This should result in restoration of normal or near normal flow (TIMI 2-3). This is confirmed on repeat angiography.
- If the flow remains poor, quick repeat aspiration should be performed and another 30 mL blood should be removed.
- Additionally, a dissection at the stent edge or at the site of filter deployment should be ruled out.
- If no dissection is noted, and flow still remains sluggish the filter can be retrieved assuming that either the filter is 'full' or there is spasm at the site.

FILTER RETRIEVAL

- After flushing it, the retrieval catheter is back loaded on the 0.014" wire.
- It is advanced out of the guiding catheter and into the stent under fluoroscopic guidance.
- Keep the deployed filter in field of view and well above the stent.
- Once through the distal end of the deployed stent, the retrieval catheter is advanced until the proximal loop of the filter (four markers on Accunet; elliptical hoop of the Spider FX) is just in the catheter.
- DO NOT attempt to contain the entire filter within the retrieval catheter. This can dislodge the trapped debris in addition to the risk of filter detachment.
- Make sure the Tuohy-Borst adapter is wide open.
- Gently pull the retrieval catheter and the filter as a unit into and through the stent.
- Once in the guiding catheter, they can be removed swiftly.
- Allow the sheath to back-bleed adequately and flush once again ensuring there is no air or debris.
- Repeat angiography should be performed to document normal flow after removal of the filter. If spasm is present and the patient's BP allows, local IA nitroglycerin can be given.
- The patient is asked to 'squeeze the toy' in the left hand to grossly assess motor function as a surrogate for distal embolization.
- Cervical ICA and intra-cranial ICA, ACA and MCA should be imaged in lateral and Towne's views.

FINAL ANGIOGRAPHY

- The 0.018" wire may still be kept in the axillary artery (especially in patients with a type II/III arch) to anchor the guiding catheter in the innominate artery while final angiographic views are obtained. This may not be necessary for type I arch patients (figure 2H).
- Sometimes, for better visulaization, an arch aortogram has to be repeated using a pigtail catheter.
- After angiography, the guiding catheter is disengaged from the innominate artery and over a 0.035" wire it is then removed.

ACCESS MANAGEMENT

- The arteriotomy is managed as appropriate.

CASE

A 62-year old right-handed male presented with severe claudication of right arm. No TIA/CVA symptoms were reported.

On exam, right upper extremity pulses were noted to be weak and there was a 60 mm Hg difference in systolic BP between the right and left arms. A bruit was heard over the right subclavian area.

The patient underwent an outpatient CTA. This study showed a type III aortic arch with a severe stenosis of the proximal innominate artery. The right SCA, CCA and ICA were noted to be free of any disease. A decision was made to intervene on the innominate artery.

Initial Diagnostic Equipment:

- 4 Fr. micropuncture kit.
- 6 Fr./11 cm sheath.
- 6 Fr./45 cm Ansel 2 sheath.
- 0.018" V-18 wire.
- 5 Fr./100 cm pigtail catheter.
- 5 Fr./125 cm Vitek catheter.
- Power injector.

Access & Diagnostic Angiography:

- Arterial access was obtained in the right CFA.
- No sedation was used.
- The patient was given IV UFH for anticoagulation. An ACT of > 250 sec was confirmed.
- Arch aortography was performed using the 5 Fr. pigtail catheter (figures 3A & 3B).
- For selective angiography of the innominate artery a 5 Fr. Vitek catheter was used (figures 3C & 3D).

Interventional Technique:

Startup equipment

Due to the acute angle between the right SCA and the CCA it was concluded that EPD delivery via brachial access would be challenging. Thus a decision was made to attempt intervention via trans-femoral approach. However, neither a guiding catheter nor a Shuttle sheath could be used to successfully engage the innominate artery and provide the necessary support for intervention. We resorted to transbrachial approach for intervention thereafter.

- 0.035" 180 cm ANSGW.
- 6.0 mm x 20 mm Powerflex balloon for pre-dilatation.
- 8.0 mm x 29 mm Genesis BES stent.
- 9.0 mm x 20 mm Powerflex balloon for post-dilatation.
- Access was obtained in the right brachial artery using a 4 Fr. micropuncture kit.
- A 6 Fr./45 cm Ansel 2 sheath was inserted in this vessel over a 0.035" ANSGWwire.
- The wire was advanced to the innominate artery and the sheath advanced over it into the distal subclavian artery.
- The dilator was removed and the sheath flushed.

Engaging the right CCA

- Attempts were made to engage the the right CCA using various catheters (IM, JR4 etc) through the Ansel sheath. These remained unsuccessful.

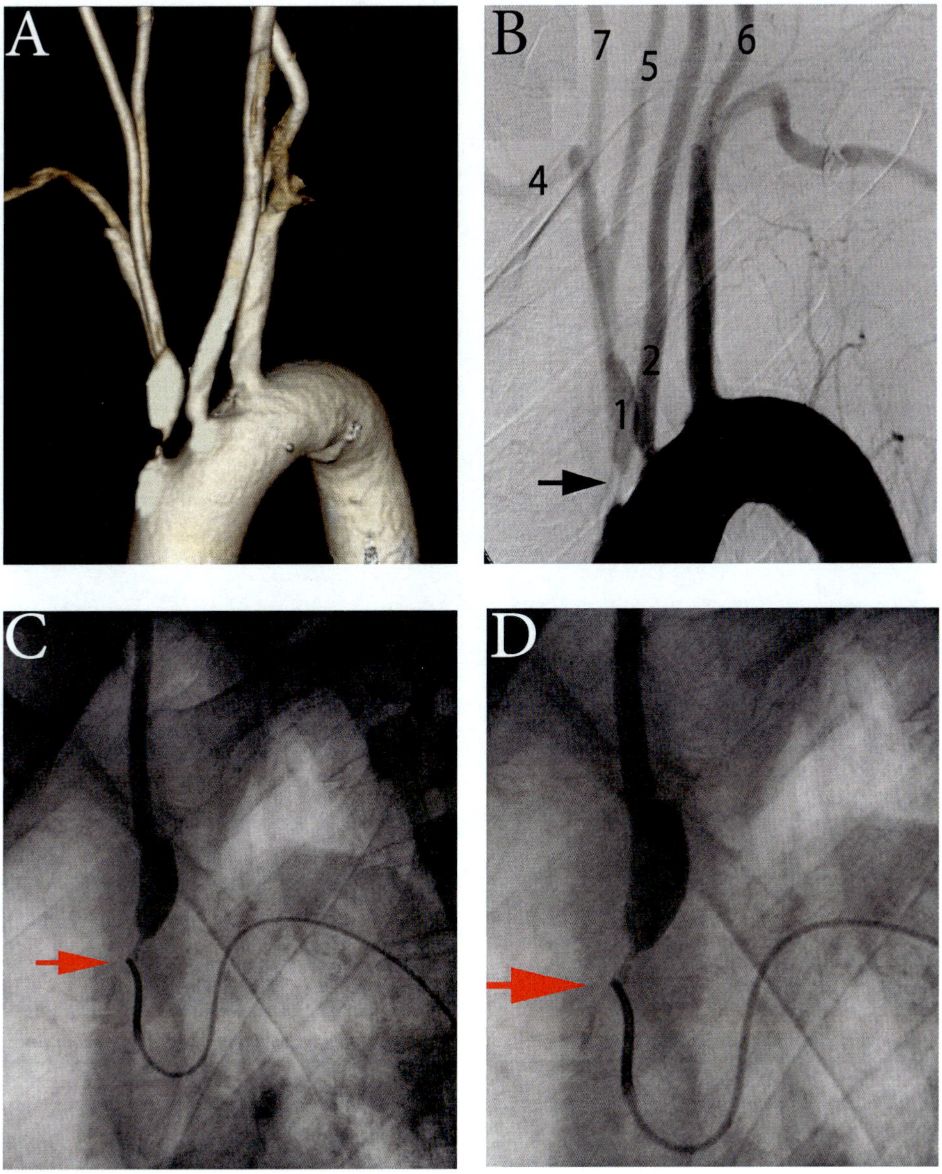

Figure 3.

Innominate artery stenting without embolic protection.

A. Arch CTA reveals a severe stenosis of the ostial innominate artery. The left CCA and SCA are patent.
B. Arch aortography reveals a type III arch. 1, Innominate artery (black arrow indicates the lesion); 2, left CCA; 3, left SCA; 4, right SCA; 5, right vertebral artery; 6, left vertebral artery;7, right CCA.
C. Selective angiography of the innominate artery using a 5 Fr. Vitek catheter (red arrow) via transfemoral access.
D. A magnified view of the selective angiogram.

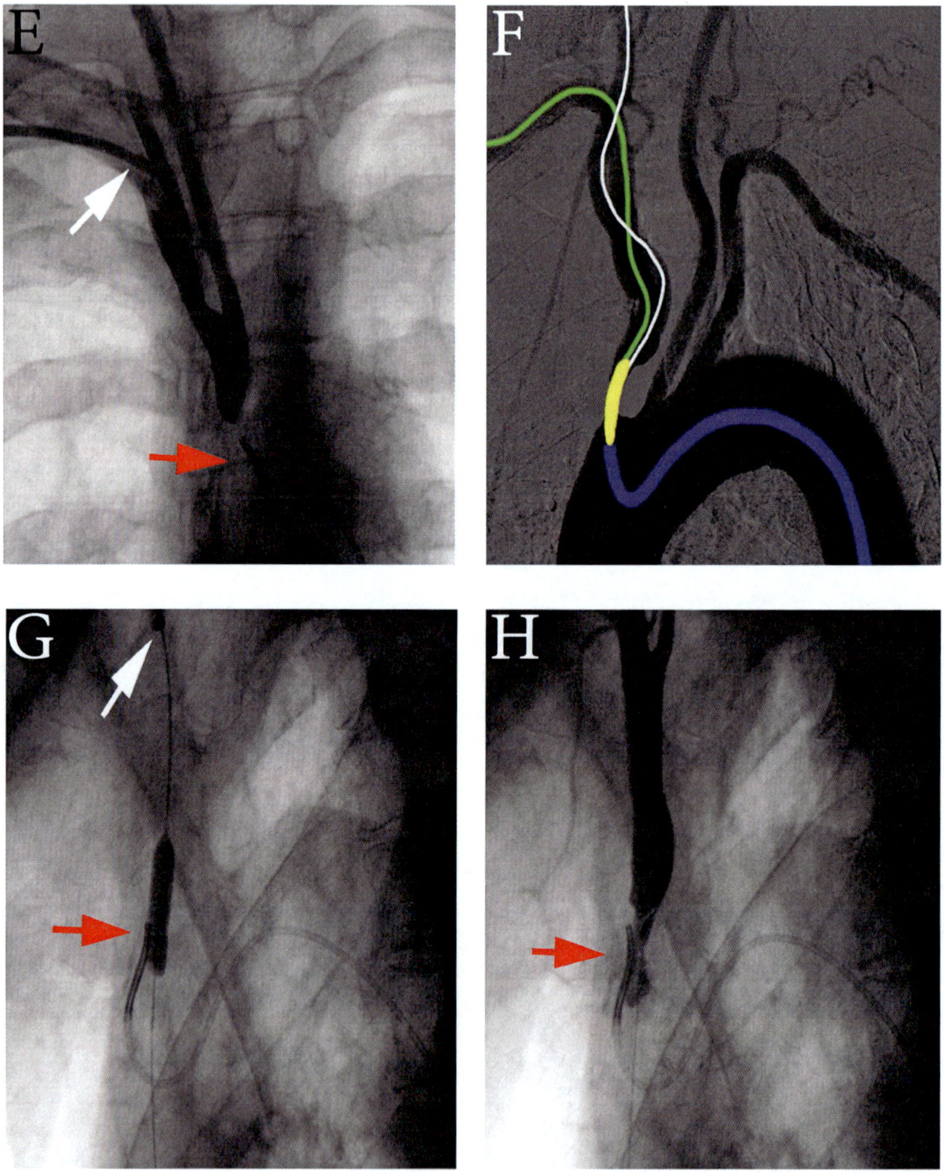

Figure 3.

Innominate artery stenting without embolic protection (continued).

E. Retrograde angiogram via right brachial artery sheath (white arrow) shows a near occlusive innominate stenosis. The Vitek catheter (red arrow) is seen engaged in the ostium from the transfemoral approach.

F. Predilatation of the innominate artery. A signficant waist is noted in the balloon. The yellow arrow indicates the V-18 wire in the ascending aorta.

G. Predilatation with good expansion of the balloon (white arrow: brachial sheath tip; red arrow: Vitek tip).

H. Retrograde angiogram after pre-dilatation (red arrow: Vitek tip).

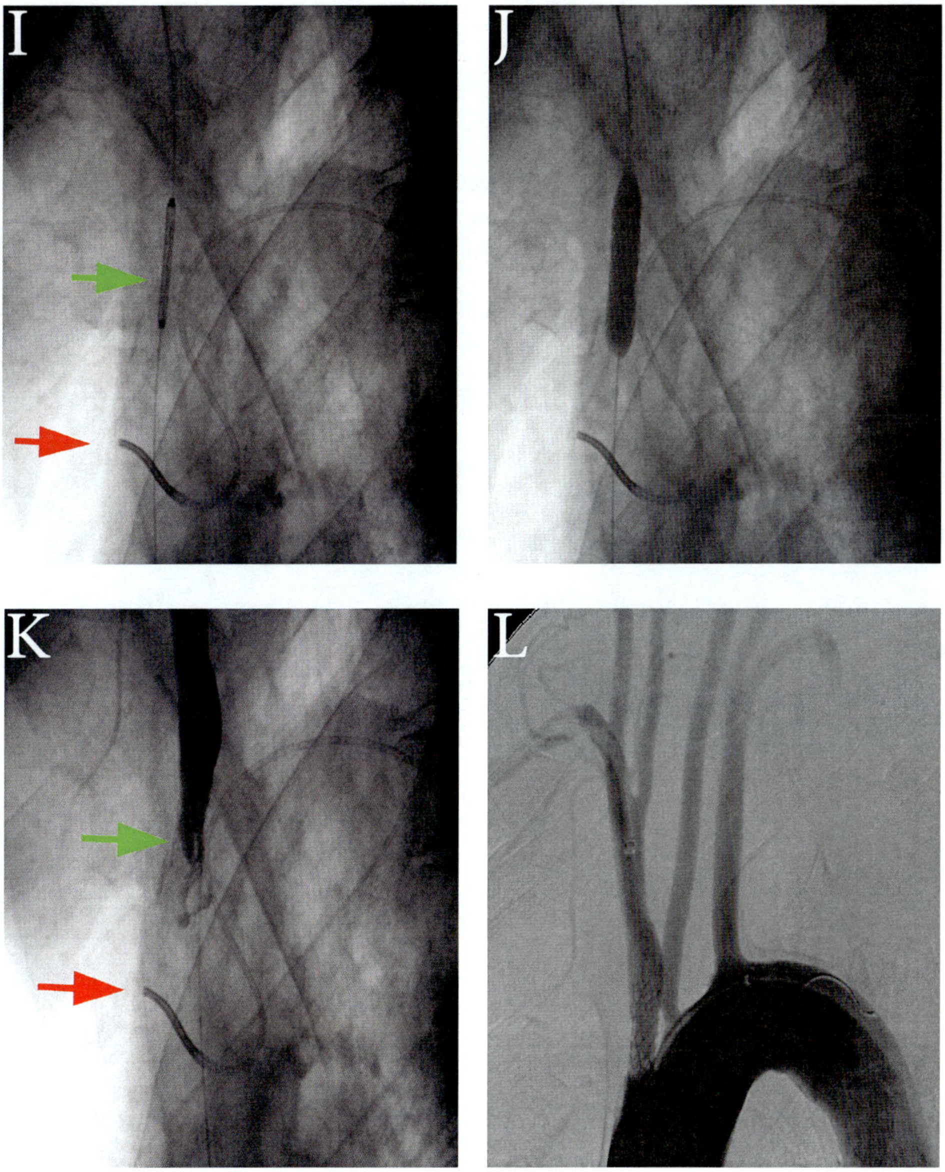

Figure 3.

Innominate artery stenting without embolic protection (continued).

I. The Vitek catheter (red arrow) is disengaged from the lesion as the stent (green arrow) is positioned in the stenosis.
J. Stent deployment.
K. Retrograde angiogram after stent deployment shows mild residual stenosis.
L. After post-dilatation, the Vitek is removed and an arch aortogram is repeated via femoral approach. This shows a widely patent innominate artery.

- A decision was thus made to proceed to innominate stenting without the use of EPD.

Wire advancement across the innominate stenosis

- A V-18 wire was advanced in a retrograde fashion across the innominate artery stenosis into the ascending aorta.
- Since retrograde injections did not allow visualization of the origin of the innominate artery, the Vitek catheter was used to engage the innominate artery via transfemoral approach, to serve as a 'landmark' for the ositum (figure 3E).

Predilatation

- A 6.0 mm x 20 mm Powerflex balloon was used to predilate the lesion (figures 3F-H).

Stent deployment

- An 8.0 mm x 29 mm Genesis BES was then positioned across the ositum of the innominate artery via the brachial sheath (figure 3I).
- Its position was confirmed using the Vitek catheter as the landmark.
- The catheter was then disengaged and the stent deployed at 8 ATM (figure 3J).

Postdilatation

- The stent balloon was removed and the stent postdilated with a 9.0 mm x 20 mm Powerflex balloon.

Final aortography

- The patient remained intact neurologically and had no defecits.
- Retrograde angiography through the sheath revealed a widely patent innominate artery with normal flow (figure 3K).
- Arch aortography was then performed using the a pigtail catheter via the transfemoral approach (figure 3L).
- This revealed an acceptable result with a widely patent innominate artery.

Arteriotomy management & patient disposition

- The sheaths were removed when the ACT was less than 180 sec.
- The patient was sent to the telemetry unit.
- He remained asymptomatic neurologically and had no deficits on exam.
- He was continued on ASA 325 mg/d and Plavix and discharged home the following day.

13

AORTIC INTERVENTION FOR OCCLUSIVE DISEASE

Aortic occlusive disease may lead to a focal, hemodynamically significant stenosis within the thoracic or abdominal aorta, diffuse severe disease, or, in more advanced cases, total occlusion of the infrarenal aorta. As a result, patients with aortic occlusive disease have symptoms of intermittent claudication that affect the calves, thighs and buttocks. These symptoms are bilateral but may be worse in one leg if there is additional ileofemoral disease on that side. Additionally, there may be erectile dysfunction in male patients.

Percutaneous intervention for aortic total occlusion is a complex procedure that requires experience with CTO recanalization and endografts.

ANATOMY

- The descending thoracic aorta becomes the abdominal aorta as it passes through the aortic hiatus in the diaphragm at the level of T12 vertebra.
- It descends in front of the vertebral column, somewhat to its left.
- In an adult male, the abdominal aorta is 25-28 mm in diameter at its origin (T12) and tapers down to nearly 20 mm in diameter at its termination at L4 level. In females, the diameters are 22-25 mm at origin and 16-18 mm at termination.

BRANCHES OF THE ABDOMINAL AORTA

The abdominal aorta gives rise to several branches before it bifurcates into the right and left common iliac arteries.

Anterior:

- Celiac trunk (T12 level).

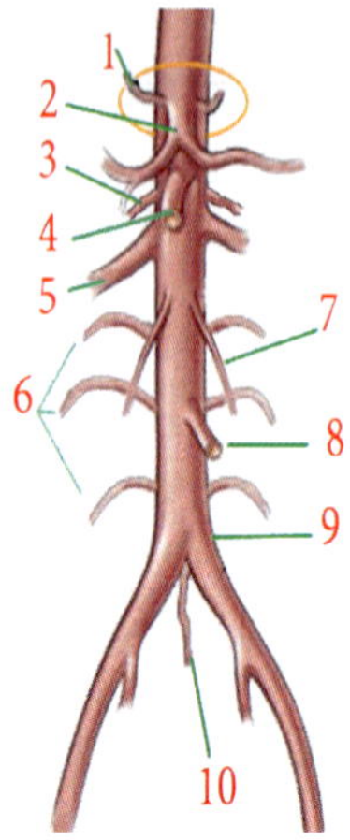

Figure 1.

Anatomy of the abdominal aorta

The orange circle represents the level of the aortic hiatus in the diaphragm.

1. Inferior phrenic artery.

2. Celiac axis.

3. Middle supra renal arteries.

4. SMA.

5. Renal artery.

6. Lumbar arteries.

7. Gonadal artery.

8. Inferior mesenteric artery.

9. Common iliac arteries.

10. Median sacral artery.

- SMA (L1 level).
- IMA arises anteriorly and somewhat to the left off the aorta (L3-L4 level).

Lateral:

These branches typically arise in pairs, one on the right and one on the left.

- Inferior phrenic arteries (T12 level)
- Middle supra-renal arteries (L1 level)
- Renal arteries, arise posterolaterally from the aorta (L1-L2 level).
- The testicular/ovarian arteries arise anterolaterally, a little below the renal arteries.

Posterior:

- Lumbar arteries, four on each side, arise posterolaterally from the aorta. A fifth pair may arise from the median sacral artery.
- The median sacral artery is a posterior branch of the abdominal aorta.

DIAGNOSIS

The history of claudication is fairly classic in patients with aortic occlusive disease. On physical examination, an abdominal bruit may be present with or without concurrent femoral bruits. The femoral pulses are commonly diminished bilaterally and may be non-palpable in cases of total aortic occlusion. However, femoral pulses may be 'normal' despite a significant aortic lesion.

ABI with pulse volume recording (PVR) or Doppler should be obtained at baseline. In patients with a patent but narrowed aorta, this test shows mild to moderate reduction in the ankle pressures and PVRs suggest 'inflow' disease. However, the results of this test may also be normal under resting conditions. Therefore, an exercise ABI should be obtained in patients with a good history of intermittent claudication and a normal resting study.

In patients with aortic total occlusion, the resting study is fairly abnormal and an exercise study is not necessary. PVR and Doppler are strongly suggestive of 'aorto-bi-iliac' occlusive disease.

STENOSIS OF THE AORTA

Stenoses of the abdominal aorta are not uncommon. The typical location for such disease is in the infra-renal aorta, particularly in the distal two thirds. The terminal aorta is particularly prone to atherosclerosis, specially in patients with dyslipidemia and tobacco abuse.

Depending on the severity of the stenosis, the patient may have varying degrees of intermittent claudication. In cases with concurrent iliac disease, the male patients may also have erectile dysfunction.

If angiography and endovascular intervention is being considered, it is a good idea to consider pre-operative imaging.

PRE-OPERATIVE IMAGING

It is recommended to obtain a pre-operative CTA in all patients suspected to have:

- Severe bilateral aorto-iliac occlusive disease.
- Severe occlusive disease of the abdominal aorta.
- Both of the above.

This study:

1. Reveals aortic, mesenteric, renal, iliac and CFA anatomy.
2. Allows measurement of the diameter and length of the diseased segments of the abdominal aorta and iliac arteries.
3. Shows the presence of plaque, calcium and thrombus in the affected segments.
4. Enables precise sizing of the angioplasty balloon and stents/stent-grafts used for the intervention.

Pre-operative imaging should allow the operator to determine the likelihood of a successful intervention and the risk of complications such as iliac or aortic dissection/perforation. A careful 'bailout' plan should also be formulated based on such imaging to deal with these potential complications. Appropriate size devices such as aortic occlusion balloons and appropriate size stents/stent-grafts should be available prior to the procedure.

DIAGNOSTIC ANGIOGRAPHY

Equipment:

- 4 Fr. micropuncture kit.
- 0.035" 260 cm angled stiff hydrophilic wire.
- 5 Fr. sheath.
- 5 Fr./100 cm pigtail or tennis racket catheter.

Access:

Site:

- Retrograde femoral access in the CFA, unilateral or bilateral.

Sheath Size:

- Initial access is obtained with a 4 Fr. micropuncture kit. Thereafter, over a 0.035" angled stiff hydrophilic wire, a 5 Fr. 20+ cm sheath is inserted.
- Advance the wire cautiously in a diseased aorta to avoid the risk of dissection.

Abdominal Aortography:

- A 4 Fr./65 cm tennis racquet (TR) catheter is advanced over the 0.035"ASGW and placed at L1 - L2 space via femoral access.
- From brachial access, a 5 Fr./100 cm pigtail catheter is used instead.
- Abdominal aortography is performed using digital subtraction at 10 - 15 cc/sec for total of 20 - 40 cc at 500 - 750 psi.
- Breath-hold should be performed.
- The field of view is set to image from the top of the flush catheter to bilateral common femoral arteries. All visceral (mesenteric and renal) arteries should be fully visualized.
- The rate of power injection depends upon the patient's body habitus, renal function and the ability to comply with instructions of breath-hold during aortography.

Selective Angiography:

Selective angiography is usually not relevant in cases of aortic occlusive disease. Sometimes, however, visceral angiography may be necessary.

Gradients:

- One pressure is recorded via the side-arm of the introducer sheath and the other through a catheter that is inserted through this sheath.
- In this case, the caliber of the measuring catheter should be at least 1 French size smaller than that of the sheath. Otherwise, artifactual dampening of the pressure waveforms may be noted.
- Obviously, if two separate access points are used for measuring the gradient, the difference in the caliber of the catheters is irrelevant.

Gradients & Vasodilators:

- In addition to NTG, Papaverine may be used for vasodilatation.
- The vasodilator is given through the sheath/catheter that is distal to/downstream from the lesion being investigated.

HEMODYNAMICS

If the aorta is not totally occluded, hemodynamic assessment with gradient measurement is strongly recommended.

- This should be done not only for lesions that appear angiographically severe but also for those that may appear to be 'non-significant'.
- Gradients should be recorded before and after provocation with IA NTG. This is particularly relevant in patients who have a good history of intermittent claudication but have less than impressive resting ABI/PVR/Doppler studies. This group of patients is very likely to have angiographically 'moderate' appearing stenoses with 'non-significant' resting gradients. After administration of 200-300 mcg of IA NTG, distal to the lesion (through the sheath), this gradient increases and becomes significant. Some operators use papaverine for the same purpose. These vasodilators should be given through a catheter/sheath distal to the lesion being investigated.
- Ideally, the gradient is measured by using simultaneous pressure recordings proximal and distal to the diseased segment, using two separate transducers.

INTERVENTION

Goal:

- To abolish or minimize the hemodynamic gradient in the aorta (resting and provoked) from the worst stenosis.
- To implant a stent at the site of the worst stenosis without any residual dissection or perforation.
- Balloon-expandable stents are preferred to self-expanding stents for focal lesions due to the precision in their placement and their radial strength.
- For long diseased segments, stent-grafts in the form of 'contra-lateral iliac limb' or 'iliac limb extension' from EVAR endograft systems can be used. These consist of self-expanding stent-grafts.

Equipment:

Wires:

- 0.035" 180 cm angled non-stiff hydrophilic wire
- 0.035" 180 cm J-tip Super Stiff Amplatz wire.

Sheaths:

- 8-12 Fr./30 cm long sheaths (depending upon the aortic diameter and that of the planned stent).

Balloons:

- 6-10 mm diameter 0.035" balloon for predilatation
- 10-14 mm diameter x 40 -60 mm length balloon.

Stents:

- Ideally, these stents should be sized on the basis of an abdominal CTA. However, if one is not available, IVUS can be used for this purpose. Alternatively, the stent(s) can be sized conservatively based on angiography.
- Palmaz XD or XL stent(s) 3110, 4010 or 5010 (tables 1 & 2; figure 2B)
- iCast balloon-expandable covered stent or its larger counterpart the Advanta stent (where its available) (figure 2C).
- Aortic endografts (figure 2D):

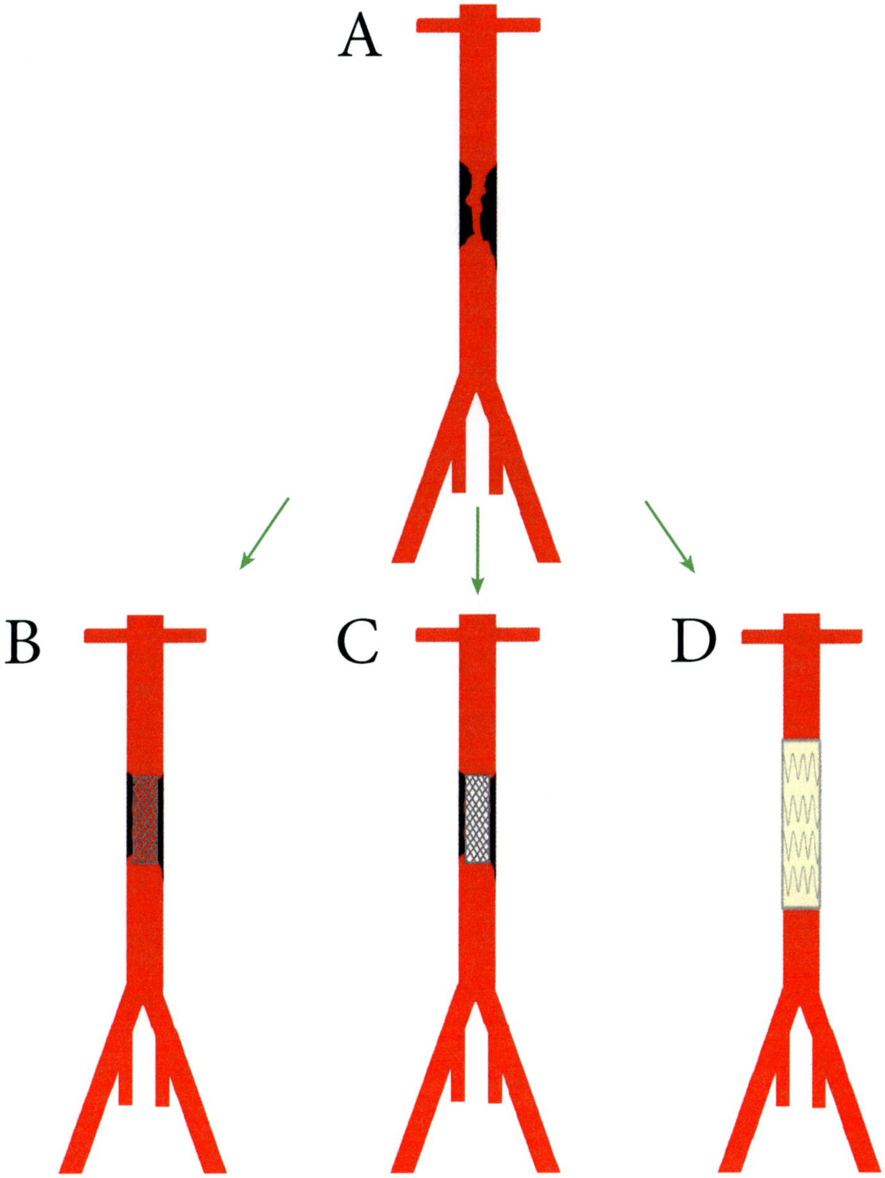

Figure 2.

Endovascular strategy for treatment of aortic stenosis.

A. Focal stenosis of abdominal aorta.
B. The above stenosis is treated with a balloon-expandable bare-metal stent (e.g.Palmaz XL).
C. The above stenosis is treated with a balloon-expandable covered stent (e.g.iCast).
D. The above stenosis is treated with a self-expanding stent-graft.
.

- ▹ 'Contralateral iliac' limb (of EVAR endografts).
- ▹ 'Iliac limb-extender' (of EVAR endografts).
- ▹ Aortic 'cuff'/'extension'.
- ▹ Thoracic stent-graft.

Access:

- Femoral access is used for these cases.
- Occasionally, radial/brachial access may be obtained for initial angiography.

TABLE 1 - PALMAZ XD STENT: • This stent can be expanded between 10 mm and 12 mm diameter			
UNEXPANDED STENT LENGTH (MM)	EXPANDED STENT LENGTH (MM)	EXPANDED STENT DIAMETER (MM)	SHEATH COMPATIBILITY (FR.)
• 19	• 17	• 10	• 10
• 19	• 16	• 12	• 10
• 29	• 27	• 10	• 10
• 29	• 25	• 12	• 10
• 39	• 36	• 10	• 10
• 39	• 34	• 12	• 10
• 59	• 55	• 10	• 10

TABLE 2 - PALMAZ XL STENT: • This stent can be expanded between 14 mm and 25 mm diameter		
UNEXPANDED STENT LENGTH (MM)	OUTER DIAMETER BEFORE EXPANSION (MM)	SHEATH DIAMETER (FR.)
• 40	• 4.6	• 12-14
• 50	• 4.6	• 12-14

Anticoagulation:

- IV UFH is used.
- For aorto-iliac CTO interventions, its preferred not to fully anticoagulate the patient until the wire is safely across the lesion and in the true lumen.
- Goal ACT > 250 sec; prior to crossing iliac CTOs, we prefer to keep the ACT < 200 seconds.
- Bivalirudin or GP IIb/IIIa inhibitors should be avoided because of the inability to reverse the effect in case of perforation.

Technique:

- After aortography and hemodynamic assessment, obtain a road map angiogram through the aortic flush catheter.
- Remove the flush catheter over a 0.035" 180 cm J-tip Super Stiff wire and leave the wire in place with the tip in the descending thoracic aorta.
- Pre-dilate the lesion with an undersized balloon.
  - ▹ Be cautious not to over-size the balloon to avoid dissection and perforation.
  - ▹ Typical diameters for predilatation balloons are 6 - 8 mm x 40 mm.
- Remove the balloon and 'shepherd' the sheath across the stenosis. If it is necessary, the sheath can be advanced over its dilator.
- Mount a Palmaz XL stent on an appropriate size balloon (see below the technique of hand

crimping a stent).

- The goal of the intervention is ***not*** to match the diameter of the treated segment to that of the healthy normal aorta. There is significant risk of perforation in that. Instead, the goal is to increase the diameter to 10 - 12 mm and abolish any hemodynamic gradient across the stenosis. The risk of restenosis is fairly low at these diameters.
- With the sheath across the lesion, advance the balloon catheter and the hand-crimped stent, into the lesion and position at the desired level.
- Withdraw the sheath enough to expose the proximal shoulder of the balloon.
- Inflate the balloon to its nominal diameter to deploy the stent.
- After removal of the stent balloon, perform an aortogram to determine the adequacy of the intervention.
- If postdilatation is required, use a somewhat undersized balloon to avoid the risk of dissection or perforation.
- Perform final hemodynamics and abdominal aortography.
- Final angiography should be performed using DSA and breath-hold technique.
- Significant edge dissection(s)/perforations should be ruled out.

AORTIC CHRONIC TOTAL OCCLUSION

Chronic total occlusion of the aorta typically occurs in the infra-renal segment, caudal to the origin of the renal arteries. In this case, the IMA is usually occluded as well and the large gut is perfused via collaterals from the celiac axis, SMA and hypogastric vessels. There may be concurrent disease (stenosis or occlusion) of the renal arteries, celiac axis and SMA.

Aortic CTO results in disabling intermittent claudication affecting the buttocks, thighs and legs, and in a minority of cases, rest pain and even tissue loss. Male patients with this condition have a high incidence of erectile dysfunction.

Traditionally, surgical revascularization is pursued for aortic CTO and consists of aorto-bifemoral or aorto-bi-iliac bypass. In patients with a hostile abdomen/aorta (previous surgery, infected aorta or horseshoe kidney) axillo-bifemoral or, rarely, thoraco-bifemoral bypass is performed.

Endovascular intervention of an aortic CTO is performed in the rare instance when the patient has disabling or limb-threatening symptoms and the risk of surgery is prohibitively high (coexisting severe CAD with LV systolic dysfunction etc.). These patients should be selected meticulously after a thorough review of the pros and cons of such an intervention and risk of associated complications. It would be prudent to document that surgery is considered extremely or unacceptably risky.

PRE-OPERATIVE IMAGING

Although ABI/PVR/Doppler assessment is very helpful in arriving at the diagnosis, high-quality, standardized imaging of the abdominal aorta should be performed in order to plan an aortic intervention. This is ideally performed using a CT angiogram (figure 3).

Some basic guidelines for the study are:

- Image from the level of the diaphragm to mid intertrochanteric region (celiac artery origin to the bifurcation of the CFA).
- Collimation should be 3 mm or less.
- Center-line measurements should be obtained for length when possible.

The study should provide the operator with the following information:

- The maximal and minimal luminal diameters of the aorta in the following locations:

  - ▹ Supra-renal aorta
  - ▹ Infra-renal aorta
  - ▹ Terminal aorta
  - ▹ Diseased segment(s)
- Presence of thrombus, plaque and calcium.
- Presence of aortic aneurysm (type, location, diameter and length).
- Diameter, length and angulation of the aorta in the diseased segment(s).
- Presence, patency and location of the celiac trunk, SMA and IMA.
- Patency, location and number of renal arteries on each side.
- Identification of the lowest renal artery.
- Distance from the lowest renal artery to the aortic bifurcation.
- Distance from the lowest renal artery to iliac bifurcation (origin of the hypogastric artery).
- Presence of CFA and iliac disease along with measurements of maximal and minimal luminal diameters.

DEVICE PLANNING

For patients with aortic occlusion, stent-grafts (covered stents) should be used for treatment of the diseased segment.

Isolated Aortic Occlusion:

- For isolated, focal aortic occlusion, an aortic 'cuff'/'extension' of an abdominal AAA repair endograft or a thoracic stent-graft can be used (figure 4A & 4B). This is placed much in the same fashion as the obsolete aortic 'tube' grafts, and requires the distal/terminal aorta to be patent and free of significant occlusive disease so the stent can be expanded without the risk of compromising flow to either iliac artery.

Aortic Occlusion with Severe Iliac Disease:

- If there is concurrent iliac occlusive disease, bifurcated modular grafting technique should be employed (figure 4C & 4D).
- This implies the placement of multiple stent-grafts in a stepwise fashion, with at least one aortic stent-graft and one iliac stent-graft on each side.

Aortic Component:

- The atherosclerotic aorta is severely and diffusely narrowed and concurrent aneurysmal disease is rare. Therefore, the main body of an aortic stent-graft designed for AAA repair (EVAR) is usually not used for the treatment of aortic occlusive disease, as there may not be enough room in the lumen of the aorta to allow adequate expansion of the stent(s) and opening and cannulation of the contralateral limb gate. The latter could lead to a situation where the contralateral gate stays compressed and under-expanded, prohibiting cannulation, necessitating surgical bypass (femoro-femoral) to the affected leg.
- Instead, the following components can be used for the treatment of the aortic segment:
  - ▹ 'Contralateral iliac' limb (of EVAR endografts).
  - ▹ 'Iliac limb-extender' (of EVAR endografts).
  - ▹ Aortic 'cuff'/'extension'.
  - ▹ Thoracic stent-graft.

Iliac Component:

- To preserve flow to both iliac arteries, stent-grafts are extended from each common iliac artery into the aortic stent, to create an inverted Y-shaped 'bifurcated modular graft'.

Overlap:

- The overlap between the aortic and iliac stents should be at least 2-3 cm to provide adequate seal, particularly in cases of aortic perforation.

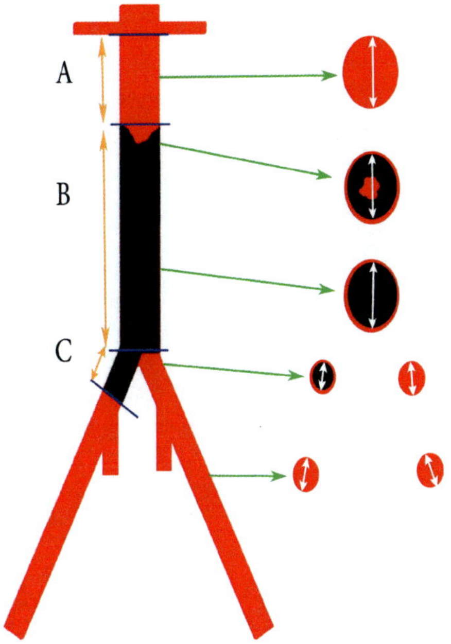

Figure 3.

CTA of the abdominal aorta

This schematic illustrates how aortic diameters and lengths are measured to plan for the intervention.

A. Distance between the lowest renal artery and the occlusion.

B. The length of the aortic occlusion.

C. Length of the iliac occlusion.

STENT SIZING

Precise sizing of the stents is critical. Oversizing of the stent may lead to vascular trauma causing dissection or even frank rupture. Additionally, when stent-grafts are used, oversizing poses the risk of pleating (bunching up) of the 'fabric' (ePTFE, etc.) within the lumen of an under-expanded stent-graft predisposing to stent thrombosis.

- The diameter of the aortic stent is chosen based on the type of the stent (tables 3 & 4).
- For balloon-expandable stents (bare or covered), sizing is done conservatively, choosing a stent 1 - 2 mm smaller in diameter than the aorta, and rarely at 1:1 to the vessel lumen.
- For self-expanding stents, the general rule of oversizing the stent by 1 - 2 mm remains true. However, caution should be used in segments of CTO where even the self-expanding stents (stent-grafts) are best sized 1:1 to the vessel.
- In order to prevent/manage iatrogenic dissection or rupture, stents should be placed meticulously in the overlap zones.
- A durable 'seal'(between the stent layers) should be obtained in the overlap zone(s).
- A key to achieving such a seal is adequate overlap and precise stent sizing.
- The iliac stents are positioned at the same level within the aortic stent, with an overlap of 2 - 3 cm, and deployed simultaneously.
- Depending on the type and diameter of the iliac stents, two configurations of the kissing stents are possible within the aortic stent. This is illustrated in figure 5.
    - ▹ If the iliac stents have only marginal contact with each other in the middle and limited apposition to the aortic stent, an 'O' configuration is achieved.
    - ▹ If, however, these stents are flattened in the middle and have linear contact with each other and additionally they are apposed to the aortic stent more or less in a uniform and circumferential fashion, they will assume a 'D' configuration.

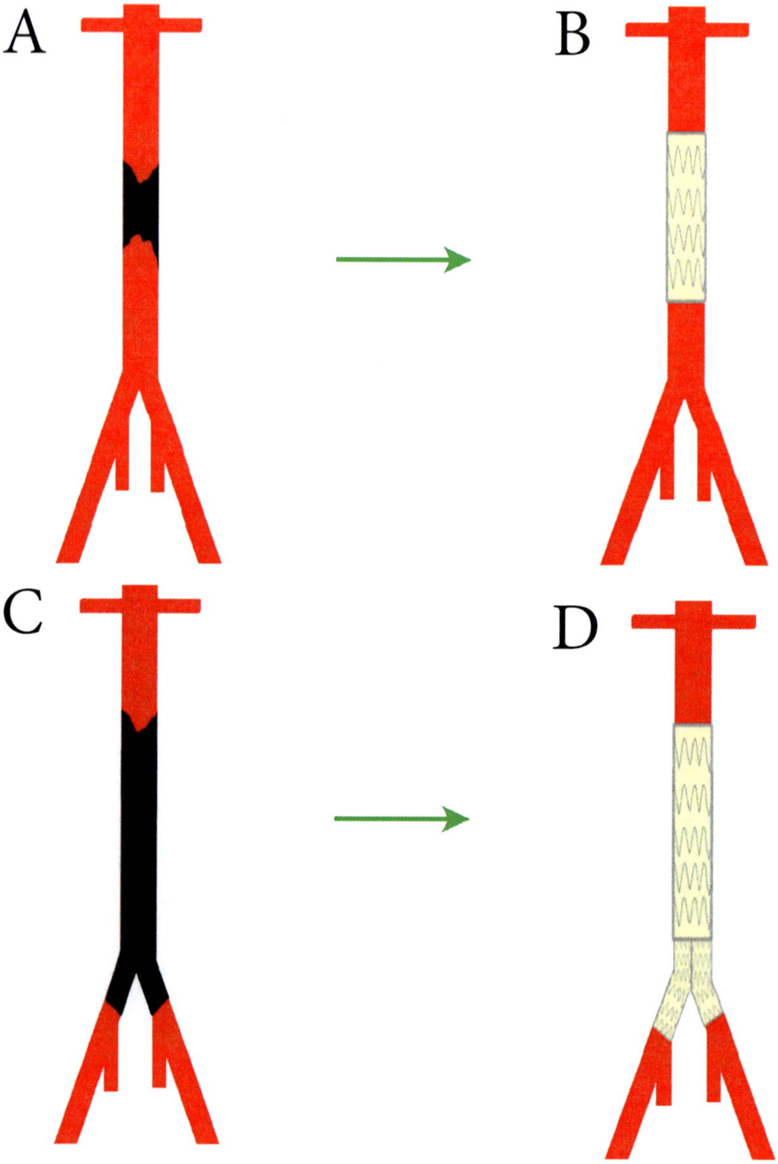

Figure 4.

Endovascular strategy for treatment of aortic occlusion.

A. Focal occlusion of abdominal aorta. Note distal aorta is disease-free.
B. The above occlusion is treated with a single piece aortic stent-graft. Note the stent-graft is away from the ostia of either iliac artery.
C. Aortic and bilateral iliac occlusion.
D. The above occlusion is treated with bifurcated modular stenting - one aortic stent-graft with overlapping iliac stent graft on each side.

A

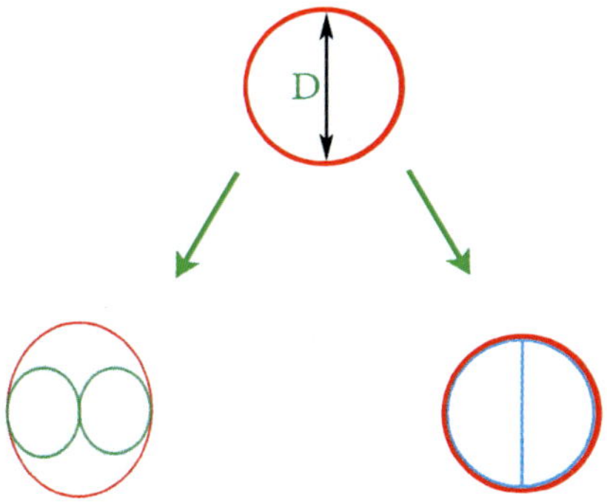

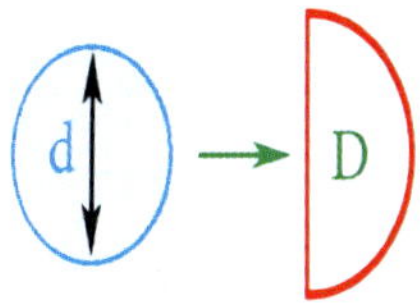

B

Circumference of a circle $= 2\pi \times (\text{radius})$
$= 2\pi \times (\text{diameter})/2$
$= \pi \times \text{diameter}$
$= \pi \times D$

Perimeter of a semi-circle = **Half the circumference + diameter**
$= (\pi \times D/2) + D$
$= (\pi D + 2D)/2$
$= D(\pi + 2)/2$
$= D\,(3.14 + 2)/2$
$= D\,(5.14)/2$
$= D \times 2.57$
$= 2.57 \times D$

Since the perimeter ($2.57 \times D$) of the red semi-circle is the same as the circumference ($\pi \times d$) of the blue circle :

$\pi \times d = 2.57 \times D$
$d = 2.57 \times D/\pi$
$d = 0.81 \times D$

Figure 5.

Sizing Aorto-iliac stents.

A. Two different configurations of aortic stent(A) placement with overlapping 'kissing' iliac stents:

- To achieve the 'O' configuration two iliac stents each with a diameter equal to the radius of the aortic stent are chosen.
- To achieve the 'D' configuration two iliac stents each with a diameter larger than the radius of the aortic stent are chosen.

B. For a circular stent to assume a semi-circular shape (each kissing iliac stent in 'D' configuration) the diameter of the iliac stent has to be carefully chosen relative to that of the aortic stent.

- Calculation of the diameter d of the (blue) stent when the diameter D of the larger (red) stent is known. Thus, under perfect conditions to achieve configuration 'D' the diameter of each kissing stent (blue circle) should be 0.8 times the diameter of the aortic stent (outer red circle).

TABLE 3 - AORTO-ILIAC STENT-GRAFT SIZING:	
• This refers ONLY to the use of 'self-expanding' stent-grafts. • Vessel diameters refer to the inner luminal diameter.	
VESSEL	STENT-GRAFT SIZING
Aorta	Oversize diameter by 1 mm (10%)
Iliac	Oversize diameter by 1 mm (10-25%)

TABLE 4 - AORTO-ILIAC BALLOON-EXPANDING STENT SIZING:	
• Vessel diameters refer to the inner luminal diameter.	
VESSEL	STENT SIZING
Aorta	Undersize diameter by 1-2 mm (10-20%)
Iliac	Size diameter 1:1

Modular stenting:
- This implies placement of more than one stent, piece by piece.

- The 'seal' between the aortic and iliac stents is relatively poor with the 'O' configuration than with the 'D' configuration.
- Iliac stent sizing to achieve the 'O' configuration is easy. Each iliac stent must have a diameter that is equal to the radius of the aortic stent (i.e. the diameter of the iliac stent is 0.5 x the diameter of the aortic stent).
- Iliac stent sizing to achieve the 'D' configuration is slightly tricky as illustrated in figure 5. The diameter of each iliac stent is greater than the radius of the aortic stent (as shown in the figure the diameter of the iliac stent is 0.8 times the diameter of the aortic stent).

DIAGNOSTIC ANGIOGRAPHY

Equipment:

- 4 Fr. micropuncture kit.
- 0.035" 260 cm angled stiff hydrophilic wire.
- 5 Fr. sheath.
- 4 Fr./100 cm pigtail or tennis racket catheter.

Access:

Site:

- Retrograde left brachial artery access is obtained. If this is not feasible, right brachial artery access is used instead.

Sheath Size:

- Initial access is obtained with a 21 gauge needle/4 Fr. micropuncture kit.
- Through the microcatheter of this kit, an angled stiff hydrophilic 0.035" wire is advanced into the descending thoracic aorta.
- The microcatheter is removed and a 5 Fr./90 cm straight sheath (e.g. Shuttle® sheath; Cook Medical, Bloomington, IN) is inserted over a 0.035" angled stiff hydrophilic wire.

- The largest caliber sheath that we like to insert in the brachial artery is 7 Fr. Although > 7 Fr. sheaths have been used at this access point, they increase the likelihood of ischemic complications to the forearm and hand.

Abdominal Aortography:

- A pigtail or tennis racquet catheter is used.
- Position the catheter at the level of L1 - L2.
- Start with proximal abdominal aortography in an AP projection.
- Digital subtraction angiography (DSA) should be used to reduce the volume of dye administration without compromising image quality. Breath-hold should be performed.
- The field of view is set to image from the top of the flush catheter to bilateral common femoral arteries. All visceral (mesenteric and renal) arteries should be fully visualized.
- The rate of power injection depends upon the patient's body habitus, renal function and the ability to comply with instructions of breath-hold during aortography.
- Injection settings are 10 cc/sec for a total volume of 20-40 cc at 500-750 psi.

Selective Angiography:

Selective angiography is not relevant in cases of aortic occlusive disease. Sometimes, however, visceral angiography may be necessary, for example, to preserve the ostia of the renal arteries or when using snorkel/periscope technique.

Snorkel or Periscope technique:

- This implies placement of a covered stent within a visceral artery, that would otherwise be 'jailed' behind an aortic stent-graft.
- When the visceral stent is deployed from a brachial/axillary approach, the aortic component points cephalad. This is called a 'snorkel'.
- If, instead, the stent is deployed from a femoral approach, the aortic component points caudad. This is called a 'periscope'.

HEMODYNAMICS

In a patient with total occlusion of the aorta, measurement of a pressure gradient across the obstruction, at baseline or otherwise, is less important.

- If gradients are to be measured, two simultaneous pressure recordings proximal and distal to the occlusion should be obtained using two separate transducers.
- Typically, one pressure is recorded via the flush catheter in the aorta from the brachial access and the other through the side arm of the arterial sheath.
- The catheter and sheath should be one French size apart in caliber to avoid a 'pseudogradient'.
- If two separate access points (brachial and femoral) are used for this purpose, the difference in the caliber of the catheters is irrelevant.

INTERVENTION

Goal:

- To restore patency of the abdominal aorta and iliac arteries bilaterally.
- To abolish or minimize the hemodynamic gradient in the aorta (resting and provoked) .
- To implant a stent/stent-graft at the site of the occlusion that may or may not extend into bilateral iliac arteries, without any residual dissection or perforation.

Equipment:

Sheaths:

- 5 Fr./90 cm sheath for brachial artery access.

- 8-14 Fr. sheath for upsizing the femoral access once the occlusion has been successfully crossed.
- Perclose Proglide® closure devices x 2 for each femoral access to preclose if the sheaths are upsized.

Wires:

- 0.035" 260 cm angled stiff hydrophilic wire; two sets.
- 0.035" 260 cm straight stiff hydrophilic wire.
- 0.035" 260 cm J-tip SuperStiff Amplatz wires x 2.
- 0.035" 150 cm straight exchange catheter (QuickCross®- Spectranetics).

Balloons:

- 6-10 mm diameter 0.035" balloons for predilatation.
- 10-14 mm diameter x 40 -60 mm length balloon for postdilatation.

Stents/Stent-grafts:

- Aortic endograft pieces (based on the measurements from the CTA).
- Covered stents - Self-expanding and balloon expandable in appropriate sizes for common and external iliac arteries.

Access:

- The choice of access site for intervention is made based on the findings of diagnostic angiography or pre-op CTA.
- More than one access site is utilized in these cases of aortic 'reconstruction', typically, bilateral CFA and (left) brachial artery access is necessary.
- Brachial artery access is used for crossing the occlusion as antegrade recanalization is usually easier and safer.
- Femoral artery access sites are utilized for delivering balloons and stents and to complete the intervention.

Anticoagulation:

- IV UFH is used. For aorto-iliac CTO interventions, its preferred not to fully anticoagulate the patient until the wire is safely across the lesion and in the true lumen.
- Goal ACT > 250 sec; prior to crossing aorto-iliac CTOs we prefer to keep the ACT < 200 seconds.
- Bivalirudin or GP IIb/IIIa inhibitors should be avoided because of the inability to reverse the effect in case of perforation.

Technique:

Probing the Proximal Cap:

- After aortography and hemodynamic assessment, obtain a road map angiogram through the aortic flush catheter (via the brachial artery access).
- Exchange the flush catheter over a 0.035" 260 cm angled stiff hydrophilic wire for a 4 Fr./120 cm angled hydrophilic catheter.
- Advance the catheter and wire to the level of the proximal cap of the occlusion. This is almost always below the origin of the renal arteries as the aorta typically occludes below that level.
- Usually, this proximal cap has a 'beak' or a fine channel at the leading edge of the occlusion cap.
- Direct the angled catheter towards this 'beak' and steer the wire into it. Bring the catheter into the 'beak' as well.
- If no 'beak' is identified, the center of the lumen at the site of occlusion is targeted with the catheter and wire to cross.
- At this point, switch out the wire for a 0.035" 260 cm straight stiff hydrophilic wire. To ob-

tain support the Shuttle® sheath sometimes has to be advanced to the point of the proximal cap.

- Once the wire is at the level of the proximal cap and within the 'beak' of the occlusion, it is advanced quickly but smoothly across the proximal cap.

Advancing Beyond the Proximal Cap:

- During advancement of the wire, try to maintain momentum.
- If the direction of wire path appears to be intra-luminal, advance the wire as far as the aortic bifurcation leading with the tip (straight) or looping it. Follow with the Glidecath on the wire.
- If the wire tip seems to be deviating away from the desired path, it can be exchanged out for an angled stiff hydrophilic wire which can then be redirected.

Crossing Into the CIA:

- If the wire appears to track into one of the iliac arteries, follow it with the hydrophilic catheter.
- Typically, alternating combinations of straight wire/angled catheter and angled wire/angled catheter allow successful crossing.
- If this approach does not work from above, advancing from either femoral artery and meeting the brachial wire/catheter 'half-way' may be an option. This may allow either wire to be externalized or snared out.
- Once the wire is successfully advanced into the CIA on one side, the catheter is tracked over it.
- Continue to advance the wire until it enters the reconstituted segment. At this point, the wire is 'externalized' through the ipsilateral femoral sheath.
- The same maneuver is repeated for the other iliac artery.

Pre-dilatation:

- Pre-dilate the lesion with an undersized balloon.
- Be cautious not to over-size to avoid a dissection or perforation.
- Typical diameters for predilatation balloons are 6 - 8 mm.

Calibrated Aortography:

- Repeat aortography should be performed after the occluded segments have been crossed and predilated.
- A calibrated (marker) pigtail catheter is used. The goal is to measure the length of the diseased segments precisely to aid in the selection of the appropriate stent lengths.
- The renal arteries, aortic termination and the origin of the hypogastric arteries should be adequately visualized.
- If possible, a smaller injection (e.g. 10 cc/sec for a total of 20 cc) should be used.

Aortic Stenting:

- The sheath is shepherded across the lesion.
- The stent catheter is advanced via the femoral sheath to the lesion and positioned at the desired level.
- The sheath is withdrawn to expose the stent.
- The stent is deployed in the usual fashion.
- As noted above, if concurrent iliac disease is present and needs to be treated then the bifurcated modular grafting technique should be used.
- The aortic stent is deployed first.
- Post-dilatation is performed with an appropriate size balloon.

Iliac Stenting:

- When indicated, the iliac stents are deployed after the aortic stent has been placed and dilated.
- Stent-grafts should be used for the reasons mentioned above.
- To maintain perfusion to both legs, kissing iliac stents are used, even if significant disease is present only on one side.

Sheath Size
- Palmaz XD stent - 10 Fr. sheath
- Palmaz XL 12-14 Fr. sheath

- After postdilatation of the aorta, the dilator for each femoral sheath is re-introduced over the wire, into its respective sheath. Under fluoroscopic guidance, the sheaths are advanced, one after the other, inside the aortic stents.
- These stents are overlapped within the aortic stent by 2-3 cm.
- Both right and left iliac stents are positioned and deployed simultaneously.
- After postdilatation of the iliac stents, the balloons are removed.
- Final hemodynamics are performed followed by final abdominal aortography.
- DSA and breath-holding technique should be used.

HOW TO CRIMP A STENT?

Pre-mounted balloon-expandable stents are not available in diameters over 12 mm. If a balloon-expandable stent is required in diameters > 12 mm it has to be manually crimped on an appropriate size balloon by the operator. Examples of bare stents that are routinely hand-crimped for this purpose include Palmaz XD and Palmaz XL (Cordis Corporation, Bridgewater NJ). Each stent is packaged with a crimping tube, a metallic tube that is meant to assist in crimping of the stent.

Manually crimped stents:
- These stents are more likely to be dislodged off the balloon during delivery than pre-mounted stents, particularly when crossing tortuous anatomy..

BALLOON PREPARATION

- After a diagnostic angiogram is performed using a calibrated catheter, an appropriate stent diameter and length is selected (see tables 1 & 2).
- The stent(s) should extend from normal to normal segment of the vessel.
- When overlapping stents are placed, the cranial stent is placed first and the caudal stent is overlapped with it.
- The balloon over which these stents are delivered is called a 'carrier' balloon (figure 6).
- The balloon should be at least as long as the unexpanded stent.
- The balloon is inflated at least once to 'wing' it. A gentle inflation (at 2-4 ATM) and deflation may be adequate.
- The surface of the inflated balloon is wiped gently with a moist piece of lint-free cloth/gauze.
- The balloon is deflated and kept on negative pressure.
- It is then re-wrapped/re-groomed with the 're-wrapping'/'re-grooming' tool provided with the balloon.
    - ▹ The balloon is loaded on a 0.035" wire.
    - ▹ The re-grooming tool is also loaded on the wire and advanced tip to hub (distal to proximal) in a counter-clockwise fashion (when looking from the tip of the balloon to its hub).

If a crimping tool (figure 6) is available it should be used. Otherwise, manual crimping is used.

CRIMPING WITH A CRIMPING TOOL

After the above preparation, the following steps are taken to crimp a stent using a crimping tool:

- The crimping tool is provided in a non-sterile state by the vendor and should be sterilized using autoclaving according to hospital policy.

- The unexpanded stent and the crimping tube (figure 7) is removed from its sterile case.
- The stent is rolled manually between the index fingers and thumbs to a smaller diameter so that it fits easily inside the crimping tube.
- Once inside the crimping tube, the stent is advanced over the carrier balloon and positioned between the markers of the balloon (figure 7A).
- The crimping tube/stent/balloon assembly is then loaded into die number 1 of the crimping tool. During this maneuver, it is imperative to ensure that the stent is still between the markers of the balloon.
- The stent is centered in die number 1, its ends outside the die.
- The handles of the crimp tool are squeezed firmly to crimp the stent (figure 7B).
- After releasing the handles, the crimping tube/stent/balloon assembly is simultaneously rotated 90°.
- After this step, the ends of the stent are sequentially crimped at 0° and then 90°.
- The crimping tube/stent/balloon are then removed from die number 1 and placed in die number 2 (centring the stent), and the above steps are repeated (figure 7C).
- With the crimping tube/stent/balloon in die number 2, the handles of the crimping tool are squeezed until they come to a fixed stop.
- The assembly is removed from die number 2 and the crimping tube is gently slid off the stent.
- The balloon is inspected to ensure the stent is still within the markers of the balloon (figure 7D).
- The stop-cock on the balloon is opened to 'neutral' position allowing contrast to enter the balloon lumen.
- The stop-cock is kept in the 'neutral' position until the stent reaches its target lesion within the body, where it is then prepped to a 'negative' hold prior to deployment.

Crimping tube:
- The crimping tube is used to crimp the stent particularly with the crimping tool.
- This tube is also used to assist in delivering the stent into the sheath, through the diaphragm so the stent does not get stuck and dislodged.

CRIMPING WITHOUT A CRIMPING TOOL

- Dipping the stent in 100% dye may make it more 'sticky' prior to loading it onto the balloon.
- Gently load the stent onto the balloon using transverse (radial) force.
- Position the stent between the two markers of the balloon (figure 8B).
- The stylet/mandrel that is provided with the carrier balloon is inserted into the tip of the balloon catheter. This provides support to the shaft during crimping.
- To crimp the stent, gently pinch onto the balloon (figure 8C). Crimp the stent tip to hub (distal to proximal).
- After crimping the length of the stent once, rotate balloon and stent 45° (together as a unit) and continue to crimp.
- Perform crimping until the stent is securely loaded onto the balloon.
- Avoid the following balloon maneuvers:
    - ▷ Rolling
    - ▷ Kinking
    - ▷ Twisting
    - ▷ Turning
- Remove the stylet.
- Do not wipe the stent.

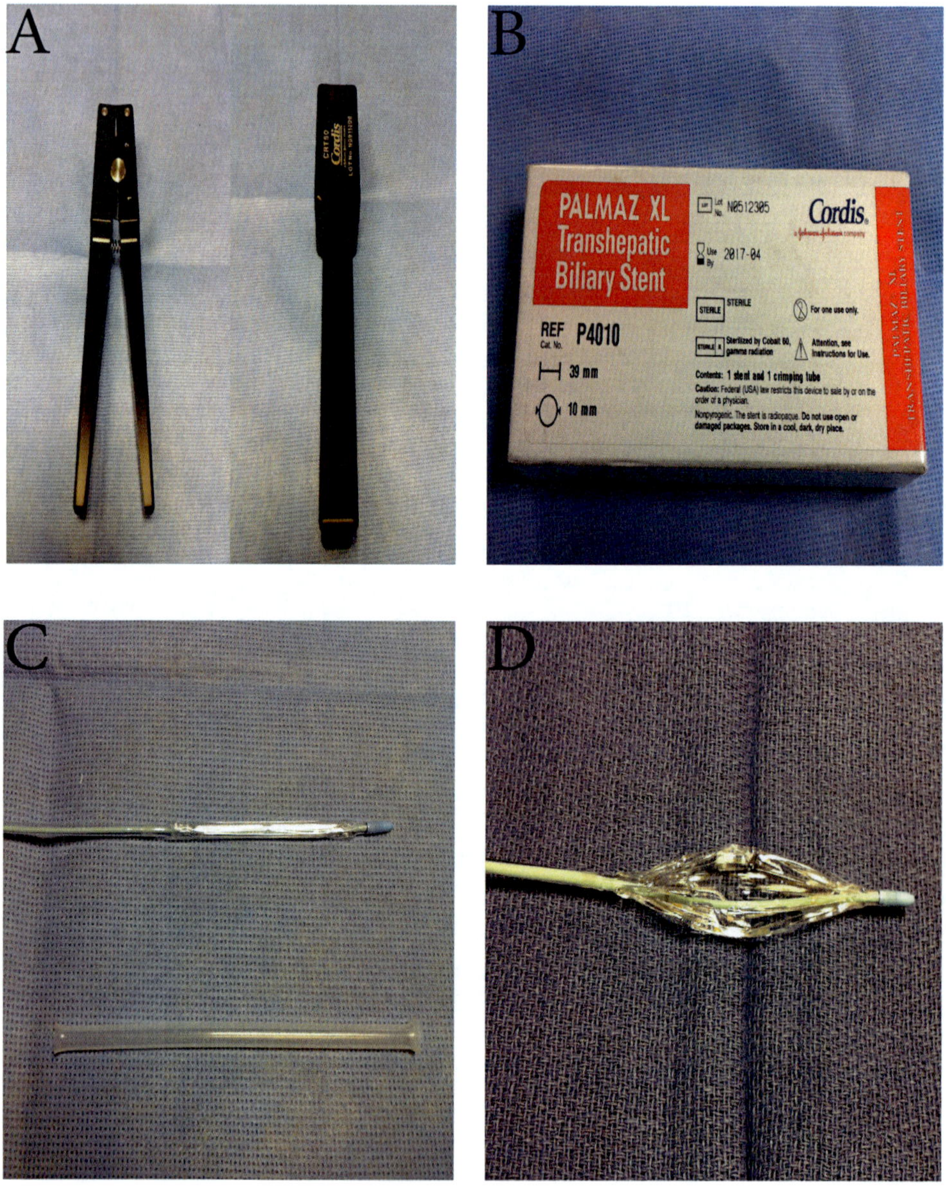

Figure 6.

Tools for crimping a stent.

A. Stent crimping tool from Cordis.
B. Palmaz XL 4010 stent in its box.
C. A carrier balloon and its re-wrapping tool.
D. The carrier balloon is noted to be 'winged' after a couple of quick inflations and deflations. This maneuver increases the profile of the balloon even after it is re-wrapped, so that it is a snug fit inside the manually crimped stent.

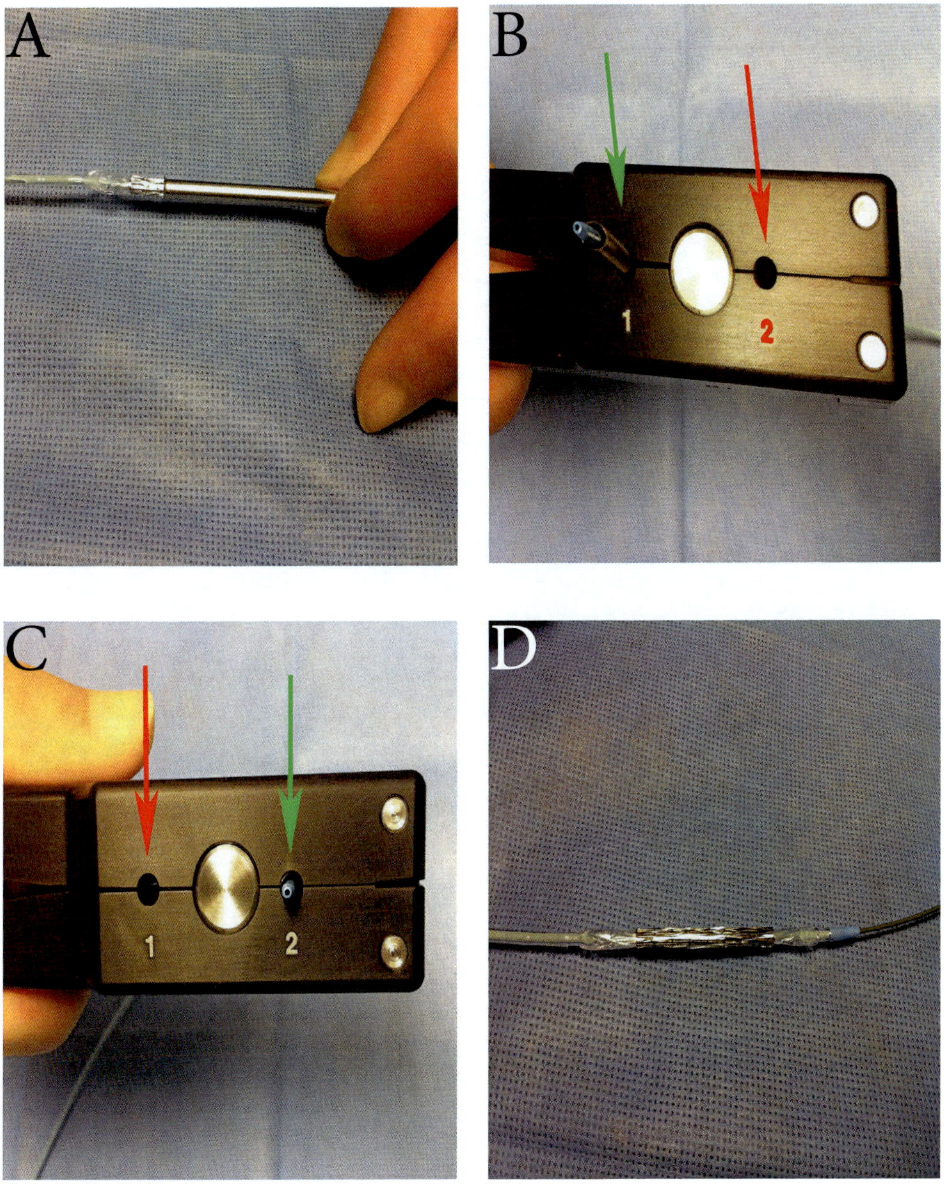

Figure 7.

Crimping a stent with a crimping tool.

A. The crimping tube is mounted on the stent loaded on the carrier balloon.
B. The crimping tube/stent/carrier balloon assembly is then loaded into die # 1 of the crimping tool.
C. After crimping in die # 1, the assembly is moved to die # 2 and crimped again. Here the distal edge of the stent is being crimped after crimping the body of the stent in die #2.
D. The crimped stent on the carrier balloon.

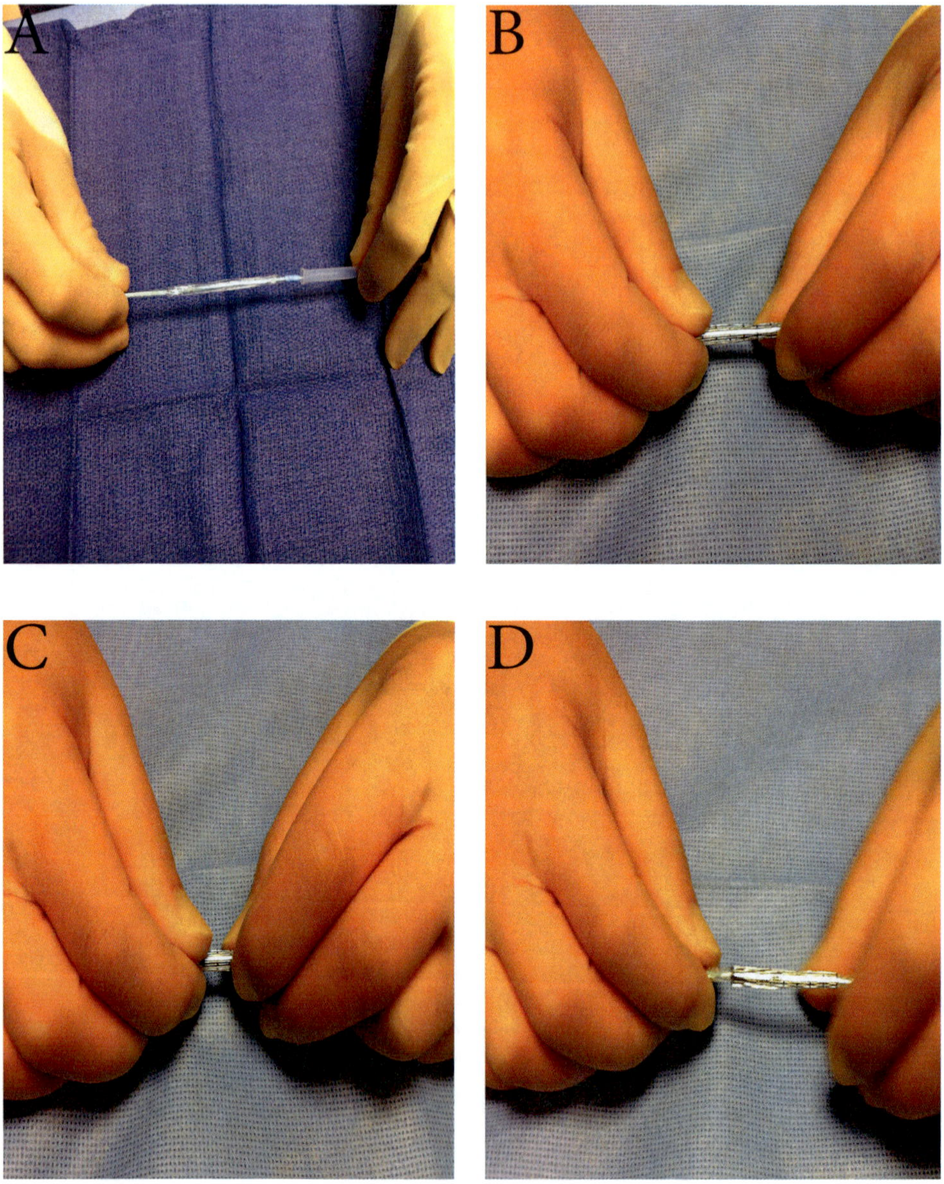

Figure 8.

Hand-crimping a stent.

A. The carrier balloon has been adequately re-wrapped.
B. The stent is loaded onto the carrier balloon gently.
C. It is pinched sequentially from distal (tip) to proximal (hub) to crimp it.
D. The crimped stent is noted on the balloon. Kinking, rolling, twisting and turning of the catheter should be avoided during crimping.

STENT DELIVERY AND DEPLOYMENT

The manually crimped stent is delivered and deployed in the following fashion, regardless of whether it was crimped with or without a crimping tool.

- The stent is loaded onto the 0.035" wire and passed through the diaphragm of the sheath with great care ensuring not to dislodge it off the balloon.
- The sheath is advanced and shepherded across the lesion under fluoroscopy.
- The stent is advanced within the sheath until it is delivered to the target lesion.
- Holding the balloon catheter in position, the sheath is withdrawn well beyond the proximal marker of the carrier balloon.
- The stent is deployed by inflating the balloon to its nominal pressure.
- Expect up to 5% reduction in the stent length from unexpanded to nominal expansion.
- The balloon is deflated and allowed to return to its lowest profile prior to removing from within the stent.

CASE 1

A 66-year old moderately obese female with hypertension, dyslipidemia, hypothyroidism and peripheral arterial disease presented with disabling intermittent claudication of lower extremities. She had bilateral calf and thigh discomfort with ambulation (less than 30 feet on level ground). Right leg symptoms were significantly worse than the left.

ABI and PVRs showed severe inflow PAD bilaterally right worse than left. Given the above findings, a CTA of abdomen, pelvis and lower extremities was obtained (figure 9).
This study showed:

- Patent suprarenal aorta.
- Patent celiac artery and SMA.
- Patent single bilateral renal arteries.
- Small diffusely diseased but patent IMA.
- Diffuse severe disease in the infra-renal aorta with multiple tandem severe stenoses.
- Severe stenosis at the ostium of the LEFT common iliac artery.
- Total occlusion of the ostium of the RIGHT common iliac artery with reconstitution in the distal RIGHT common iliac artery.
- Patent bilateral hypogastric arteries and external iliac arteries.
- RIGHT external iliac artery was undersized compared to the LEFT likely due to chronic under-filling.
- Patent bilateral CFA with a relatively low bifurcation on the RIGHT and a relatively high bifurcation of the LEFT.

Access:

- Arterial access was obtained in the left brachial artery using a 4 Fr. micropuncture kit in a modified Seldinger fashion.
- A 5 Fr./11 cm Cordis Britetip sheath was inserted in the artery over a 0.035" angled stiff Glidewire wire.
- The patient was given 2000 units of IV heparin.

Diagnostic Angiography:

- A 5 Fr./100 cm marker (calibrated) pigtail catheter was advanced via the LEFT brachial sheath into the abdominal aorta and abdominal aortography was performed (figure 10A).
- The findings of the CTA were confirmed on this study.

Intervention:

Left CFA access

- The pigtail catheter was then exchanged out for a QuickCross Extreme catheter which was advanced over a 0.035" angled stiff 260 cm Glidewire into the LEFT external iliac artery.
- A 4 Fr. micropuncture kit was used to obtain access in the LEFT CFA under fluoroscopic/roadmap guidance.
- A 5 Fr./23 cm Brite-tip sheath was inserted in the LEFT CFA over a 0.035" J-wire that was advanced through the microcatheter in the LEFT CFA.

Externalization of brachial wire via LEFT CFA sheath

- The angled stiff Glidewire and the QuickCross Extreme catheter from brachial access, were maneuvered into the tip of the LEFT CFA sheath.
- This catheter and wire were externalized through the LEFT CFA sheath (figure 10B).
- The wire was clamped outside the left CFA sheath and the catheter was removed via the LEFT brachial artery.

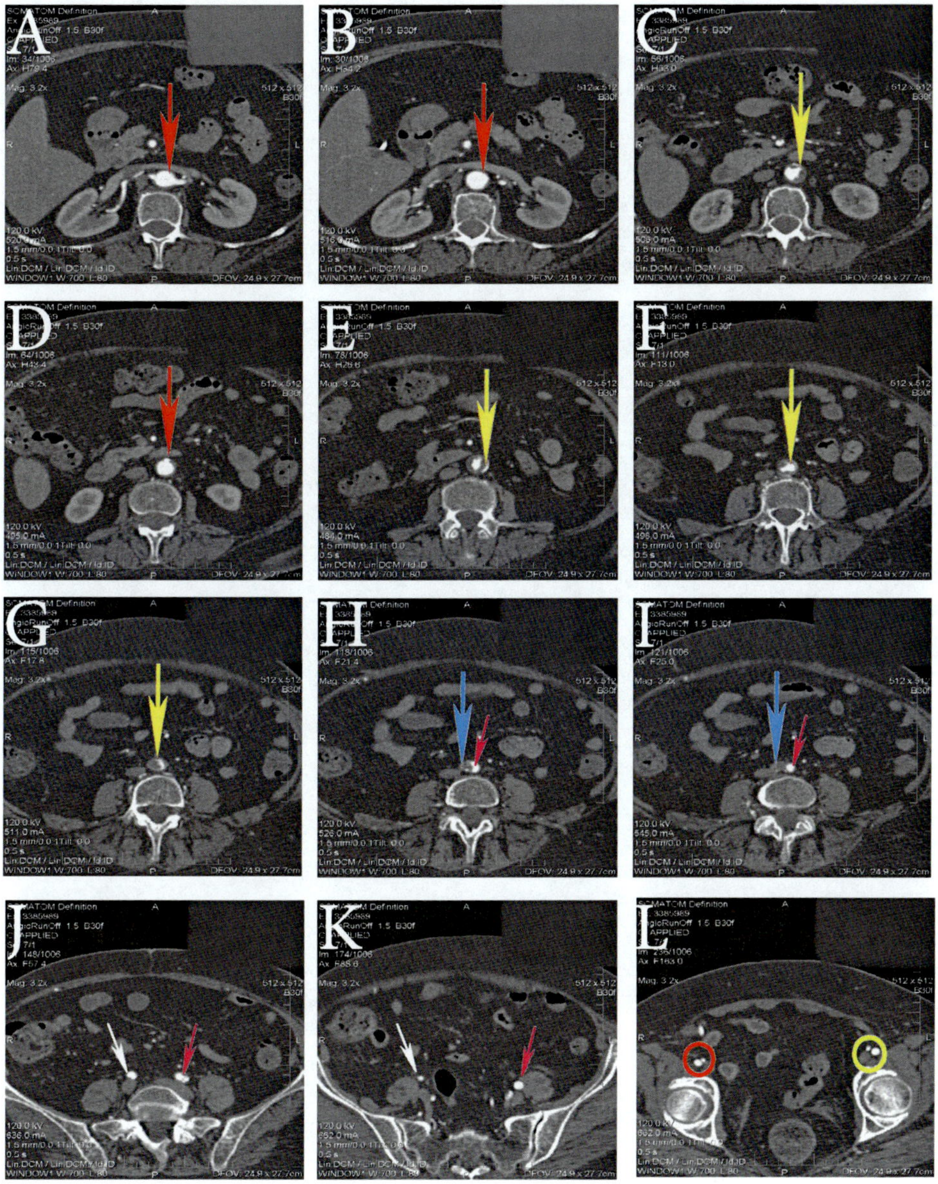

Figure 9.

Pre-operative CTA.

Red arrows: Patent segments of abdominal aorta.
Yellow arrows: Severely diseased segments of abdominal aorta with thrombus/plaque.
Blue arrows: Occluded right common iliac artery.
Magenta arrows: Patent left common iliac artery.
White arrows: Reconstituted right common (J) and external (K) iliac artery.
Red Circle: Patent right common femoral artery.
Yellow Circle: Patent left common femoral artery.

Calibration aortography

- The marker pigtail catheter was advanced now via the LEFT femoral sheath into the abdominal aorta and another aortogram was performed to measure the distance between the lowest renal artery and aortic bifurcation (figure 10C).
- This distance was noted to be close to 10 cm.
- This catheter was then removed over a 0.035" J-tip 180 cm Superstiff Amplatz wire.

Preclosure and upsizing the LEFT CFA sheath

- A 6 Fr. sheath was then inserted over this wire back into the LEFT CFA.
- Preclosure was performed using two Perclose® devices in a standard fashion.
- Thereafter the subcutaneous track was dilated with an 8 Fr., 10 Fr. and finally a 12 Fr. dilator.
- A 12 Fr./30 cm sheath was then inserted over this wire into the distal LEFT external iliac artery.

PTA of LEFT CIA/EIA

- Angioplasty was performed of the LEFT common and external iliac arteries using a 6.0 mm x 40 mm balloon (figure 10D).

Aortic stent-graft placement

- The dilator was re-introduced and the 12 Fr. sheath was advanced under fluoroscopic guidance into the distal abdominal aorta.
- A Gore Excluder® 16 mm/12 mm/7 cm iliac extender was then prepared and advanced into the sheath to the level below the renal arteries. Care was taken to ensure that the proximal (hub) end of the stent-graft was above the terminal aorta and would not cover access to the right CIA ostium.
- The sheath was withdrawn and the stent was deployed under roadmap guidance (figure 11B).
- The stent was then postdilated using an 8.0 mm x 20 mm Fox balloon (figure 11C).

Upsizing the LEFT brachial artery sheath

- The LEFT brachial artery sheath was then exchanged out for a 6 Fr./90 cm Shuttle sheath.
- This sheath was advanced over an angled stiff Glidewire into the abdominal aorta and its tip was brought within the stent-graft. The dilator was removed and the sheath was flushed.
- The patient was given an additional 3000 units of IV heparin.

Crossing the RIGHT CIA Occlusion

- A 6 Fr. multipurpose coronary guiding catheter was advanced through the Shuttle sheath into the distal abdominal aorta. A roadmap was obtained and the guiding catheter was pointed towards the ostium of the occluded RIGHT CIA.
- The wire was exchanged out for a straight stiff 260 cm Glidewire. This was used to cross the proximal of the occluded RIGHT common iliac artery.
- The wire was easily advanced into the RIGHT CFA (figure 12A).
- The guiding catheter was exchanged out through the Shuttle. A 0.035" straight QuickCross 150 cm catheter was advanced through the Shuttle sheath into the RIGHT CFA over the straight wire.
- The wire was removed and a confirmatory angiogram was performed (third order selective angiogram) (figure 12B).

Right CFA access

- Arterial access was then obtained in the RIGHT CFA using a 4 Fr. micropuncture kit.
- A 0.035" J-tip wire was advanced into the RIGHT common iliac artery, below the occluded segment.
- To provide support for sheath placment this wire was exchanged out through a 4 Fr./65 cm Glidecath for a 0.035 inch J-tip 260 cm Superstiff Amplatz wire.
- The Glidecath was removed and an 8 Fr./23 cm Cordis Britetip sheath was advanced into the RIGHT CFA and was brought into the distal RIGHT EIA.

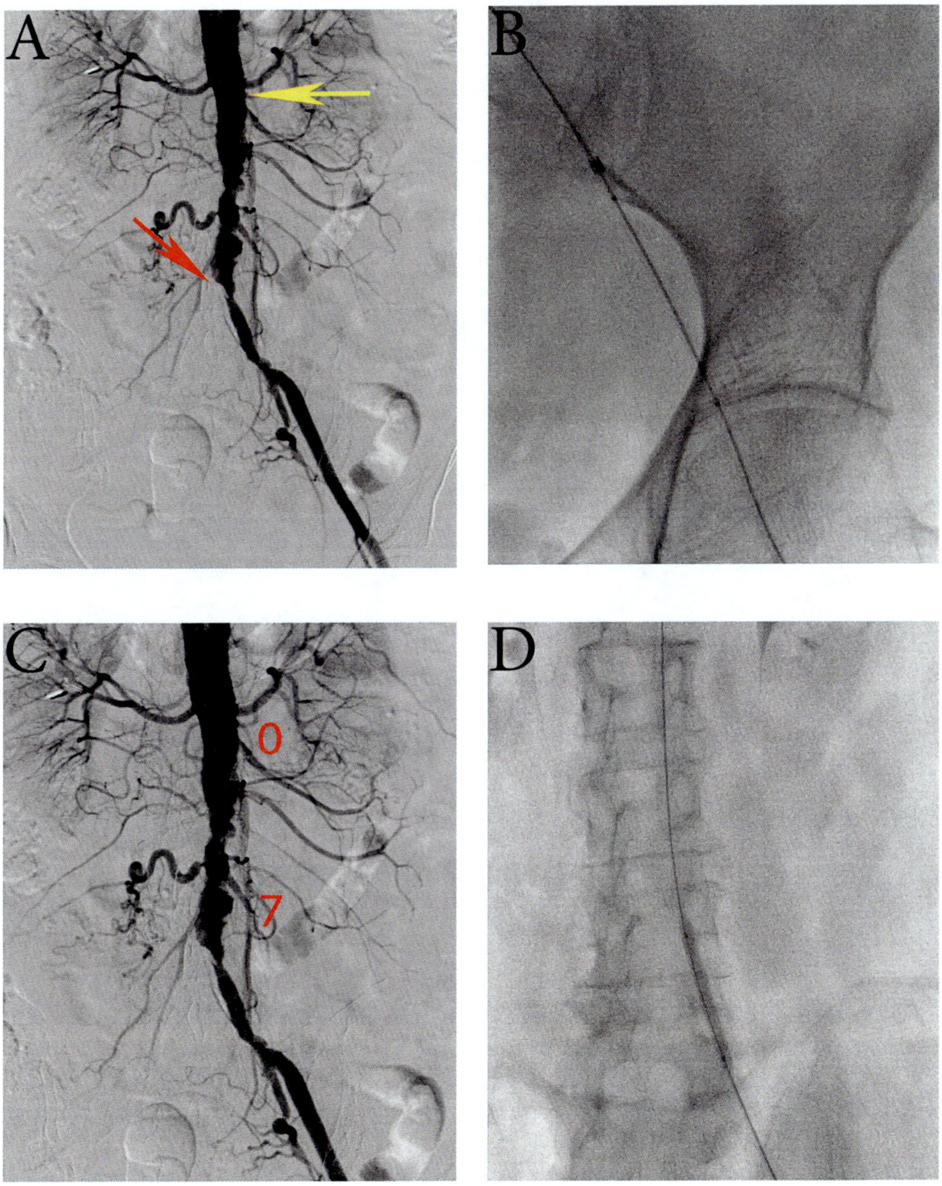

Figure 10.

A. Abdominal aortogram via left brachial artery. Yellow arrow indicates lower margin of the lowest renal artery (left). The red arrow indicates ostial occlusion of the right common iliac artery.
B. Externalization of ASGW and a QuickCross Extreme catheter via left CFA sheath.
C. Abdominal aortogram with a calibrated pigtail catheter. The '0' indicates the planned level of the distal (upper) end of the 7 cm long aortic stent-graft (16/12/7) and the '7' indicates the level where the proximal (lower) end of this stent would land.
D. Angioplasty of the left common iliac artery.

Externalization of RIGHT CIA wire

- An angled stiff Glidewire was advanced through the QuickCross (QC) catheter via the Shuttle sheath into the RIGHT external iliac artery.
- This wire was maneuvered into the RIGHT CFA sheath and externalized.

Reversal of RIGHT CFA wire

- The QC catheter was removed from the brachial sheath, flushed and reintroduced over this wire via the RIGHT CFA sheath into the descending thoracic aorta.
- After removing the wire, a confirmatory injection was performed through the catheter. The wire was then re-introduced.

Removal of the Shuttle sheath from the LEFT brachial artery

- The Shuttle sheath was removed from the LEFT brachial artery and a 6 Fr./11 cm sheath was inserted in its place. The dilator was removed and the sheath was flushed.
- An ACT was checked and additional heparin was given.

PTA of RIGHT CIA

- The 260 cm Superstiff Amplatz wire was introduced via the RIGHT CFA sheath through the QC catheter into the descending thoracic aorta.
- The catheter was then removed.
- The RIGHT common iliac artery was then predilated with a 6.0 mm x 40 mm balloon (figure 12D).
- Along with this dilator, the RIGHT CFA sheath was then brought into the distal abdominal aorta. The dilator was removed and the sheath was flushed.

Placement of bilateral iliac stents

- A Gore Viabahn 9 mm x 50 mm stent was then positioned via the RIGHT CFA sheath into the distal abdominal aorta overlapping with aortic stent by 2 cm.
- A 9 mm x 50 mm Viabahn stent was introduced via the LEFT CFA sheath and positioned at the level of the RIGHT side Viabahn stent.
- Under fluoroscopic guidance, the 2 stents were deployed simultaneously with the proximal edges overlapping with aortic stent by 2 cm and their distal 2 cm overlapped with their respective common iliac arteries.
- Both the stents were postdilated in a kissing fashion with an 8 x 20 mm and 8 x 30 mm Fox balloons (figure 13A).
- The RIGHT CIA stent was extended by overlapping with another 8 x 50 mm Viabahn stent. The RIGHT hypogastric was not covered.
- The LEFT CIA stent was overlapped with a 10 x 60 mm Absolute stent as disease extended into the EIA as we intended to keep the IIA open.
- Both the stented segments were then postdilated with 8 mm balloons including the overlap.

Post-dilatation of the aortic stent-graft

- A 12 mm x 40 mm Atlas balloon was then introduced via the LEFT CFA sheath in the aortic stent, above the Viabahn stents.
- The aortic stent was postdilated at 6 atmospheres x 3 (figure 13B).
- The RIGHT external iliac artery was noted to be undiseased but relatively "undersized" due to being chronically underfilled.
- No hemodynamic gradient was noted between the abdominal aorta and the distal RIGHT external iliac artery or the LEFT external iliac artery.

Final Angiography:

- Patent infra-renal aorta with patent aortic, bilateral CIA and EIA stents (figure 13C).
- The RIGHT CFA sheath was closed with a single Perclose device without issues.
- The LEFT CFA sheath was closed with the 'preclosed' devices x 2, without any issues.

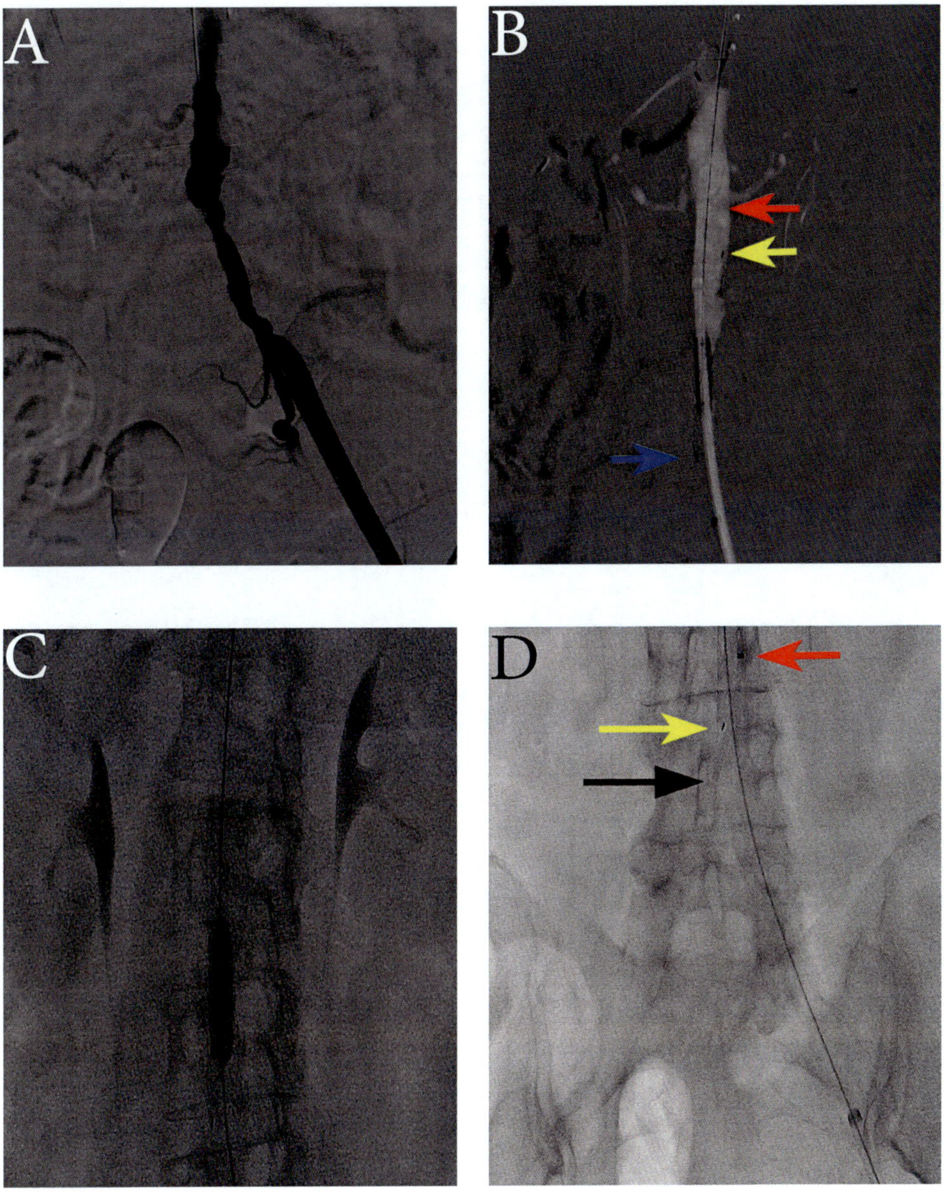

Figure 11.

A. Angiogram of the left CIA after PTA of the stenosis in the proximal segment.
B. Road-map aortogram after placement of the aortic stent. Red arrow - lowest renal artery; yellow arrow - distal marker of the aortic stent; blue arrow - proximal marker of the aortic stent.
C. Post-dilatation of the aortic stent.
D. Red arrow - tip of the Shuttle sheath; yellow arrow - proximal marker of the aortic stent; black arrow - tip of the multipurpose guiding catheter.

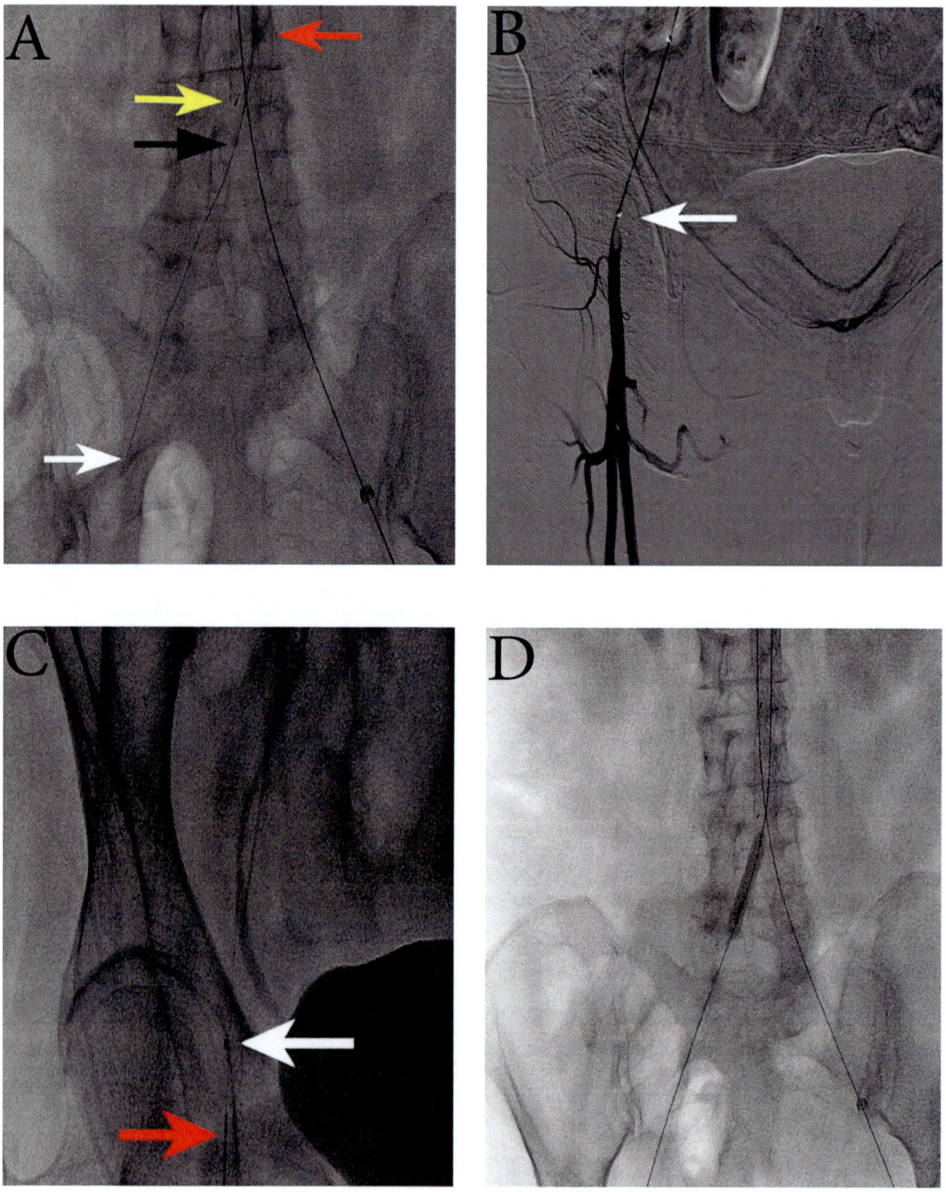

Figure 12.

A. Red arrow - tip of the Shuttle sheath; yellow arrow - proximal marker of the aortic stent; black arrow - tip of the multipurpose guiding catheter; white arrow - tip of the straight stiff Glidewire crossing into the right EIA.
B. Selective angiogram of the right CFA via a QuickCross catheter advanced via the left brachial artery.
C. Retrograde access being obtained in the right CFA using a 21 G needle (red arrow). The tip of the QuickCross catheter (white arrow) is seen in the proximal CFA.
D. After externalization and reversal of the wire in the aorta via the right CFA, PTA of the right CIA ostium is performed.

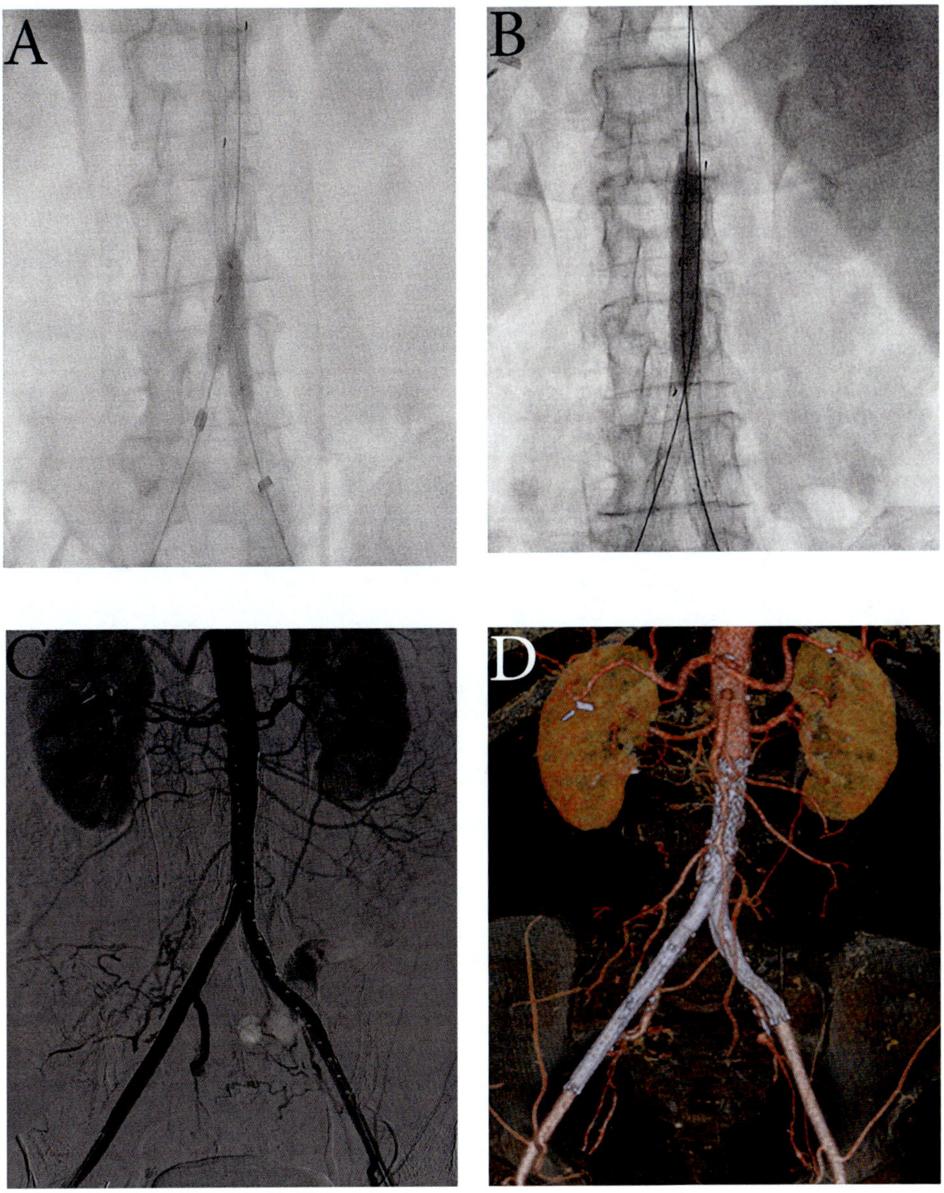

Figure 13.

A. Kissing postdilatation of bilateral common iliac arteries.
B. Post-dilatation of the aortic stent with the 12 mm x 40 mm Atlas balloon.
C. Final aortogram via the left CFA shows a patent aorta and widely patent bilateral common iliac and external iliac arteries.
D. An abdominal CTA with 3D-reconstruction confirming the above findings at 30-days post intervention.

Post-op Care:

- The patient was monitored overnight in the hospital and discharged home the following day.
- She was maintained on ASA 325 mg/d and clopidogrel 75 mg/d.

Follow-up:

- At 30-day follow-up, the patient was completely asymptomatic and had total resolution of her claudication symptoms in both legs.
- She underwent a repeat CTA (figure 13D) which showed widely patent stents in the abdominal aorta as well as bilateral iliac arteries. Her physiologic studies (ABI/PVR) were also within normal limits.
- She was advised to maintain her dual anti-platelet therapy indefinitely.

CASE 2

A 51-year old male with hypertension, dyslipidemia, bilateral prior carotid endarterectomy, known CAD with prior CABG and re-do CABG presented with severe chronic claudication of both legs now progressing to nocturnal rest pain in both feet. He did not have any tissue loss or ulceration.

His ABI/PVR showed severe aorto-iliac disease in both legs. Given this, a CTA of his abdomen, pelvis and legs was obtained. This study showed total occlusion of the infra-renal aorta and bilateral common and proximal external iliac arteries. He had prominent collateral flow to his bilateral CFA via lumbar and anterior abdominal wall collaterals (Winslow's pathway - superior and inferior epigastric arteries).

His pre-operative workup revealed severe LV systolic dysfunction. A recent cardiac cath had shown only a patent LIMA bypass graft to the LAD. His other grafts and native coronary arteries were occluded.

Given the above, his risk for surgical revascularization of aorto-iliac disease was considered to be prohibitive. However, due to his advanced symptoms endovascular intervention was considered.

Access:

- Before the start of the procedure, appropriate size aortic and iliac stent-grafts and aortic oc-

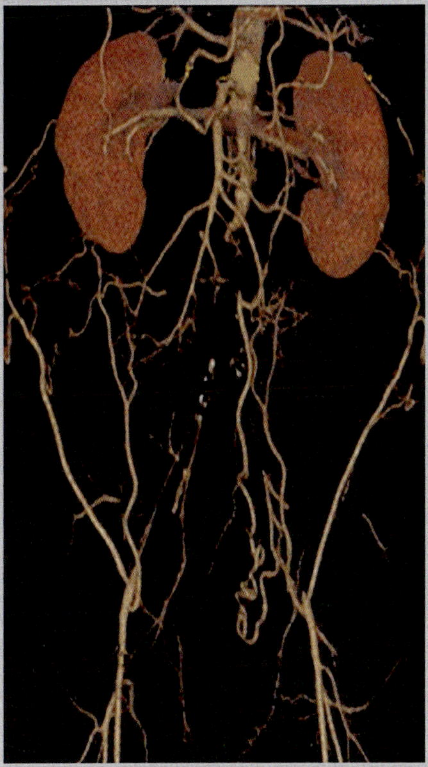

clusion balloon was kept handy.

- Arterial access was initially obtained in the right brachial artery (due to a patent LIMA the left brachial artery was not considered first).
- A 4 Fr. micropuncture kit was used. A 5 Fr./11 cm Cordis Britetip sheath was inserted in the artery over a 0.035" ASGW.

Diagnostic Angiography:

An abdominal aortogram was performed with a 5 Fr. pigtail catheter. On abdominal aortography the following were noted (figure 14):

- Patent suprarenal aorta.
- Patent celiac artery and SMA.
- Patent single bilateral renal arteries with bilateral ostial disease.
- Occluded inferior mesenteric artery, collateralized via SMA.
- The infra-renal aorta tapered down to a total occlusion.
- Prominent lumbar arteries were noted, collateralizing the hypogastric arteries bilaterally.
- Bilateral common arteries were occluded at the ostium and remained occluded throughout.
- The right EIA was severe diseased but patent as a 'string' like channel.
- The left EIA was totally occluded throughout.
- The right IIA was occluded at its origin and reconstituted distally via lumbar collaterals.
- The left IIA was also occluded at the ostium but reconstituted distally via superior rectal artery, a branch of the inferior mesenteric artery.
- Prominent circumflex iliac and inferior epigastric collaterals were noted on both sides.

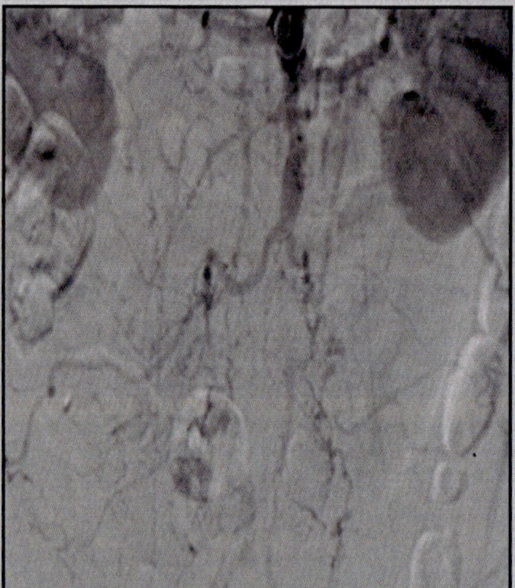

Figure 14.

Abdominal aortography via right brachial artery.

- Bilateral CFA were noted to be widely patent and disease free.

Upgrading the arterial access

- Under road-map guidance arterial access was obtained in bilateral CFA with a micropuncture kit.
- A 5 Fr./23 cm CBT sheath was inserted over an ASGW.
- The sheath tip was kept in the CFA on both sides.
- The right CFA arteriotomy was preclosed with two Perclose™ devices.
- The left CFA arteriotomy was preclosed in the same fashion.
- In a sequential fashion, a SSGW was advanced first from the right CFA sheath, retrograde towards the aorta. The wire encountered no difficulty up to the point of aortic bifurcation. Thereafter, it did not track cephalad on a straight path and appeared subintimal.
- The same was repeated on the left, with an identical result.
- It was then decided to try antegrade cannulation of the aorta, via right brachial artery access.

Intervention:

Antegrade traversal of the aortic occlusion

- The sheath in the right brachial artery was exchanged out for a 5 Fr./90 cm Shuttle sheath. This was parked just distal to the renal arteries.
- A QuickCross (QC) catheter was then advanced through the sheath to the level of the aortic occlusion.
- A SSGW was advanced across the proximal cap without much difficulty (figure 15).

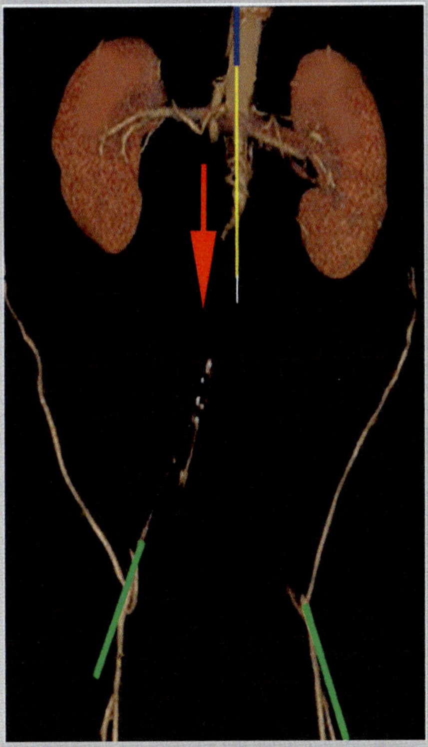

Figure 15.

Schematic showing antegrade traversal of aortic occlusion. Blue: Shuttle sheath. Green: Femoral sheaths. Red arrow: indicates the antegrade direction of wire traversal.

- The wire/catheter combination was advanced without difficulty to the level of the aortic bifurcation.

Crossing the distal cap into the right EIA

- At the level of the aortic bifurcation, the catheter was exchanged out for an angled Glidecath (GC). The SSGW was then advanced through the Glidecath, directing the wire towards the right CIA. With some effort the wire/catheter combination was advanced across the distal right CIA and into the right EIA (figure 16A).

Externalization of the wire via the right CFA

- The SSGW was then exchanged out for an ASGW which was maneuvered into the right CFA sheath and externalized.
- The Glidecath was removed via the right brachial artery (figure 16B).

PTA of the right EIA, CIA and aorta

- A 4 mm x 100 mm balloon was used to perform PTA along the entire length of the occluded right EIA, CIA and infra-renal aorta (figure 17A).

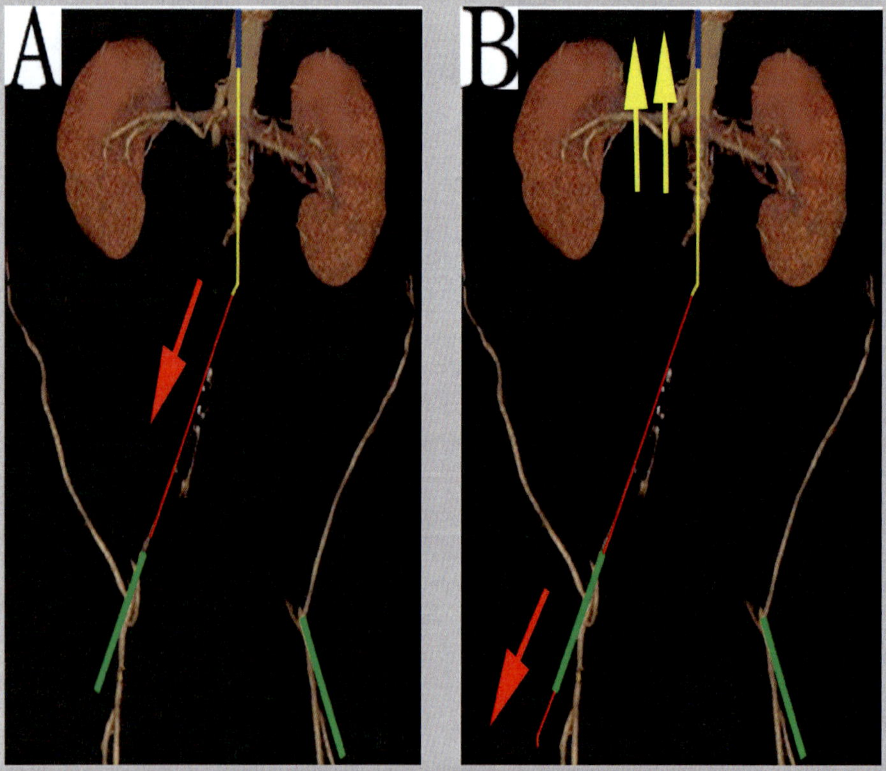

Figure 16.
A. The SSGW supported by an angled GC via the right brachial artery, crosses the right CIA and EIA occlusion (red arrow).

B. The wire is switched out for the ASGW which is externalized (red arrow) via the right CFA and the GC then removed via the right brachial sheath (double yellow arrows).

Reversal of the right CFA wires

- The QC was then advanced over the ASGW via the right CFA into the aorta and then inside the Shuttle sheath. This confirmed that the wire was in the true lumen in the aorta (figure 17B).
- The ASGW was then pulled out of the body.
- A 0.014"/300 cm Platinum Plus wire was advanced via the right CFA sheath, into the QC which was still in place.
- A 0.018"/300 cm V-18 wire was also advanced simultaneously through the same catheter in the same fashion (figure 18A).

Externalization of the V-18 via the right brachial artery

- The V-18 wire was externalized via the right brachial artery Shuttle sheath. Its back end was fixed outside the right CFA sheath.
- The Shuttle was pulled back a few inches, into the thoracic aorta. This exposed the tip of the Platinum Plus wire (figure 18B). This wire was then advanced up to the level of the left subclavian artery. The back end of this wire was fixed outside the right CFA sheath.

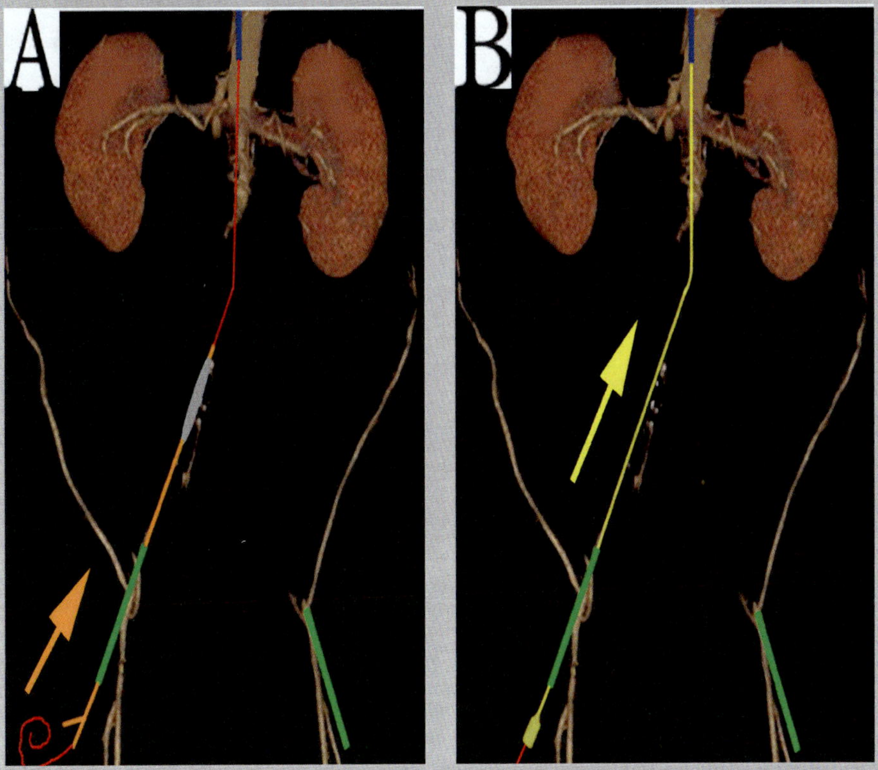

Figure 17.
A. PTA of the right iliac arteries via the right CFA.

B. The QuickCross catheter is inserted over the ASGW via the right CFA into the aorta, to switch the ASGW for two wires, a 0.014" and a 0.018" wire, both 300 cm long.

Advancing a GC catheter via the right brachial into the left iliac artery

- The QC catheter was then removed from the right CFA (figure 19A).
- The angled GC was introduced via the right brachial artery, over the V-18 wire.
- It was advanced until its tip reached the aortic bifurcation (figure 19B).

Crossing the left iliac artery occlusion

- The V-18 wire was removed from the body.
- It was replaced with a SSGW which was introduced into the GC via the brachial artery.
- The SSGW was advanced via the GC into the terminal aorta, till the level of the aortic bifurcation.
- This catheter was pointed towards the ostium of the left CIA and the occlusion crossed in the standard fashion (figure 20A).
- The GC was then advanced over this wire into the left EIA (figure 20B).

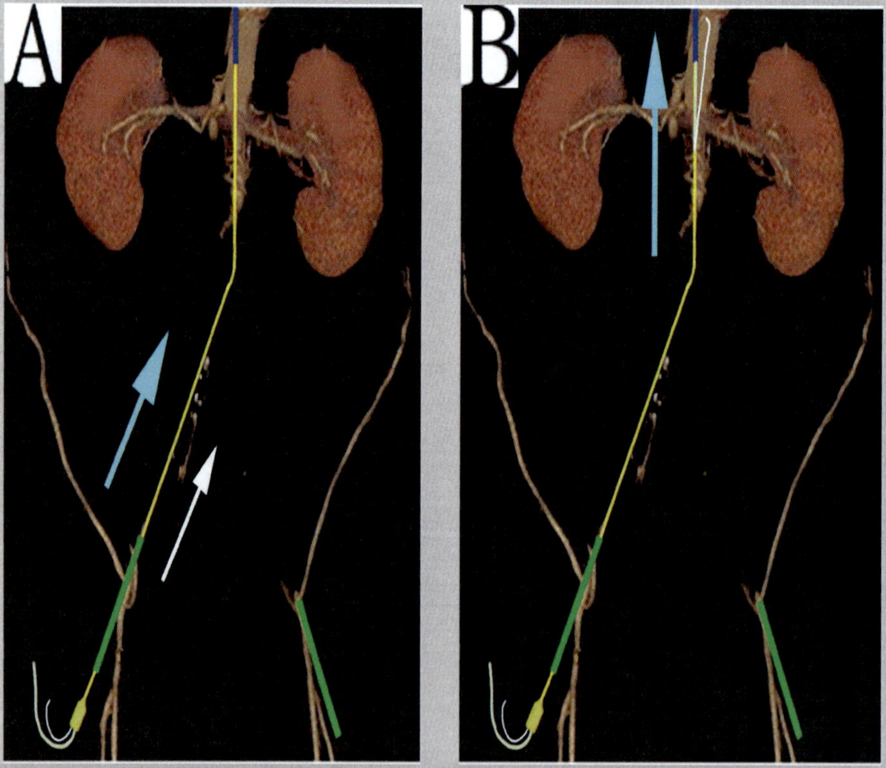

Figure 18.
A. A 0.014"/300 cm Platinum Plus wire (white arrow) and a 0.018"/300 cm V-18 wire (blue arrow) are advanced through the right CFA QuickCross catheter.
B. The Platinum Plus is kept in the descending thoracic aorta, outside the Shuttle, while the V-18 is externalized via the right brachial artery (blue arrow).

Externalization of the wire via the Left CFA sheath

- The left EIA SSGW was then removed via the brachial artery.
- It was replaced with an ASGW.
- This wire and the GC were then externalized via the left CFA sheath (figure 20B).

Reversal of the left CFA wire

- Through the externalized tip of the GC, the ASGW was introduced into the thoracic aorta (figure 21A).
- The back end of this wire was fixed outside the left CFA sheath. The tip of the wire was brought in the descending thoracic aorta at the same level as the Platinum Plus via the right CFA.

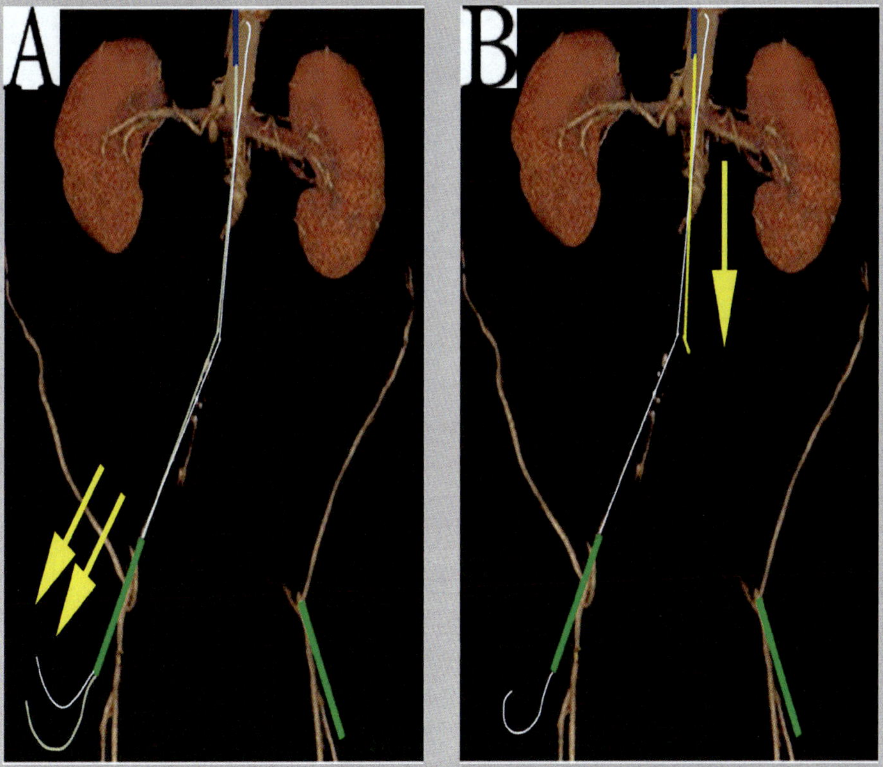

Figure 19.

A. The QuickCross catheter is removed from the right CFA (double yellow arrows). The two wires are kept in their positions.

B. The angled Glidecath is next introduced over the externalized V-18 wire via the right brachial artery Shuttle sheath. It is brought to the level of the aortic bifurcation over the V-18 wire, which is then removed .

- The GC was then removed via the right brachial artery (figure 21B).

PTA of the aorta and bilateral CIA

- At this point, two wires were across the aorto-iliac occlusion - a 0.014" Platinum Plus wire via the right CFA sheath and ASGW via the left CFA sheath (figure 21B).
- The tips of both these wires were in the descending thoracic aorta.
- PTA of both CIA and the aorta was then performed using 0.035" wire compatible 6.0 mm x 60 mm balloon (figure 22A).
- Before removal of the right CFA balloon, the Platinum Plus wire was exchanged through it for a 0.035" Super Stiff Amplatz wire. The tip of this wire was parked in the descending thoracic aorta.

Up sizing the CFA sheaths

- The 5 Fr. sheath in the right CFA was exchanged out for a 12 Fr./30 cm sheath.
- This sheath was advanced into the infra-renal abdominal aorta.
- The left CFA sheath was exchanged out for an 11 Fr./23 cm Britetip sheath.
- This sheath was parked in the proximal left CIA (figure 22 B).

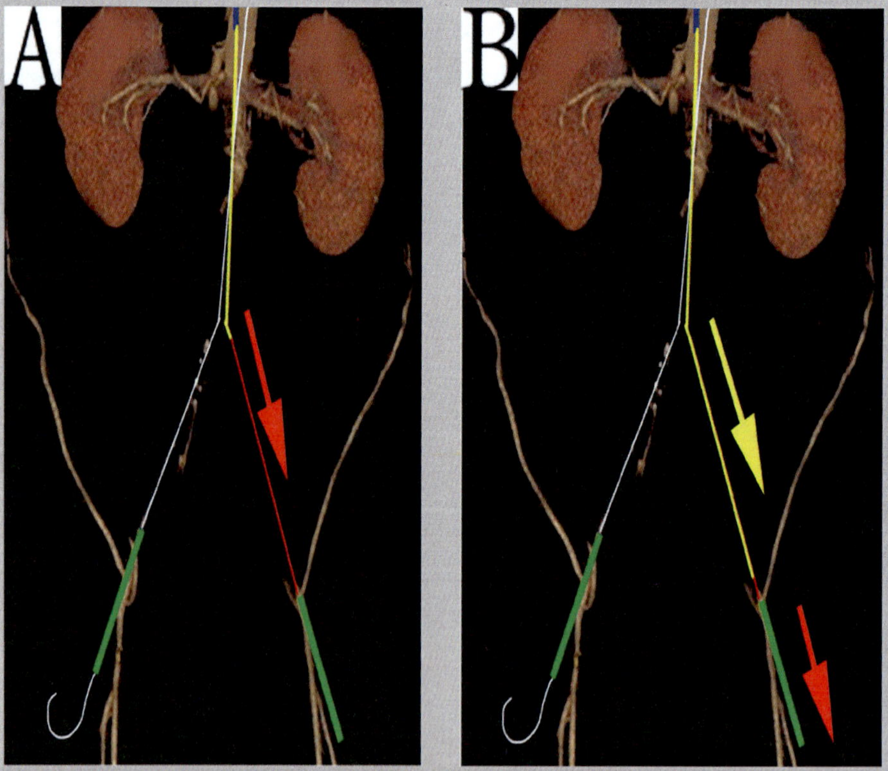

Figure 20.
A. The left CIA occlusion is crossed, similar to the right, with a SSGW supported by an angled GC (red arrow).
B. The GC is advanced over the wire into the left EIA (yellow arrow). The wire is switched out for an ASGW which is then maneuvered through the sheath and externalized via the left CFA sheath (red arrow).

Calibrating aortography via the left CFA

- A marker pigtail catheter was used to perform calibrating aortography via the left CFA sheath (figure 23A).
- This provided another opportunity to measure precisely the lengths of the occluded segment of the aorta and left common iliac artery.
- The aortic length between the lowest renal artery and the aortic bifurcation was noted to be 11 cm.

Positioning of aortic endograft

- Based on measurements from the pre-op CTA a Gore Excluder 'contra-lateral' limb' prosthesis measuring 16 mm/14.5 mm/70 mm (proximal diameter/distal diameter/length) was selected for stenting the aortic CTO segment (figure 23B).

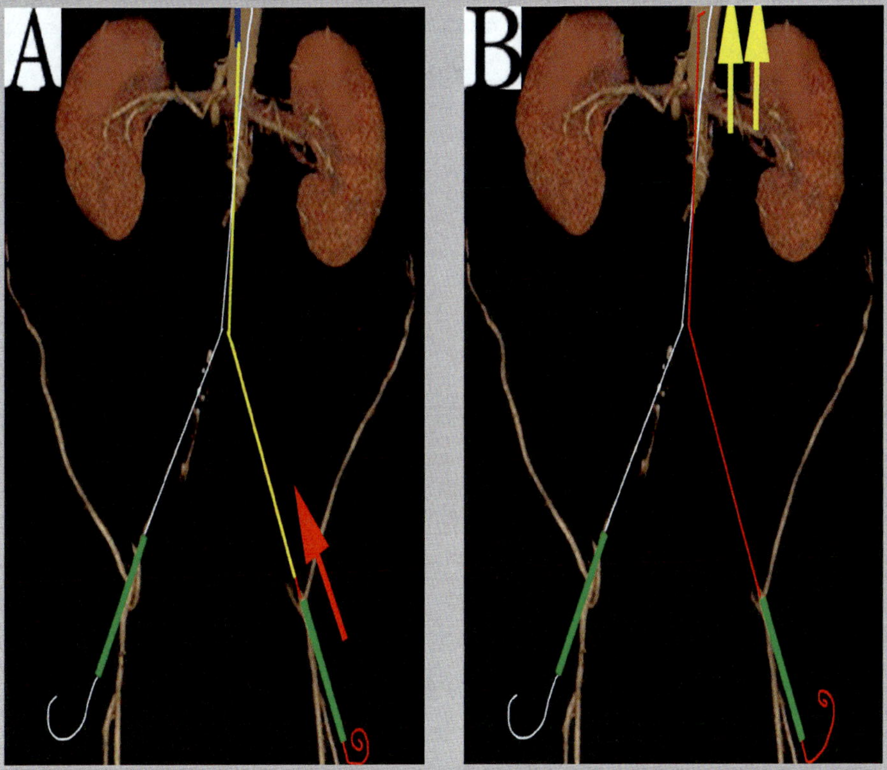

Figure 21.
A. An ASGW is introduced through the externalized tip of the left CFA GC (red arrow).
B. Once the wire tip is in the descending thoracic aorta, the GC is removed via the right brachial artery (double yellow arrows).

- Prior to introducing this stent-graft, the 12 Fr. sheath in the right CFA was advanced with its dilator into the infra-renal aorta.
- The dilator was removed and the sheath flushed.
- The pigtail via the left CFA was still in position for guidance in precise positioning.
- After the Excluder prosthesis was inserted in the sheath and positioned at the site of desired deployment, the sheath was pulled back into the right CIA.
- The stent-graft was deployed in the standard fashion.

Removal of pigtail catheter

- An ASGW was introduced via the pigtail catheter through the left CFA. This catheter was still behind the aortic stent-graft.
- The catheter was withdrawn without trouble from behind the stent-graft.
- The ASGW was then used to cannulate the aortic stent-graft. The pigtail was advanced over this wire into the aortic stent-graft, the wire withdrawn within the catheter. The catheter was 'spun' within the stent-graft to confirm intra-luminal position of the catheter.
- The ASGW was exchanged out for a 0.035" Super Stiff Amplatz wire through the pigtail catheter. This wire was then advanced into the descending thoracic aorta.

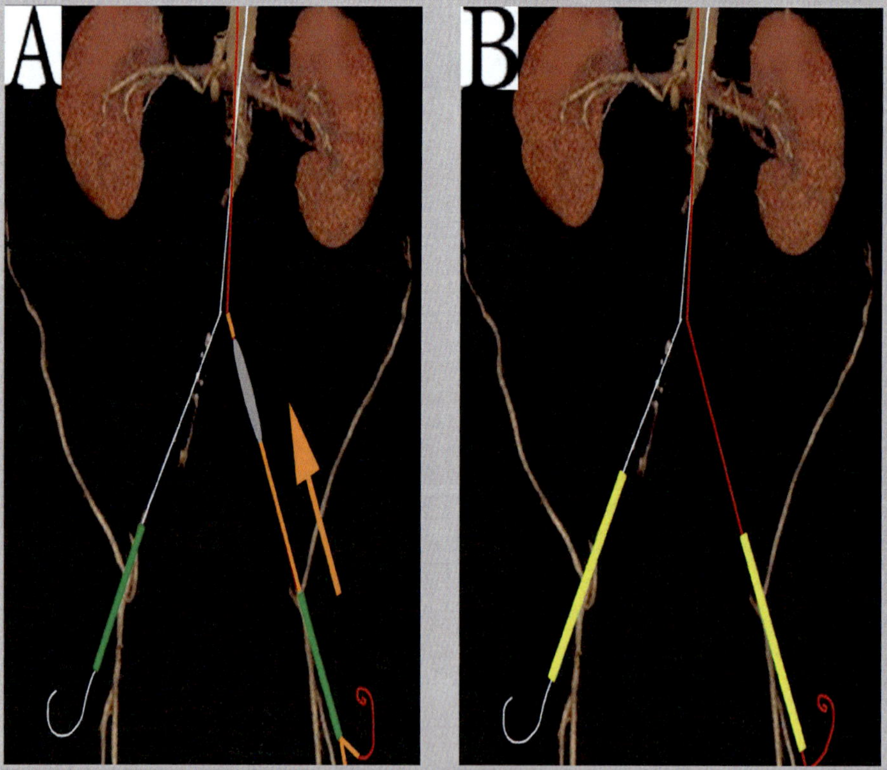

Figure 22.
A. PTA of the left iliac arteries via the left CFA sheath (orange arrow).
B. Both the right and left CFA sheaths, which were previously preclosed with two Perclose devices on each side, are upsized to 12 Fr. and 11 Fr. respectively.

- The pigtail catheter was then removed.

IVUS of the stent-graft

- To re-confirm the presence of the this SuperStiff Amplatz wire within the lumen of the aortic stent-graft and not behind it, IVUS was performed using the PV catheter.
- The IVUS catheter was then removed.

Post-dilation of the aortic stent-graft

- The aortic stent-graft was then postdilated with a 14 mm x 40 mm Atlas balloon.

Iliac stent-grafts

- Based on the concepts discussed earlier in the chapter, an 11 mm diameter Viabahn stent-graft was chosen for each CIA(14.5 mm x 0.8 ~ 11 mm).
- An 11 mm x 10 cm Viabahn was introduced on each side via the respective CFA sheath (shown schematically in figure 24A).
- These stent-grafts were positioned within the aortic stent-graft in the non-tapered distal segment.

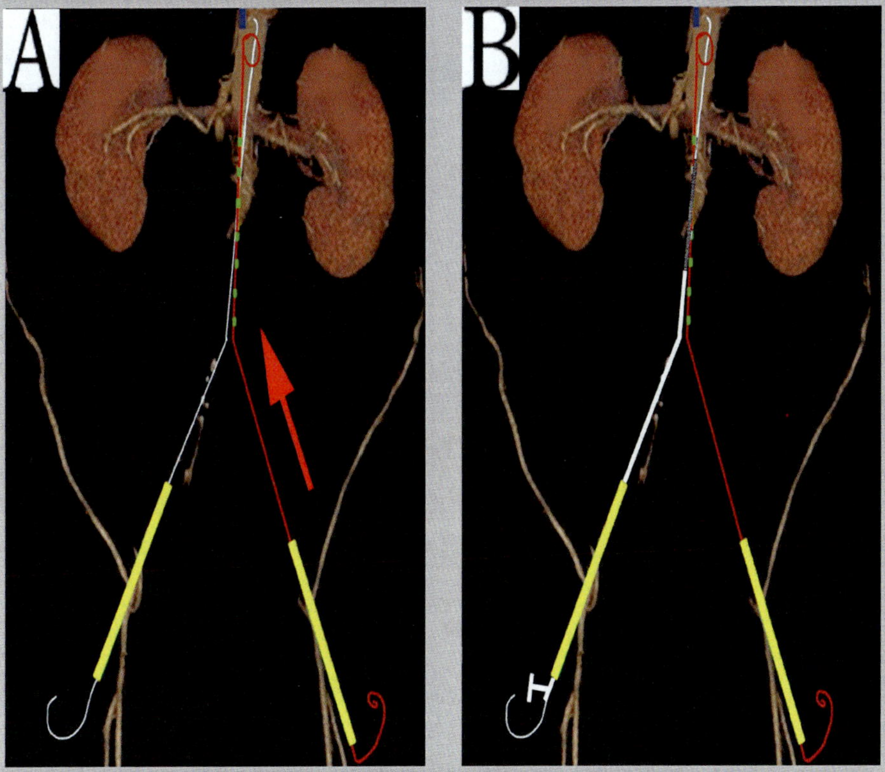

Figure 23.
A. A marker pigtail catheter is inserted via the left CFA (red arrow) into the abdominal aorta and a calibration aortogram is performed.
B. A Gore Excluder contralateral limb (with the dimensions 16 mm proximal diameter x 14.5 mm distal diameter x 10 cm length) is positioned in the infra-renal aorta via the right CFA sheath.

- The position of the two stent-grafts was matched at the top, their respective sheaths withdrawn into the EIA bilaterally, and they were deployed in a standard fashion (figure 24B).
- These were then postdilated in a 'kissing' fashion with 10 mm x 40 mm balloons.
- For radial strength, the inside of the proximal end of each of these Viabahn stent-grafts was lined with a 9 mm x 39 mm Genesis balloon-expandable stents. These again were deployed in a kissing fashion.

External iliac stenting

- The Viabahn stents extended into the CIA/proximal EIA bilaterally.
- Bilateral IIA were known to be occluded at their ostia, hence we deemed it OK to cover them with covered stents.
- The distal EIA was then treated with 8 mm x 100 mm Absolute stents, overlapped with the above.
- The overlap zone was postdilated with 8 mm x 40 mm balloons.
- Simultaneous pressure recordings were made from the supra-renal aorta and bilateral external iliac arteries.
- This showed no hemodynamic gradient between the supra-renal aorta and either external iliac artery.

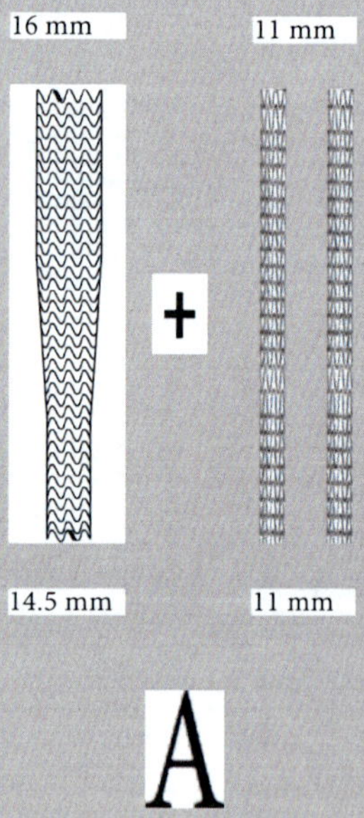

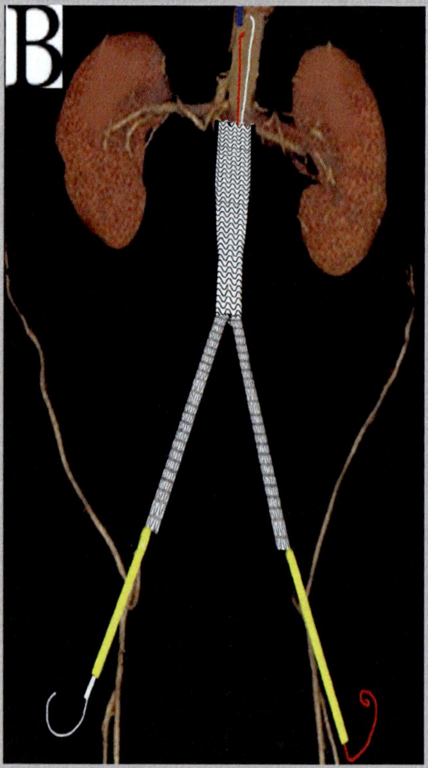

Figure 24.
A. A Gore Excluder limb (16 mm x 14.5 mm x 70 mm) is used in the aorta and a Viabahn stent-graft 11 mm x 10 cm is used in each CIA, overlapped inside the aortic stentgraft.
B. Final configuration of the aortic and iliac stentgrafts.

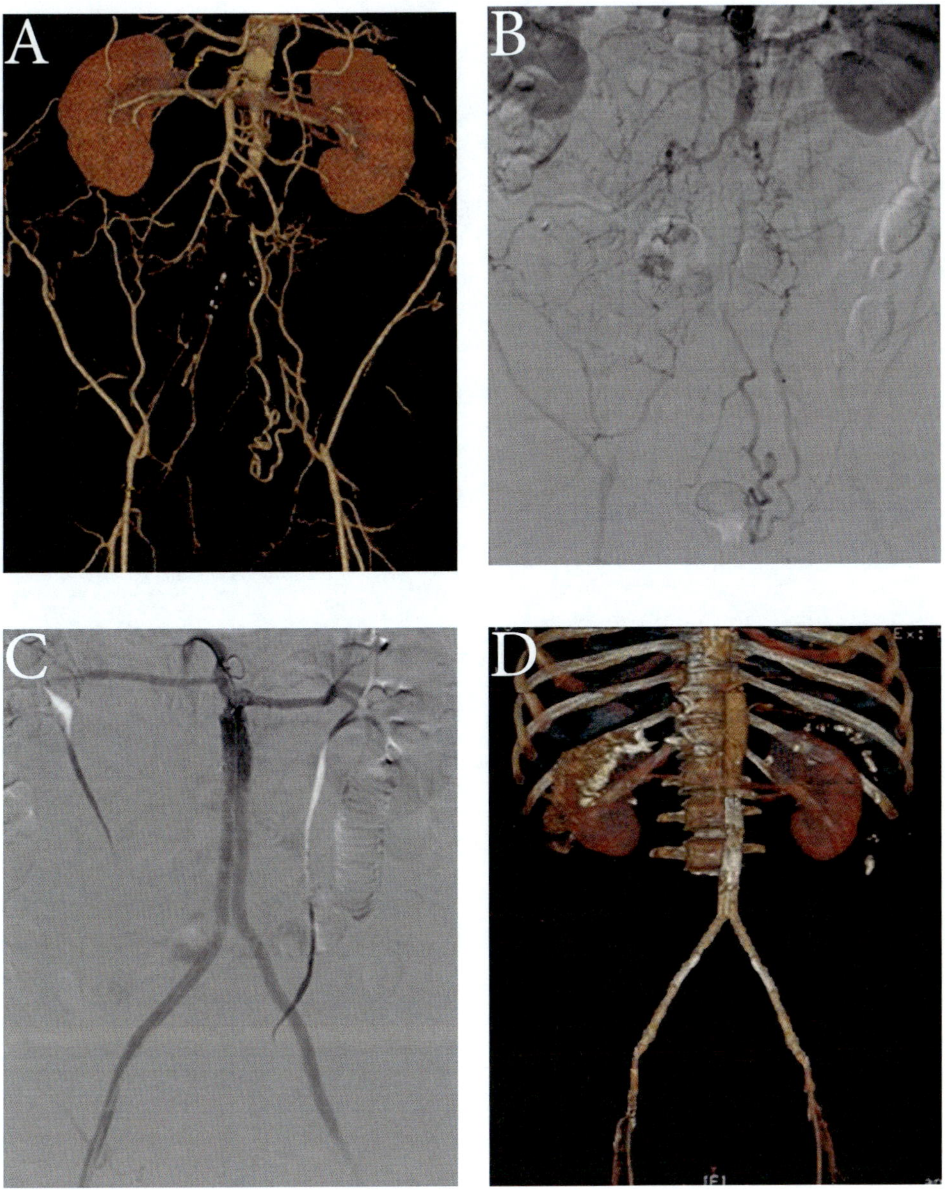

Figure 25.

A. Abdominal CTA prior to the intervention.
B. Abdominal aortography via the right brachial arery.
C. Final abdominal aortography reveals bilateral patent renal arteries, patent infra-renal aorta as well as bilateral CIA and EIA.
D. A CTA was obtained at 30-day followup and shows the aortic and iliac stent configuration with excellent flow.

Final Angiography:

- Abdominal aortography was then performed using a pigtail catheter via the right brachial artery.
- This showed widely patent infra-renal aorta with widely patent bilateral CIA, EIA and CFA (figure 25C).

Arteriotomy Closure:

- Bilateral CFA sheaths were then removed and using the pre-closure technique, the already partially deployed Perclose devices were used to close bilateral CFA arteriotomies.
- Right brachial artery sheath was switched out for a short 6 Fr. sheath. This was sewn in place and removed later when the ACT was < 180 seconds.

Post-op Care:

- The patient was monitored overnight in the hospital and discharged home the following day without any adverse outcomes.
- He had 2+ pedal pulses bilaterally.
- He was maintained on ASA 325 mg/d and clopidogrel 75 mg/d to be continued indefinitely.

Follow-up:

- At 30-day follow-up, the patient was completely asymptomatic and had total resolution of his claudication symptoms in both legs.
- He underwent a repeat CTA (figure 25D) which showed widely patent stents in the abdominal aorta as well as bilateral iliac arteries.
- He underwent ABI/PVR which were completely normal.
- He was advised to maintain his dual anti-platelet therapy indefinitely.

14

ENDOVASCULAR ABDOMINAL AORTIC ANEURYSM REPAIR

Abdominal aortic aneurysm (AAA) is one of the leading causes of death in the Western world. At any age, this condition is more common in men than women with a ratio of 5:1. With advancing age the incidence increases signficantly. It is reported that in men between the ages of 45 and 54 years the incidence of AAA is 1.3%. This increases to 12.5% as the subjects approach the ages of 75 to 84 years. The corresponding prevalence in women is 0% and 5.2%, respectively.

Risk factors for development of an AAA include:

- Tobacco use
- COPD
- Family history in a first degree relative
- Hypertension
- Dyslipidemia
- Age
- Male gender

An AAA can be can be supra-renal, juxtarenal or infra-renal. The focus of this chapter is on infra-renal AAA therapy.

DEFINITION

- An aneurysm by definition implies a dilated segment of a blood vessel such that its diameter is at least 50% larger than that of the adjacent normal segment.
- Dilatation to an extent of < 50% diameter increase is called ectasia.
- The diameter of the normal abdominal aorta is 18 - 22 mm.
- Dilation of the abdominal aorta to ≥ 3.0 cm is considered aneurysmal.
- Since women have a smaller aortic diameter at baseline, it has been suggested that the threshold should be lowered for them.

- Abdominal aortic aneurysms most commonly occur in the infra-renal segment of the aorta.

ANATOMY OF AAA

An AAA consists of the following anatomical segments:

- Proximal neck - this is the segment of non-aneurysmal aorta between the lowest renal artery and the aneurysm sac.
- Sac - this is the aneurysm proper, between the proximal and distal necks.
- Distal neck/cuff - this is the aortic segment between the sac and the aortic bifurcation.

Most AAAs have luminal thrombus due to turbulent flow at the their periphery.

NATURAL HISTORY OF AAA

- Untreated aneurysmal aorta continues to expand over time.
- Growth rates are variable and depend on the diameter of the AAA. Average annual expansion rates are:
    - ▹ 1 - 4 mm for AAAs < 4.0 cm in diameter.
    - ▹ 4 - 5 mm for those 5.0 - 6.0 cm in diameter.
    - ▹ 7 - 8 mm for those > 6 cm in diameter.
- Continued expansion poses a significant risk for rupture which is proportionate to the aneurysm diameter (table 1).

Figure 1.

The anatomy of an AAA.

A. Supra-renal aorta.
B. Right renal artery.
C. Left renal artery.
D. Proximal neck of AAA.
E. Sac.
F. Distal neck of AAA.
G. Right CIA.
H. Left CIA.
I. Right IIA.
J. Left IIA.
K. Right EIA.
L. Left EIA.

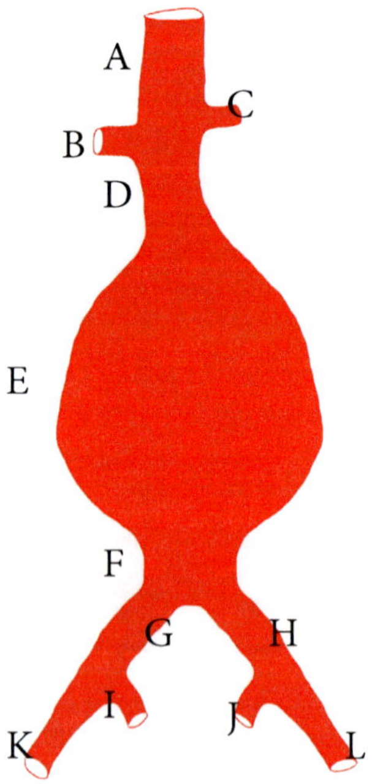

TABLE 1 - ABDOMINAL AORTIC ANEURYSM SIZE & RUPTURE RISK		
Diameter (cm)	Annual Rupture Risk	Eventual Rupture Risk
5.0 - 5.9	4%	20%
6.0 - 6.9	7%	40%
7.0	20%	50%

INDICATIONS FOR REPAIR

The two factors that strongly influence the decision to repair an AAA are symptoms and size. This is true for open as well as endovascular aortic aneurysm repair (EVAR).

SYMPTOMS

- Patients with symptomatic AAAs must undergo prompt repair regardless of aneurysm diameter unless their life expectancy and prognosis is otherwise poor.
- Symptoms of AAA may be due to:
  - Rupture/imminent rupture
  - Peripheral embolism
  - Thrombotic occlusion of the aorta
  - Mass effect on adjacent tissues

Rupture/Imminent Rupture:

- This is the most feared complication of an AAA with an overall mortality rate of 90%, including those patients who make it alive to the hospital.
- Rupture is most commonly associated with sharp sudden onset abdominal or back pain which may radiate towards the flank.
- Depending on how well the adjacent tissues contain the rupture, hypotension occurs and may lead to presyncope or syncope. These along with abdominal pain should be a red flag for a ruptured AAA.
- On examination, hypotension, ranging from mild to hypovolemic shock, abdominal tenderness and a pulsatile mass may be felt.
- Ruptures that are well-contained may cause symptoms for days to weeks before they are finally diagnosed. Early on symtpoms are subtle and non-specific in these cases.

Peripheral Embolism:

- Embolization of thrombus from the AAA sac is a potentially limb threatening event and warrants repair even if no signficant tissue loss has resulted from the index event.

Thrombotic Occlusion:

- Rarely, the lumen of the aorta thromboses causing an acute aortic occlusion.
- This is a surgical emergency as viability of the lower extremities is threatened.

Ruptured AAA:

AAA may rupture anteriorly or posteriorly:

- 20% of AAAs rupture anteriorly into the peritoneal cavity. These are more likely to be fatal due to the extent of free hemorrhage.
- 80% of AAAs rupture in the retroperitoneal cavity. These are more likely to become 'contained ruptures' due to the early tamponade provided by adjacent tissues.

Mass Effect:

- More common than acute aortic occlusion is compression of adjacent tissues by the enlarging aorta. This can lead to bowel obstruction, hydronephrosis, frequent micturition and chronic back pain.
- Since these findings typically occur in an aneurysm that is rather large in size or rapidly expanding, it is usually an indication to repair.

SIZE

- The diameter of an AAA is the strongest predictor of rupture risk as shown in table 1.
- In the absence of symptoms, a diameter of ≥ 5.5 cm is considered an appropriate size for elective repair, provided the patient is a good risk.
- For asymptomatic women, this threshold is lower so that at a diameter ≥ 4.5 cm elective repair may be offered.
- However, repair may be justified at a smaller size in the presence of other factors:

Symptoms:

- As noted above, the presence of symptoms takes precedence over other factors and the absolute diameter of the AAA in this case becomes irrelevant.

Rapid enlargement:

- An AAA is considered to be rapidly enlarging if its diameter is increasing at > 0.5 cm/6 months or > 1.0 cm/12 months.
- Rapid enlargement at this rate projects a high risk of rupture and should prompt early repair.

Saccular aneurysm:

- Saccular AAAs, due to their higher risk of rupture, should be considered for repair when their diameter approaches twice that of the adjacent normal aorta.

Eccentric expansion with wall thinning:

- Asymmeteric expansion of an AAA may lead to thinning of the wall, increasing the risk of rupture.
- Rapid eccentric expansion and wall thinning, even when below the above thresholds should lead to early repair.

Patient preference:

- Patient participation in decision-making is critical. Since most patients with > 4.0 cm AAA will eventually undergo repair, it is reasonable to consider endovascular repair at diameters < 5.5 cm for suitable risk patients.
- Additionally, for patients with high emotional stress due to an unrepaired AAA, earlier repair may be reasonable.
- Similarly, earlier repair should be considered for patients that are likely to be lost to follow-up. Although in this latter group, EVAR may not be the best option if the patient is likely to be non-compliant with close post-procedure imaging and clinical follow-up (life long).

Female patient:

- Women have a smaller aortic diameter at baseline. Thus it is believed that a treatment threshold of 5.5 cm may be too high for women.
- This combined with the fact that risk of rupture is 4 times as high among women as men, suggests that in female patients elective AAA repair should be considered at a lower diameter, perhaps 4.5 - 5.0 cm.

EVAR

- In 1991 Juan Parodi and colleagues first reported a successful case of EVAR as a treatment alternative for patients with infra-renal AAA and high surgical morbidity.

- Since then the procedure has evolved and the technology has improved signficantly.
- Due to these improvements, the risks of the procedure have diminished and it is now being employed increasingly in the treatment of patients with acceptable surgical risk as well.
- The goal of EVAR is to prevent future aneurysm rupture. This is acheived by successful exclusion of an AAA such that over time the aneurysm sac shrinks in diameter.
- The procedure consists of inserting stent grafts into the infra-renal aorta through transfemoral access.
- For a vast majority of patients, transfemoral access is obtained in both right and left CFA.
- EVAR has been shown to reduce the morbidity of AAA repair compared to open surgical repair. It also reduces the length of hospital stay and confers early mortality benefits.
- However, like most other procedures, EVAR is not for every patient who is a candidate for AAA repair.
- This is primarily due to anatomical constraints that have to be satisfied to achieve a successful and durable exclusion of a AAA using a stent graft.

THE ANATOMY OF A STENT GRAFT

- The conventional stent graft is constructed from a metal skeleton (stent) attached to a graft.
- The stent may be on the inside or outside of the graft.
- Contemporary stent grafts have a bifurcated modular design.
- A bifurcated stent graft consists of:
    - ▹ A main body.
    - ▹ A contralateral limb.
- Ancillary components that may be necessary include:
    - ▹ An aortic cuff.
    - ▹ Iliac extenders.

MAIN BODY

The main body has several components:

Supra-renal Stent:

- Most stent grafts approved in the U.S. for EVAR now have a supra-renal stent at the cranial end.
- The supra-renal stent is a bare stent that is meant to provide better fixation and prevent graft migration.
- Typically, the stent has fixation 'barbs' that are atraumatic.
- Below the supra-renal stent, all segments of the stent graft are lined with a graft.

Trunk Piece:

- Caudal to the supra-renal stent is a single trunk piece that is designed to be positioned in the proximal neck.
- The diameter of the trunk piece is chosen based on the diameter of the proximal aortic neck.

Flow Divider:

- At the caudal end of the trunk the stent graft bifurcates, dividing into the two iliac limbs - a long 'ipsilateral' limb and a short 'contralateral stub'.

Ipsilateral Limb:

- This piece is continuous cranially with the trunk piece of the main body.
- Caudally, the ipsilateral limb is deployed in the ipsilateral common iliac artery.
- The diameter of the caudal end of the ipsilateral iliac limb is chosen based on the segment of

the CIA where the seal is intended.

- The length of the ipsilateral limb is chosen to cover the distance between the lowest renal artery and the ipsilateral hypogastric artery. In cases where the ipsilateral common iliac artery is aneurysmal, the iliac limb may have to be extended to achieve seal in the EIA instead. When this is being considered, the ipsilateral hypogastric artery is coil embolized to prevent future retrograde flow leading to a type II endoleak (vide infra).

Contralateral Stub:

- This is the shorter limb of the main body.
- It is designed to receive the contralateral iliac piece.
- The caudal end of this piece is called 'the gate'. This is prominently highlighted with radio-opaque markers to facilitate engagement with a wire via the contralateral femoral artery access.

'Ipsilateral' vs. 'contralateral':

- The side on which the 'main body' of the stent graft is introduced is often referred to as the 'ipsilateral' side.
- The opposite side is thus referred to as the 'contralateral' side.
- This classification is ***not*** in reference to the side of origin of the lowest renal artery.

CONTRALATERAL ILIAC LIMB

- The contralateral iliac limb is inserted via the contralateral femoral access.
- It is designed to fit cranially inside the contralateral stub of the main body, and caudally to attach in the contralateral common iliac artery.
- The diameter of the cranial end is 16 mm regardless of the diameter of the caudal end.
- The diameter of the caudal end is chosen based on the segment of the contrlateral CIA where the seal is intended.

ILIAC LIMB EXTENSION

- Iliac limb extensions are available if additional distal stent graft is needed.
- The caudal end diameter is chosen based on the diameter of the distal landing zone.

AORTIC EXTENSION

- Similar to the iliac extension, aortic extension pieces are available to allow further treatment at the level of the proximal neck.
- These are overlapped with the trunk of the bifurcated graft and deployed using the same precautions as the main body.

CONFIGURATIONS OF STENT GRAFTS

BIFURCATED AORTIC STENT GRAFTS

- Ideally, the stent graft should extend from just caudal to the lowest renal artery in the aorta to just above each hypogastric artery.
- For most patients, a bifurcated modular stent graft is used (figure 3).
  - Main body inserted through the 'ipsilateral' CFA.
  - Contralateral iliac limb inserted through the contralateral CFA and overlapped cranially with the contralateral stub of the main body.

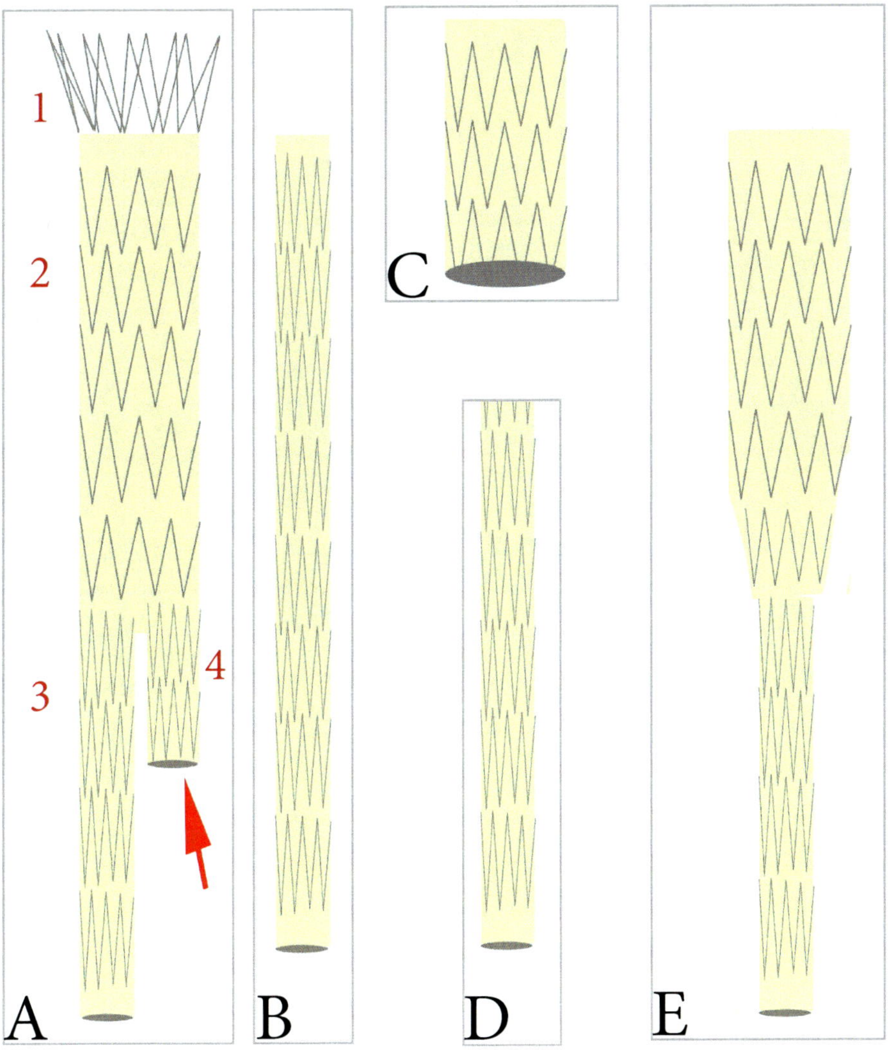

Figure 2.

The anatomy of an EVAR stent graft.

A. Main body of a bifurcated stent graft.
1- Supra-renal stent.
2- Trunk piece.
3- Ipsilateral limb.
4- Contralateral stub.
The red arrow points to the open 'contralateral gate'.
B. Contralateral iliac limb.
C. Aortic extension.
D. Iliac extension.
E. An AUI stent graft.

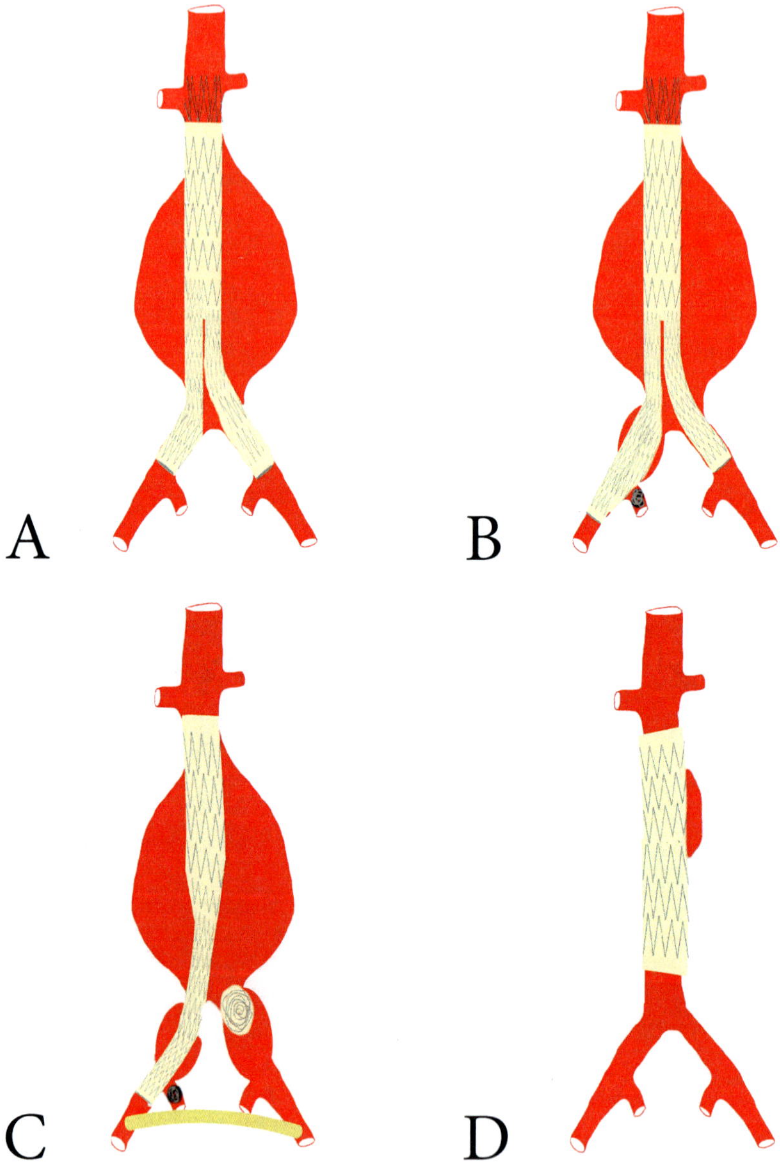

Figure 3.

Configurations of EVAR.

A. Bifurcated stent graft with bilateral iliac landing zones in the CIA.
B. Bifurcated stent graft with unilateral CIA aneurysm and the ipsilateral landing zone extended in the EIA after IIA coiling.
C. AUI stent graft with landing zone in the EIA. Ipsilateral IIA and contralateral CIA have been embolized. There is a femoro
femoral bypass graft in place to perfuse the contralateral leg.
D. A PAU treated with a 'tube' graft concept using a TEVAR prosthesis.

NON-BIFURCATED AORTIC STENT GRAFTS

- Rarely, anatomic factors exist that preclude the use of a bifurcated device.

Aorto-Uni-iliac Stent Graft:

- For example, severe obstructive disease in one iliac artery may not allow the placement of the contralateral limb. In these cases, an aorto-uni-iliac (AUI) graft may have to be used (figure 2).
  - ▹ The stent graft is introduced via one CFA.
  - ▹ The contralateral iliac artery if not already totally occluded, is occluded with an occlusion plug.
  - ▹ The contralateral leg is perfused via a surgical femoro-femoral bypass.

Aortic 'Tube' Grafts:

- These grafts consist of a single, non-tapering, non-bifurcating cylindrical stent graft. Examples include the stent grafts utilized are in the thoracic aortic endovascular aneurysm repair (TEVAR).
- Aortic tube grafts are considered inadequate for the treatement of AAAs and are not used anymore for this indication.
- However, these grafts may be used for the treatment of penetrating aortic ulcers or some saccular aneurysms.

APPLICATIONS OF EVAR

ISOLATED AAA

- Patients with isolated AAA and without CIA aneurysms that are > 2 cm in diameter can undergo EVAR with a bifurcated modular stent graft (figure3A).

AAA WITH UNILATERAL ILIAC ANEURYSM

- Patients with an infra-renal AAA and a unilateral CIA or hypogastric artery aneurysm that needs to be treated (diameter > 2 cm) can usually safely undergo EVAR (figure 3B).
- First, however, the hypogastric artery needs to be iatrogenically occluded with vascular coils/plugs to minimize the risk of retrograde flow via this vessel into the AAA.
- Next, the patient undergoes EVAR with a bifurcated stent graft. The iliac limb on the side of the iliac aneurysm is extended so the landing zone is in the EIA.

AAA WITH BILATERAL ILIAC ANEURYSMS

- In patients with an infra-renal AAA and bilateral CIA aneurysms the above approach of hypogastric artery embolization and extension of the iliac landing zones to the EIA is not feasible as occlusion of bilateral hypogastric arteries can lead to bowel (colorectal) and bladder ischemia.
- Ideally, these patients should undergo open repair.
- Alternatively, the insertion of an AUI device may offer an endovascular solution (figure 3C).
  - ▹ The hypogastric artery is embolized on one side.
  - ▹ An AUI device is used through that side with the stent graft extended to the EIA.
  - ▹ The contralateral CIA is 'occluded' using an occlusion plug.
  - ▹ A femorofemoral bypass graft is performed during the same procedure, to perfuse the contralateral leg.

TABLE 2 - FDA APPROVED STENT GRAFTS FOR EVAR.

Device	Manufacturer	Aortic Prosthesis Diameter (mm)	Aortic Delivery Sheath (Fr.)	Iliac Prosthesis Diameter (mm)	Iliac Delivery Sheath (Fr.)
AFX	Endologix	22 - 34	19	13 - 25	11- 19
Endurant	Medtronic	23 - 36	18 - 20	10 - 28	14 - 16
Excluder	Gore	23 - 35	18 - 20	12 - 27	12 - 18
Ovation	Trivascular	20 - 34	14 - 15	10 - 22	13 - 15
Zenith	Cook Medical	22 - 36	18 - 22	8 - 24	14 - 16

PENETRATING AORITC ULCER

- A penetrating aortic ulcer (PAU) is a degenerative atherosclerotic disease affecting the aorta that leads to erosion of the plaque through the intima into the media.
- Along with aortic dissection and intramural hematoma, the PAU constitutes the acute aortic syndromes.
- PAUs are more common in the descending thoracic aorta than in the abdominal aorta.
- PAUs may be precursors of saccular aneurysms.
- A PAU is usually picked up incidentally on radiologic studies ordered for other indications.
- An axial slice would demonstrate a discrete disruption in the aortic wall circumference, possibly filled with thrombus.
- In patients with abdominal/back pain (i.e. symptomatic patients), a PAU should be fixed urgently to prevent aortic rupture.
- Endovascular treatment with a stent graft is the ideal method of repair (figure 3D).
- Due to the focal nature of the disruption, there is usually adequate proximal and distal landing zone for the placement of a stent graft.
- A non-bifurcated stent graft is used in a overwhelming majority of cases.
- All the same precautions are used as during traditional EVAR.

PRE-OPERATIVE IMAGING

- High resolution computerized tomography, both non-contrast and contrast enhanced (CT/CTA) serves as the 'gold standard' for preoperative imaging. This study allows appropriate patient selection, precise stent graft sizing and planning of the procedure.
- CT images are acquired from just above the celiac trunk to the level of the CFA bifurcation.
- Images should be acquired at ≤ 3 mm collimation.
- Diameter of the proximal aortic neck, its length, and the diameters and lengths of bilateral CIA 'landing zones' and EIA/CFA determine if EVAR can be safely and successfully performed in a certain patient.
- This study allows visualization and distinction between the vessel lumen, calcium, plaque and thrombus in the vessel wall.
- All axial slices should be reviewed in detail to determine the anatomy of the aorta, its branches and the ileofemoral arteries.
- Multiplanar reformatting (MPR) allows simultaneous display of the vessel in axial, sagittal and coronal planes.
- Using MPR, a curvilinear line can be drawn parallel to the vessel long axis ('center line').

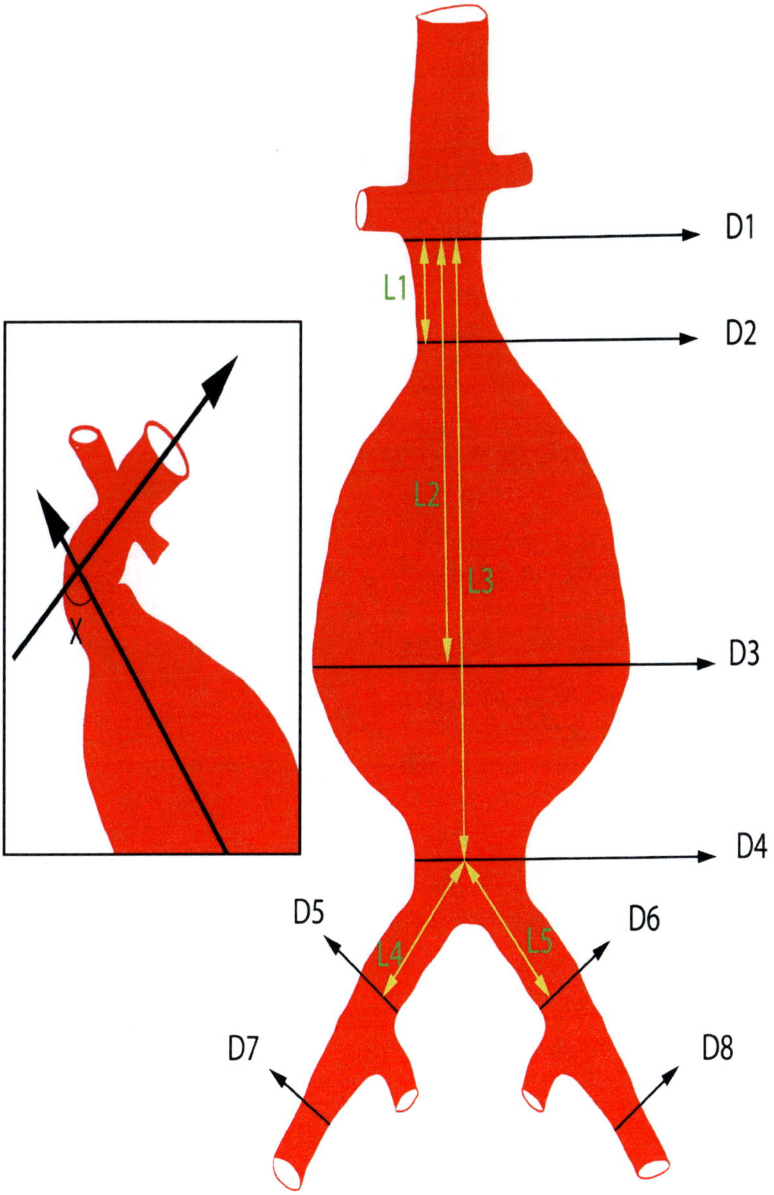

Figure 4.
Pre-operative CTA Measurements for EVAR.

D1/D2. Proximal neck diameters.
D3. Sac maximal diameter.
D4. Distal neck diameter.
D5/D6. CIA diameters.
D7/D8. EIA diameters.

L1. Proximal neck length.
L2. Distance from lowest renal artery to the maximal sac diameter.
L3. Distance from lowest renal artery to the aortic bifurcation.
L4/L5. Distance from aortic bifurcation to the hypogastric arteries.

Inset: AAA with an angulated proximal neck. The angle of the neck to the AAA body is denoted by the letter X.

This allows more accurate measurement of length in the long axis.

- Transverse measurements (diameter) should be made perpendicular to the long axis of the vessel using the center line.
- 3-D reconstructions allow determination of precise C-arm angulation and rotation during the procedure so that the lowest renal artery and the hypogastric arteries are optimally visualized.

PATIENT SELECTION

Other than the factors listed above (symptoms, size, life expectancy etc.), the etiology of the aneurysm and aorto-iliac anatomy determine if EVAR is feasible and safe for a particular patient.

ANATOMIC PRE-REQUISITES

Iliac Arteries:

- CFA and iliac artery diameters (figure 4, D7 and D8) should be adequate to allow safe delivery of the devices (at least 5.0 mm - 7.6 mm of luminal diameter).
- EVAR can be successfully performed in patients who have iliac arteries ranging in diameter from 8 - 25 mm.
- The length of the CIA should be at least 3 cm with a distal fixation length of ≥ 15 mm.
- Significant iliac/CFA disease should not be present as it may not only preclude safe delivery of device(s) but also impede graft outflow and risk graft limb thrombosis on the affected side.
- If there is a CIA aneurysm (diameter > 20 mm), it may require the aneurysm be covered as well and that the seal zone be extended to the EIA. In this case, the hypogastric artery on the side of the CIA aneurysm should be coil embolized pre-emptively to reduce the chance of a type II endoleak due to retrograde flow from this vessel into the AAA sac.
- If bilateral CIA aneurysms are present, EVAR with a bifurcated system should be avoided - coil embolization of bilateral hypogastric arteries will not only lead to buttock claudication and erectile dysfunction but also pelvic visceral (bladder and rectal) ischemia. If open repair is too risky, then an aorto-uni-iliac (AUI) device should be considered in these patients along with a surgical femorofemoral bypass graft.
- Similarly, if one hypogastric artery is severely diseased/occluded and there is a CIA aneurysm on the other side, an AUI device or open repair should be considered.

Proximal Neck:

- Morphology of the proximal neck is the most common factor that renders a patient ineligible for EVAR.
- The following criteria apply:

Length:

- Proximal neck length (figure 4, L1) should be at least ≥ 10 - 15 mm to allow for adequate fixation and seal.
- Some newer devices allow treatment of a shorter neck (e.g. it is claimed that the Trivascular Ovation stent graft can be used to treat an AAA with a proximal neck length of 7 mm).

Diameter:

- An AAA with a proximal aortic neck diameter (figure 4, D1) of 19 - 32 mm can be safely treated with EVAR with adequate seal.
- A large number of devices are available commercially, and depending on the specific device being used the possible treatment diameters vary (range 15 mm - 32 mm).

Angulation:

- Infra-renal neck angulation (figure 4, inset X) should be < 60°.
- A more angulated neck increases the chances of poor fixation and seal and increase the risk of early graft migration.

Shape:

- The shape of the proximal neck should be cylindrical and not conical with a reverse taper (figure 5).
- A conical neck with a reverse taper implies an increase in the diameter of the neck, cranial to caudal, by at least 4 mm over a length of 10 mm.
- The presence of a conical neck with a reverse taper increases the likelihood of graft migration.

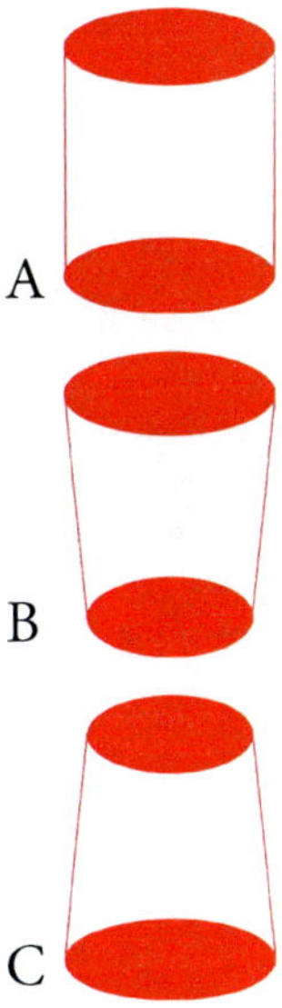

Figure 5.

Shape of the Proximal Neck.

A. Cylindrical.
B. Taper.
C. Reverse taper.
Shapes A & B are desirable.

Aneurysm Sac:

- The aneurysm sac should have an adequate central luminal diameter (at the level where the contralateral gate is expected to open) to allow easy cannulation and placement of the contralateral limb.
- Lack of adequate luminal diameter may exclude the use of a bifurcated device.

Distal Neck/Cuff:

- The distal aortic neck should be devoid of severe obstructive disease, to allow delivery of the sheaths and devices into the aorta without any issue.
- If present, angioplasty and rarely stenting may have to be performed prior to EVAR. At the conclusion of the procedure, the aortic bifurcation should be re-dilated with kissing balloons to allow adequate expansion, and to prevent kinking and hence graft thrombosis.

ANATOMIC EXCLUSIONS

By extension of the above, the presence of the following anatomic features preclude a successful EVAR.

- Supra-renal or juxta-renal aneurysms.
- Reverse taper/conical neck (> 4 mm increase in diameter over a 10 mm length).
- Presence of significant mural thrombus/calcium/plaque (defined as ≥ 2 mm in thickness or ≥ 25% of vessel circumference of aortic neck/iliac artery, or ≥ 50% of the length of the iliac artery) at the proximal or distal fixation sites as it may compromise adequate seal.
- Presence of a dominant IMA in the presence of severely diseased celiac trunk and SMA.
- Bilateral common iliac artery aneurysm/ectasia that may require exclusion of bilateral hypogastric arteries.

ETIOLOGIC EXCLUSIONS

Endovascular stent grafts should not be used for the repair of the following types of AAAs:

- Mycotic aneurysms.
- Inflammatory aneurysms.
- Degenerative aortic disease (e.g. Marfan's, Ehlers-Danlos syndromes, etc.).
- Connective tissue disorders.
- Pseudoaneurysm (resulting from previous graft placement).

PATIENT COUNSELING

- Risks and benefits of ongoing medical therapy, open surgical repair and endovascular repair should be discussed in detail.
- The possibility of future re-intervention/surgical repair(s) should be discussed.
- The need for life-long regular follow-up with interval imaging and importance of compliance should be emphasized.
- The commitment for the the above should be obtained prior to performing EVAR.

EQUIPMENT

- The equipment used for EVAR is shown in table 3.

FLUSH CATHETER

- A 5 Fr./65 cm calibrated flush catheter (pigtail, Omniflush) with one cm markers over a 20 cm length is used.

WIRES

- 0.035" 180 cm Wholey wire x 2.
- 0.035" 180 cm Superstiff Amplatz wire x 2.
- 0.035" 180 cm ASGW.

OTHER CATHETERS

- 5 Fr./65 cm angled Glidecath or 5 Fr./65 cm Kumpe catheter.

SHEATHS

- 4 Fr. micropuncture kit.
- 6 Fr./11 cm and 8 Fr./11 cm sheaths.
- 'Ipsilateral' sheath:
  - ▹ 14 - 20 Fr./30 cm.
- 'Contralateral' sheath:
  - ▹ 12 - 18 Fr./30 cm .

DILATORS

- It is critical to have dilators of the correct size available.
- A set of Coons taper dilators (Cook Medical, Bloomington, IN) is invalueable. These dilators are relatively economical and do not have a hydrophilic coating. They are 0.038" wire compatible and are 20 cm in length.
- Dilators with a hydrophilic coating are superior but more expensive.

BALLOONS

Molding Balloon:

These balloons (table 4) are used to 'mold' the stent graft to the anatomy of the aorto-iliac arteries. Most of the balloons in this category are also aortic 'occlusion' balloons and are useful in occluding aortic flow in the event of AAA rupture. These balloons are inflated manually with a 60 cc syringe. An in-delfator is not used for these balloons. When the balloon begins to elongate during inflation its an indication to stop. Continued inflation may lead to aortic/iliac rupture.

TABLE 3 - EQUIPMENT FOR EVAR:	
Access kit	• 4 Fr. micropuncture kit.
Dilators	• Coons tapered dilators (10 - 16 Fr.).
Sheaths	• 6 Fr. /11 cm and 8 Fr./11 cm sheaths if using percutaneous access. • 12 -18 Fr. /30 cm for contralateral limb. • 14 - 20 Fr./30 cm for main body.
Catheters	• 5 Fr./65 cm calibrated flush catheter. • 5 Fr./65 cm angled Glidecath. • 5 Fr./65 cm Kumpe catheter.
Wires	• 0.035" (180 cm) ▹ Angled stiff Glide wire. ▹ Wholey wire. ▹ Superstiff Amplatz.
Balloons	• Occlusion/moulding: ▹ Coda (32 and 40 mm). ▹ Reliant. ▹ Equalizer. • Angioplasty: ▹ 12 mm x 40 mm.
Stent grafts	• Based on operator preference and patient anatomy one of the several available devices can be used. ▹ Main body. ▹ Contralateral limb. ▹ Aortic extension. ▹ Iliac extension(s).
Miscellaneous	• Inflation device. • 60 cc luerlock syringes x 2. • Palmaz XD stent. • Sterile mineral oil.

TABLE 4 - MOLDING BALLOONS

Balloon	Manufacturer	Sizes (mm)	Delivery sheath (Fr.)
Coda	Cook Medical	32, 40	14
Equalizer	Boston Scientific	20, 27, 33, 40	14/16
Reliant	Medtronic, Inc.	46	12
Trilobe	Gore	34, 42	18

Angioplasty Balloon:

These balloons may be used for 'molding' the contralateral limb and iliac limbs/extenders. Additionally, kissing inflation may be required to post-dilate the stent grafts in case of moderate to severe obstructive disease at the aortic bifurcation. Typically, a 12 - 14 mm x 40 mm balloon is sufficient for this purpose. Inflation should be performed cautiously. An in-deflator is used but the inflation pressure is kept fairly low at 2 - 4 ATM.

DEVICE SIZING

The general principles of stent graft sizing for the treatment of aneurysmal disease are as follows:

- The main body/trunk diameter of the bifurcated stent graft is oversized by 10-20% with respect to the measured inner diameter of the aorta at the proximal seal zone (proximal neck).
- The main body length is based on the distance between the lowest renal artery and the ipsilateral hypogastric artery.
- The ipsilateral and contralateral iliac limb diameters are oversized by 10-25% of the measured inner diameter at the distal seal zones.

ACCESS

EVAR is performed via transfemoral route and requires bilateral CFA access. This may be obtained either by surgical cutdown or percutaneously with 'preclosure'. Depending upon the anatomy of the ileofemoral arteries, operator preference and experience, one or both techniques may be used for access. Whenever feasible, we prefer to use percutaneous technique for access as it results in early ambulation and faster recovery of the patient.

CUTDOWN

- Surgical cutdown to expose the CFA/terminal EIA is the most commonly used method for obtaining arterial access for EVAR.
- This is essential in the presence of inadequate CFA lumen or when the vessel is calcified precluding preclosure.
- Using 5-0 Prolene, a purse-string suture is placed in the anterior wall of the vessel.
- Access is then obtained through this purse-string enclosed CFA using modified Seldinger technique.
- Prior to upsizing for the large delivery sheaths, the arteriotomy is dilated with serial dilators of progressively larger sizes.
- At the conclusion of the procedure, the arteriotomy is closed using standard surgical technique.

PERCUTANEOUS

- This method is applicable only when the CFA is disease-free and has minimal calcification.
- 'Preclosure' is performed after obtaining initial arterial access.
- We obtain access using a 4 Fr. micropuncture kit. U/S guidance may be used.
- Using a number 11 blade, a skin incision is made at the site of the needle entry and the track is dilated.
- Before a sheath is inserted, we obtain an angiogram through the microcatheter to confirm that the arterial puncture is in the CFA and not in a smaller/branch vessel.
- A 0.035" Wholey wire is used to introduce a 6 Fr. sheath.
- This wire provides better steerability through the aneurysm.
- Two Perclose® Proglide (Abbott Vascular, Abbott Park, IL) devices are prepared in a standard fashion.
- The 6 Fr. sheath is removed over the Wholey wire and the first Perclose device is inserted. We choose to keep this device oriented at 10 O'clock position (figures 6A & 6B).
- Once the sutures are deployed, the wire is re-introduced as the device is being removed.
- The sutures are secured with a pair of 'hemostats' and wrapped in a piece of wet gauze. The knot is not secured at this point.
- The second Perclose is then inserted over the wire in the same fashion except it is oriented to

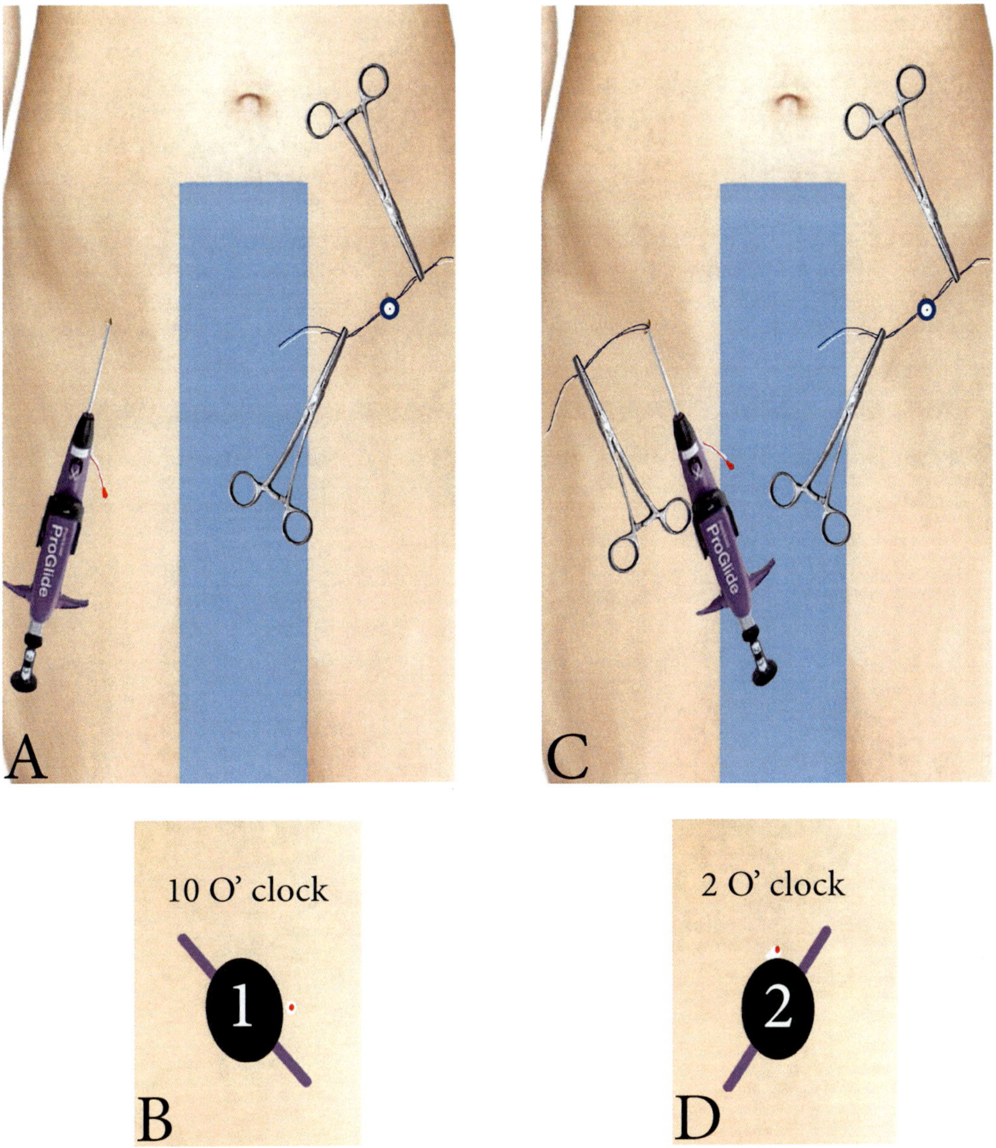

Figure 6.

Groin 'Preclosure' technique using two Perclose™ Proglide devices.

A. The right CFA arteriotomy being 'preclosed' with a Perclose Proglide device. The left CFA arteriotomy has already been preclosed. The two sets of sutures have been secured by the hemostats and an 8 Fr. sheath is in place in the left groin.

B. The first Perclose device (1) is oriented at 10 O'clock position at the time of insertion and deployment.

C. After deployment of the first Perclose device in the right CFA, its sutures have been secured with a hemostat. The second Perclose device is inserted and its sutures are about to be deployed.

D. The second Perclose device (2) is oriented at 2 O'clock position at the time of insertion and deployment.

a 2 O'clock position, orthogonal to the first one (figures 6 C & 6D).

- Again, as the device is being removed, the Wholey wire is re-inserted through it. The sutures are secured in a similar fashion with a separate pair of 'hemostats'.
- An 8 Fr. sheath is inserted over the Wholey wire into the arteriotomy.
- The dilator is removed. The sheath is flushed.
- A pigtail/Glidecath is then advanced over the Wholey wire into the descending thoracic aorta to exchange the wire out for a J-tip SuperStiff Amplatz wire.
- Over this wire, the subcutaneous track is serially dilated with dilators of progressively larger diameters until the final sheath, adequately sized for the chosen device, can be inserted safely.
- The final sheath is advanced into the abdominal aorta under fluoroscopic guidance. The tip of the sheath is advanced to L1-L2 vertebral level. The dilator is removed and the sheath flushed.
- If contralateral CFA access is planned in a similar fashion, the above steps are repeated.

AORTOGRAPHY

- After both arterial sheaths are in place, the flush catheter is introduced via the side ('ipsi' side) that will be used for main body delivery.
- Using a calibrated (i.e. marker) catheter, length is measured from the lowest renal artery to the 'ipsilateral' hypogastric artery.
- This confirms the length of the main body device to be used.
- The catheter is then removed over a wire and introduced on the contralateral side and positioned just cranial to the highest renal artery.

CATHETER

- A calibrated flush catheter such as a pigtail or Omniflush is used.
- The markers are 1 cm apart and are used for measuring length. These measurements are required to measure the total treatment length on each side (lowest renal to the hypogastric artery on the 'ipsilateral/main body' side and from the contra-gate to 'contralateral' hypogastric artery).

CATHETER POSITIONING

- The catheter is positioned just above the highest renal artery.

INJECTION

- The flush catheter should be kept connected to the power injector.
- If the renal function allows, initial aortogram should be performed at 10-30 cc/sec at 500 psi.
- If the GFR is < 60 mL/min., 50:50 dilution may be used.
- Breath-hold technique should be used to optimize image quality.
- The field of view should include the renal arteries as well as the hypogastric arteries (at least the 'ipsilateral' hypogastric artery).
- Based on the aortic anatomy, cranial/caudal and LAO angulation may be required to avoid foreshortening of the proximal segments of the renal arteries.
- This prevents inaccurate estimation of the origin of these vessels.
- The necessary gantry angulation can be estimated from the pre-operative CTA.

TECHNIQUE

- After confirmation of the length measurements, the appropriate size main body stent graft is opened and prepared.
- Prior to introduction of the main body in the sheath, it is important to align it so that the contra gate will face in the desired direction.
- This direction of the contra gate positioning is dictated by the contralateral wire bias in the sac.
- If necessary, place the graft under the C-arm (outside the patient) to visualize and identify the contra gate and its position in relation to the handle of the device.
- It is then introduced in this orientation via the ipsilateral CFA sheath into the aorta.

MAIN BODY POSITIONING

- The main body is advanced to the level of the lowest renal artery (marked on the screen).
- The position of the stent graft should be such that the fabric portion of the graft is just caudal to the lowest renal artery.
- The sheath on this side is withdrawn to a level below the contra gate.
- The image detector should be kept perpendicular to the centerline of the infrarenal aorta to avoid parallax errors.
- Another aortogram is performed with the injector to accurately align the stent graft with the lowest renal artery.
- Prior to initiation of deployment, it should be confirmed that the contralateral gate is within the aorta and above the aortic bifurcation, and not in the CIA.

MAIN BODY DEPLOYMENT

- Deployment is initiated.
- Depending on the brand/generation of the device being used, the position of the leading edge of the stent graft may be adjustable even after partial deployment.
- Another injection should be performed to ensure proper position of the graft with respect to the renal arteries.
- If necessary, the graft may be gradually pushed up or pulled down.
- Deployment is carried out until the contra gate is open.
- The rest of the main body is not deployed at this point.

CONTRALATERAL GATE CANNULATION

- The pigtail catheter is disconnected from the injector.
- An ASGW is introduced in the catheter to straighten out its loop.
- The catheter and wire are withdrawn from behind the stent graft and pulled below the level of the contra gate.
- The wire is used to enter into the contra gate.
- This usually requires the assistance of a catheter. The pigtail or another catheter such as an angled GC or Kumpe may be used for this purpose.
- Once the wire is believed to be inside the gate, the catheter is advanced up the main body into the trunk of the stent graft.
- The operator must confirm that the wire/catheter is actually inside the stent graft and not behind it.
- The most common maneuver used to confirm this is to rotate the pigtail catheter in the trunk of the stent graft.
    - ▹ If the catheter can be easily rotated 360° then in all likelihood it is inside the stent graft.
    - ▹ If instead the catheter cannot be rotated freely, it must be withdrawn and the maneuver

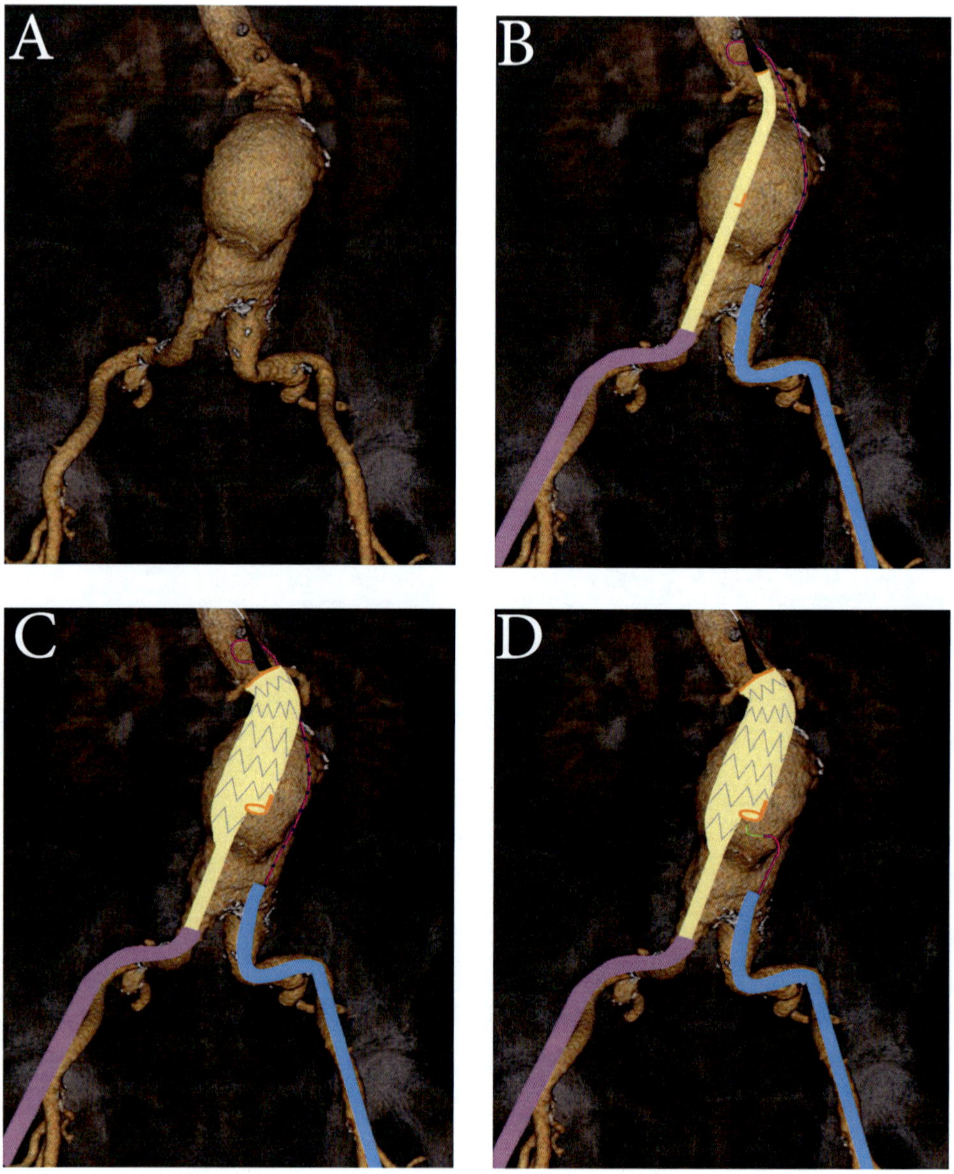

Figure 7.

Steps in EVAR.

A. Large infra-renal AAA.
B. The ipsi (purple) and contra (blue) sheaths are inserted. The main body is positioned. There is a marker flush catheter, inserted via the contra sheath.
C. The main body is partially deployed, without deploying the ipsi limb. The flush catheter still behind the graft.
D. The flush catheter is straightened out over an ASGW and withdrawn into the sac.

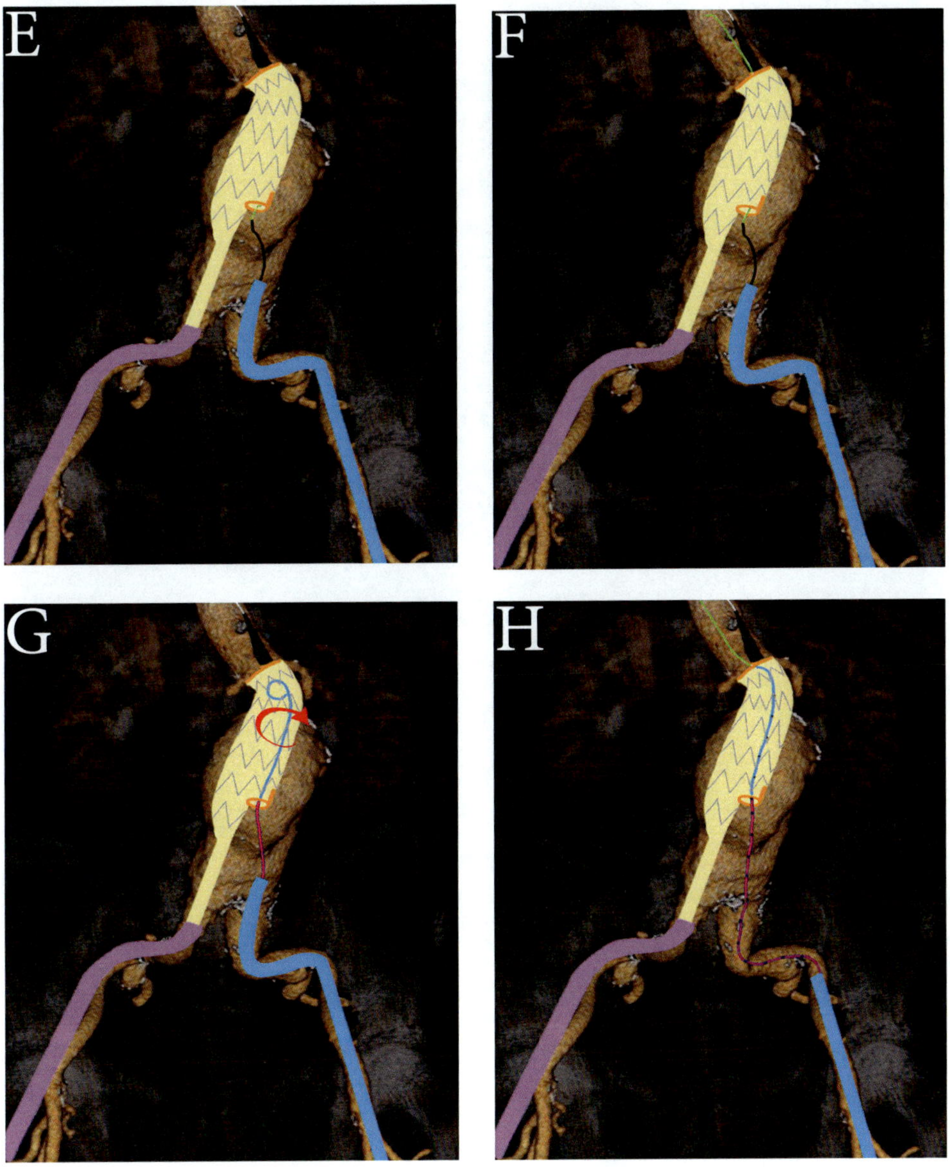

Figure 7 (continued).

Steps in EVAR.

E. The contra gate is engaged using an ASGW (green).
F. The gate is cannulated and the ASGW is advanced through it into the supra-renal aorta.
G. The support catheter is exchanged out. The marker flush catherer is advanced into the trunk of the main body
and is rotated to confirm positioning within the graft.
H. Over the ASGW the catheter is withdrawn to measure length from the contra gate to the contralateral hypogastric
artery.

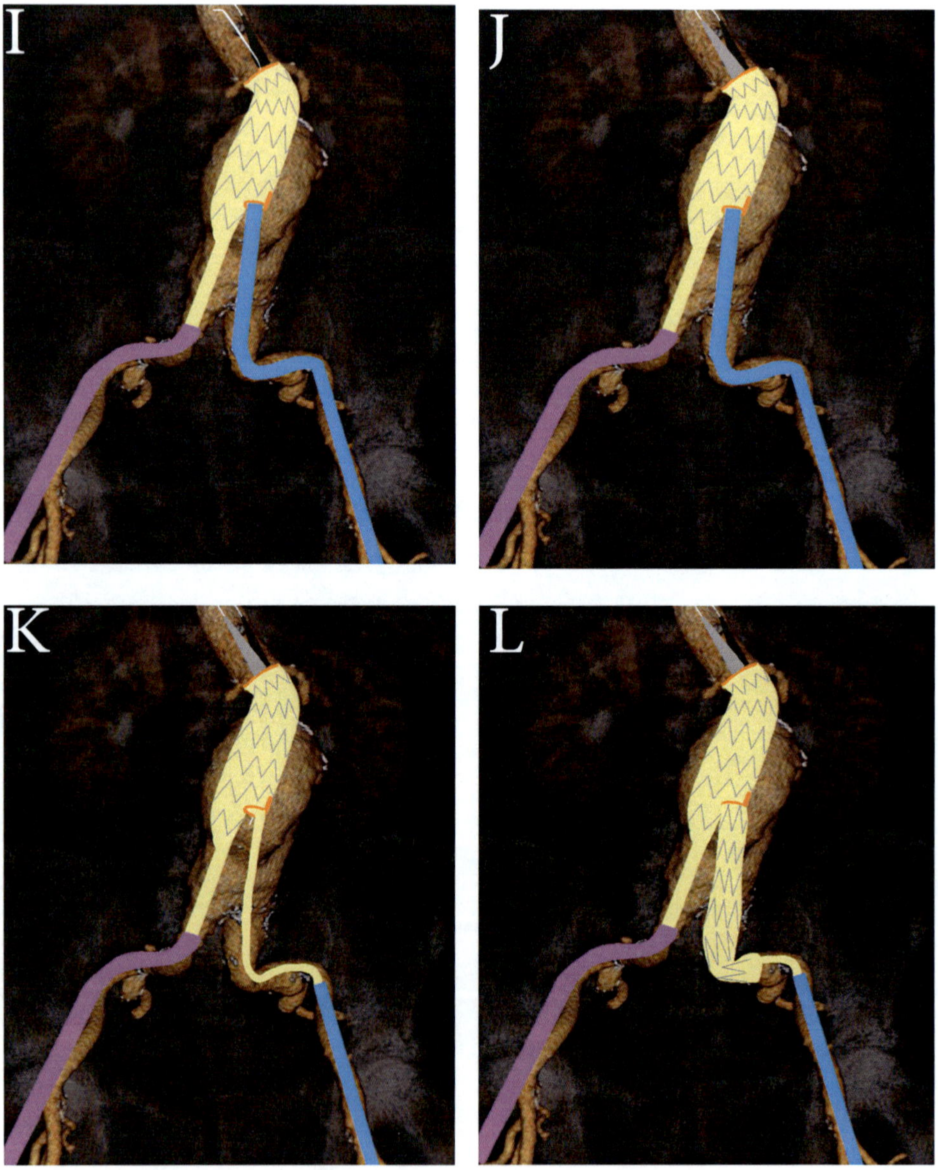

Figure 7 (continued).

Steps in EVAR.

I. The ASGW is switched out for a SS Amplatz wire, over which the contra sheath is advanced into the contra gate.
J. The contra device is advanced into the contra sheath.
K. The contra sheath is withdrawn into the EIA, exposing the contra limb.
L. The contra limb is deployed.

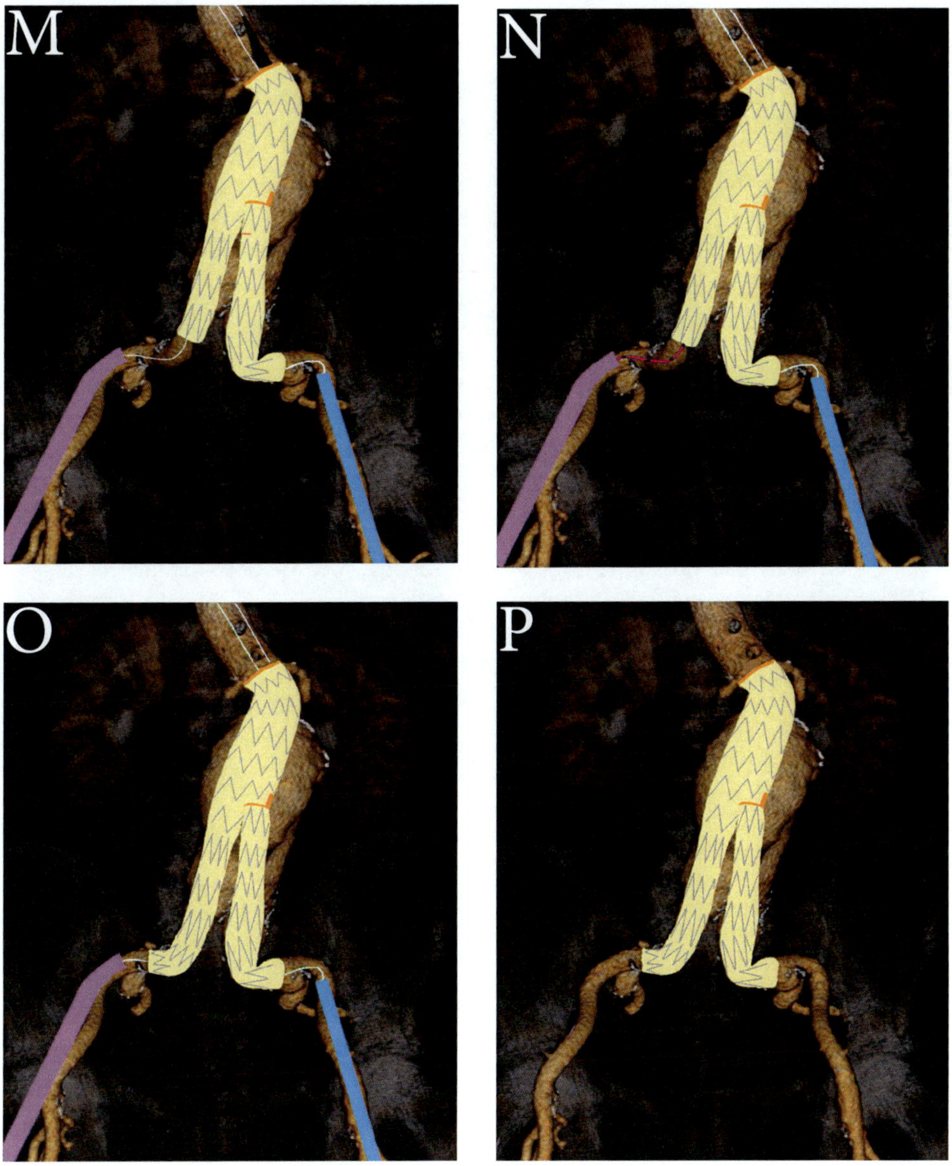

Figure 7 (continued).

Steps in EVAR.

M. The contra limb delivery catheter is removed. Ipsi limb deployment is completed.
N. The marker catheter is inserted on the ipsi side to measure the length between the ipsi limb and the ipsilateral
hypogastric artery.
O. After post-dilatation of the stent grafts final aortography is performed with the sheaths still in place.
P. Once a satisfactory final results is obtained, the sheaths are removed and the arteriotomies closed appropriately.

repeated.
- ▹ The operator may choose alternative confirmatory methods such as a contrast injection inside trunk or IVUS.

CONTRALATERAL LENGTH MEASUREMENT

- After confirming the position of the pigtail, length measurement between the contra gate and the contralateral hypogastric artery is performed.
- An oblique projection of 20° or so is used to optimally visualize the origin of the hypogastric artery (RAO 20° for the left hypogastric artery and LAO 20° for the right).
- The hypogastric position should be marked on the screen for easy reference.

CONTRALATERAL STENT DEPLOYMENT

- The appropriate length contralateral piece is prepared.
- Prior to introducing this piece into the sheath, the sheath should be advanced inside the contra gate to facilitate smooth delivery of the graft.
- For this purpose, the pigtail catheter is removed from the sheath, the dilator is introduced into the sheath and under fluoroscopic guidance the sheath is advanced inside the contra gate.
- The dilator is removed and the contralateral piece is introduced through the contralateral sheath.
- The entire graft must be above the level of the hypogastric artery so that this vessel is not covered inadvertently.
- Once the marker on the contralateral graft is lined up with that on the contra gate indicating adequate overlap, the sheath is withdrawn to expose the entire length of the contralateral stent graft.
- The contralateral piece is deployed.
- The delivery catheter is removed.

COMPLETION OF IPSILATERAL GRAFT DEPLOYMENT

- After deployment of the contralateral piece, the main body deployment is completed.
- The delivery catheter is removed.

IPSILATERAL LENGTH MEASUREMENT

- If indicated (e.g. when the ipsi limb does not cover an adequate length of the CIA), an iliac limb extension may be required.
- The marker catheter is introduced on the ipsi side.
- The length is measured between the ipsi limb and ipsilateral hypogastric artery in the same fashion as on the contralateral side.
- The appropriate size iliac limb extension is deployed.
- The catheter is removed.

POSTDILATATION

- After deployment of all the components of the stent graft, post-dilatation is undertaken.
- The idea is to 'mold' the stent graft to the aorta/iliac arteries at the seal and overlap zones rather than to stretch and expand the stent graft.
- Typically, an aortic occlusion balloon such as the Coda, Reliant or Equalizer is used to post-dilate.
- A 60 cc syringe is used for balloon inflation/deflation.
- The inflations should only be performed within the stent graft to avoid aortic/iliac rupture.
- In the event there is obstructive disease affecting the aortic bifurcation/CIA, kissing inflations should be performed in this area.

- The second balloon chosen is a 12 - 14 mm diameter x 40 mm length and inflated with an in-deflator.

FINAL AORTOGRAPHY

- After removal of the balloon catheters, repeat aortography is performed.
- The flush catheter is positioned just above the renal arteries.
- DSA and breath hold must be used to rule out the presence of endoleaks (continued contrast flow into the sac).
- During aortography, blood should be rapidly aspirated through both ipsi and contralateral sheaths using a 60 cc syringe on each side. This allows room for the contrast to runoff into the iliac arteries as the sheaths are partially occlusive in the femoral arteries.
- After the injection, the blood may be returned to the patient.
- The aortogram is carefully evaluated for the presence of an endoleak.
- If no significant endoleak is present, the repair is considered successful.
- The flush catheter is removed.

ACCESS CLOSURE

- Arterial closure is performed as appropriate.
- It may be appropriate to give the patient protamine to reverse the effect of IV UFH.
- If used, a test dose must be given prior to the entire dose.

Percutaneous Closure:

- If percutaneous technique has been used for access, we perform closure with the wires still in situ.
- On each side the respective dilator is loaded on the wire.
- An assistant applies manual compression to groin just above the arteriotomy as the sheath is withdrawn, leaving the wire in place.
- The knot on the 10 O'clock sutures is tightened first and delivered to the arteriotomy. The sequence is repeated with the 2 O'clock sutures.
- If hemostasis appears satisfactory, the wire is removed and the knots are tightened.
- The suture is then cut in the tract, 1 - 2 mm above the arteriotomy.
- Despite adequate hemostasis, manual compression should be applied for a duration of 5 - 10 minutes.
- The same is repeated on the other side.
- A sterile dressing is applied.

Open Femoral Repair:

- If open access was used for arterial access, then surgical repair of the vessel is performed on one or both sides.
- A sterile dressing is applied.

POST-OP CARE

- We usually transfer the patient to the ICU overnight.
- If general anesthesia has been used, we aim to awaken the patient and extubate them as early as possible.
- Clear liquid diet is initiated as early as possible and advanced to regular as tolerated.
- If percutaneous access had been used, we encourage the patient to get out of bed and ambulate with assistance on the second day. Many patients may even be able to ambulate independantly and be discharged home on post-op day 1.

- Patients do not need dual antiplatelet therapy after EVAR unless they other indications (e.g. CAD).

FOLLOW-UP CARE

- All patients should be seen in follow-up within the first 2 - 4 weeks of EVAR.
- Access sites are examined to ensure proper healing.
- A follow-up CTA is performed around this time to evaluate:
  - ▷ Device integrity.
  - ▷ AAA size.
  - ▷ Presence of any endoleaks.
  - ▷ Patency of vessels.
- After EVAR, patients need to be followed up longitudinally with periodic imaging (table 5). Although CTA is the best modality for this purpose, the adverse impact of cumulative radiation dose and contrast dye make this a less than desirable long term option.
- During the first post-op year, a CTA should be obtained at 1 month, 6 months and 12 months post-op. Thereafter, annual imaging is suggested. If significant endoleaks or other issues are present, more frequent imaging may be warranted.
- However, a combination of other modalities such as abdominal X-ray, aortic duplex U/S, non-contrast CT and MRI/MRA may obviate the need for an annual CTA in stable patients.

TESTS

CT/CTA:

- This is the study of choice as it provides the most detailed information about the device and vessel anatomy.
- However, it should be used in conjunction with other modalities to increase the imaging interval and to reduce radiation and contrast exposure.
- For the contrast run, imaging should start 1 cm proximal to the celiac trunk and terminate at the bifurcation of the CFA with a collimation of < 3 mm.
- The non-contrast should start at the level of the diaphragm and terminate at the level of the proximal femur.
- Collimation is kept at < 3 mm.

X-ray:

- This modality is used to monitor device integrity (e.g. wire separation, fracture, kinking) and component uncoupling/migration.
- Four views are obtained: AP, cross-table lateral, RPO 30° and LPO 30°.
- The entire device should be captured on each view.
- The table-to-film distance should be recorded and replilcated in every follow up study.

Duplex US:

- An aortic duplex should be performed to measure maximal aneurysmal diameter, detect endoleaks, confirm stent patency and detect stenosis.

MRI/MRA:

- This modality is used rarely. Its primary indication is in patients who are unable to tolerate iodinated contrast due to allergic history.

TABLE 5 - FOLLOW-UP IMAGING SCHEDULE				
Interval	CTA	KUB	Duplex US	MRI/MRA (only if allergic to iodinated contrast)
Pre-op	X	-	-	X
1 month	X	X	-	X
6 months	X	X	X	X
12 months	X	X	X	X
24 months	-	X	X	X
36 months	X	X	X	X
48 months	-	X	X	X

FACTORS INDICATING NEED FOR RE-INTERVENTION

The following findings if present suggest a need for re-intervention or conversion to open surgical repair in the appropriate patient:

- AAA diameter growth by ≥ 0.5 cm over imaging interval, even without endoleak.
- Persistent endoleak (especially types I and III).
- Development of a new endoleak (especially types I and III).
- Stent graft migration (with inadequate seal zone).
- Worsening renal function due to compromise of renal artery flow.

ENDOLEAKS

DEFINITION

- Persistent blood flow into the aneurysmal sac despite endovascular repair with a stent graft is called an 'endoleak'.
- Endoleaks are the most common 'complication' that occurs after EVAR.
- An endoleak may lead to continued expansion of the sac with eventual rupture.

TYPES OF ENDOLEAK

Five main types of endoleak are described.

Type I:

- A type I endoleak results from incomplete apposition of the stent graft to the arterial wall.
- This is usually detectable easily on contrast angiography.
- It is considered a high-pressure endoleak and should be treated promptly.

Type I endoleaks are further classified based on the location of the endoleak.

Type IA:

- This type I endoleak results from poor proximal apposition of the stent graft in the aorta.

Type IB:

- This type I endoleak results from poor distal apposition of the stent graft in the iliac arteries.

Type IC:

- When a type I endoleak results from incomplete apposition of an iliac occluder in patients in whom an AUI has been used, it is categorized as type IC.

Type II:

- A type II endoleak results from the retrograde flow into the sac via one of the aortic branches, such as the IMA, a lumbar artery, hypogastric artery or an accessory renal artery.

- This is the commonest type of endoleak and is considered a low-pressure endoleak.
- Most of these endoleak resolve spontaneously over a few months.
- However, rarely, a type II endoleak may lead to continued aneurysm expansion and thus need to be treated to avoid rupture.

Type IIA:

- Single vessel inflow into the sac.

Type IIB:

- Two or more vessel inflow into the sac.

Type III:

- A type III endoleak occurs when blood enters into the sac through the body of a stent graft.
- Usually this is due to uncoupling/separation of the components of the stent graft (e.g. iliac limb separation) or due to rupture/tear of graft material. These factors are used to classify a type III endoleak:

Type IIIA:

- A type IIIA endoleak results from fractures or holes in the stent graft.

Type IIIB:

- A type IIIB endoleak is due to junctional separation of modular components.

Type IIIC:

- This endoleak is postulated to be due to suture holes in the graft.

Type IV:

- When an endoleak is present immediately after EVAR and no obvious type I, II or III source is noted, it is called a type IV endoleak.
- It is postulated that this endoleak is through the graft fabric due to graft porosity.
- Type IV endoleaks are transient and resolve when the patient's anticoagulation is reversed at the end of the procedure.
- They are not dectected on follow-up imaging and do not lead to sac expansion or rupture.

Type V:

- A type V endoleak is said to be present when there is continued sac expansion on follow-up imaging, without any other evident endoleak.
- A type V endoleak is also called 'endotension'.

TREATMENT

- Most endoleaks are transient and do not require treatment.
- However, persistent endoleaks may contribute to aneurysm expansion and should be treated.

Type I:

- Early type I endoleaks should be treated during the procedure.
- Balloon dilatation/molding is the first step. The molding balloon should be introduced not only through the 'ipsi' side but also via the contralateral side.
- If balloon molding fails to produce the desired result, placement of a Palmaz stent within the aortic stent graft trunk should be considered.
- If enough room is present between the renal arteries and the aortic trunk margin an aortic extension may be feasible.
- Rarely, however, a type I endoleak may be present despite the above interventions. These should be evaluated at follow-up imaging at 30 days post-op. Many will disappear.
- Persistent type I endoleaks require further treatment. This may consist of placement of additional stents as noted above or glue injection or embolization.
- If a type IB endoleak is present balloon molding of distal seal zones should be performed.

- If this is unsuccessful and both hypogastric arteries are patent, the ipsilateral hypogastric artery should be coiled and the stent graft extended to the EIA.
- If despite endovascular re-intervention the endoleak persists, open conversion is necessary to avoid rupture.

Type II:

- Most type II endoleaks resolve spontaneously.
- Persistent type II endoleaks that cause sac expansion require treatment in the form of embolization of the feeding vessels such as the lumbars or the IMA in a retrograde fashion.
- In principle, both inflow and outflow vessels should be treated.
- For IMA type II endoleaks, the SMA is cannulated and the middle colic artery is selected. A microcatheter is advanced into this vessel and tracked into the marginal artery and then the IMA in a retrograde fashion.
- Lumbar artery type II endoleaks are addressed by cannulating the hypogastric arteries and selecting the iliolumbar branches. Retrograde cannalization of the lumbar arteries is then achieved.
- Translumbar coil embolization of the sac (direct aortic puncture with the patient lying in the prone position) using fluoroscopic/CT guidance may lead to thrombosis of the inflow or outflow of the endoleak and cure the problem.
- Complications of embolic therapy include bowel ischemia and colonic infarct.
- If re-intervention fails and the sac continues to expand, surgical conversion becomes necessary.
- Laproscopic or robotic ligation of the feeder arteries may be feasible in select cases.

Type III:

- Similar to a type I endoleak, a type III endoleak must be treated promptly.
- Treatment usually consits of endovascular 'relining' of the stent graft with additional prostheses.
- Failed repair should lead to surgical conversion.

Type IV:

- This type of endoleak is transient and resolves after EVAR.
- It does not require any further treatment.

Type V:

- This type of endoleak may be treated initially with re-intervention but if persistent it requires open repair.

CASE

- Mr. D. is a 54-year-old male reformed smoker with known CAD. He presented with syncope and symptoms of congestive heart failure. After admission, recurrent ventricular tachycardia was noted and considered to be the cause of his presentation. His TTE showed severe left ventricular dilatation with severe systolic dysfunction with an estimated LV ejection fraction of 10%. On coronary angiography he was found to have severe triple vessel CAD and later shown to have no significant myocaridal viability.
- For evaluation of this puslatile mass in his abdomen, the patient underwent an abdominal CTA which showed a large (7.2 cm diameter) infra-renal AAA without any evidence of acute rupture.
- A normal suprarenal aorta was noted with a patent celiac axis and SMA.
- Single renal arteries were noted bilaterally, right approximately 4 mm lower than the left.

Pre-op CTA findings:

- The CTA (figure 8) measurements were as follows:
  - ▹ Proximal neck cranial diameter (D1) = 25 mm.
  - ▹ Proximal neck caudal diameter (D2) = 25 mm.
  - ▹ Proximal neck length (L1) = 15 mm.
  - ▹ Length from lowest (right) renal artery to right hypogastric artery (L3 + L4) = 180 mm.
  - ▹ Length from lowest (right) renal artery to left hypogastric artery (L3 + L5) = 187 mm.
  - ▹ Right CIA diameter (D5) = 13 mm.
  - ▹ Right CIA landing zone length = 40 mm.
  - ▹ Left CIA diameter (D6) = 13 mm.
  - ▹ Left CIA landing zone length = 47 mm.
  - ▹ Right EIA diameter (D7) = 8 mm.
  - ▹ Left EIA diameter (D8) = 8 mm.
  - ▹ Right CFA mildly calcified, diameter = 7 mm.
  - ▹ Left CFA moderately calcified, diameter = 7 mm.

Indications for EVAR:

- Large asymptomatic infra-renal AAA.
- A Medtronic Endurant™ stent graft system was chosen for this case.

Access:

- Given the patient's favorable ileofemoral anatomy, it was decided to obtain percutaneous access and perform preclosure bilaterally prior to EVAR.
- Since the left CFA was noted to have more calcium, it was decided to use the right CFA for main body delivery and the left for contrlateral devices.
- The patient was given conscious sedation with IV Fentanyl.
- Local anesthesia was used. Given his poor LV systolic function, general anesthesia was avoided.
- Using a 4 Fr. micropuncture kit, arterial access was obtained in the right CFA (figure 9A).
- After confirmation of the arteriotomy in the desired portion of the CFA, a Wholey wire was advanced into the aorta (figure 9B).
- A 6 Fr. /11 cm sheath was inserted.
- The same was repeated for the left CFA.
- The right CFA 6 Fr. sheath was removed over the wire and preclosure was performed using two Perclose Proglide devices as previously outlined.
- An 8 Fr./11 cm sheath was inserted thereafter in this CFA.

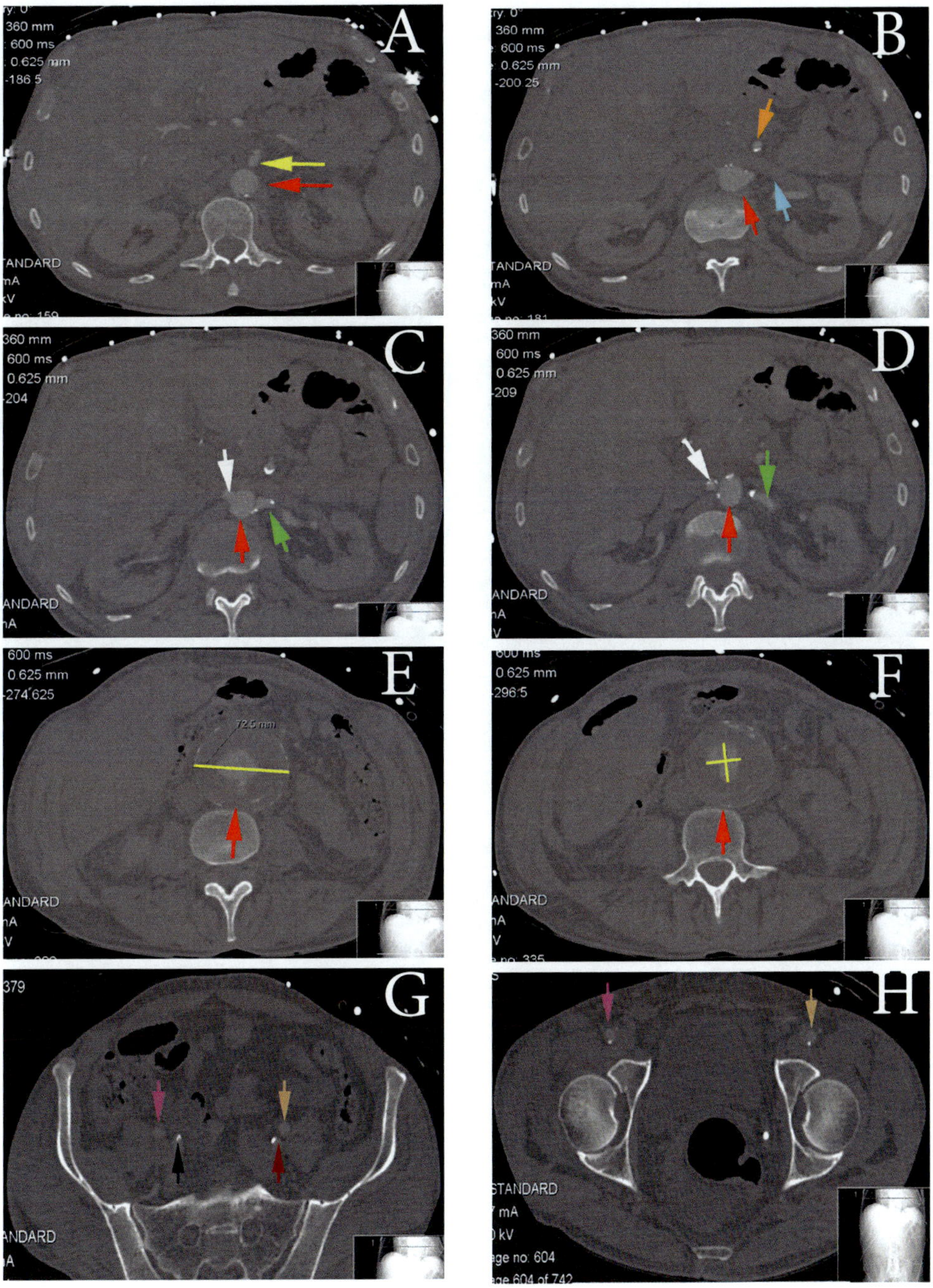

Figure 8.
Pre-operative CTA.

Red arrow: aorta; Yellow arrow: celiac axis; Orange arrow: SMA; White arrow: right renal artery; Green arrow: left renal artery; Blue arrow: left renal vein; Pink arrow: right EIA; Brown arrow: left EIA; Black arrow: right hypogastric artery; Mauve arrow: left hypogastric artery.

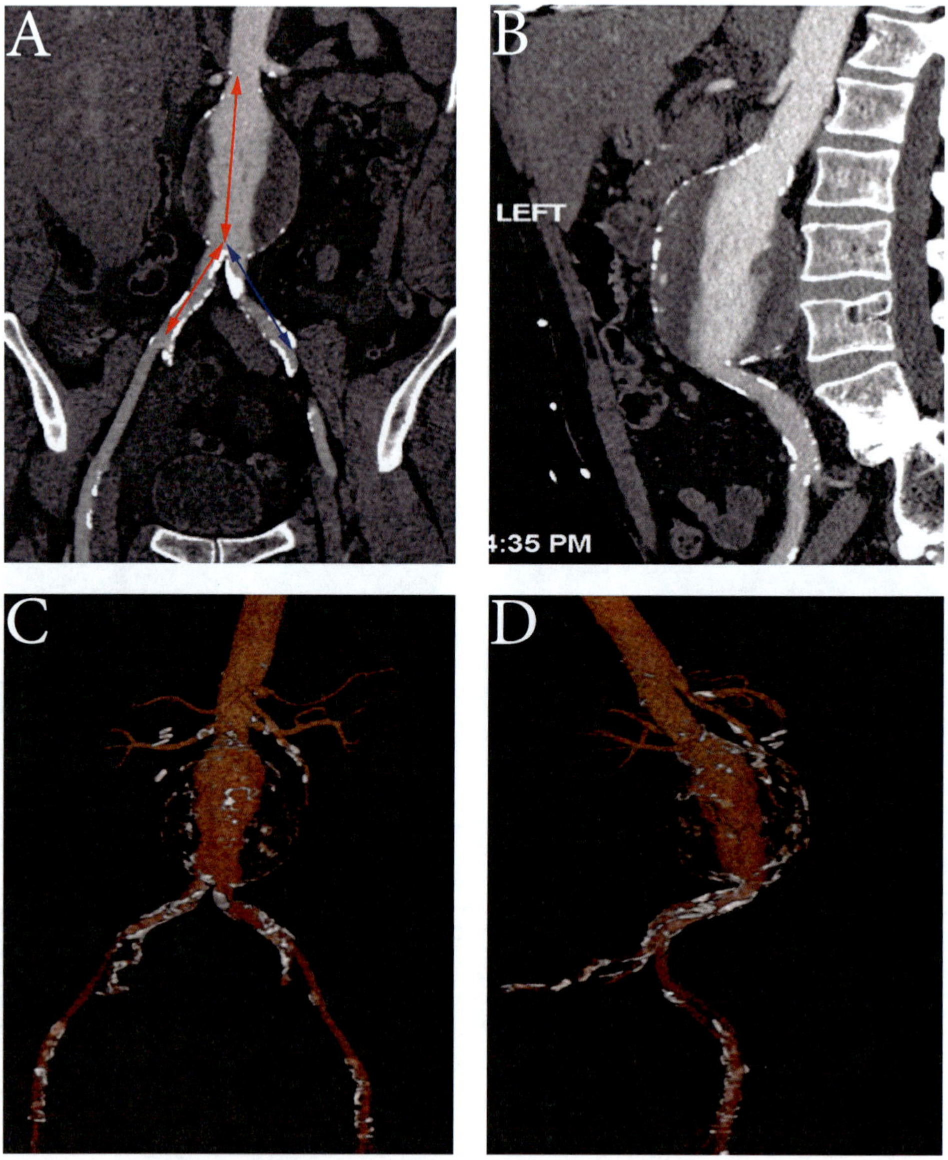

Figure 8.
Pre-operative CTA.

Coronal, sagittal and 3D reconstructions of the abdominal aorta showing thrombus in the sac and moderate calcification of bilateral CIA.

- Through a 4 Fr. angled GC the Wholey wire was exchanged out for a 0.035" 180 cm J-tip Amplatz SuperStiff wire.
- The GC and the sheath were then removed over this wire.
- Using dilators of progressively larger caliber (10 - 18 Fr.), the arteriotomy track was adequately dilated (figures 9D - 9F). The dilators were advanced under fluoroscopic guidance.
- After final dilatation, an 18 Fr./30 cm sheath was inserted in the right CFA and it was advanced under fluoroscopic guidance until its tip was in the infra-renal aorta.
- The same procedure was repeated for left CFA to preclose. A 14 Fr./30 cm sheath was inserted in the left CFA (figures 9G - 9I) and its tip was advanced into the distal abdominal aorta.
- Once both the sheaths were in place, the patient was given 6000 U IV UFH.

Diagnostic Aortography:

- A marker flush catheter was introduced via the right CFA into the abdominal aorta. It was flushed and connected to a power injector.
- Using DSA, an abdominal aortogram was obtained (figure 9K) to measure the length from the lowest renal artery to the right hypogastric artery.

Interventional Technique:

- With the contralateral 0.035" Amplatz SuperStiff wire in place, the desired orientation of the contralateral gate was determined.
- Since the contralateral (left CFA) wire was noted to be coursing up along the right side of the sac, it was decided to keep the contralateral limb oriented to that position so it could be cannulated later without difficulty.

Selection of the main body

- Given the diameter of the proximal aortic neck, a 28 mm diameter main body was selected with a length of 145 mm and an ipsilateral limb diameter of 16 mm.
- This was prepped in a standard fashion.
- Prior to introduction into the ipsilateral sheath, the stent graft was positioned under fluoroscopy on the patient's abdomen to confirm the orientation of the contralateral gate with respect to the handle of the device.

Introduction of the marker catheter on the contralateral side

- The marker catheter was removed over the ipsilateral 0.035" Amplatz SuperStiff wire.
- The catheter was now introduced via the left CFA over the contralateral wire and advanced into the supra-renal aorta.
- The wire was removed, the catheter flushed and connected to the power injector.
- An abdominal aortogram was performed (figure 9J).

Introduction of the main body

- The main body was then inserted via the right CFA and advanced into the peri-renal aorta.
- With the marker catheter connected to the injector, an aortogram was performed at 10 cc/sec for a total of 10 cc at 500 psi (figure 9K).

Deployment of the main body

- After confirming the position of the stent graft and the right (lower) renal artery, main body deployment was initiated (figures 9L & 9M).
- The deployment was continued until the contralateral gate was opened (figure 9N).

Removal of the marker catheter from behind the stent graft

- With the aortic portion of the main body deployed and the contra gate open, the marker catheter was removed from behind the stent graft after straightening the catheter out with an ASGW.

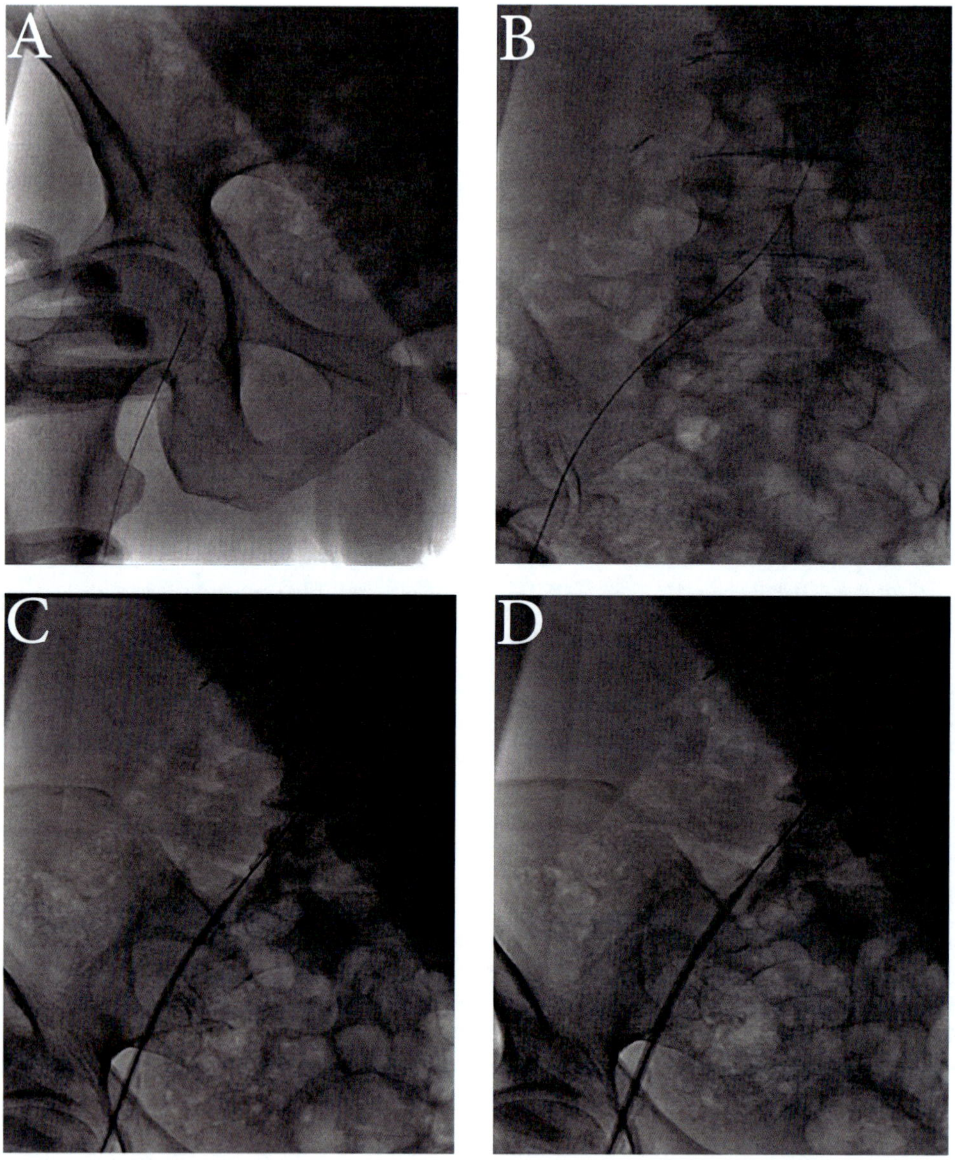

Figure 9.

EVAR Stepwise.

A. Right CFA access with a 21 G needle.
B. A Wholey is advanced into the sac.
C. A 6 Fr. sheath is inserted in the right CFA.
D. After preclosure and wire exchange, the right CFA track is being dilated with a 10 Fr. dilator.

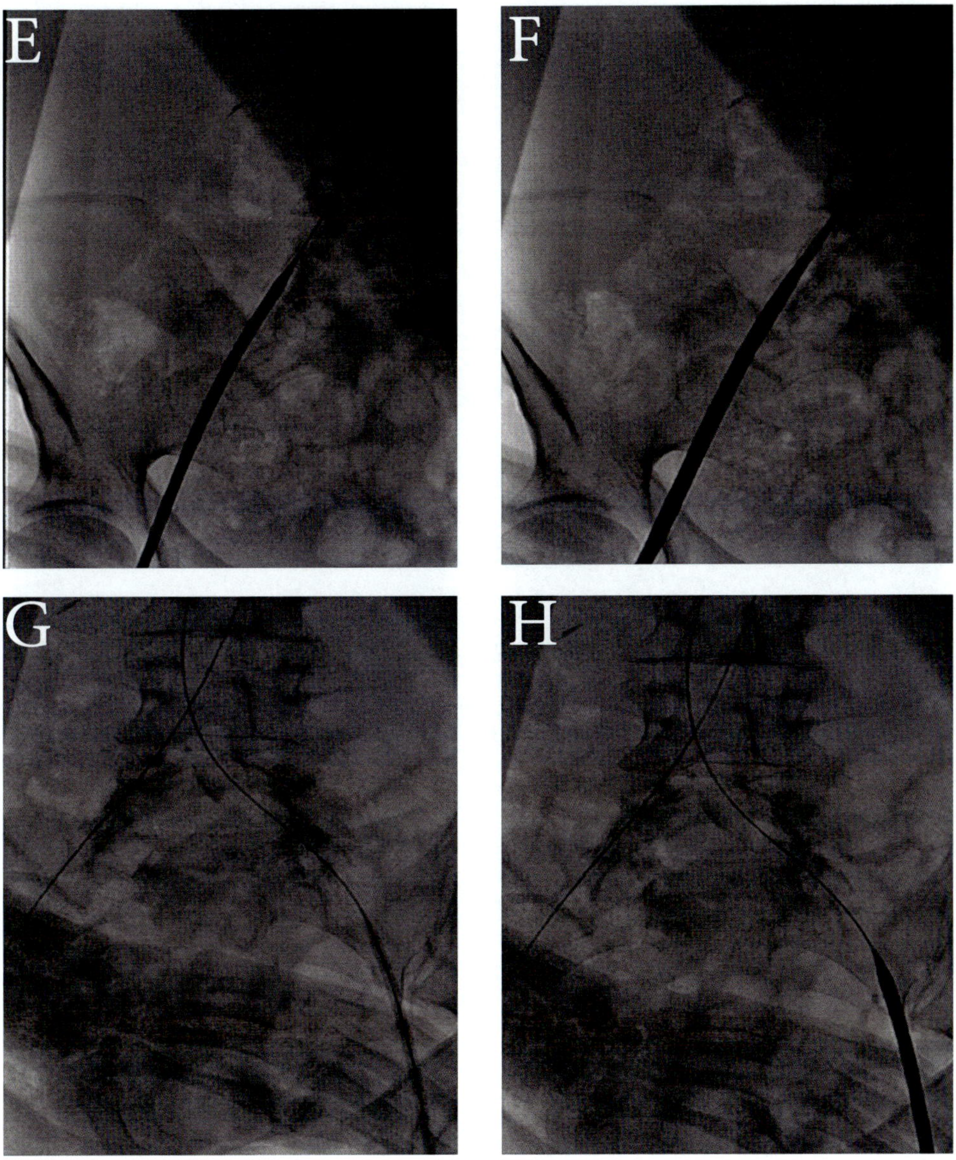

Figure 9 (continued).

EVAR Stepwise.

E. The right CFA track is being dilated with a 14 Fr. dilator.
F. The right CFA track is being dilated with an 18 Fr. dilator.
G. The left CFA track is being dilated with an 10 Fr. dilator.
H. The left CFA track is being dilated with an 14 Fr. dilator.

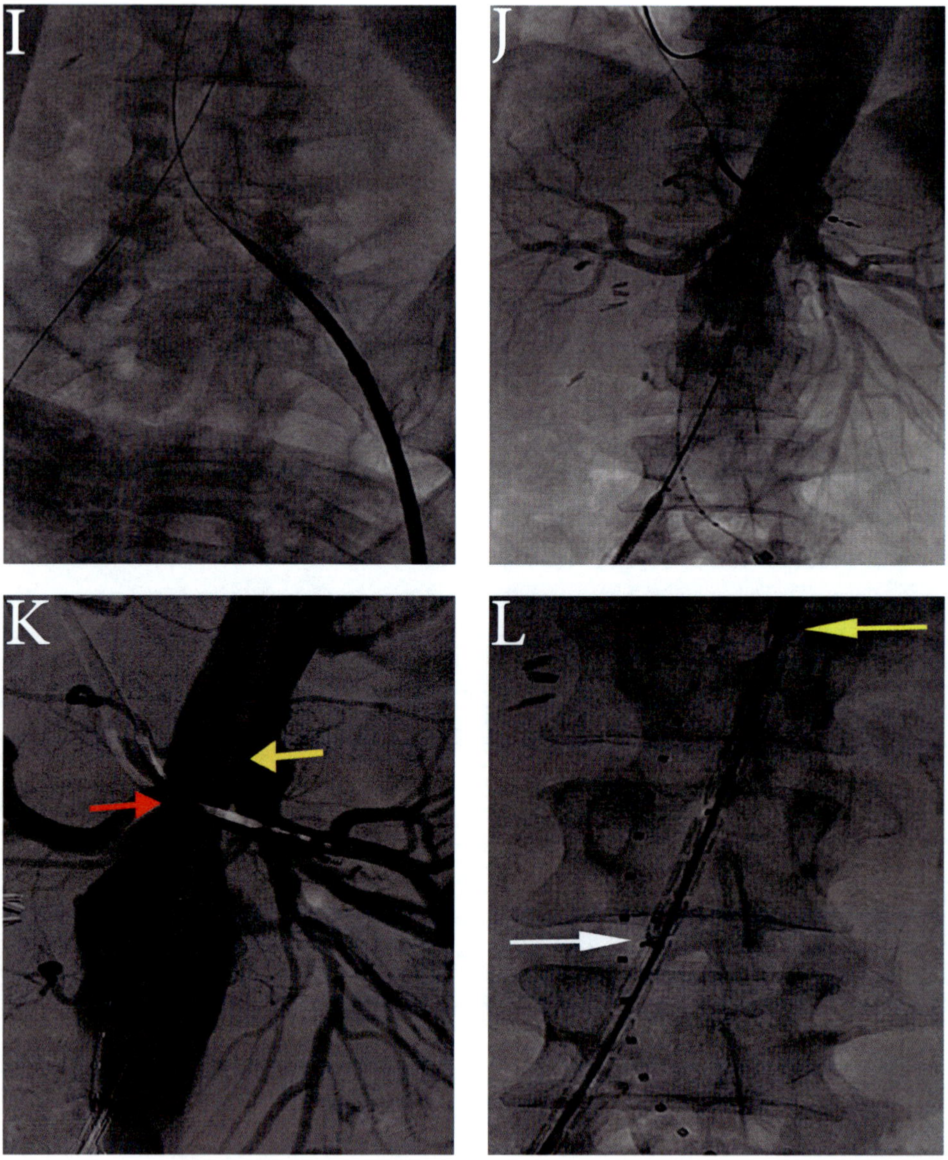

Figure 9 (continued).

EVAR Stepwise.

I. A 14 Fr. sheath is inserted via the left CFA and it is advanced into the distal abdominal aorta.
J. The marker catheter has been switched to the left side. The main body has been introduced via the left CFA and a positioning aortogram is performed.
K. A positioning aortogram performed under DSA. The yellow arrow shows the top of the graft material on the main body and the red arrow shows the lower margin of the lowest (right) renal artery.
L. A magnified image of the distal main body shows the contra gate (white arrow tip). Note that the contra gate is aligned along the plane of the marker catheter.

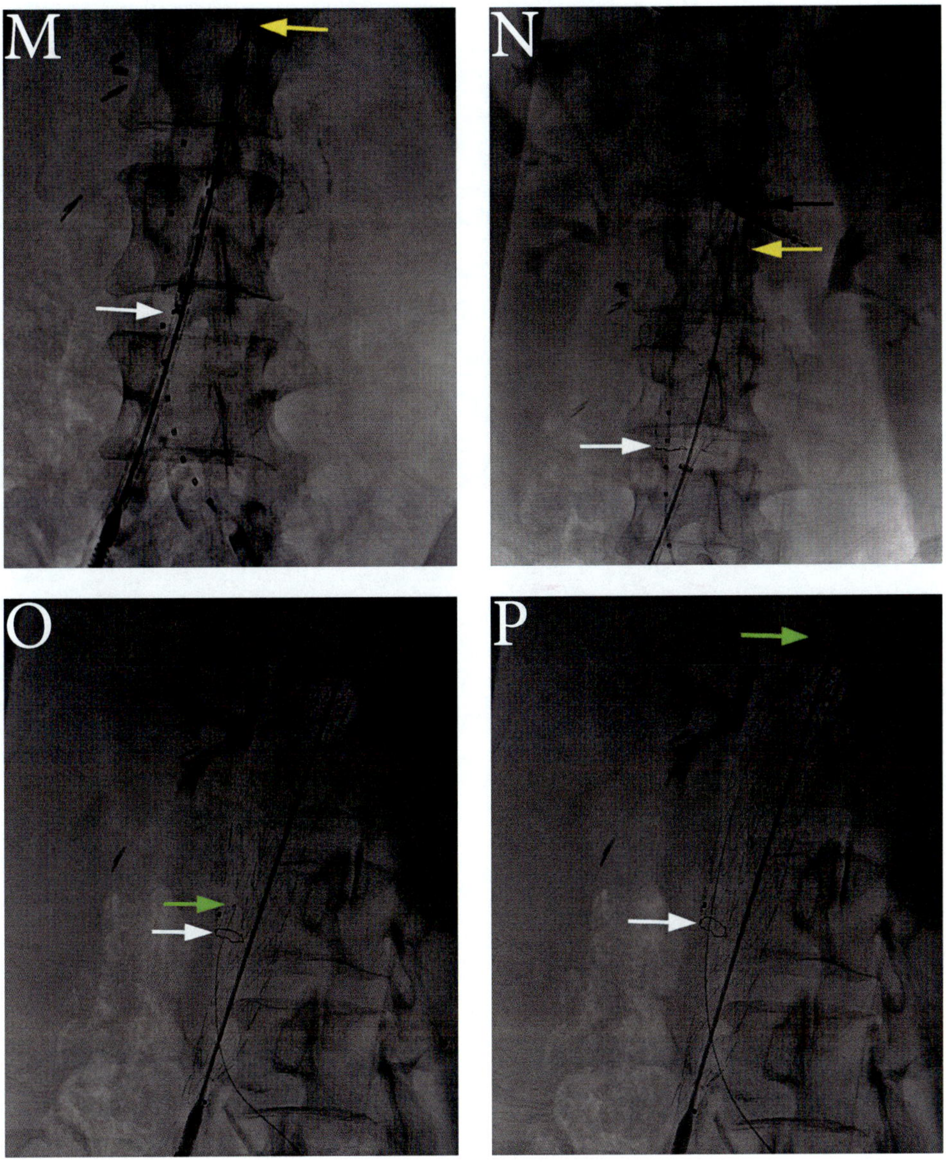

Figure 9 (continued).

EVAR Stepwise.

M. Deployment of the main body has begun.
N. The supra-renal stent and the trunk has been deployed. The main body is deployed up to the point the contra gate is open.
O. The marker catheter has been withdrawn from behind the stent graft and removed. An ASGW (green arrow) has been advanced into the gate.
P. The wire has been advanced into the trunk piece.
Yellow arrow: Proximal graft markers; white arrow: contra gate; Black arrow: supra-renal stent;

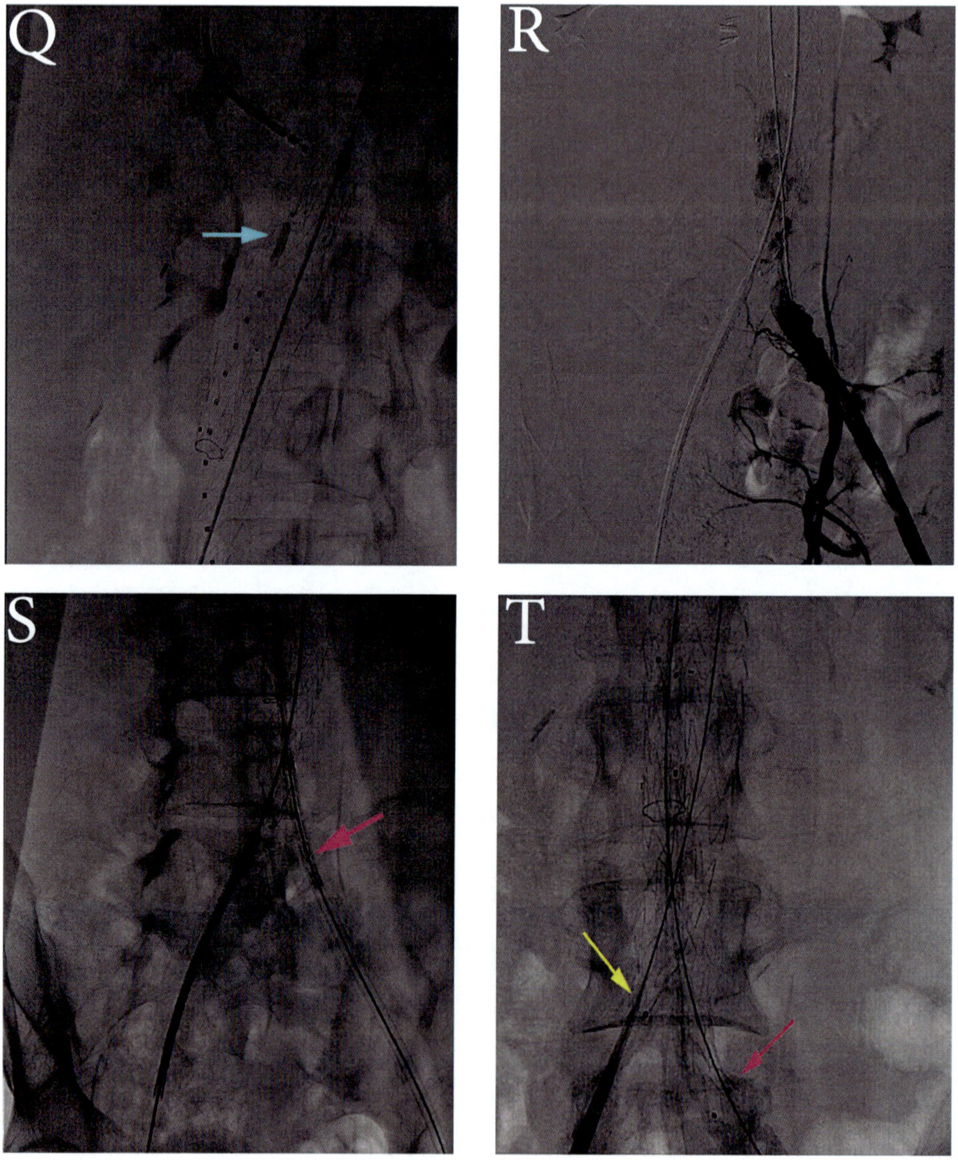

Figure 9 (continued).

EVAR Stepwise.

Q. The flush catheter is advanced over the ASGW into the trunk and rotated to confirm presence within the graft lumen.
R. An angiogram is performed to measure the distance between the contra gate and the left hypogastric artery.
S. The contralateral limb is being positioned. The purple arrow marks the caudal end of the graft.
T. Deployment of the ipsilateral limb is completed (yellow arrow).

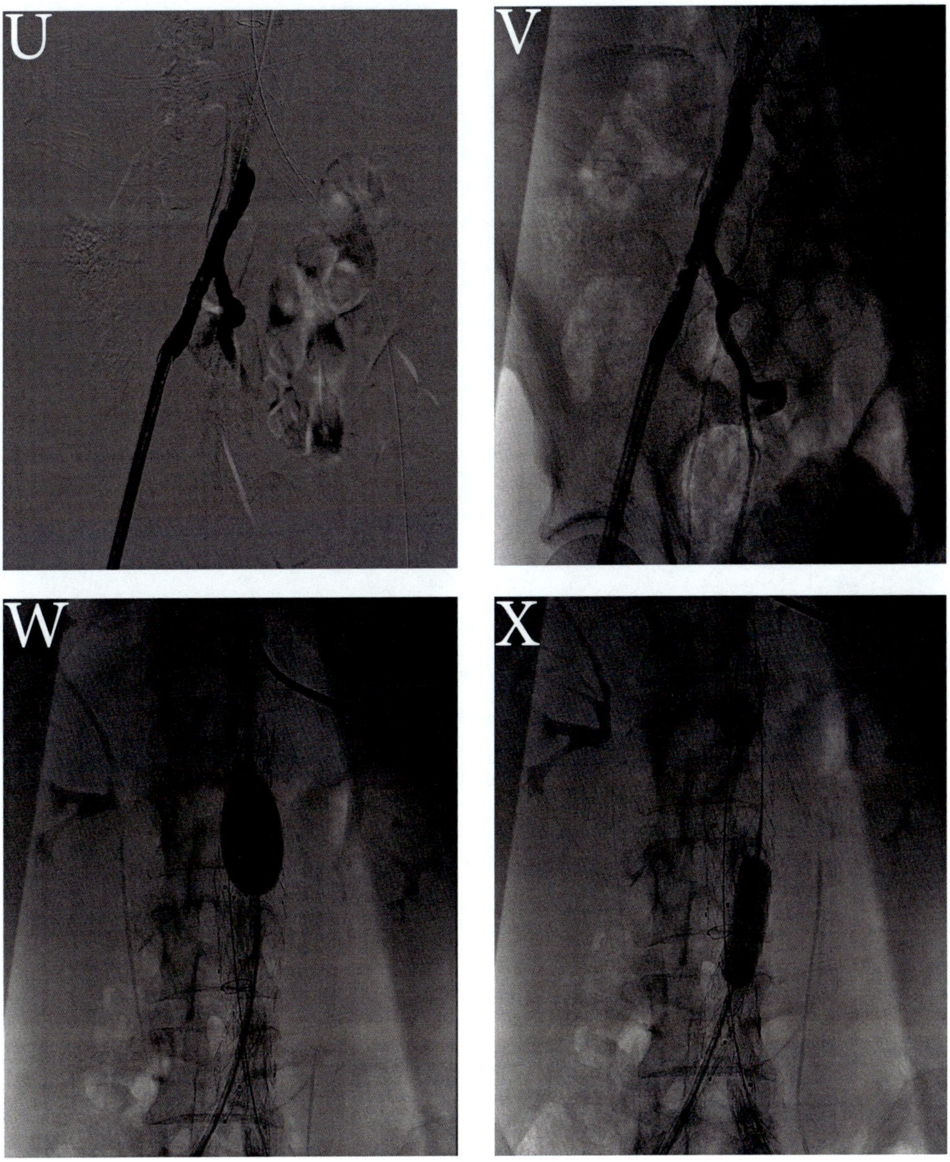

Figure 9.

EVAR Stepwise.

U. An angiogram is performed to measure the length from the ipsi limb to the right hypogastric artery.
V. Another angiogram is obtained as the ipsi iliac extender is being positioned.
W. The Coda balloon is used to post-dilate the trunk piece.
X. The ipsi limb is dilated with the same balloon.

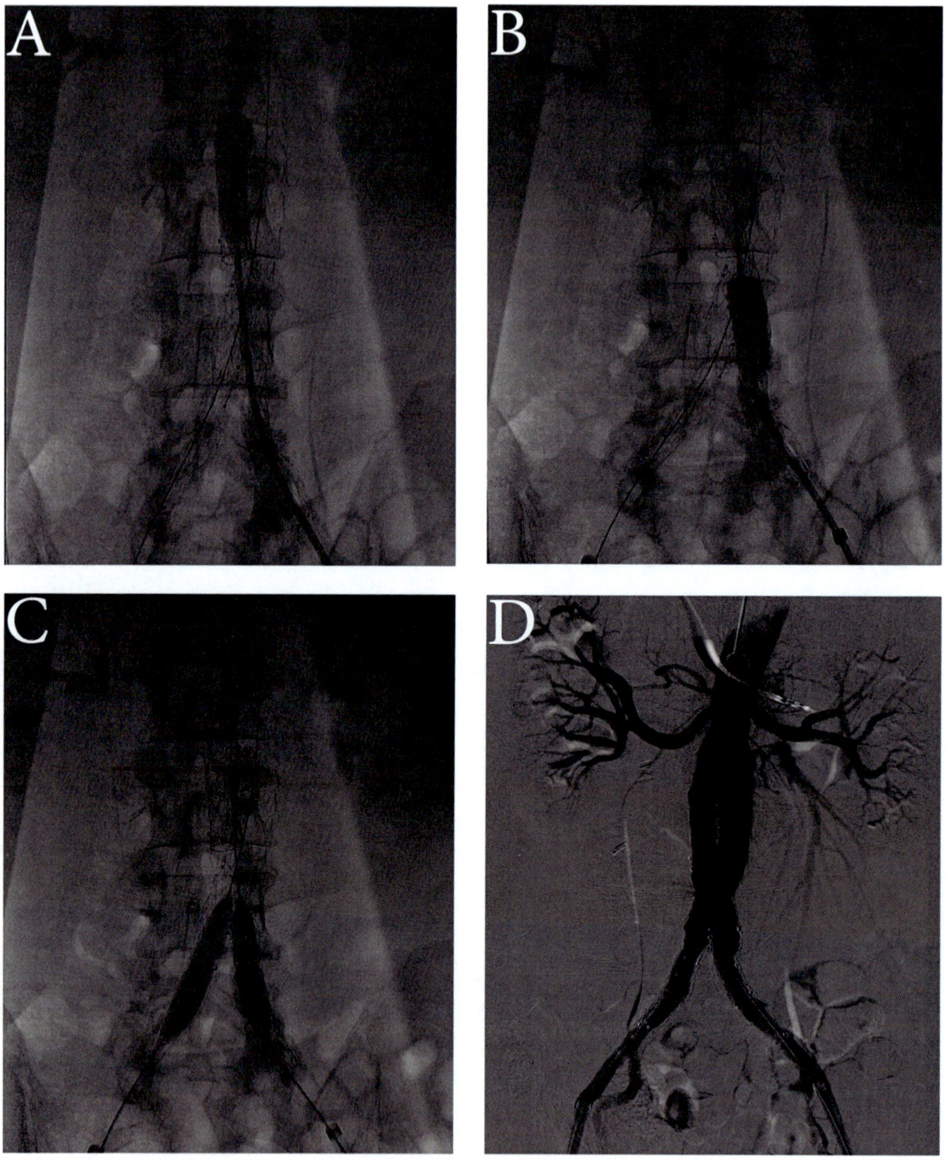

Figure 10.

EVAR Stepwise.

A. The contra limb proximal overlap zone is post-dilated with the same Coda balloon.
B. The contra limb proximal overlap zone is post-dilated with the same Coda balloon.
C. The aortic bifurcation is post-dilated with a kissing inflation of two 12 mm x 40 mm Atlas balloons.
D. Final aortogram demonstrating successful exclusion of the AAA with preserved bilateral renal and hypogastric arteries. No endoleaks are noted.

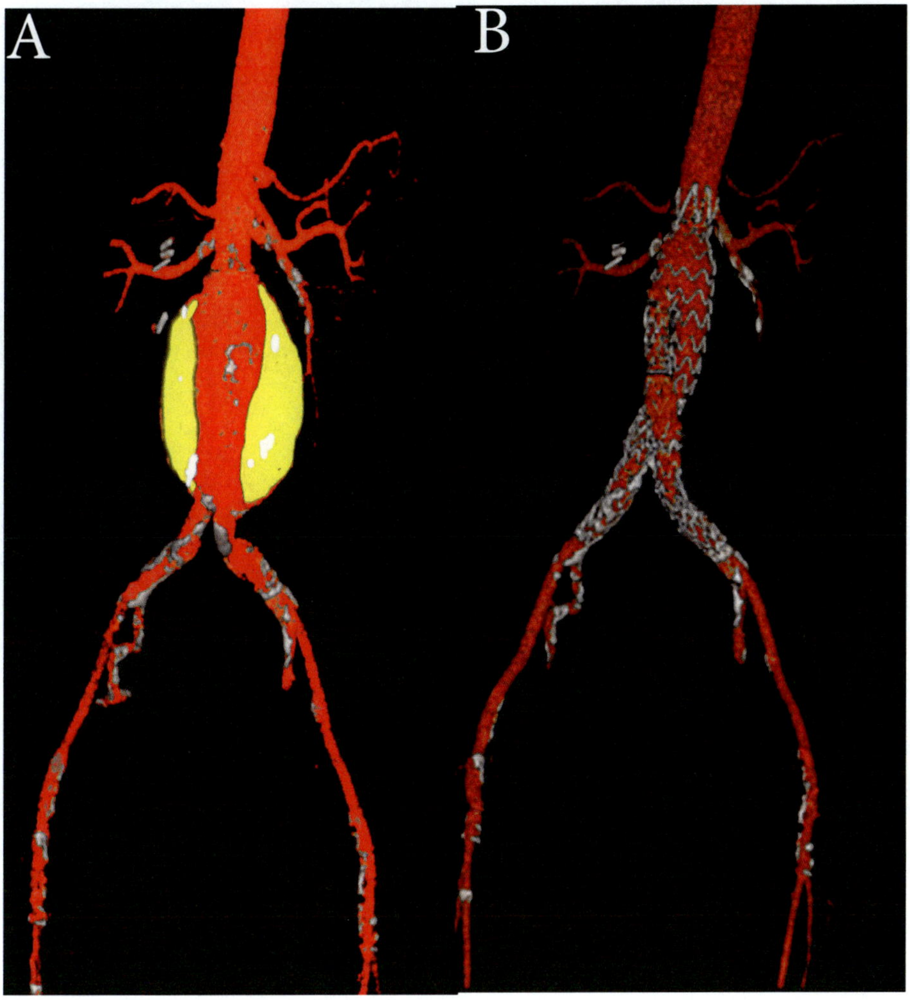

Figure 11.

30-day follow-up CTA.

A. Pre-op CTA with 3D recontruction.
B. Post-op follow-up CTA with 3D reconstruction.

- The catheter was withdrawn to a level below the contra gate and exchanged out for a 5 Fr. Kumpe catheter.
- The wire was then used to cannulate the contra gate without difficulty (figures 9O & 9P).
- The Kumpe catheter was exchanged out for the flush marker catheter again.

Confirmation of the intra-graft position of the marker catheter

- The catheter was advanced into the trunk of the main body over the ASGW.
- Once in the trunk, the wire was pulled into the catheter and the catheter was freely rotated 360 ° (figure 9Q) to demonstrate that it is not behind the stent graft but inside its lumen.

Contralateral length measurement

- The marker catheter was then positioned to measure the length between the contra gate and the contralateral hypogastric artery origin.
- An angiogram was obtained via the left CFA sheath which had been withdrawn into the left EIA, in an RAO 20° projection. This clearly showed the origin of the left hypgastric artery (figure 9R).
- The origin of the artery was marked on the screen.

Contralateral iliac limb selection

- The contralateral limb was selected in a straight 16 mm diameter and a 124 mm length.
- Prior to the introduction of the contralateral limb into the sheath, the dilator was introduced into the sheath and under fluoroscopic guidance, the sheath was advanced into the contralateral stub.
- The dilator was removed and the sheath flushed.
- The contralateral limb was then inserted via the left CFA sheath into the contra stub.
- The sheath was withdrawn to below the left hypogastric artery origin.

Deployment of the contralateral limb

- After confirming adequate overlap with the contra stub, the contralateral iliac limb was deployed successfully into the left CIA (figure 9S).
- The delivery system was removed.

Deployment of the contralateral iliac extension

- It was deemed that the length of the landing zone in the left CIA was not adequate at this point. So an iliac limb extension (16 mm diameter and 93 mm length) was used to overlap proximally within the left iliac limb, for a length of at least 3 cm, and within the distal CIA caudally, just cranial to the origin of the left hypogastric artery.
- The device was deployed successfully into the left CIA.
- The delivery system was removed.

Deployment of the ipsilateral limb

- At this point, the deployment of the ipsilateral limb was completed (figure 9T) and the delivery system removed via the right CFA.

Extension of the ipsilateral limb

- Similar to the left CIA stent graft, it was determined that the right CIA landing zone length was inadequate.
- It was decided to extend this caudally into the distal CIA.
- Using the marker catheter, now introduced via the right CFA, the distance between the ipsilateral limb and the right hypogastric artery (figures 9U & 9V) was measured.

Postdilatation

- After deployment of the ipsilateral limb, the delivery system was removed.
- A Coda balloon was used to perform post-dilatation of the proximal neck as well as all the overlap zones on both sides (figures 9W, 9X, 10A & 10B).
- After this, using two 12 mm x 40 mm Atlas balloons, post-dilatation of the aortic bifurcation was performed (figure 10C).

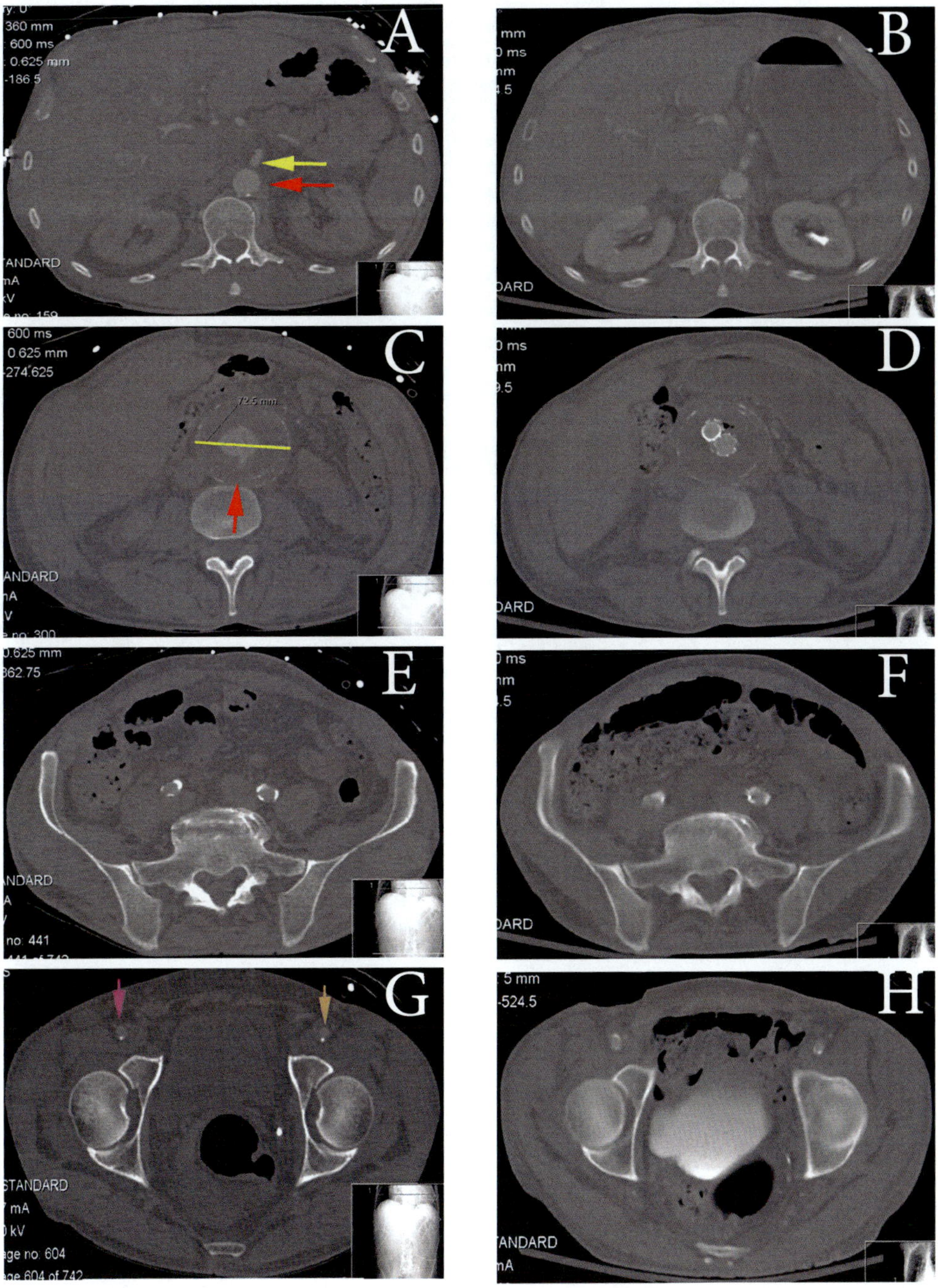

Figure 12.

30-day follow-up CTA.

Comparable pre-op and post-op CTA axial slices show successful EVAR with no endoleaks and preserved supra-
renal aorta, CIA and EIA bilaterally.

Final Aortography:

- Final aortography was performed at 10 cc/sec for 30 cc at 500 psi.
- DSA with breath hold was used for image optimization (figure 10D).
- Simultaneous with the aortogram, using a 60 cc syringe on each CFA sheath, as piration was performed to allow contrast runoff into the iliac arteries and CFA.
- A satisfactory post-EVAR aortogram was noted without any evidence of endoleaks.

Arteriotomy Closure:

- Using standard technique, the arteriotomies were closed bilaterally in a sequential fashion, ipsilateral before contralateral.
- After hemostasis on the ipsi side, an angiogram of the ipsilateral (right CFA) iliac and common femoral arteries was performed via the contralateral CFA sheath.
- Similarly, to the right, the left CFA sheath was removed and closure performed.
- He had palpable pedal pulses in both feet.

Hospital Course:

- The patient was transferred to surgical ICU.
- At 24 hours, he ambulated without difficulty.
- At post-op day seven, he underwent implantation of a ventricular assist device.
- On post-op day 14, the patient was discharged home uneventfully.

Follow-up:

- At 30-day follow-up visit, the patient was stable and doing well.
- He had undergone a CTA for follow-up that showed successful exclusion of the AAA (figures 11 & 12) without under any evidence of endoleaks.

15

CATHETER-DIRECTED THROMBOLYTIC THERAPY

Systemic (intravenous) thrombolytic therapy has a well-established role in the management of acute ST segment elevation myocardial infarction (STEMI), ischemic stroke and massive pulmonary embolism. However, one of the major factors that limits the success of this therapy is the risk of bleeding complications including hemorrhagic stroke.

In cases of acute arterial and venous thrombosis outside the above vascular beds, systemic thrombolytic therapy is not recommended. The reason is two-fold:

1. The thrombus burden in such occlusions is usually significantly higher resulting in poor penetration of the systemically given thrombolytic agent.
2. The risk of bleeding complications remains high and makes the risk/benefit ratio unfavorable.

Catheter-directed thrombolytic (CDT) therapy allows delivery of the thrombolytic agent to the affected segment of the culprit vessel, enhancing drug penetration into the thrombus. The significantly reduces the total dose of the thrombolytic agent thus administered (up to 50% - 75%), thereby reducing the likelihood of bleeding complications.

PRINCIPLES OF CDT

CDT consists of delivering thrombolytic therapy through a catheter to the affected blood vessel.

- The delivery catheter usually has several dozen sideholes to allow for uniform drug delivery through the length of the occlusion.
- The duration for which the catheter is placed in the thrombosed segment of the blood vessel is called the dwell time.

- The dwell time for CDT ranges from 20 minutes to 24 hours. The average duration is 12 hours.
- After the infusion is complete, the patient returns to the endovascular suite for removal of the catheter and angiography is repeated.
- Definitive intervention is then performed.

INDICATIONS

CDT is used for the following indications:

- Acute limb ischemia.
- Massive ileofemoral or proximal femoral DVT with high risk of limb gangrene.
- Submassive and massive pulmonary embolism (see Chapter 15).

CONTRAINDICATIONS

Contraindications to thrombolytic therapy should be ruled out. These are shown in table 1. While these conditions are recognized as contraindications to systemic thrombolysis, many of them are applicable to CDT as well.

TABLE 1 - CONTRAINDICATIONS TO THROMBOLYTIC THERAPY
ABSOLUTE
• Active bleeding or bleeding diathesis (other than menses). • History of intracranial hemorrhage. • Presence of structural intracranial disease (e.g. AV malformation) • Known structural cerebral vasuclar lesion. • Known intracranial neoplasm. • Ischemic stroke within prior 3 months. • Suspected aortic dissection. • Significant closed head or facial trauma within 3 months.
RELATIVE
• Age > 75 years • Recent major non-vascular surgery or trauma (<10 days). • Severe uncontrolled hypertension (SBP > 180 mm Hg and DBP >110 mm Hg). • Recent internal bleeding (during prior 2-4 weeks). • Pregnancy/postpartum status. • Active peptic ulcer disease. • Prolonged CPR (> 10 min.). • Dementia. • Remote ischemic stroke. • Major surgery within 3 weeks.

THROMBOLYTIC AGENTS

- Various thrombolytic drugs have been used for CDT.
- Streptokinase (SK) was the original drug used for CDT and was later replaced by urokinase (UK). The latter, in turn, was largely replaced by r-tPA (recombinant tissue plasminogen activator; *Activase*®- Genentech, San Francisco CA) and its analogues reteplase (RPA; *Retavase*® - Cornerstone Therapeutics, Cary, NC) and tenecteplase (TNK; *TNKase*®- Genentech, San Francisco, CA).
- UK is no longer available in the US.
- Currently, in North America the thrombolytic agent of choice is alteplase (r-tPA; *Activase*®)
- Neither SK nor UK are fibrin-specific and thus have a lower fibriniolytic efficacy. Alteplase and its analogues are significantly more fibrin-specific and potent and hence serve as first-line agents for CDT. In regions where these drugs are not readily available, UK is preferred as SK is antigenic and re-exposure to this drug may result in an allergic reaction ranging from mild rash to anaphylaxis. Thus, if r-tPA or its analogues are not a feasible choice, then UK should be considered.

TABLE 2 - CHARACTERISTICS OF THROMBOLYTIC AGENTS			
AGENT	FIBRIN SPECIFIC	HALF LIFE (MIN.)	DOSE
Urokinase (UK)	No	13	60,000 IU/hr. - 240,000 IU/hr.
Streptokinase (SK)	No	23	5,000 U/hr.
Alteplase (r-tPA)	Yes	5	0.25-2 mg/hr.
Reteplase (RPA)	Yes	15	0.25-1.0 U/hr.
Tenecteplase (TNK)	Yes	25	0.25-1.0 mg/hr.

ADJUNCTIVE DRUGS

Aspirin:

For Arterial/graft Thrombosis, Aspirin Is Used As An Antiplatelet Agent.

UFH:

- Unfractionated heparin (UFH) is used in conjunction with thrombolytic agents to minimize their prothrombotic effects locally.
- The goal PTT is 45-60 seconds to keep the risk of bleeding low.
- A typical initial dose is 300-750 U/hr. Thereafter, the infusion rate is adjusted to keep the PTT within the above goal.
- If UFH is not administered, the indwelling sheath and catheter remain at risk of thrombosis around them. Thus, a portion of the UFH dose is given through the sheath (typically half the initial dose is given through the sheath).
- When UFH dose is adjusted to target the goal PTT, ideally only the peripheral IV portion of the infusion is changed and not the infusion through the sheath.

MODES OF CDT DELIVERY

CDT has undergone an evolution since its inception. Several different modes of CDT have been developed that differ primarily in how the thrombolytic agent is given. CDT can be instituted with or without adjunctive aspiration/mechanical thrombectomy (tables 3 and 4).

PROTOCOL

- A bolus dose of 5 mg of r-tPA is given through the infusion catheter prior to starting the infusion.
- r-tPA infusion is initiated through the catheter at 1-2 mg/hour for a duration of 12-24 hours. Total dosage should not exceed 40-50 mg to keep the bleeding complications low.
- Low dose unfractionated heparin (UFH) is given concurrently to achieve a PTT of 45-60 seconds. Part of the UFH infusion is given through the vascular sheath through which the infusion catheter is inserted and the rest through a peripheral IV.
- The limb that has the introducer sheath in situ is immobilized (e.g. with a knee immobilizer) to restrict patient motion and reduce the risk of access site bleeding.
- The patient should be monitored post procedure in the ICU with close monitoring.
- Frequent (once an hour) neuro-cognitive checks are routinely per-

Activase® (r-tPA):

Activase® (Alteplase or r-tPA) is a tissue plasminogen activator produced by recombinant DNA technology (r-tPA)

100 mg = 58 million IU.

Retavase®:

Retavase® is a non-glycosated deletion mutein of tPA. It is produced by recombinant technology in E. Coli.

Its potency is expressed in units (U) which is specific for Retavase® and is not comparable with the units used for other lytic agents.

Do not administer r-tPA and UFH through the same catheter due to the risk of precipitation of drug.

How to make a 'home-made' infusion catheter:

- A 5 Fr. catheter such as the MPA is selected
- Multiple side holes are created at 5-10 mm intervals using a 27-gauge needle.
- A Tuohy-Borst adapter is attached to the hub of the catheter.
- Over a 0.035" wire, the catheter is advanced into the thrombosed segment.
- The end-hole at the tip of the catheter is plugged with a beaded wire.

TABLE 3 - CDT MODES WITHOUT ADJUNCTIVE THROMBECTOMY	
Intra-thrombus lacing	• Initial delivery of thrombolytic agent within the thrombosed segment.
Continuous infusion	• Intra-thrombus delivery of thrombolytic agent via a multi-sidehole catheter that is connected to an infusion pump. • Drug is delivered at a constant rate for the duration of the infusion.
Step-wise infusion	• This consists of treating the thrombosed segment in a graded fashion. • The catheter is first positioned in the proximal half of the thrombus and CDT is initiated. • With dissolution of the thrombus, the catheter is advanced distally in a 'step-wise' fashion.
Pulse-spray	• Small volumes of concentrated thrombolytic agent are injected through a multi-sidehole catheter within the thrombus. • The catheter tip is placed within, and not distal to the thrombosed segment. • It is expected that a forceful injection of the drug will physically fragment the thrombus to some extent, increasing the surface area over which the drug acts. • This technique is expected to reduce the duration of thrombolytic therapy compared to continuous infusion.
Ultrasound-assisted CDT	• This mode is nearly identical to the *continuous infusion* method with the exception that it utilizes the EkoSonic® MACH 4e catheter from EKOS® Corp instead of a regular infusion catheter. • This 6 Fr. catheter is inserted like an infusion catheter. • r-tPA is given in 1-2 mg/hr doses along with concurrent UFH. • Parameters for PTT are identical. • The ultrasound waves transmitted through the catheter increase the penetration of r-tPA and reduce the time and dosage of thrombolytic therapy.

TABLE 4 - CDT MODES WITH ADJUNCTIVE THROMBECTOMY	
CDT with aspiration thrombectomy	• This mode works best for occlusions related to a smaller thrombus burden. • Aspiration thrombectomy is performed using one of the several commercially available catheters. • CDT is performed thereafter in a standard fashion. • 4 and 5 Fr./0.035" wire compatible; 45 cm/90 cm/ 135 cm length.
Trellis® assisted CDT	• This device performs mechanical thrombectomy in a 'contained' segment of the thrombosed vessel. • The catheter is inserted over a wire which is removed once the device is in position. • The thrombolytic agent is then given after inflation of proximal and distal balloons. • The sinusoidal wave inside the device works to break the thrombus, leading to an increase in surface area for the thrombolytic agent to act.
Power-pulse spray	• This technique requires the use of the Angiojet device. • The Angiojet catheter is prepped in a standard fashion. • The device is then switched to the 'Power-pulse' mode. • The saline infusion bag on the device console is replaced with a 250 mL bag of normal saline with 25 mg of r-tPA mixed in it. • The catheter is inserted in the thrombosed segment of the vessel and, while stepping on the pedal, the catheter is advanced at 1 mm/sec. • One to two runs of antegrade and retrograde passes are made with the catheter. • A total dose of 20-25 mg of r-tPA is used. • The catheter is then removed. • The patient is monitored for 20-25 minutes to allow the thrombus to be 'marinated' with r-tPA. • The device infusion bag is switched back to normal saline. • The catheter is re-flushed and the device reintroduced. • 'Thrombectomy' mode is activated and several passes are made through the occluded segment. • Repeat angiography and definitive intervention is performed.

formed to rule out an intracranial hemorrhage (<4% risk). If any neurologic symptoms are noted, the infusion should be stopped immediately and a head CT scan should be obtained urgently to rule out intracranial hemorrhage.

- Labs are checked periodically (once every 6 hours). These include complete blood count (CBC), coagulation profiles (PT/INR/PTT), fibrinogen levels (traditionally monitored to indicate enhanced fibrinolysis. The infusion must be stopped if the levels are < 100 mg/dL due to an increased risk of bleeding).
- The various CDT catheters and other equipment required for the procedure are listed in tables 6 & 7.

TABLE 5 -EVALUATION & MONITORING DURING CDT	
Baseline blood-work	• Type and screen • Complete blood count • PT/PTT/INR • Fibrinogen level • CK/CKMB • Chem 8 • Two peripheral IVs 18 G (saline locks)
Vital signs	• Q 15 min. x 1 hr.; Q 30 min. x 2 hr.; Q 1 hr. x 24 hr.
Nursing	• Bed rest x 24 hr. • IV fluid 50-100 cc/ hr. • Foley catheter to gravity. • Keep SBP 100-160 mm Hg; keep DBP < 85 mm Hg. • Keep O2 sat. > 92%.
Bleeding	• Check access site once every hour. • Neurologic checks to rule out intracranial bleeding.
Avoid	• New venipuncture x 12 hr. • Avoid IM injections. • Avoid placement of nasogastric tube x 24 hr. • No warfarin x 12 hr.
Cardiac	• Continuous telemetry. • BP assessment every 15 min. • O2 saturation monitoring. • 12-lead EKG once/24 hr.
Follow-up blood work (once every 6 hours)	• Complete blood count. • PT/PTT/INR. • Fibrinogen level. • CK/CKMB. • Chem 8.

TABLE 6 - THROMBOLYTIC INFUSION CATHETERS	
Fountain catheter	• Merit Medical. • 4 and 5 Fr./45 cm/90 cm/135 cm catheter length. • 0.035" wire compatible. • 5 cm/10 cm/20 cm/30 cm/40 cm/50 cm treatment lengths.
Pulse*Spray® Infusion system	• AngioDynamics. • 3-5 Fr./0.035" wire compatible; 45 cm/90 cm/135 cm length. • 2 cm/5 cm/10 cm/20 cm 30 cm/ 40 cm/50 cm treatment lengths.
SpeedLyser® catheter	• AngioDynamics • 4 and 5 Fr./0.035" wire compatible; 45 cm/90 cm/ 135 cm length.
Uni*fuse™ catheter	• AngioDynamics. • 4 and 5 Fr./0.035" wire compatible; 45 cm/90 cm/ 135 cm length.
EkoSonic® MACH 4e Endovascular System	• EKOS® Corp. • 5.4 Fr./0.035" wire compatible;106 cm/135 cm catheter lengths. • 6 cm/12 cm/18 cm/24 cm/ 30 cm/40 cm/50 cm treatment lengths for 106 cm catheter. • 12 cm/30 cm/40 cm/50 cm treatment lengths for 135 cm catheter.

TABLE 7 - EQUIPMENT FOR CDT	
Sheaths	• 5 Fr. /23 cm • 5/6 Fr. /45 cm Ansel 1
Wires	• 0.035" 260 cm Angled Stiff Glidewire
Flush Catheter	• TR
Selective Catheters	• IM
Infusion Catheters	• See table 7
Medications	• r-tPA • RPA • TNK • UK • SK • UFH

ACUTE LIMB ISCHEMIA

Acute limb ischemia (ALI) typically results from a thrombotic occlusion of a lower extremity artery. This may occur in the native, previously untreated vessel due to an embolus or due to thrombosis of a previously treated segment from graft/stent thrombosis. Regardless of the etiology, the result is severe reduction in the perfusion to the lower extremity such that the viability of at least a portion of the lower extremity is threatened.

PATIENT SELECTION

For limb salvage it is imperative that a prompt diagnosis is made and followed by timely restoration of perfusion. In the absence of any contraindications, local delivery of thrombolytic agents works remarkably to 'dissolve' the thrombus and to restore downstream flow.

Facors outlined in table-8 are considered when evaluating patiehts with ALI and determining suitability for CDT.

EQUIPMENT

- Micropuncture kit.
- 0.035" 260 cm angled stiff Glidewire (ASGW).
- 5/6 Fr. sheath.
- TR catheter.
- IM/cross-over catheter.
- 4 Fr./65 cm Glidecath.
- Thrombolytic agent.
- Infusion catheters (table 7).
- Embolic protection device:
  - ▹ Spider® FX
  - ▹ Filterwire® EZ
- Mechanical thrombectomy devices:
  - ▹ Angiojet®
  - ▹ Trellis®

PROCEDURE OUTLINE

- Abdominal aortography.
- Selective angiography.
- Insertion of a cross-over sheath (for contralateral distal iliac/ femoro-popliteal occlusion).
- Predilatation with a small low-profile balloon.
- Infusion catheter placement.
- Bolus thrombolytic agent administration.
- Continuous infusion of the thrombolytic agent.
- Repeat selective angiography after 12 hrs.
- Adjunctive mechanical thrombectomy (dependent on thrombus burden) if necessary.
- Definitive intervention of underlying obstructive disease.

TABLE 8 -ACUTE LIMB ISCHEMIA CLASSIFICATION & TREATMENT STRATEGIES

STAGE	PROGNOSIS	SENSORY LOSS	MOTOR WEAKNESS	ARTERIAL DOPPLER SIGNAL	VENOUS DOPPLER SIGNAL	TREATMENT RECOMMENDATION
I	Viable unthreatened limb	None	None	Audible	Audible	Non-invasive imaging Angiography
IIa	Threatened	Minor	None	Inaudible	Audible	Angiography
IIb	Immediately Threatened	Beyond toes; Rest pain	Mild to moderate	Inaudible	Audible	Emergent surgical or hybrid repair
III	Non-viable	Anesthetic	Paralysis	Inaudible	Inaudible	Primary amputation

TECHNIQUE

Arterial Access:

- Employing ultrasound guidance using a 4 Fr. micropuncture kit, arterial access is obtained in the contralateral CFA.
- A 5 Fr./23 cm straight sheath is inserted over an ASGW.

Abdominal Aortography:

- A 4 Fr./65 cm tennis racquet (TR) catheter is advanced over the 0.035"ASGW and placed at L1 - L2 space.
- Abdominal aortography is performed using digital subtraction at 10 - 15 cc/sec for total of 20 - 40 cc at 500 - 750 psi.

Selective Angiography:

- This step is not relevant for proximal iliac artery occlusion but more so for the distal iliac and femoropopliteal occlusions.
- After aortography, the TR catheter is exchanged out for a 5 Fr./65 cm IM catheter (or another appropriate catheter based on the configuration of the aortic bifurcation).
- The common iliac artery (CIA) ostium is engaged with the IM catheter and the ASGW is advanced into the PFA.
- Once the wire is anchored well, the IM catheter is exchanged for a 4 Fr./65 cm Glidecath.
- This catheter is advanced to the distal EIA.
- Selective angiography of the symptomatic limb is then performed through this catheter.

Placement of Cross-over Sheath:

This step is not relevant for aortic/proximal iliac artery occlusion but for the distal iliac and femoropopliteal occlusions.

- The ASGW is re-inserted into the 4 Fr./65 cm Glidecath which is then removed.
- The 5 Fr. sheath in the contralateral CFA is exchanged out for a 5/6 Fr. Ansel 1 sheath.
- Under fluoroscopic guidance, this sheath is advanced over the aortic bifurcation and the tip is parked into the distal EIA/CFA.

Predilatation:

- Predilatation may not always be necessary. It is used only if the thrombus is somewhat subacute. The goal is to 'crack' the thrombus so that the surface area of the clot is increased to improve the action of the thrombolytic agent.
- The ASGW is directed into and across the segment with thrombosis.
- A low profile, undersized predilatation balloon (e.g. 3-4 mm diameter for ileo-femoral arteries) is advanced into the occluded segment and gentle PTA (2-4 ATM) is performed.
- The balloon is deflated and removed.

For aortic and proximal iliac cases a straight sheath is used.

Placement of the Infusion Catheter:
- An infusion catheter of choice is then advanced across the lesion, so that the tip is barely beyond the distal cap of the occlusion.
- The wire is removed and the catheter flushed.
- A bolus of r-tPA (5 mg) is then given through this catheter. Subsequently, a continuous infusion through the catheter is initiated at 1-2 mg/hr for a total of 12-24 hrs.
- If r-tPA is being used, it is imperative to give UFH in addition.
- The goal is to achieve a PTT of 45-60 seconds and the dose is typically 300-750 units/hour. A portion of this UFH can be given through the side-arm of the arterial sheath and the rest through a peripheral IV line.

Repeat Angiography:
- The patient returns to the lab at a pre-specified interval (maximum 12-24 hours).
- They are cleaned and prepped in a standard sterile fashion.
- The sheath and the infusion catheter from the initial procedure are cleaned using sterile technique.
- Local anesthetic is infiltrated in the groin around the original access site.
- Over a 0.035"ASGW, the infusion catheter is removed.
- Similarly, the cross-over sheath is exchanged for a new clean sheath. It may be prudent to upsize the sheath by 1 Fr. size to prevent bleeding around the original arteriotomy.
- The sheath is advanced under fluoroscopic guidance over the aortic bifurcation and to the distal EIA.
- A selective angiogram is performed to reassess for residual thrombus.
- If residual thrombus persists, can use adjunctive mechanical thrombectomy. Prior to this an embolic protection device should be used.

Definitive Intervention:
- If CDT was successful in 'cleaning' up the vessel, which it almost always is, an underlying problem in the vessel around/distal to the site of thrombosis is searched for.
- Unless, the underlying issue is resolved, this problem will recur.
- Thus, the treatment of the underlying problem (stenosis) is imperative.

Post-lytic Care:
During and after CDT, the patient needs to be monitored very closely (in an ICU setting) for potential complications. These complications include:
- Bleeding (access site, gastrointestinal and intra-cranial),
- Hyperperfusion and compartment syndrome.

Acute compartment syndrome
This surgical emergency is characterized by high pressure in an 'enclosed' osteo-fascial compartment.

Fascial compartments of the leg:
- There are four fascial compartments in the lower leg (figure 1A). These along with their respective neurovascular bundles are:
  - ▹ Anterior (ATA, ATVs and peroneal nerve).
  - ▹ Lateral (superficial peroneal nerve).
  - ▹ Deep posterior (PTA, PTVs, peroneal artery and veins and tibial

nerve).
- ▹ Superficial posterior (sural nerve and greater & lesser saphenous veins).

Causes:

- Revascularization (esp. surgical or post-lytic after ALI).
- Trauma (hemorrhage).
- Burns.

Clinical features:

- These include severe pain and sensory loss (partial or total).
- The leg usually appears swollen, tense and tender. Distal pulses are reduced.

Diagnosis:

- A compartment pressure (CP) > 30 mm Hg or the difference between the systemic diastolic BP and compartment pressure (DBP- CP) < 30 mm Hg is confirmatory.
- The compartment pressure can be measured using a pressure gauge like the one shown in figure 1B.

Treatment of acute compartment syndrome

- Prompt decompression (within the first few hours) is essential to prevent irreversible muscle and nerve damage.
- Prompt fasciotomy of the lower leg is performed for decompression.

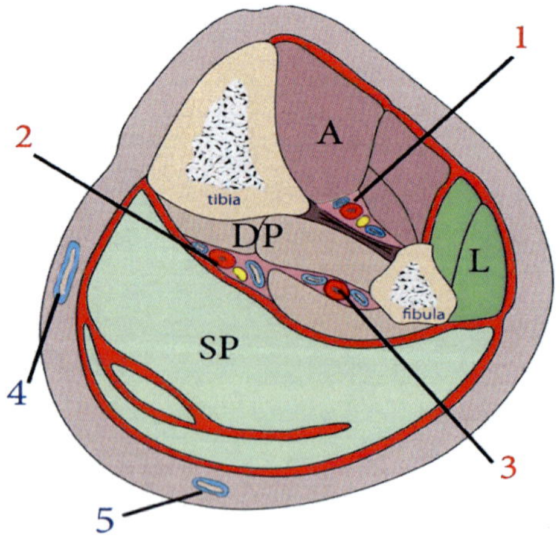

Figure 1A.
Compartments of the (left) leg. A, Anterior. L, Lateral. DP, Deep posterior and SP, superficial posterior.

The interosseus membrane separates the anterior (A) comparment from the deep posterior (DP) compartment.
1. Anterior tibial vessels and peroneal nerve.
2. Posterior tibial vessels and tibial nerve.
3. Peroneal vessels.
4. Greater saphenous vein.
5. Lesser saphenous vein..

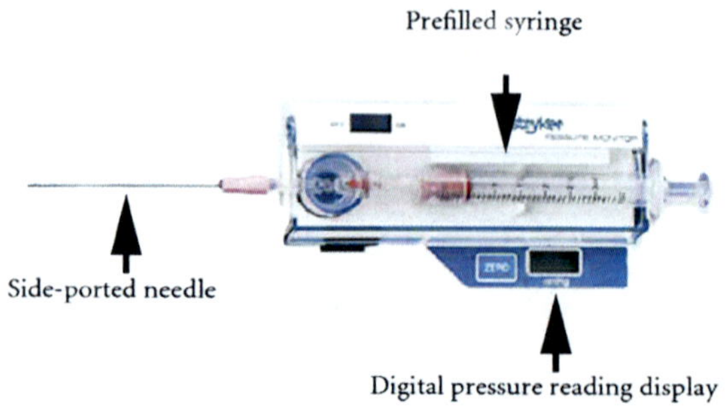

Figure 1B.

Compartment pressure measurement tool.

The compartment pressure can be measured using a hand-held tool such as the gauge shown here from Stryker Corporation, Kalamazoo, MI.

ILEOFEMORAL DEEP VEIN THROMBOSIS

Proximal deep vein thrombosis (DVT) is thrombosis that involves either the popliteal, femoral and/or iliac veins. Up to 90% of acute PEs originate from proximal DVTs and are associated with a higher mortality compared to distal DVT's.

For patients with ileofemoral DVT, treatment options that should be considered are catheter-directed thrombolytic therapy or surgical thrombectomy. Phlegmasia cerulea dolens, the result of extensive proximal venous thrombosis, is the only widely accepted indication for use of thrombolytic therapy in treatment of lower extremity DVTs.

There is also data to suggest that treatment of proximal DVT with thrombolytic therapy decreases the risk of post-thrombotic syndrome. However, systemic thrombolytic therapy is seldom used for lower extremity DVT because of the increased risk of major bleeding. Most patients are unwilling to accept the risks of thrombolytic therapy to prevent post-thrombotic syndrome.

EQUIPMENT

- 4 Fr. micropuncture kit
- 0.035" 260 cm angled stiff Glidewire (ASGW)
- 5/6 Fr. sheath.
- 4 Fr./65 cm Glidecath.
- Thrombolytic agent.
- Infusion catheters (table 7).
- Mechanical thrombectomy devices:
  - Angiojet
  - Trellis

PROCEDURE OUTLINE

- Jugular venous access.
- IVC filter placement.
- Patient in prone position.
- Popliteal vein access.
- Selective venography.
- Infusion catheter.
- Administration of thrombolytic agent.
- Repeat venography.

TECHNIQUE

Venous Access:

- Using ultrasound guidance, obtain access in the right internal jugular vein.

Venography:

- Advance a 0.035" wire (J-tip, ASGW or other wire) into the IVC.
- Insert a flush catheter and perform IVC venography in a standard fashion.
- An optional IVC filter is then deployed in a standard fashion.
- After removal of the delivery system of the filter, the patient is placed in the prone position.
- Prepare the ipsilateral popliteal fossa using sterile technique and access

Phlegmasia alba dolens:

Also known as the 'milk leg' as it was seen in pregnant women in the last trimester.

- It is characterized by painful swelling of the extremity (usually leg) that is associated with blanching.
- It results from massive proximal DVT.
- Only deep veins are thrombosed and usually greater saphenous vein is patent.
- Peripheral pulses are present.
- If not treated in a timely fashion can progress to phlegmasic cerulea dolens ("painful blue leg").

Phlegmasia cerulea dolens:

- This results from progression of phlegmasic alba dolens.
- It is a vascular emergency.
- Massive thrombosis of deep and superficial veins obstructs venous outflow, resulting in a swollen, painful, bluish extremity and threatening venous gangrene.
- May result in compartment syndrome, limb loss and death.

the popliteal vein under ultrasound guidance with a 4 Fr. micropuncture kit.
- Insert an 8 Fr. short sheath over a 0.035"ASGW.
- Perform popliteal venogram.

Intervention:
- Insert the infusion catheter over the 0.035"ASGW through the 8 Fr. sheath.
- Remove the wire and administer a bolus of r-tPA (5 mg) and start a continuous infusion at 1-2 mg/hr.

Repeat Venography:
- Position the patient in the prone position.
- Insert a 0.035"ASGW through the infusion catheter and place it in the SVC.
- Remove the existing popliteal sheath and insert a new 8 Fr. short sheath.
- Perform repeat venography to assess the results of lytic therapy.
- If there is residual thrombus, adjunctive mechanical thrombectomy may be used.

Post-Op Care:
- Patient is admitted to the ICU for close observation.
- Continuous pulse oximetry, ECG and non-invasive blood pressure monitoring is used.
- Frequent neurologic assessments are performed to detect intracranial bleeding secondary to thrombolytic therapy.
- Blood work including CBC, PT/PTT, D-dimer and fibrinogen should be monitored during therapy.

CASE 1

- Mr. C is a 77-year-old male with known PAD. Seven years prior to his presentation he had undergone left femoral-posterior tibial artery bypass grafting with an in situ vein. He had also undergone left iliac stent placement in the past.
- He now presented with severe rest pain in the left foot x 2 days. The pain came on suddenly and had been constant. He was on dual anti-platelet therapy.
- Physical exam was remarkable for non-palpable pedal pulses in both feet. The pulse in the graft was not palpable. On Doppler, a very faint monophasic signal was noted in both the left DP and PT. No Doppler signal was noted in the graft. The right pedal signals were also monophasic but stronger. The motor and sensory function were normal.
- Noninvasive arterial testing revealed a resting ABI of 0.68 in the right leg and 0.20 in the left leg. PVR suggested severe femoro-popliteal and infra-popliteal disease in the left leg. The left big toe waveform was 'flat' (non-pulsatile).
- An arterial duplex showed no flow in the bypass graft.
- A diagnosis of ALI was made, the patient was started on IV UFH and he underwent angiography.

Access & Diagnostic Angiography:

- For initial angiography, we used a 5 Fr./23 cm Cordis Brite-tip sheath.
- ACT was checked and an additional 2000 units IV UFH was given.
- Abdominal aortography was performed.
- This was followed by selective left iliac and infra-inguinal angiography with runoff.

Findings:

- No obstructive disease was noted in the aorto-iliac arteries.
- The left CFA was noted to be disease free.
- The bypass graft was noted to be completely occluded just beyond the proximal anastomosis (figure 2A).
- The left PFA was noted to have the appearence of a recanalized vessel with prior thrombosis.
- The native SFA and popliteal arteries were both noted to be totally occluded throughout their course (figures 2B & 2C).

Interventional Technique:

Exchanging the right CFA sheath

- An ASGW was advanced into the left PFA.
- Over this wire, the selective catheter and right CFA sheath were removed and a 6 Fr./45 cm Ansel 1 sheath was inserted. This sheath was advanced over the aortic bifurcation and into the distal left EIA.
- This was done under fluoroscopic guidance.
- Additional 4000 units of IV UFH was given to achieve a therapeutic ACT of >250 sec.

Cannulation of the SVG bypass graft

- A road-map was obtained.
- A 4 Fr./65 cm angled Glidecath was used to cannulate the ostium of the bypass graft.
- The ASGW was then maneuvered into the graft through this catheter. The wire was advanced without difficulty to the level of the distal anastomosis.
- The angled GC was removed and the straight 100 cm GC was advanced into the graft, over the ASGW.
- The wire was removed and angiography of the distal graft with runoff was performed (figure 2D).

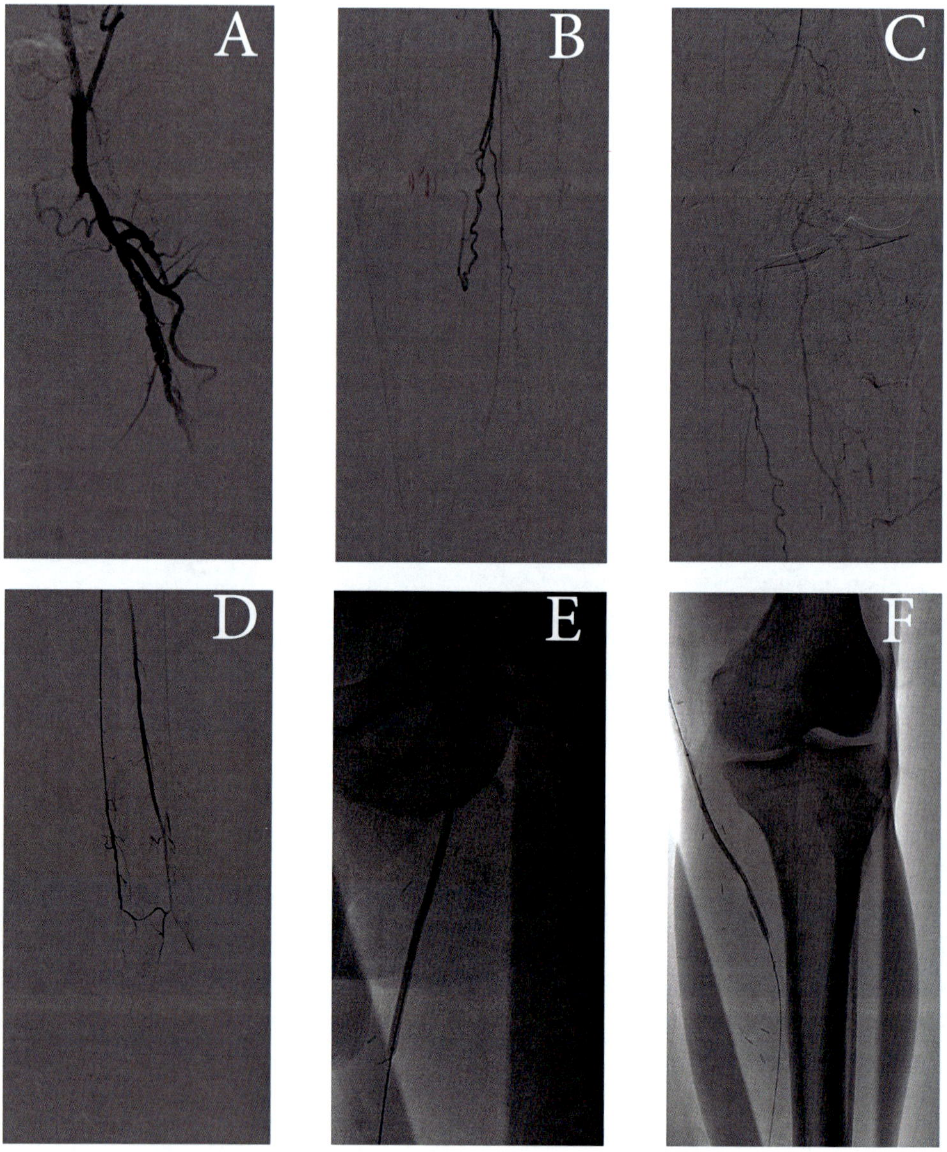

Figure 2.

A. Selective angiogram of the left CFA showing totoal occlusion of the bypass graft at the proximal anastomosis.
B. No flow was seen in the native SFA or ATK popliteal artery.
C. The BTK popliteal artery was also occluded.
D. A selective angiogram through the GC placed in the distal bypass graft.There was diffuse disease in the distal PT and it was noted to be occluded in the distal third of the leg.
E & F. PTA of the bypass was performed to increase the surface area of the thrombus for r-tpA to act up on.

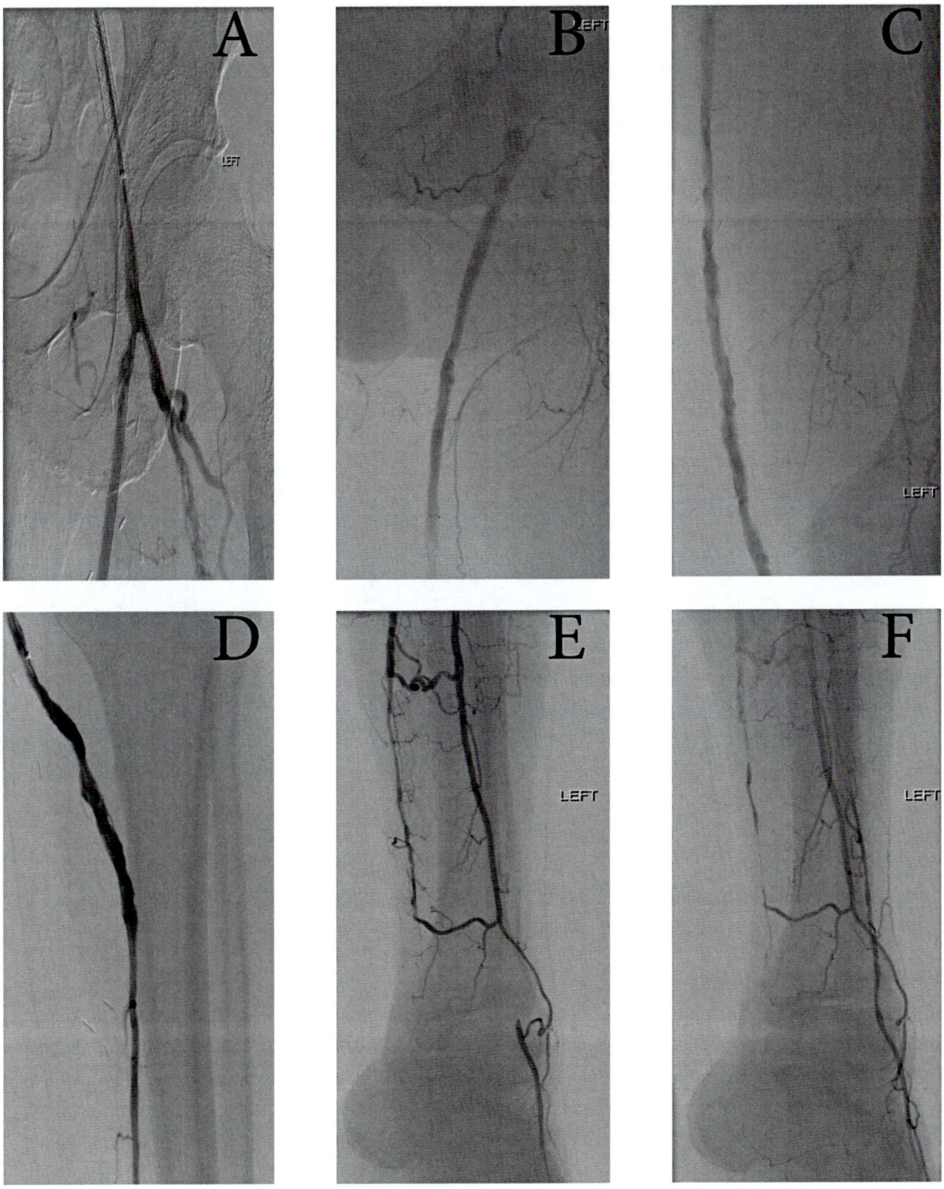

Figure 3.

A. The graft was noted to be patent with normal flow in it.
B & C. Mild residual thrombus was noted throughout the graft.
D. The distal anastomosis was noted to be patent as well.
E. DIffuse disease was noted in the distal PT with collaterals to the peroneal artery. This in turn collateralized the DP.
F. Retrograde flow was noted via the DP into the AT.

PTA of the occluded bypass graft

- The GC was removed and a 4.0 mm x 80 mm balloon was used to performed PTA of the entire length of the bypass (figures 2E & 2F).
- This was followed by removal of the balloon.
- A therapeutic ACT was confirmed.
- It was decided to place an EKOS catheter for ultrasound assisted lysis of the bypass graft.

Placement of an EKOS catheter

- A 5.4 Fr./106 cm long EKOS catheter with a treatment length of 50 cm was used.
- This catheter was placed over the ASGW.
- The distal end of the catheter was placed in the graft close to the distal anastomosis.
- The proximal end was placed just proximal to the proximal anatomosis.
- Once in position, the wire was removed and the catheter was connected to the console in a standard fashion.
- Infusion of r-tPA was started at 1 mg/hour.

Repeat angiography:

- The patient underwent repeat angiography the following monring.
- The EKOS catheter was removed over an ASGW and an angiogram was performed after exchanging the sheath.
- The bypass graft was noted to be patent with only mild residual thrombus (figures 3A-D).
- There was flow in the PT, which collateralized the peroneal artery and then occluded prior to reaching the ankle (figure 3E).
- The peroneal artery collateralized the DP which then provided retrograde flow up the AT (figure 3F).
- Angiography and hemodynamics identified a signficant stenosis at the proximal anastomosis of the graft. This was deemed to be the culprit stenosis. Duplex confirmed the velocity at > 540 cm/sec.
- The patient was kept on IV UFH.
- Three days later the patient underwent surgical revision of the anastomosis with patch angioplasty.

Follow-up:

- At 30-day follow-up, the patient remained asymptomatic.
- His surgical incision site had healed well.
- He had 1+ palpable DP pulse in the left foot.
- His resting ABI and PVRs improved considerably.
- He was maintained on ASA 325 mg/d and Plavix 75 mg/d indefinitely.
- At a recent 9-month follow-up visit he continued to be asymptomatic and had a normal arterial duplex of the bypass graft.

CASE 2

- Mr. R, a 72-year-old male, was evaluated for a non-healing ulcer on his right lower leg. The ulcer was traumatic in nature but did not heal over 6 months. He was an active smoker with a history of worsening intermittent claudication in the right calf. His exam showed diminished popliteal and pedal pulses in the right foot.
- Non-invasive arterial studies showed severe femoro-popliteal occlusive disease in the right leg.
- He underwent angiography via left CFA. This showed severe disease in the right CIA/EIA, a long CTO of his right SFA (figure 4A), reconstituting above the Hunter's canal. His popliteal artery had no significant obstructive disease. He had three vessel runoff.
- He underwent PTA/stenting of his right SFA (figure 4B) with placement of two overlapping covered stents (Viabahn 6 mm x 150 mm, from ostial SFA to the distal SFA).
- He was maintained on dual antiplatelet therapy. At 3 month follow-up, he had stopped smoking, his ulcer had completely healed and he had no residual claudication. A resting ABI was normal at 0.99 in the right leg and 1.00 in the left. An arterial duplex of the right SFA showed patent stents with normal flow velocities.
- Twelve months after the intervention, he presented with new onset numbness in his right foot and severe calf claudication at minimal ambulation (< 30 feet) that started approximately two weeks prior. He was compliant with his dual anti-platelet therapy.
- Upon questioning he admitted to recurrent calf claudication x 2 months.
- Dependant rubor and elevation pallor was noted in the right foot. His right popliteal and pedal pulses were only Dopplerable. No skin breakdown or tissue necrosis was present. Motor function was intact.
- Noninvasive arterial testing revealed a resting ABI of 0.40 in the right leg and 0.99 in the left leg. PVR suggested severe inflow and femoro-popliteal disease in the right leg. The waveform was non-pulsatile at the ankle.
- The patient was scheduled for urgent angiography with a clinical diagnosis of stent thrombosis.

Access & Diagnostic Angiography:

- Using a 4 Fr. micropuncture kit arterial access was obtained in the left CFA.
- For initial angiography, we used a 5 Fr./23 cm Cordis Brite-tip sheath.
- IV UFH 5000 units were given. ACT was kept at > 250 seconds.
- Abdominal aortography was performed.
- This was followed by selective right iliac and infra-inguinal angiography with runoff.

Findings:

- No obstructive disease was noted in the aorto-iliac arteries.
- The right CFA was noted to have mild disease (figure 4C).
- The right PFA was noted to have severe ostial stenosis (figure 4D).
- The right SFA (figure 4D) was noted to be occluded at the ostium (with no flow in the stented segment). Distal reconstitution occurred via PFA collaterals in the Hunter's canal (figure 4E).
- The popliteal artery was patent throughout its course (figures 4E).
- There was three vessel runoff.

Interventional Technique:

We decided to use a Trellis™-6 (Covidien, Mansfield, MA) peripheral infusion system (see chapter 3) for thrombolysis.

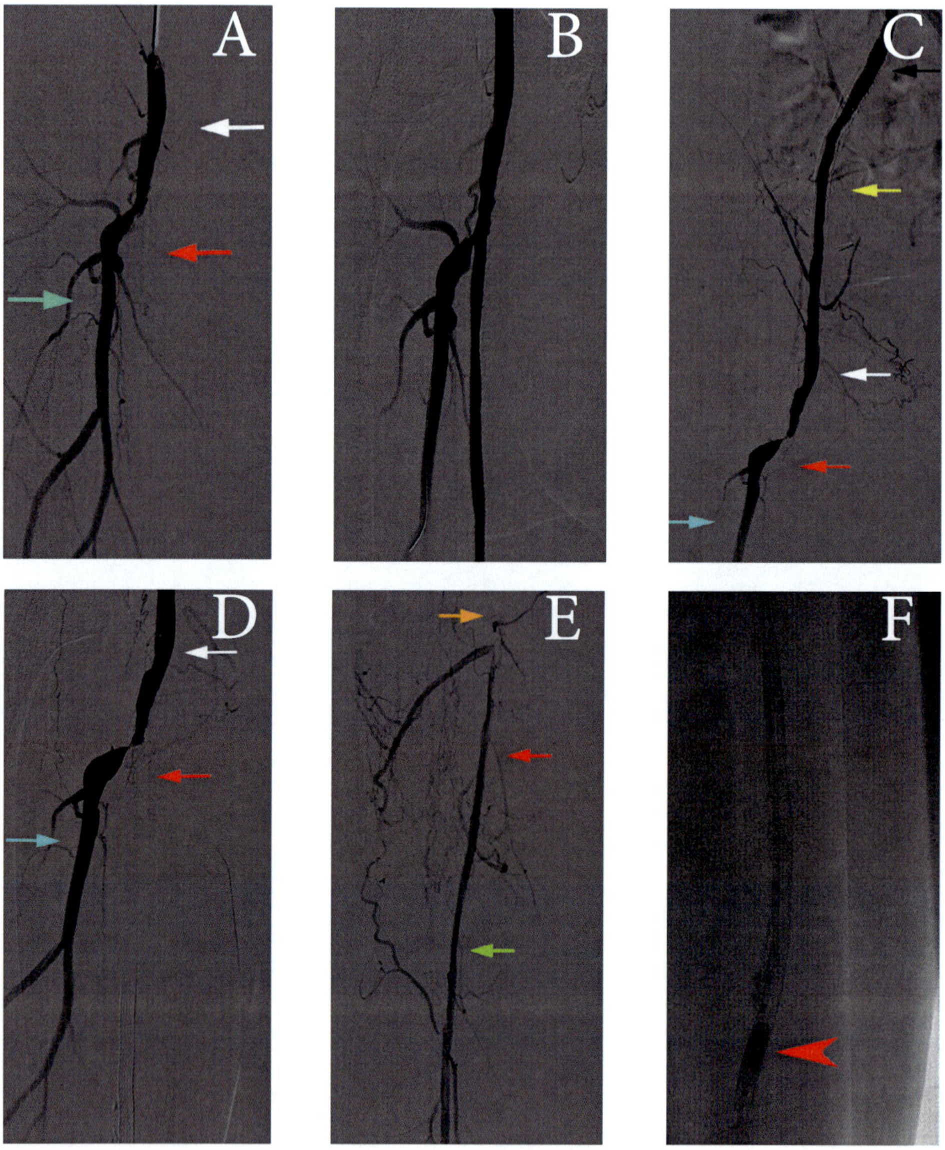

Figure 4.

A. Right CFA angiogram showing original occlusion of the right SFA at the time of the first angiogram 1 year ago.
B. Right CFA angiogram after SFA intervention 1year ago.
C. Right CIA angiogram 12 months after the first intervention (current presentation).
D. An upclose view of the right CFA, PFA and proximal SFA.
E. Distal reconstitution of the right SFA via PFA collaterals.
F. The Trellis- 6 device is seen within the SFA, with the distal balloon (red arrow head) inflated within the distal edge of the SFA stent.

Arrow key: black= CIA; yellow=EIA; white=CFA; blue=PFA; red=SFA; orange=distal SFA occlusion; green=popliteal.

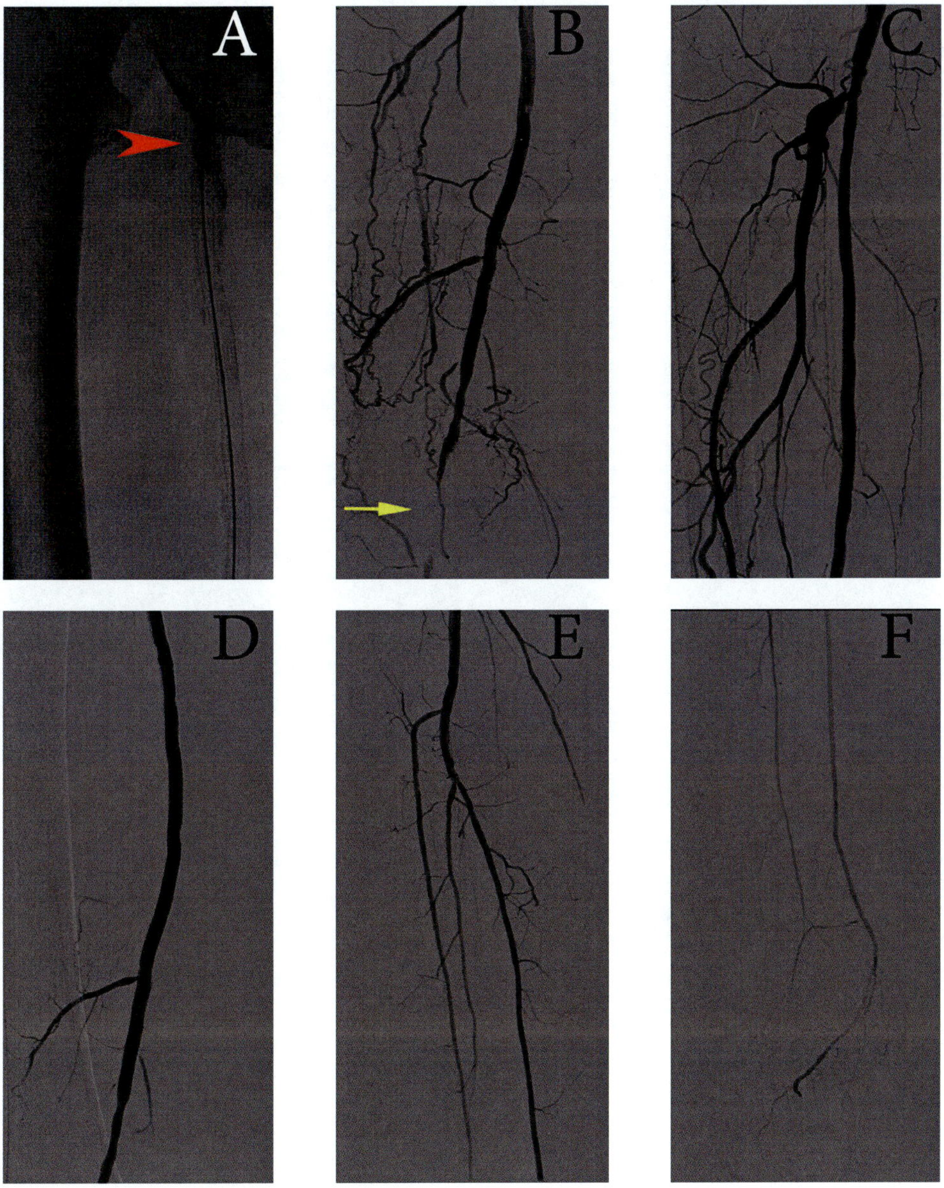

Figure 5.

A. The Trellis with the proximal balloon (red arrow head) inflated within the proximal edge of the SFA stent.
B. A filling defect (yellow arrow) at the site of the Spider, suggesting captured thrombus within the filter.
C. Patent CFA, PFA and proximal/mid SFA.
D. Patent distal SFA. The lesion at the distal edge of the Viabahn has been treated with a Zilver PTX stent.
E. Preserved runoff showing lack of distal embolization due to the Spider.
F. Angiogram of the foot showing a preserved dominant PT.

Exchanging the left CFA sheath

- An ASGW, supported by a 4 Fr. angled Glidecath, was advanced into the right popliteal artery, through the SFA stents without much resistance.
- Over this wire, the selective catheter and CFA sheath were removed and a 6 Fr./45 cm Ansel 1 sheath was inserted. This sheath was advanced over the aortic bifurcation and into the distal right EIA.
- This was done under fluoroscopic guidance.

Placement of Trellis™ catheter

- A Trellis-6/30 cm device was then placed in the right SFA.
- It consists of proximal and distal occlusion balloons which are inflated to isolate the treatment zone.
- An infusion port on the catheter allows administration of drug (tPA) in the treatment zone.
- A total of 10 mg IA tPA was administered through the device.
- Treatment was initiated in a standard fashion (figures 4F and 5A).
- The treatment zone included the entire length of the stented SFA.
- The treatment duration was 15 minutes.

Placement of Spider EPD through the Trellis catheter

- After removal of the dispersion wire, with the balloons still inflated, a 6 mm Spider™ EPD was advanced through the wire port of the catheter in a 'bare back' fashion (figure 5B).
- This was advanced beyond the tip of the catheter into the popliteal artery and deployed in a standard fashion.
- With the filter deployed, the balloons were deflated and the catheter removed.

PTA/stenting of distal SFA

- Angiography at this point showed near total resolution of thrombosis in the SFA stents.
- PTA of the distal edge of the stented segment was performed with a 6.0 mm x 60 mm balloon.
- This was followed by placement of a 6.0 mm x 80 mm drug-eluting Zilver PTX (Cook Medical, Bloomington, IN) at this distal edge.
- The stent was post-dilated with the same balloon.
- The ostial segment of the SFA was treated with a 6.0 mm x 60 mm Absolute stent (Abbott Vascular, Abbott Park, IL).
- The ostial PFA stenosis was treated with PTA using a 5.0 mm balloon.

Repeat angiography:

- Final angiography revealed a widely patent SFA (figures 5C & 5D) with normal flow.
- The ostium of the PFA also appeared improved with normal flow in it.
- Preserved runoff was noted (figures 5E & 5F).

Follow-up:

- At 30-day follow-up, the patient remained asymptomatic.
- He had 2+ palpable DP pulse in the right foot.
- His resting ABI and PVRs normalized.
- He was maintained on ASA 325 mg/d and Plavix 75 mg/d indefinitely.

16

IVC FILTER PLACEMENT & RETRIEVAL

The morbidity and mortality of pulmonary embolism has been recognized for over a century. Initial preventive efforts focused on the surgical plication or ligation of the femoral vein and later the inferior vena cava (IVC). These surgeries, however, were not very well tolerated.

In 1967, Dr. Mobin Uddin, a cardiovascular surgeon, developed a precursor to the modern day IVC filter. The Mobin Uddin Umbrella (Edwards Laboratories, Santa Ana CA), as it was called, was the first percutaneously inserted IVC interruption device, introduced through the right internal jugular vein. A decade later, multiple improvements of the initial concept were developed. Although the original IVC filters were all permanent, in contemporary practice, optional (i.e. retrievable) filters are more popular.

Prior to embarking upon the delineation of the technique for deployment and retrieval of IVC filters it would be fruitful to discuss the anatomy of the IVC.

ANATOMY OF THE IVC

- The IVC starts at the level of L4 - L5, along the vertebral column and to the right of the aorta.
- It is formed by the confluence of the right and left common iliac veins.
- It terminates superiorly, just above the diaphragm, in the right atrium.
- The IVC is divided into three different parts - supra-renal, renal and infra-renal.
- The diameter of the normal infra-renal IVC in an adult is 25-27 mm. If the diameter exceeds 28 mm the IVC is often called 'mega cava'.
    - Most IVC filters are designed for deployment in patients with an IVC diameter < 28 mm. In patients with a 'mega cava', the only filter that is large enough to adequately appose the IVC is the Bird's Nest filter (BNF) (Cook medical, Bloomington, IN). In fact, it can be used in an IVC up to 40 mm in diameter.

VARIANTS

Variations in the anatomy of the IVC are not uncommon and are usually asymptomatic. It is very

> ***Tributaries of the IVC:***
> The IVC receives the following veins:
> - Right gonadal
> - Lumbar
> - Renal
> - Supra-renal
> - Right inferior phrenic
> - Hepatic

important to be familiar with these variants and to be able to recognize them on venography prior to IVC filter placement.

Circumaortic Left Renal Vein:

- The prevalence of this variant is 8-9% (figure 1B).
- In this case, there are two left renal veins and one right.
- The left renal veins have a common pedicle at the hilum of the kidney and thereafter they diverge.
- The superior left renal vein passes anterior to the aorta to join the IVC.
- The inferior left renal vein passes posterior to the aorta to join the IVC
- The separation between the openings of the two left renal veins is usually 1-2 cm.
- It is critical to recognize this variant prior to placement of an IVC filter. Optimal protection can only be achieved if the filter is placed caudal to the inferior left renal vein. Placing the filter in the segment between the two renal veins may allow blood to shunt from the infra-renal cava into the inferior renal vein and via the pedicle to the superior vein and back into the IVC. This circuitous path may thus lead to continued thrombo-embolization to the lungs.

Double IVC:

- The prevalence of duplicated IVC is up to 3% (figure 1C).
- In this condition there are two separate IVCs, one on the right of the aorta and one the left.
- The right IVC is a continuation of the right common iliac vein. The left IVC, likewise, is a continuation of the left common iliac vein.
- The right IVC receives the right renal vein. The left IVC receives the left renal vein, which usually courses anterior to the aorta to join the right IVC.
- The course of the supra-renal IVC is usually normal.

Retro-aortic Left Renal Vein:

- The prevalence of this variant is 2% (figure 1D).
- The IVC is normal in its entire course.
- The left renal vein passes posterior to the aorta to drain into the IVC.

Azygous Continuation of the IVC:

- The prevalence of this variant is 0.6% (figure 1E).
- The IVC is normal in its infra-renal and renal segments.
- The supra-renal IVC does not continue to the liver and does not pass through a caval hiatus in the diaphragm.
- Instead, the supra-renal IVC continues as the azygous vein, behind the diaphragmatic crura, to the right of the aorta. This azygous vein drains normally into the SVC.

Left Sided IVC

- A left sided IVC has a prevalence of up to 0.5% (figure 1F).
- This IVC is also formed by the confluence of the right and left common iliac veins.
- It lies to the left of the abdominal aorta up to the level of the left kidney.
- The left IVC joins the left renal vein, crosses the aorta anteriorly, as the left renal vein normally does, and unites with the right renal vein. Beyond this confluence of the two renal veins, the IVC continues its normal supra-renal course.
- The obvious clinical significance of this anomaly is potential difficulty in placement of an IVC filter via a right femoral or jugular approach, due to the angulated path to the infra-renal cava. Left femoral approach offers a more natural path for filter placement in this case.

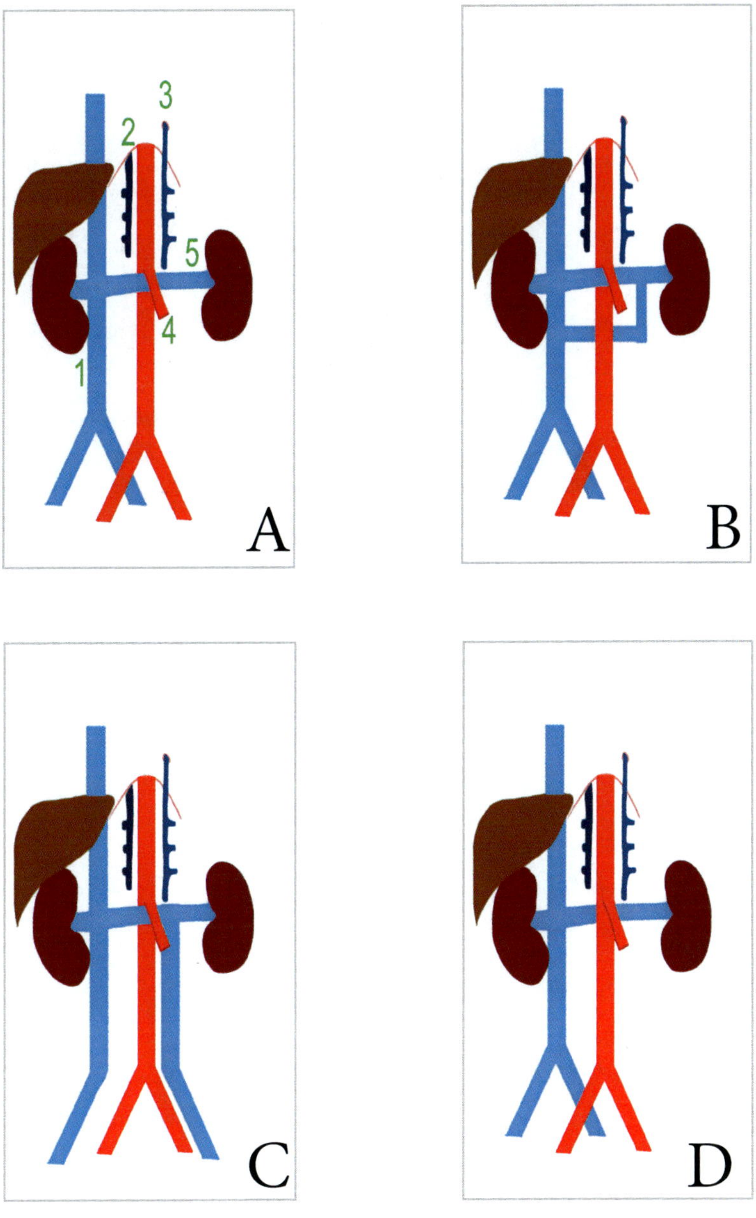

Figure 1.

Anatomical variants of the IVC.
1. IVC; 2. Azygous vein; 3. Hemiazygous vein; 4. SMA; 5. Left renal vein.

A. Normal IVC.
B. Circum-aortic left renal vein.
C. Double IVC.
D. Retro-aortic left renal vein.

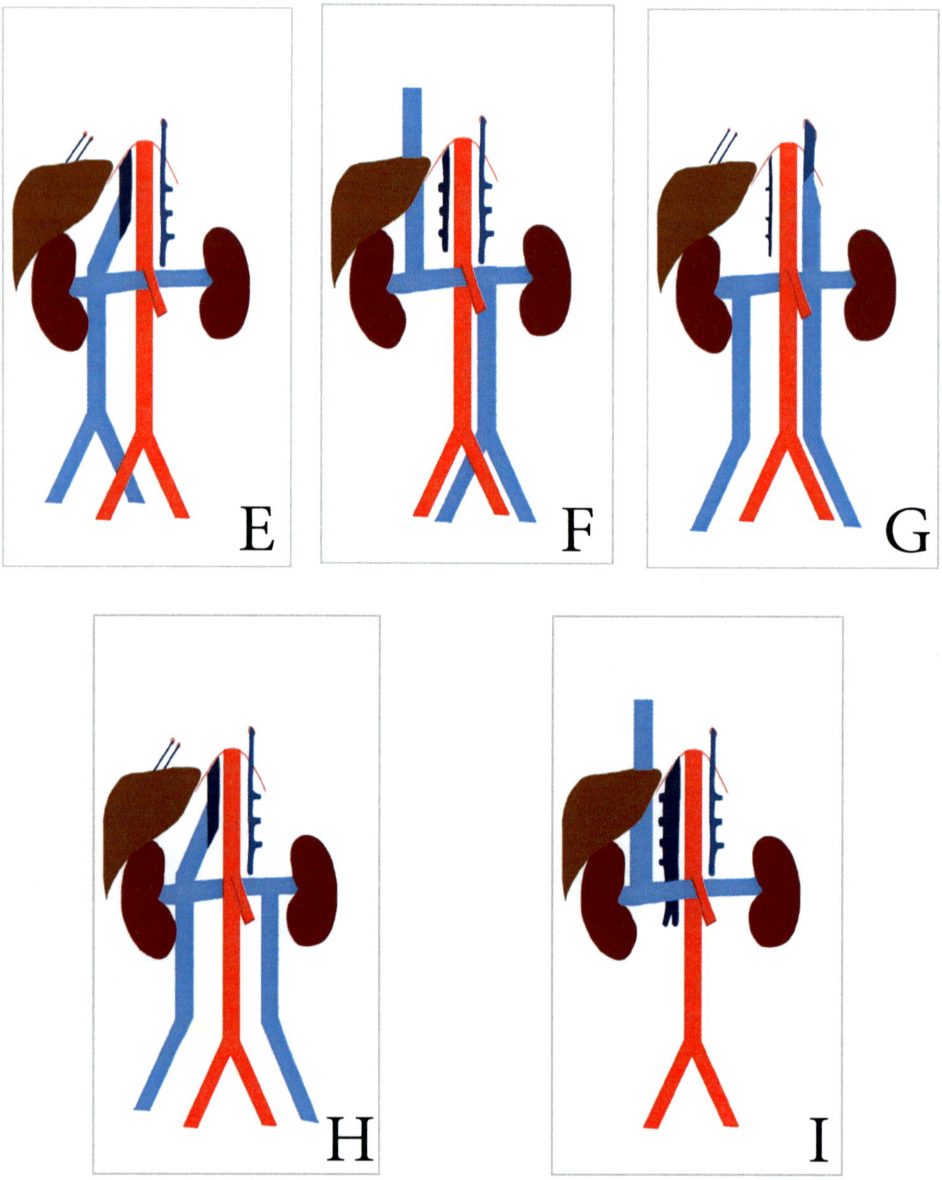

Figure 1 (continued).

Anatomical variants of the IVC (continued).

E. Azygous continuation.
F. Left-sided IVC.
G. Double IVC with retro-aortic right renal vein and hemi-azygous continuation.
H. Double IVC with retro-aortic left renal vein and azygous continuation.
I. Absent infra-renal IVC.

Double IVC with Retro-aortic Right Renal Vein and Hemiazygous Continuation:

- The prevalence of this variant is 0.1% (figure 1G).
- In this case multiple anomalies are present -
  - ▹ Double IVCs,
  - ▹ The right renal vein draining into the left IVC
  - ▹ The absence of the normal supra-renal IVC
  - ▹ The hepatic veins drain directly into the right atrium.
- There are two separate IVCs up to the level of the renal veins (L1-2), one to the right of the aorta and the other to its left.
- The right IVC joins the right renal vein which then crosses the aorta posteriorly to join the left IVC (L1).
- The left IVC then continues as the hemiazygous vein and enters the thorax through the left diaphragmatic crus (T10-11).
- The azygous vein is rudimentary and is joined by the hemi-azygous vein in the normal fashion (T9).
- The azygous vein then drains into the SVC in a normal fashion (T4).
- As the supra-renal IVC is absent, the hepatic veins drain directly into the right atrium.

Double IVC with Retro-aortic Left Renal Vein and Azygous Continuation:

- The prevalance of this variant is 0.1% (figure 1H).
- Again, there are multiple anomalies present.
  - ▹ There are two separate IVCs up to the level of the renal veins (L1-2), one to the right of the aorta and the other to its left.
  - ▹ The left IVC joins the left renal vein which then crosses the aorta posteriorly to join the right IVC (L1).
  - ▹ The right IVC then continues as the azygous vein through the aortic hiatus (T12).
  - ▹ The azygous vein receives the hemiazygous vein (T9) and drains into the SVC normally (T4).
  - ▹ As the supra-renal IVC is absent, the hepatic veins drain directly into the right atrium.

Absent Infra-renal IVC with Preserved Supra-renal IVC:

- There is controversy about the embryologic basis for this anomaly. It is suggested that it results from prenatal IVC thrombosis (figure 1I).
- The lower extremity and pelvic veins drain via ascending lumbar veins into the azygous vein on the right and hemiazygous on the left. These continue cephalad with their normal courses.
- The renal veins merge to form the renal segment of the IVC which then ascends into the supra-renal IVC normally.

THE ANATOMY OF AN IVC FILTER

- IVC filters are made of metal (steel, nitinol, etc).
- The filter design usually has a conical construction with or without non-conical components. The OptEase™ (Cordis, Rahway NJ) for example, has a proximal cone and a distal inverted cone, connected through six vertical struts. The Bird's Nest™ filter (Cook Medical, Bloomington IN), an exception, is a 'nest' of wire.
- The tip of the cone is usually designed to point cephalad.
- The cone is made of legs/struts with fixation barbs on them.
- These barbs may be present in the cranial, caudal or both directions.
- Optional filters have a hook on the cranial (e.g. Celect™, Cook Medical, Bloomington IN) or caudal (e.g. OptEase™) end.
- This hook is used to grasp the filter with a snare at the time of retrieval.

INDICATIONS FOR FILTER PLACEMENT

The most common indications for IVC filter placement include:

- Acute DVT/PE with contraindication to anticoagulation.
- Recurrent PE in the setting of therapeutic anticoagulation.
- Extensive ileo-femoral DVT with pre-existing pulmonary embolism.
- Patients with acute PE and limited cardiopulmonary reserve.
- Pre-operatively for gastric bypass surgery.
- Multi-trauma patients.

ACCESS

SITE

- IVC filters are designed to be placed in the infra-renal IVC.
- Right common femoral vein (CFV) is the most logical choice for access site as it offers the shortest 'straight-line' path to the IVC. However, certain factors may preclude this choice:
  - ▹ CFV or iliac thrombosis.
  - ▹ Low IVC thrombosis.
  - ▹ Left sided IVC.
- In patients with lower extremity DVT, it is advisable to perform a common femoral vein (CFV) duplex US (intra-operatively or at least the same day) to assess for the presence of thrombus in the proposed access site.
- Most IVC filter delivery systems have a shaft length of only 40-80 cm. These are, therefore, suitable for delivery only via femoral or internal jugular (IJ) access.
- However, if antecubital (e.g. basilic vein) or subclavian vein access is necessary, filters with longer delivery systems are available (e.g. OptEase® with 90 cm delivery system).

SHEATH SIZE

- We use a micro-puncture kit for access.
- Thereafter, a 0.035" 150 cm J-wire is used to remove the micro-catheter from the right CFV.
- The delivery sheath for the particular filter is introduced over this wire.
- Under fluoroscopic guidance, this sheath is advanced to the level of L5-S1.
- The wire is removed, the dilator flushed and a small dye injection is used to make sure the sheath tip is positioned below the confluence of the common iliac veins and that there is no proximal thrombus.
- For jugular access, we universally use US guidance. This method lowers the likelihood of inadvertent arterial sticks and other complications.

DIAGNOSTIC VENOGRAPHY

VENACAVOGRAM

- The IVC filter is ideally deployed in the infra-renal vena cava. This segment lies between the confluence of the common iliac veins caudally, and the renal veins cranially.
- The goal of the cavogram is to determine caval anatomy and to identify these veins as angiographic landmarks.
- This step also allows for measurement of the IVC diameter.

CATHETER

- The cavogram can be performed either through the delivery system (sheath + dilator) of the

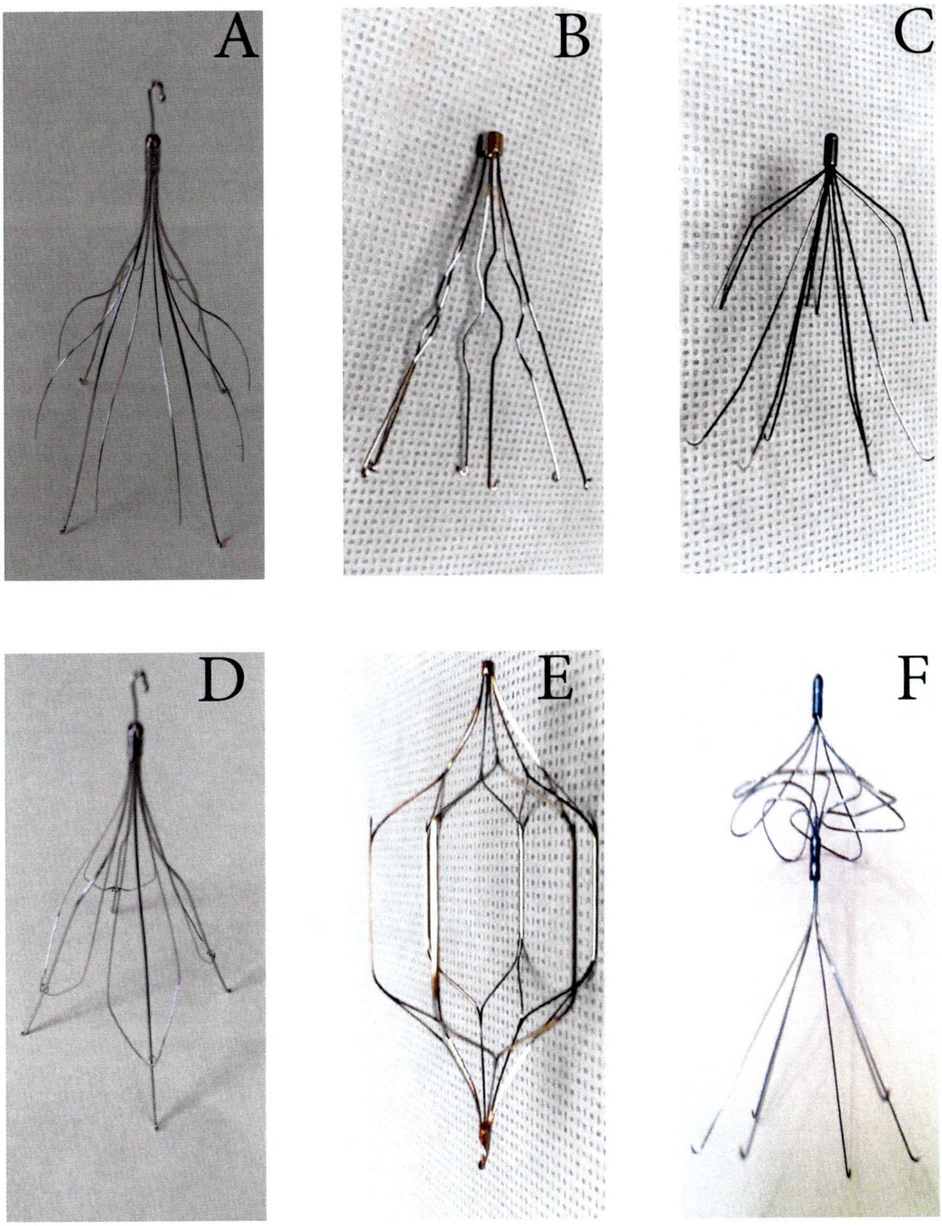

Figure 2.

Examples of IVC filters.

A. Celect.
B. Greenfield.
C. Bard G2.
D. Tulip.
E. OptEase.
F. Simon Nitinol.

particular IVC filter or by using a flush catheter such as the Omniflush®catheter (Angiodynamics, Latham, NY).

- Ideally, a 5 Fr. flush catheter is used.
- Use of a marker catheter or measuring tape is recommended for calibration purposes. This allows for measurement of the diameter of the IVC.
- Many filter delivery kits now come with a dilator that has two radio-opaque bands 3 cm apart (end to end). These allow for easy reference for IVC diameter measurement.

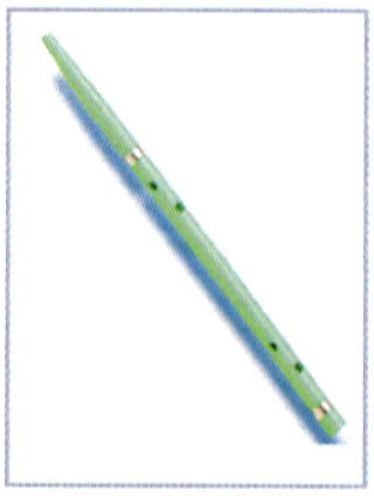

Figure 3.

Dilator for the OptEase™ filter delivery sheath.

The two bands are 3 cm apart, outer end to outer end and can be used to measure the diameter of the IVC.

CATHETER POSITIONING

- The catheter is placed in the distal right common iliac vein or 'hooked' in the left common iliac vein. This allows for the ability to visualize a left or duplicate IVC if present.

DYE INJECTION AND IMAGING

- An AP projection is used.
- We use DSA for optimal imaging.
- The patient is instructed to take in a deep breath and bear down (perform a Valsalva maneuver) at the time of imaging. This distends the IVC and allows for more accurate diameter measurement.
- A power injector is used. A total of 40 ml of 70:30 diluted contrast (use 50:50 dilution in patients with renal dysfunction) is injected at 20 ml/sec at 500-900 psi.

ANATOMICAL LANDMARKS

- Once a cavogram is obtained, identify the common iliac vein confluence and bilateral renal vein origins.
  - ▷ The contralateral iliac vein and the renal veins are recognized by inflow of the unopacified blood.
  - ▷ The left renal vein is usually lower than the right.
  - ▷ Determine if two renal veins are present on the left instead of one. This suggests the presence of a circumaortic left renal vein and requires the IVC filter be placed caudal to the lower left renal vein.
  - ▷ If the renal veins cannot be identified with a cavogram, selective cannulation with a Cobra 2 (C2) catheter should be considered.
  - ▷ If the left common iliac vein is not identified, a duplicate IVC should be ruled out.
- Measure the diameter of the infra-renal IVC. Ideally this should be < 28 mm. Most IVC filters are designed for an IVC diameter < 28 mm.
- If the IVC diameter is > 28 mm (a condition called 'mega cava'), then only a Bird's Nest filter (BNF) may be used, as previously mentioned.
- The other filters are designed to fit an IVC < 28 mm in diameter and, therefore, would be too small and not well apposed to the wall. This would predispose the filter to "tilting" or cephalad migration.

Figure 4.

Venacavogram.

Yellow arrow, right suprarenal vein; blue arrow, right renal vein; red arrow, right common iliac vein; black arrow, left renal vein; green arrow, left common iliac vein. The blue oval indicates region of the IVC between the confluence of the common iliac veins and the lowest renal vein. This is the region where an IVC filter is placed.

TECHNIQUE OF FILTER DEPLOYMENT

FILTER POSITIONING & DEPLOYMENT

- Each filter has its own deployment mechanism which dictates a specific technique. However, certain common principles apply.
- Once a satisfactory cavogram has been performed, the origin of the lowest renal vein and the confluence of the iliac veins should be marked on the screen.
- After the cavogram, the flush catheter is removed and the wire and dilator are reinserted into the sheath.
- The sheath is advanced up to the anticipated location for filter deployment.
- The filter delivery system is loaded into the sheath. It is critical at this point to make sure that the filter orientation matches the access site being used (femoral vs. jugular).
- The filter is advanced until the tip of the filter matches that of the sheath.
- The tip (cephalad end) of the filter should be placed just below the level of the origin of the lowest renal vein.
- The feet (caudad end) of the filter should be placed above the confluence of the iliac veins.
- The sheath is then withdrawn using a simple 'pin and pull' technique.
- Once the filter has been successfully deployed, perform another cavogram. Also, document a cineangiographic 'snap-shot' of the filter in place (without DSA) for future reference.

Positioning of the IVC filter:
The IVC filter is positioned caudal to the lowest renal vein. This prevents:
- Thrombosis of the renal vein, in case of filter thrombosis
- Placement of the tip or cephalad end of the filter in the renal vein which may lead to tilting of the filter and risk migration.

OPTEASE™ PLACEMENT

The OptEase™ is an optional filter.

FILTER DESIGN & MATERIAL

- The OptEase™ filter is made of nickel titanium alloy (nitinol).
- It has a proximal and distal basket that are connected by six vertical struts.
- Fixation barbs are present only at the cranial end.
- A centrally placed hook is present at the caudal end of the filter. This is used for snaring during retrieval.

FILTER KIT

The filter kit (figure 5) consists of:

- An OptEase™ filter inside the filter introducer cartridge with labels for femoral ('FEMORAL') insertion on one side and jugular ('JUGULAR') insertion on the other side (figure 5 B & C). For femoral insertion, the hook is at the proximal end of the introducer, for jugular insertion the cartridge is flipped so that the hook is at the distal end. The cartridge is introduced into the sheath with the arrows pointing towards the valve.
- A 6 Fr./55 cm Cordis Britetip® delivery sheath for femoral/jugular route (for antecubital routes this sheath is 90 cm in length).
- A 66 cm long dilator for the sheath with side holes and 2 radio-opaque markers at a distance of 30 mm from each other (end to end).
- A 68 cm filter pusher (a white, solid obturator). It has a non-hollow tip, a black marker closer to the hub and a small round bump proximal to the marker.

TECHNIQUE

OptEase™ Orientation:
The hook should ALWAYS be at the caudal (foot) end of the delivery sheath so that after deployment it points towards the iliac veins and not the renal veins. Otherwise, the sheath and filter should be removed together and discarded. The procedure is resumed then with a new filter (figures 5 B&C).

Cavogram:

- Obtain venous access in the right CFV.
- Introduce a 0.035" J-tip wire (150 cm) into the suprarenal IVC.
- Introduce the filter delivery sheath (6 Fr./55 cm Britetip™)into the proximal IVC.
- Remove the wire.
- A cavogram can be performed through the dilator of the sheath.
- Alternatively, perform a venacavogram using the technique outlined above and confirm normal caval anatomy.
- Identify the renal veins and confluence of the common iliac veins. Mark these on the screen.

Sheath Advancement:

- Reintroduce the wire.
- Advance the sheath over the wire so that the radio-opaque tip sits caudal to the lowest renal vein.
- Remove the wire (and dilator, if not already removed) from the sheath.
- Purge any air from the sheath and flush it.

Filter Preparation:

- Flush the filter cartridge with (heparinized) saline.
- Orient the filter cartridge appropriate to the selected access site (figure 5 B & C) 'FEMORAL' vs. 'JUGULAR'.
- Following the arrows, insert the filter cartridge into the valve of the sheath. Stabilize the cartridge and using the filter pusher advance the filter out of the cartridge into the sheath.
- Once the filter is inside the sheath, the cartridge can be pulled back on the pusher, proximal to the bump.

Filter Positioning:

- Without moving the sheath, advance the filter in the sheath to the point where the black mark on the pusher is at the valve. This indicates that the filter tip is just inside the radio-opaque marker on the sheath (caudal to the lowest renal vein). The filter is still entirely within the sheath and its position can be adjusted if necessary.
- Confirm the hook is at the caudal end of the filter.
- The undeployed OptEase™ is 67 mm in length and, depending on diameter of the IVC, may shrink to 54 mm.
- To position accurately, use the center of the undeployed filter as a reference. At deployment, the position of the center does not change, the length shrinks on either end.

Filter Deployment:

- Hold the pusher stationary and under fluoroscopic guidance withdraw the delivery sheath on the pusher.
- This maneuver exposes the distal anchoring hooks of the filter and deploys it fully.

Final Cavogram:

- Perform a final cavogram to confirm satisfactory filter placement.

TABLE 1 - EQUIPMENT FOR IVC FILTER PLACEMENT	
Sheaths	• 6-11 Fr.
Wires	• 0.035" 150 cm
Flush Catheters	• 5 Fr. Omniflush®

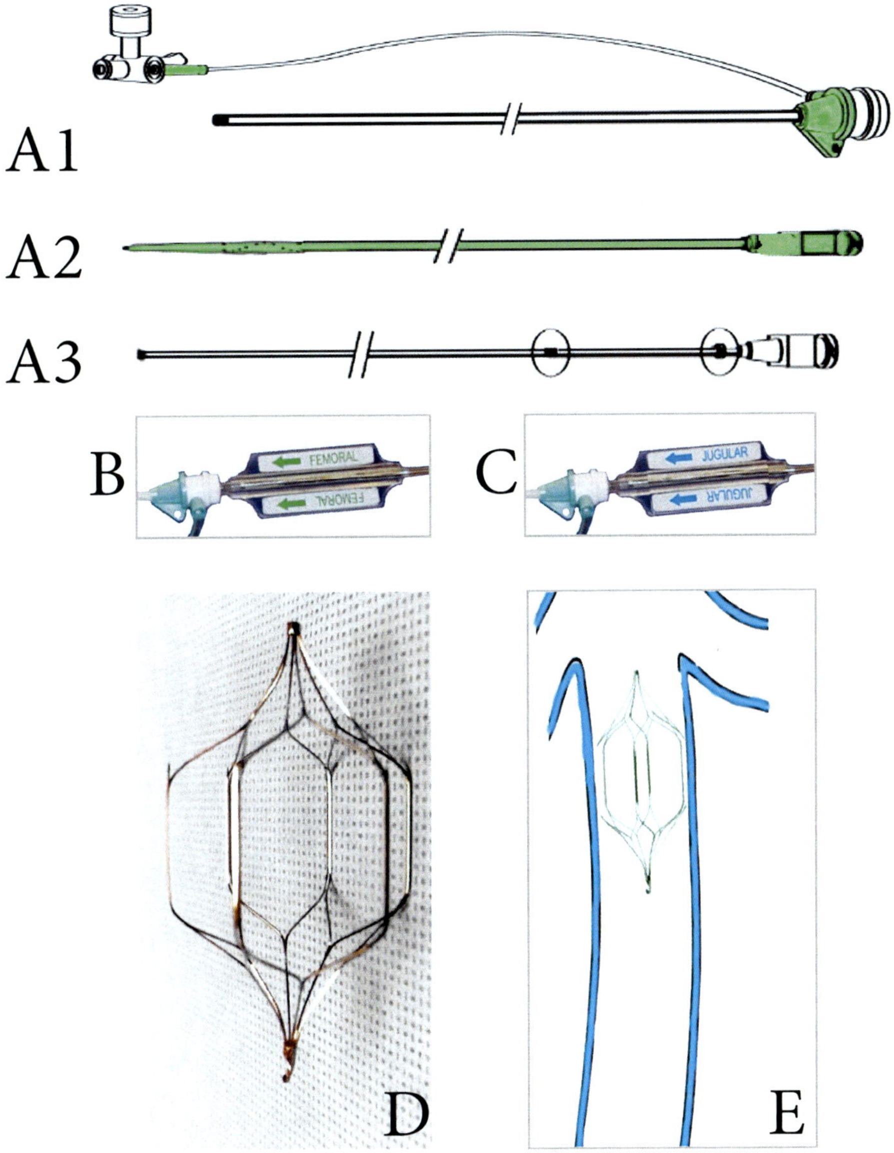

Figure 5.
The OptEase™ IVC filter.

A. The OptEase™ delivery system consists of a 6 Fr. Brite tip sheath (A1), a dilator (A2) and a filter pusher (A3).
B. The filter itself is in a cartridge with arrows on one side indicating the direction for femoral insertion.
C. And on the other side for jugular insertion.
D. An OptEase™ filter has a cranial and a caudal cone connected by six vertical struts. The barbs are present on the cranial cone.
E. A schematic of a deployed OptEase™ caudal to the renal veins.

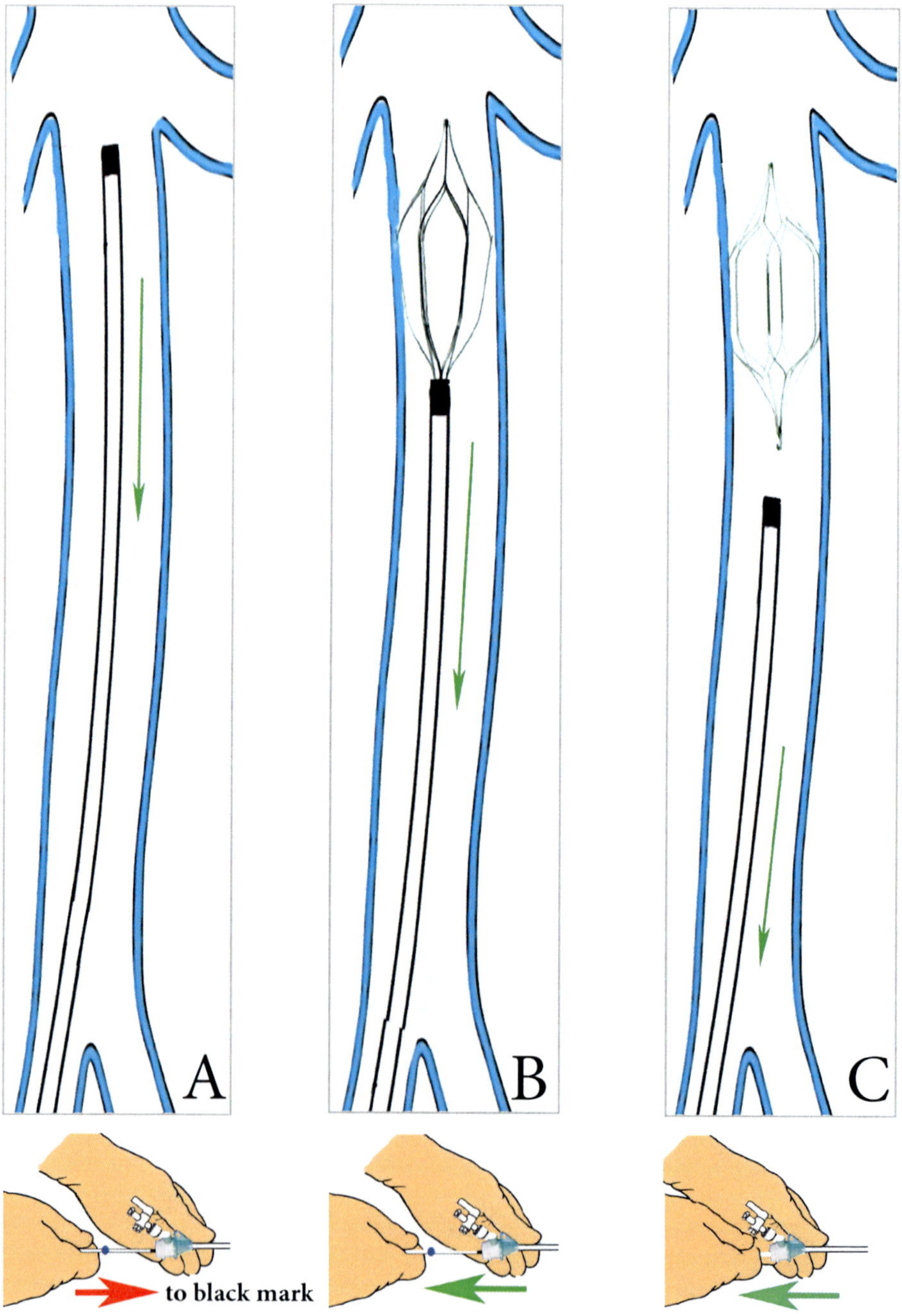

Figure 6.
Steps for deployment of an OptEase™ filter.

A. Advance the pusher to the black mark. Confirm accurate positioning of the sheath tip in the IVC.
B. FIx the pusher and withdraw the sheath.
C. Continued withdrawal results in completion of the deployment.

CELECT™ FEMORAL PLACEMENT

The Celect is an optional filter and may be delivered via femoral or jugular routes, labeled 'FEM' or 'JUG' respectively.

FILTER DESIGN & MATERIAL

- The Celect™ filter is made of Conichrome alloy (cobalt, nickel and chromium).
- It has four primary legs and eight secondary legs.
- Anchors, rather than fixation barbs are present only at the caudal end of the primary legs.
- A centrally placed hook is present at the cranial end of the filter. This is used for snaring during retrieval.

FILTER KIT

The filter kit (figure 7) consists of:

- A pre-loaded Celect filter with labels for femoral ('FEM') insertion.
- The hook is always on the cephalad end of the filter.
- A 10 Fr./20 cm radio-opaque hydrophilic dilator.
- A NavAlign™ 7 Fr. /65 cm delivery sheath.
- A measuring dilator for the sheath with eight side holes and 2 radio-opaque markers at a distance of 30 mm from each other (end to end).

TECHNIQUE

Cavogram:

- Obtain venous access in the right CFV.
- Introduce a 0.035" J-tip wire (150 cm) into the suprarenal IVC.
- Introduce the filter delivery sheath (7 Fr./65 cm) into the proximal IVC.
- Remove the wire.
- A cavogram can be performed through the dilator of the sheath (maximum flow rate is 20 mL/sec at 1000 psi).
- Alternatively, perform a venacavogram using the technique outlined above and confirm normal caval anatomy.
- Identify the renal veins and confluence of the common iliac veins. Mark these on the screen.

Sheath Advancement:

- Reintroduce the wire.
- Advance the sheath over the wire so that the radio-opaque tip sits caudal to the lowest renal vein.
- Remove the wire (and dilator, if not already removed) from the sheath.
- Purge any air from the sheath and flush it.

Filter Advancement:

- Introduce the pre-mounted filter on the delivery catheter into the hub of the sheath and advance into the sheath.
- Advance until a tactile bump is felt (when the circular 'bump' marker reaches the valve of the sheath) (figure 8A).
- At this point, the entire filter, including the hook, is inside the sheath.

Filter Positioning:

- Stabilize the delivery catheter, withdraw the sheath past the bump, until the hub of the sheath connects to the handle of the delivery catheter (figure 8B).
- This exposes most of the filter, however, the primary legs are still constrained and attached to

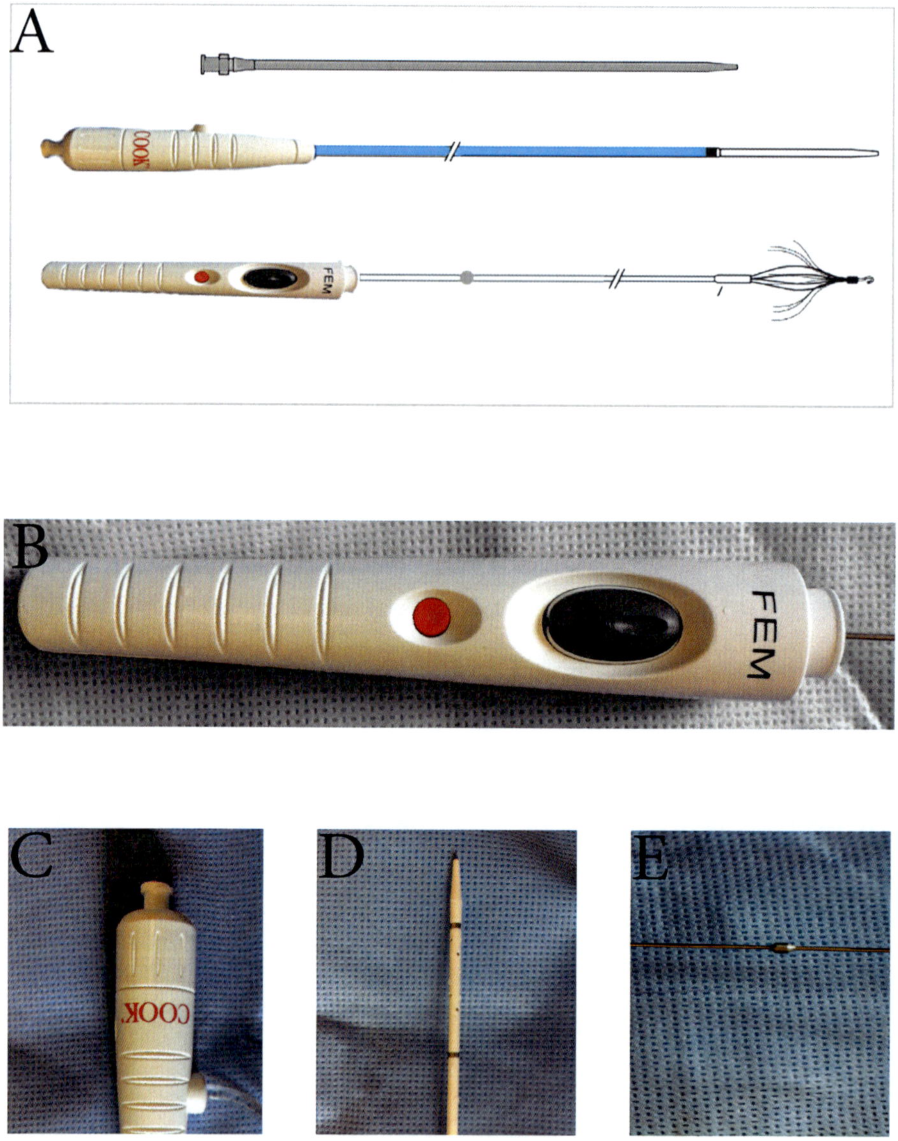

Figure 7.

The Celect™ FEMORAL IVC filter kit.

A. 10 Fr./20 cm radio-opaque hydrophilic dilator, 7 Fr./65 cm delivery sheath with radio-opaque tip and flush dilator with markers (3 cm apart) and a pre-mounted Celect™ filter on NavAlign™ delivery system.
B. A magnified view of the handle of the NavAlign™ system. 'FEM' indicates the filter is designed for femoral introduction, the red button is the safety button and the gray button is the release button.
C. The hub of the delivery sheath with its introducer dilator locked in place.
D. The tip of the measuring dilator with two markers 3 cm apart and flush side-holes.
E. The circular 'bump' on the delivery catheter.

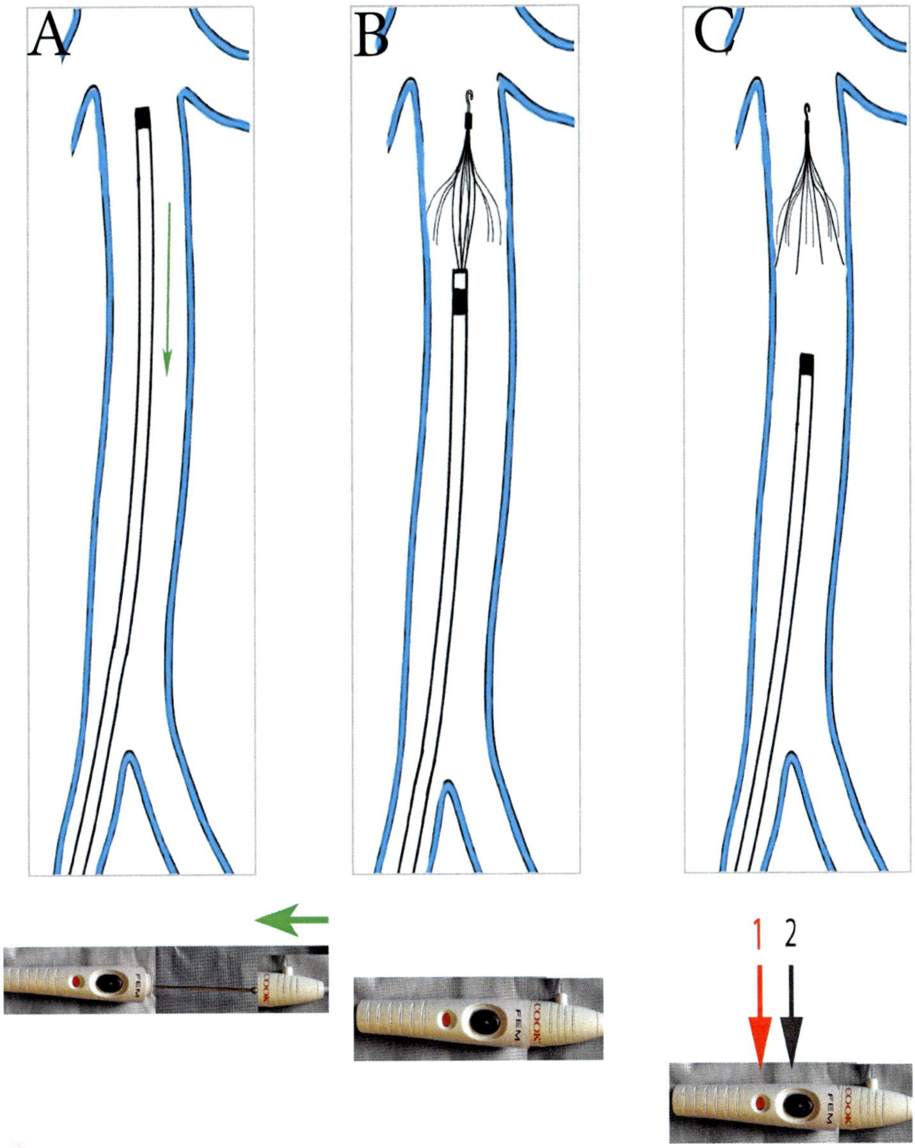

Figure 8.

The Celect™ IVC filter femoral deployment.

A. Once the 'bump' on the delivery catheter is at the hub of the sheath, the sheath is withdrawn past the bump.
B. Stabilizing the catheter, the sheath withdrawal is continued until the sheath connects with the delivery catheter handle.
C. The RED button is depressed first to 'unlock' the filter, followed by the GRAY/BLUE button for release.

the distal end of the delivery catheter.

Filter Deployment:

- Confirm that the tip of the delivery catheter (constraining the primary legs) is outside the sheath.
- Press the RED safety button to prepare the filter for release (figure 8C).
- Press the GRAY/BLUE button to release the filter.
- Confirm filter deployment fluoroscopically.

Final Cavogram:

- Perform a final cavogram to confirm satisfactory filter placement.

CELECT™ JUGULAR PLACEMENT

The Celect 'JUG' filter is exactly the same as the Celect 'FEM' IVC fitler but with a different delivery system.

JUGULAR FILTER KIT

The Celect jugular kit (figure 9) consists of:

- A Celect filter with labels for jugular ('JUG') insertion on the handle, pre-loaded in a protection sheath.
- The hook is always on the cephalad end of the filter.
- A 10 Fr./20 cm radio-opaque hydrophilic dilator.
- A NavAlign™ 7 Fr. /65 cm delivery sheath.
- A measuring dilator for the sheath with eight side holes and 2 radio-opaque markers at a distance of 30 mm from each other (end to end).

TECHNIQUE

Cavogram:

- Obtain venous access in the right IJV.
- Introduce a 0.035" J-tip wire (150 cm) into the proximal IVC.or proximal common iliac vein.
- Introduce the filter delivery sheath (7 Fr./65 cm) into the proximal IVC.
- Remove the wire.
- A cavogram can be performed through the dilator of the sheath (maximum flow rate is 20 mL/sec at 1000 psi).
- Alternatively, perform a venacavogram using a pigtail catheter and confirm normal caval anatomy.
- Identify the renal veins and confluence of the common iliac veins. Mark these on the screen.

Filter Introduction:

- Advance the filter introducer with the protection sheath to the hub of the delivery sheath.
- Using clockwise motion twist the protection sheath onto the sheath hub (figure 10A).
- This maneuver brings the filter (tip of the legs) to the radio-opaque marker band on the tip of the sheath.

Sheath Withdrawal:

- Holding the filter introducer, the delivery sheath and the connected protection sheath are withdrawn under fluoroscopic guidance until the handle and the protection sheath are connected (figure 10B).
- At this point the filter is completely outside the sheath but still connected to the delivery system.

Filter Positioning:

- It should be ensured that the hook of the filter is caudal to the renal veins.
- If the filter is not in the desried position, the introducer sheath is advanced over it to the anchors. The filter is then repositioned.
- The introducer is then withdrawn again, connecting the protection sheath hub to the handle.

Filter Deployment:

- Confirm that the hook of the filter is caudal to the renal veins.
- Press the RED safety button to prepare the filter for release (figure 10C).
- Press the GRAY/BLUE button to release the filter.
- Confirm filter deployment fluoroscopically.

Final Cavogram:

- Perform a final cavogram to confirm satisfactory filter placement.

BIRD'S NEST™ FILTER PLACEMENT

- This is the only filter that is currently available in the U. S. for use in patients with 'mega cava'. The cava diameter should be < 40 mm for this filter to be successfully placed.
- This is a permanent filter.
- This filter is deployed through a series of controlled steps.

FILTER MATERIAL & DESIGN

- This filter is made of 100 cm 'nest' of stainless steel filtering wires.
- The filter has a hook-wire design with two "V" shaped struts.
- It has junction points on each "V"-shaped strut serve as fluoroscopic markers.

THE FILTER KIT

The filter kit consists of:

- A 12 Fr. delivery sheath with dilator.
- An 11 Fr./40 cm long preloaded filter catheter (40 cm delivery system for femoral approach and 75 cm for jugular approach).
- Wire guide pusher.

TECHNIQUE

Cavogram:

- Obtain venous access in the right CFV using a modified Seldinger technique with a 4 Fr. micro-puncture kit.
- Using a 0.038" wire, upsize to a 5 Fr./11 cm sheath.
- Advance this wire in the supra-renal IVC.
- Advance a flush catheter to the right common iliac vein.
- Perform a venacavogram using the same technique as outlined before and confirm normal caval anatomy ('mega cava' is OK).
- A power injector is used. A total of 40 ml of 70:30 diluted contrast (use 50:50 dilution in patients with renal dysfunction) is injected at 20 ml/sec at 500-900 psi.
- Identify the renal veins and confluence of the common iliac veins and mark them on the screen.

Sheath Exchange:

- Remove the flush catheter over the 0.038" wire. Keep this wire close to the diaphragm (to provide a steady rail for the delivery sheath).

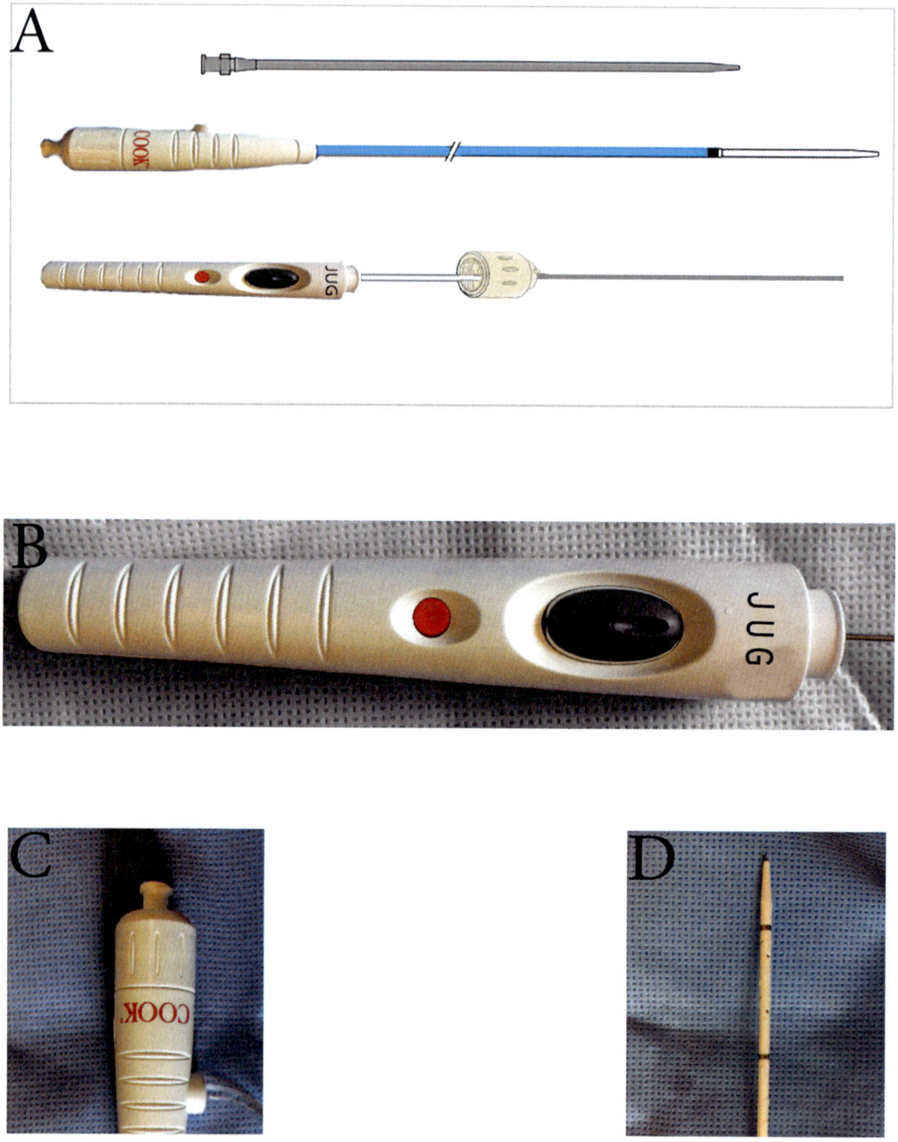

Figure 9.

The Celect™ JUGULAR IVC filter kit.

A. 10 Fr./20 cm radio-opaque hydrophilic dilator, 7 Fr./65 cm delivery sheath with radio-opaque tip and flush dilator with markers (3 cm apart) and a pre-mounted Celect™ filter on NavAlign™ delivery system.
B. A magnified view of the handle of the NavAlign™ system. 'JUG' indicates the filter is designed for jugular introduction, the red button is the safety button and the gray button is the release button.
C. The hub of the delivery sheath with its introducer dilator locked in place.
D. The tip of the measuring dilator with two markers 3 cm apart and flush side-holes.

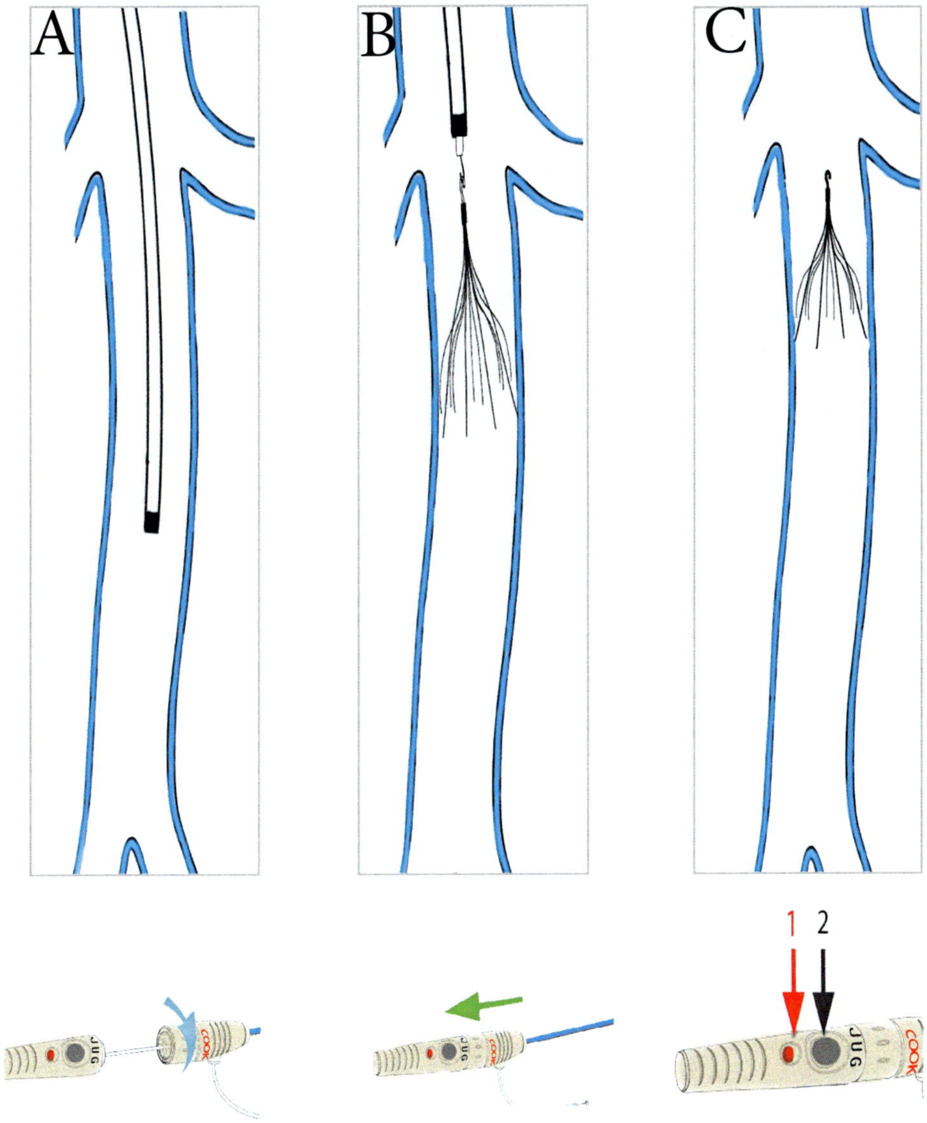

Figure 10.

The Celect™ IVC filter jugular deployment.

A. The filter in the protection sheath is advanced into the sheath and twisted clockwise to connect to the hub of the sheath.
B. Keeping the handle stationary, the protection sheath is withdrawn until it connects with the delivery catheter handle.
C. If the filter position is desirable, the RED button is depressed first to 'unlock' the filter, followed by the GRAY/BLUE button for release.

- Exchange the existing 5 Fr. sheath for the delivery sheath (use serial dilators as needed) and advance it over the wire to its hub.

Filter Preparation:

- Prime the lumen of the filter catheter with contrast via its sidearm. This aids in subsequent injections for filter positioning.
- Remove the dilator and wire from the sheath.
- Insert the filter catheter into the sheath and Luer lock it into position (in this position the tip of the filter catheter extends approximately 1.5 cm beyond the tip of the sheath).

Filter Positioning:

- Position the filter catheter tip just proximal (caudal) to the lowest renal vein.
- Perform a hand injection through the sidearm of the filter catheter. If necessary adjust the filter position.

Filter Deployment:

First Assembly Withdrawal:

- Loosen the Tuohy-Borst valve, hold the pusher stationary and withdraw the filter catheter/ sheath assembly to the black mark on the pusher (figure 12A).
- This maneuver exposes the distal anchoring hooks of the filter while the junction point of the hook wires remain within the tip of the catheter.
- Usually this is clearly visualized under fluoroscopy.
- At the end of the filter, the V-shaped hook wire struts are readily visible, but filter wires are too fine to be identified fluoroscopically.

First Assembly Advancement:

- Gently advance the filter catheter/sheath assembly once, 1-3 **mm**, to secure the hooks to the vena cava wall (figure 12B).

Second Assembly Withdrawal:

- Holding the pusher stationary, withdraw the filter catheter/sheath assembly 1-3 cm (avoid pulling the assembly into the iliac vein; figure 12C).

First Pusher Advance:

- Hold the assembly steady and advance the pusher with a steady motion until the junction point of the proximal hook wires is seen under fluoroscopy at the tip of the filter catheter. This step pushes out the filter wire (figure 12D).
- Advance gradually to avoid wire kinking within the filter catheter.

Second Assembly Advance:

- Advance the assembly so that the proximal junction point is in close proximity to or overlaps the junction point of previously positioned hook wires (figure 12E). This allows compression of the filter wires and places the second pair of hooks in the IVC.

Third Assembly Withdrawal:

- Maintain forward pressure on the pusher and withdraw the assembly. Proximal hook wires exit the catheter springing open to engage with the caval wall (figure 12F). This can be seen well under fluoroscopy.

Filter Disengagement:

- Maintain pusher position (figure 12G), remove red safety lock from the handle.
- Hold the pusher steady.
- Fully depress the thumb grip of the pusher forward into the cannula. This will separate the filter which can be visualized fluoroscopically.
- Release pressure on the thumb grip. Remove the pusher out of the filter catheter (figure 12H).

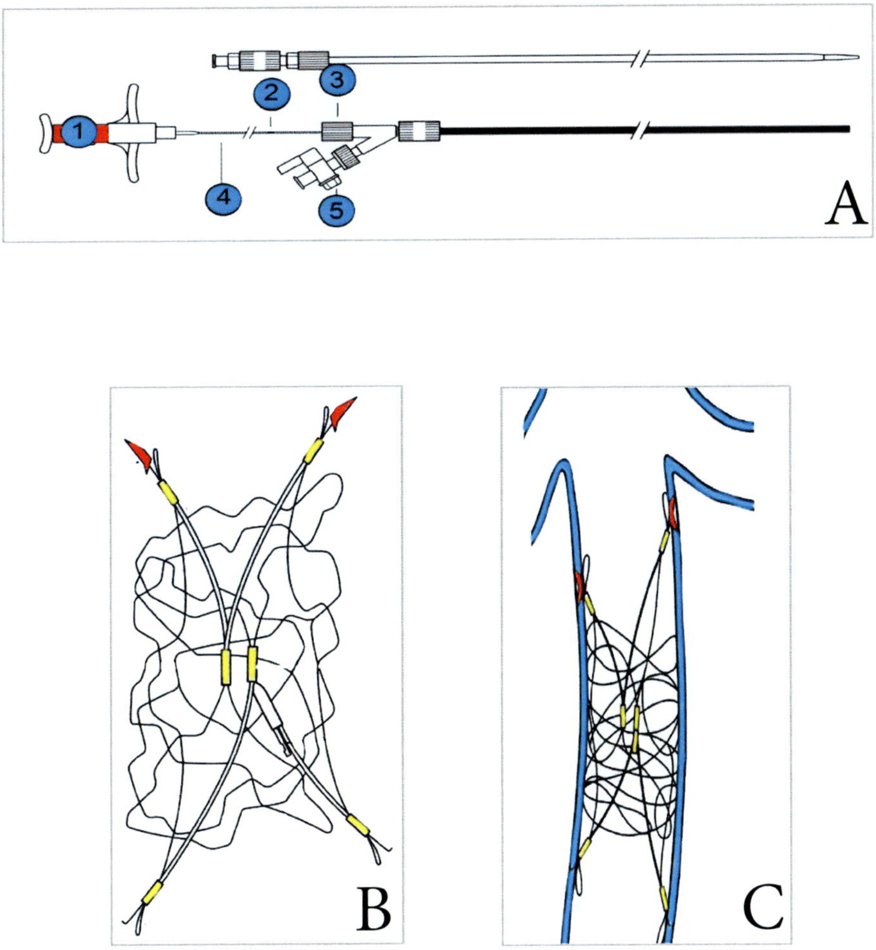

Figure 11.
Components of Bird's Nest filter (modified from Cook).
A. The filter delivery system - (1) Safety lock, (2) Black mark on the pusher, (3) Tuohy-Borst adapter, (4) Filter pusher and (5) Stopcock.
B. A Bird's Nest filter consists of two V-shaped struts, proximal and distal fixation barbs and steel wire.
C. A schematic of a deployed Bird's Nest filter.

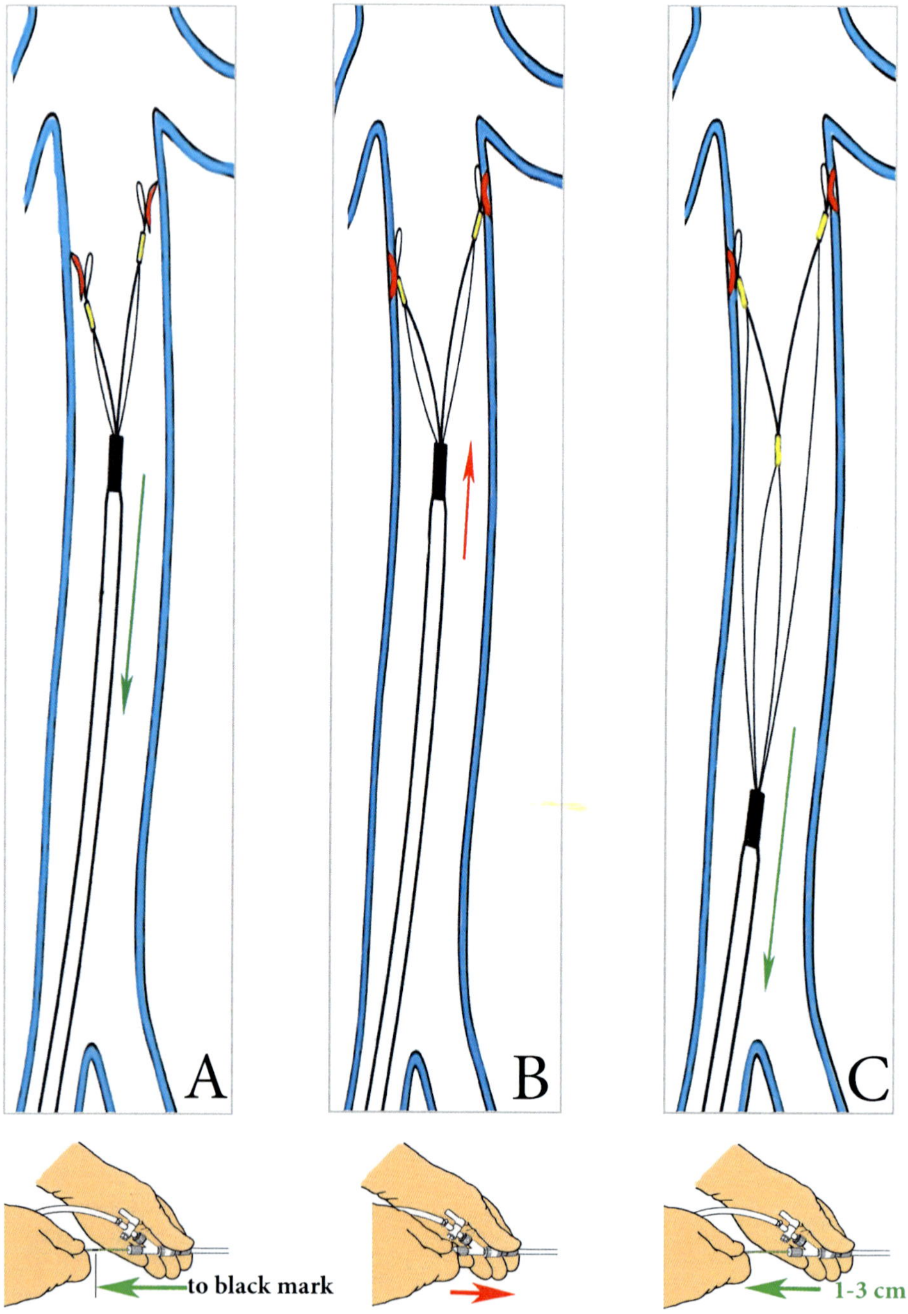

Figure 12.

Steps for the deployment of the Bird's Nest filter.
Refer to the text for details.

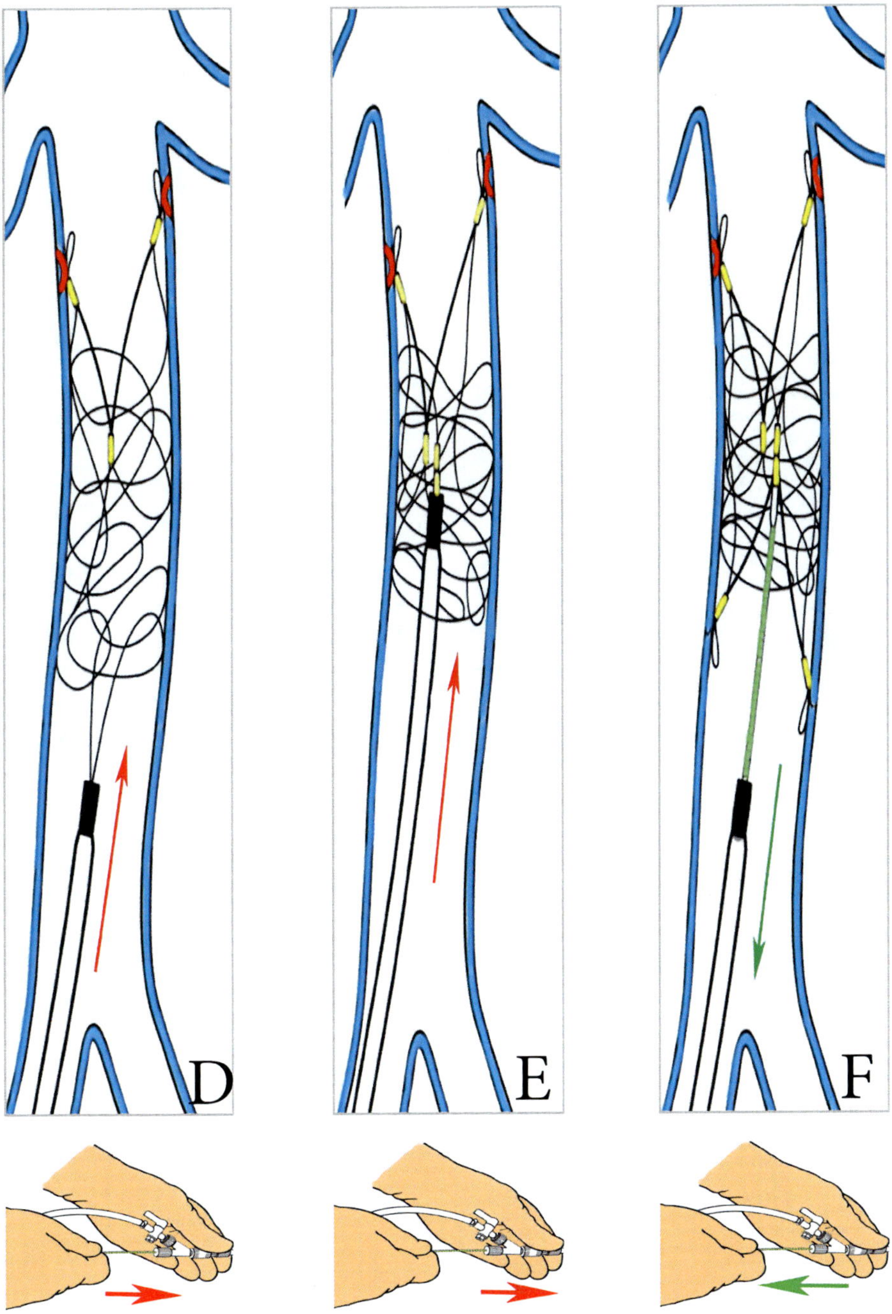

Figure 12.

Steps for the deployment of the Bird's Nest filter (continued).
Refer to the text for details.

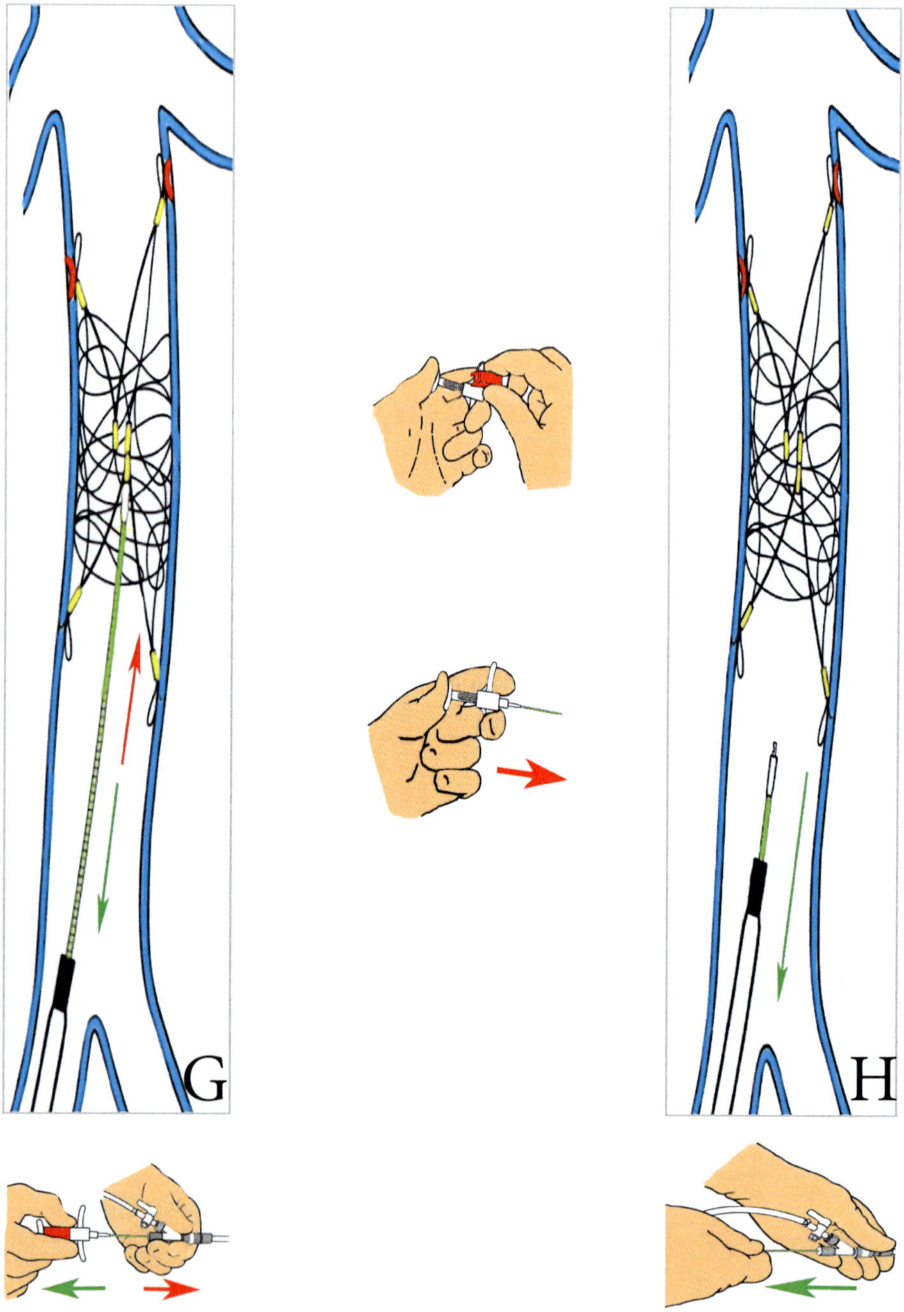

Figure 12.

Steps for the deployment of the Bird's Nest filter (continued).
Refer to the text for details.

IVC FILTER RETRIEVAL

Optional IVC filters are designed to be retrieved within a pre-specified time window. Although, there are several reports of successful late retrieval without significant complications, its best to abide by the time frame that is recommended by the manufacturer.

CONTRAINDICATIONS

- Significant ileo-femoral DVT burden with active PE
- Presence of significant thrombus within the filter (> 25% of the filter cone).

PRE-RETRIEVAL WORKUP

- A venous duplex of the lower extremities to confirm absence of significant ileo-femoral thrombus burden.
- A CTA of the abdomen with venous phase should be considered. This can provide critical information about the presence of thrombus within the filter, position of the filter, any tilt, position of the 'legs' of the filter and filter integrity.

TECHNIQUE

- Although, each optional filter has its own specific retrieval technique and perhaps equipment, the general principles remain the same.
- Most IVC filters are removed via the IJ, except the Optease® which is retrieved via the femoral approach.
- Right internal jugular vein is the preferred route.
- Access is obtained in a standard fashion, ideally under ultrasound guidance.
- A 0.035" J-tip wire is advanced through the sheath, beyond the filter.
- A flush catheter is advanced over this wire into the IVC, distal to the filter.
- A cavogram is performed.
- Absence of significant thrombus is confirmed.
- The flush catheter is exchanged out for the retrieval sheath of the particular filter.

Subsequent steps can be demonstrated by the following examples.

OPTEASE™ RETRIEVAL

Access:

- The OptEase™ filter is removed via CFV access as the hook is located at the caudal end of the filter.

Equipment:

- A special 10 Fr./80 cm OptEase™ retrieval catheter (figure 13) is available for removing this filter.
- Alternatively, an 8 Fr./55 cm Brite tip® sheath can be used for this purpose.
- 260 cm J-tip 0.035" wire.
- A 15-20 mm Amplatz Gooseneck snare.

Technique:

- Obtain right CFV access in a standard fashion.
- Insert the appropriate sheath (8 Fr. vs 10 Fr.).
- Do a cavogram first to rule out significant thrombus (>25%)inside the filter or below it.
- Advance the sheath/catheter into the IVC.
- Load the snare into the sheath/catheter in a standard fashion.
- Catch the hook of the OptEase™ with the snare. Advance the snare catheter to tighten it around the hook.

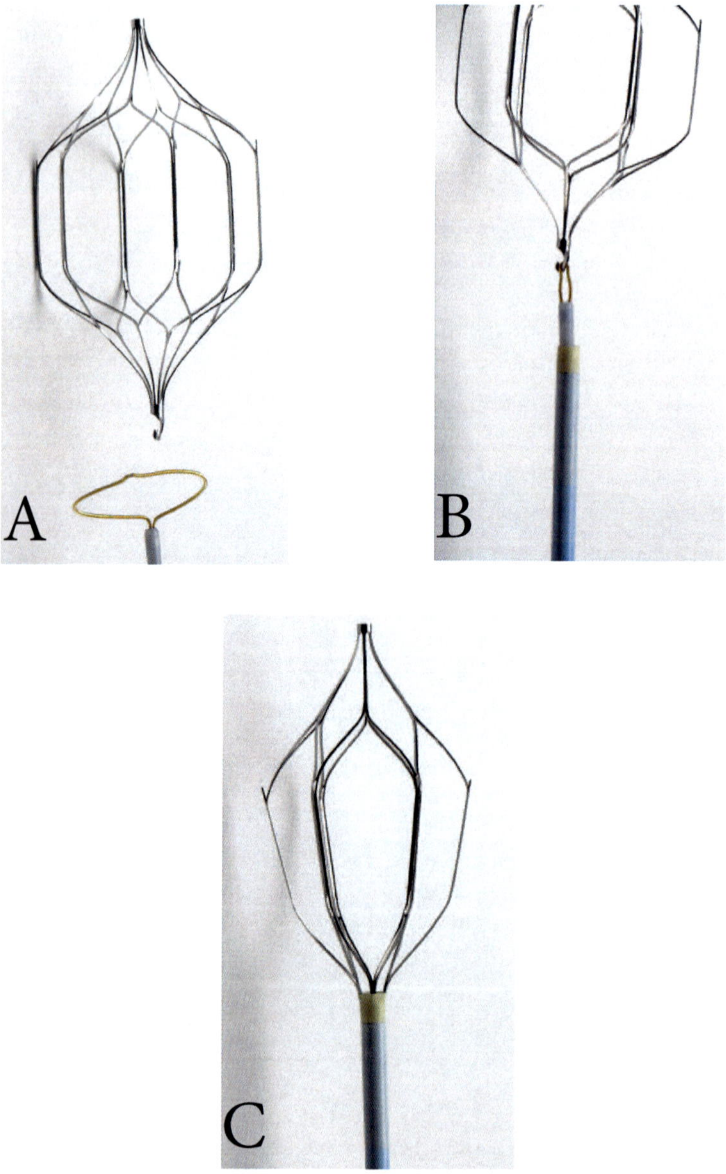

Figure 13.

OptEase™ filter retrieval.

A. The snare is advanced so it surrounds the filter hook.
B. The snare catheter is advanced while maintaining traction on the snare.
C. The catheter/sheath is advanced completely over filter.
The entire assembly should be withdrawn and removed from the body.

- Advance the sheath/catheter on top of this combo and the IVC filter, while maintaining steady backward traction on the snare.
- Pull gently but firmly. Maintain steady traction on the snare.
- Once the entire filter is inside the sheath/catheter, the snare is pulled back with the filter still in its grasp.
- The filter can usually be pulled out of the sheath without too much difficulty.
- Repeat cavogram at the end to document removal of the filter and to assess the IVC.

CELECT™/TULIP™ RETRIEVAL

Access:

- The Celect™ filter is removed via IJ access as the hook is located at the cephalad end of the filter.

Equipment:

- A special retrieval system is available for removal of the Celect filter (figure 12).
- It consists of an 11 Fr./60 cm sheath along with its dilator (red/salmon) and a 6.3 Fr. retrieval loop (Amplatz Gooseneck type snare) along with its catheter (black).
- A 6 Fr. /11 cm sheath.
- A 150 cm 0.035" J-tip wire.
- A 5 Fr. pigtail catheter.

Technique:

- Obtain right IJ access in a standard fashion and insert a 6 Fr. sheath.
- Remove the dilator, flush the sheath.
- Advance the pigtail catheter over the wire, caudal to the IVC filter, under fluoroscopic guidance.
- Do a cavogram first to rule out significant thrombus (>25% of the filter) inside the filter or below it.
- Remove the pigtail over the wire, under fluoroscopic guidance. NEVER pull this catheter through the filter without first straightening it over a wire. Otherwise, it will get caught in the filter and lead to filter migration.
- Exchange the sheath out for the retrieval sheath.
- Remove the dilator and wire.
- Introduce the snare/loop retrieval system.
- Advance it so it fully expands and surrounds the filter.
- Once the hook has been engaged by the snare, put upward traction on the snare. Do not pull hard to disengage the filter.
- Advance the snare catheter over the hook and onto the filter. This will collapse the filter, which disengages from the IVC.
- Advance the sheath over the snare/filter/catheter combination.
- Once the filter is inside the sheath, remove the entire assembly as a unit.

TABLE 2 - EQUIPMENT FOR IVC FILTER RETRIEVAL:	
Sheaths	• 8 Fr. /55 cm Brite tip (OptEase™) • 11 Fr./60 cm
Wires	• 0.035" 150 cm to 260 cm, J-tip
Catheters	• 5 Fr./100 cm pigtail catheter
Snares	• 2-3 cm Amplatz Gooseneck snare (OptEase™) • CloverSnare™ (Celect™/Tulip™) • Loop snare system (Celect™/Tulip™)

Figure 14.

Celect filter retrieval.

A. Celect™ retrieval kit consists of a 'loop' (snare), dilator for the sheath (salmon/red), catheter for snare delivery and an 11 Fr./60 cm sheath. The same kit can be used to retrieve a Tulip™ filter.

B. The sheath, snare catheter and the dilator are inserted initially in a telescoping fashion.

C. After removal of the dilator the snare is used to engage the hook of the filter.
D. Once the hook has been engaged, traction is applied and the snare catheter advanced over the hook.

E. The filter collapses.

F. The sheath is advanced over the filter.

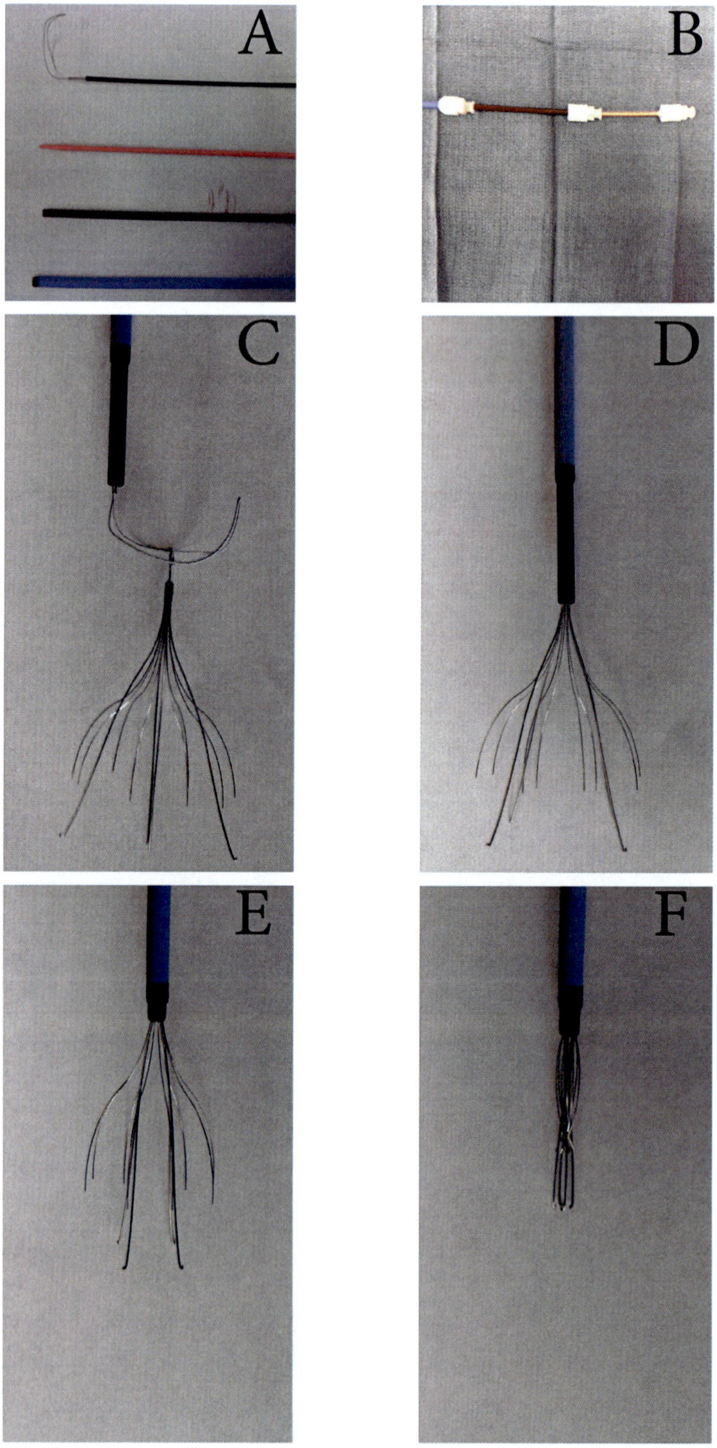

Figure 15.

The new four loop CloverSnare™ designed to improve Celect/Tulip retrieval.

17

PULMONARY EMBOLISM

Pulmonary embolism (PE) carries significant morbidity and a high risk of death. This risk is directly proportional to the thrombo-embolic burden in the pulmonary circulation (acute or chronic).

PE is classified into three different groups based upon the thrombo-embolic burden, hemodynamic impact, RV insult and symptomatology. We use the definitions proposed by the American Heart Association and the American College of Cardiology for this purpose.

MASSIVE PE

Acute PE with sustained hypotension (SBP < 90 mm Hg for at least 15 minutes or requiring inotropic support, not due to a cause other than PE, such as arrhythmia, hypovolemia, sepsis or LV dysfunction), pulselessness, or persistent profound bradycardia (HR < 40 bpm with signs or symptoms of shock).

Symptoms include chest pain, SOB, dizziness, presyncope and syncope.

SUBMASSIVE PE

Acute PE without systemic hypotension (SBP ≥ 90 mm Hg) but with either RV dysfunction or myocardial necrosis.

- RV dysfunction is defined as the presence of at least one of the following factors:
    - ▹ RV dilatation (apical four-chamber view RV end diastolic diameter divided by LV end diastolic diameter >0.9) or RV systolic dysfunction on echocardiography.
    - ▹ RV dilatation (four-chamber view RV end diastolic diameter divided by LV end diastolic diameter >0.9) on CTA axial slice analysis.
    - ▹ Elevation of BNP (>90 pg/mL)
    - ▹ Elevation of N-terminal pro-BNP (>500 pg/mL)
    - ▹ EKG changes consistent with RV strain (new complete or incomplete right bundle branch block, anteroseptal ST elevation or depression or anteroseptal T wave inversion).
- Myocardial necrosis is defined as either of the following:

- ▷ Elevation of troponin I (>0.4 ng/mL) or
- ▷ Elevation of troponin T(>0.1 ng/mL).

Symptoms include chest pain, SOB, dizziness, presyncope and syncope.

LOW-RISK PE

Acute PE with absent clinical markers of adverse prognosis that define massive and submassive PE.

Symptoms include chest pain, SOB or dizziness.

Patients with low-risk PE are treated safely with systemic anticoagulation with IV UFH with transition to oral warfarin therapy.

Patients with submassive PE tend to be treated less aggressively than they probably should be. Traditionally, these patients have been treated with anticoagulation and thrombolytic therapy is not offered. We believe that this group of patients is at a high risk of recurrent episodes and death compared to those with low-risk PE. As such, these patients should be identified promptly, stratified appropriately and be considered for catheter directed thrombolytic therapy (CDT) or systemic fibrinolysis.

Patients with massive PE must be treated emergently. Traditionally, surgical embolectomy and systemic thrombolytic therapy have been used for the management of these patients. With the advent of a device called AngioVac (see below) interventional management of massive PE is now possible in select cases.

ANATOMY OF PULMONARY ARTERY

- The main pulomnary artery (PA) arises from the right ventricle.
- It is 5 cm in length and 3 cm in diameter.
- The main PA bifurcates into the right and left pulmonary arteries, each with an average diameter of 2 cm.

RIGHT PULMONARY ARTERY

- The right PA divides into two main branches, an ascending and descending branch.
- The ascending branch supplies the right upper lobe and divides into three branches:
  - ▷ Apical posterior
  - ▷ Apical anterior
  - ▷ Apical superior
- The descending branch of the right PA supplies the right middle and lower lobe.
- This descending branch divides into several branches:
  - ▷ Middle lobe artery
  - ▷ Lateral basal
  - ▷ Anterior basal
  - ▷ Posterior basal
  - ▷ Medial basal

LEFT PULMONARY ARTERY

- The left PA is shorter and bifurcates early into an ascending and descending branch to supply the left upper and lower lobes respectively.
- The ascending branch gives rise to three branches:

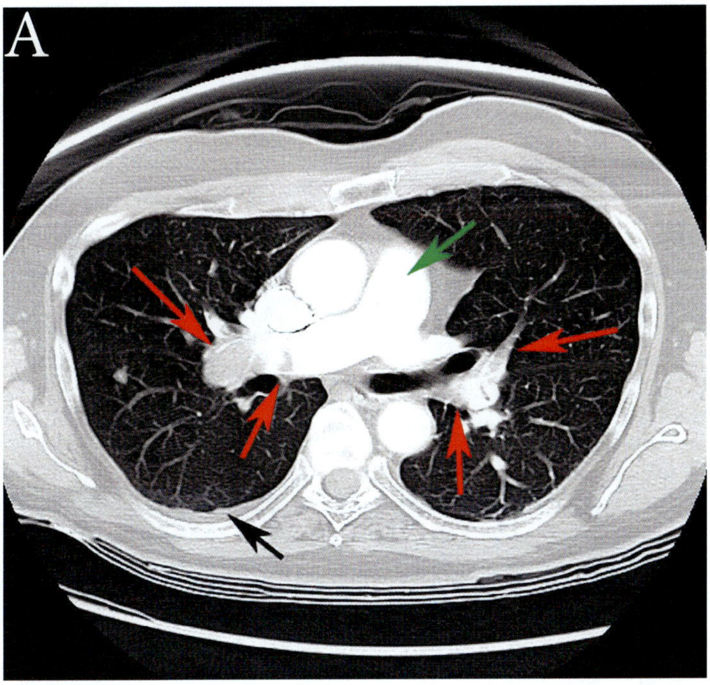

Figure 1A.

An axial slice of a CT pulmonary angiogram of a patient with submassive pulmonary embolism shows significant thrombii (red arrows) in both right and left pulmonary arteries.

The main pulmonary artery (green arrow) is noted to be free of any thrombus.

A very small pleural effusion is also present on the right side (black arrow).

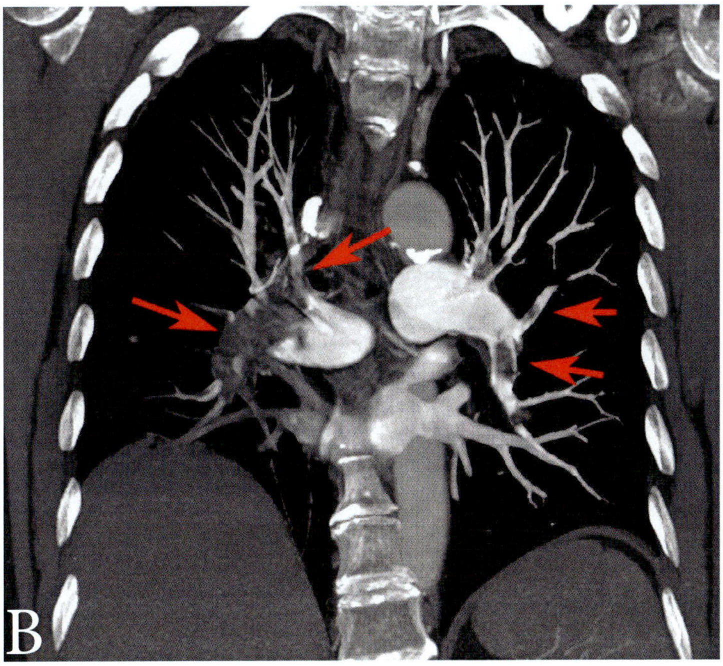

Figure 1B.

A coronal slice of a CT pulmonary angiogram of a patient with submassive pulmonary embolism. This study shows significant thrombi (red arrows) in both right and left pulmonary arteries.

- ▹ Apical posterior
- ▹ Superior
- ▹ Anterior

- The descending branch is a continuation of the left PA and gives rise to several branches:
  - ▹ Superior lingual branch
  - ▹ Inferior lingual
  - ▹ Lateral basal
  - ▹ Anteromedial basal
  - ▹ Posterior basal

DIAGNOSIS OF PE

All patients suspected of having a PE should undergo the following tests:

- Serum troponin, BNP and D-dimer levels.
- A standard 12-lead EKG.
- A transthoracic echocardiogram.
- CTPA (if the renal function allows) or alternate pulmonary artery imaging (figures 1A &1B).
- V/Q scan, a less desirable alternative, should be considered in patients if they cannot undergo CTPA and if significant baseline ventilatory defects are not present.

In patients with submassive PE, troponin and BNP are elevated but not always proportionate to the thrombus burden in the pulmonary circulation. Echocardiographic findings suggestive of submassive PE include:

- RV dilatation
- RV systolic dysfunction typically with apical sparing
- RV/LV end-diastolic diameter ≥ 0.9 (figure 1C)
- Moderate pulmonary hypertension
- IVC dilatation

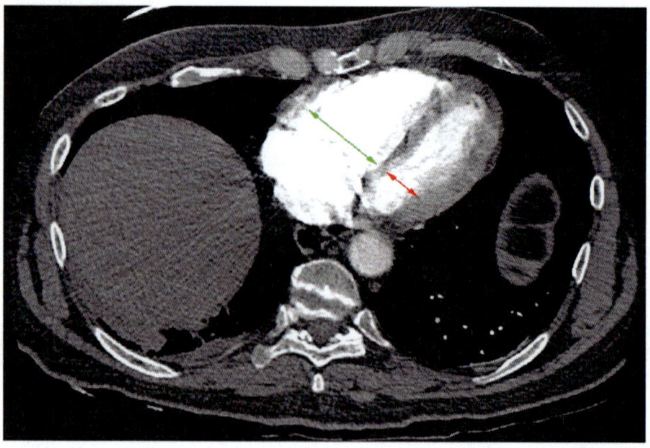

Figure 1C.

An axial slice of a CTA showing RV (green arrows) and LV (red arrows) dimensions. This shows RV diameter >> LV diameter.

TABLE 1 - TESTING FOR PATIENTS WITH PULMONARY EMBOLISM	
Imaging	• Echocardiogram (transthoracic) • CTPA • Lower extremity venous duplex US
Blood tests	• Baseline coagulation profile and CBC • Serum troponin • Serum BNP • Plasma D-dimer • Hypercoaguability workup when indicated
Invasive	• Venacavogram • Right heart catheterization • Pulmonary angiography
Infusion	• EKOS Ekosonic catheter 12 cm/106 cm • 4 Fr./100 cm infusion catheter (e.g. Fountain catheter) with 5-10 cm infusion length
IVC Filter	• Retrievable filter such as Celect (Cook Medical, Bloomington, IN) or Optease (Cordis, NJ)
Drugs	• r-tPA • Urokinase (or Streptokinase) where r-tPA is not readily available • IV UFH

CDT FOR SUBMASSIVE PE

OUTLINE OF THE CDT PROCEDURE

1. Placement of an optional IVC filter.
2. Measurement of right heart hemodynamics (right heart catheterization).
3. Pulmonary angiography.
4. Insertion of an infusion catheter into the pulmonary artery on the affected side (bilateral if there is bilateral PE).
5. Administration of 'intra-arterial' thrombolytic therapy (bolus and infusion).

EQUIPMENT

- A 4 Fr. micropuncture kit.
- Sheaths: 6 Fr./11 cm and 8 Fr./11 cm; one each.
- Wires: 0.035" j-tip 260 cm wire.
- A 6 Fr. right heart catheter (RHC).
- A 5 Fr./100 cm non-angled pigtail catheter.
- Retrievable IVC filter suitable for femoral or jugular deployment.
- EKOS catheter 6 Fr. /12 cm infusion length/102 cm shaft length x 2.
- r-tPA.

ACCESS

- Arterial access should be avoided even for invasive BP monitoring as administration of thrombolytic therapy may lead to hemostasis issues at the access site.
- Venous access is obtained in the upper or lower extremity veins as necessary. If bilateral PE is suspected two separate venous access sites are needed as a catheter is inserted in each pulmonary artery.
- Since, in our practice an IVC filter is placed before embarking on pulmonary intervention, right common femoral vein access is chosen for at least one of the access sites.

Duration of CDT
CDT should be continued for < 24 hr.
- The therapy can be instituted for PE that is up to two weeks old.

- A micropuncture kit is used under ultrasound guidance to obtain access to minimize bleeding complications that may follow thrombolytic therapy.
- An 8 Fr./11 cm sheath is placed in this vein and is used as a conduit to advance the IVC filter delivery sheath.
- A 6 Fr./11 cm sheath is adequate for placement in the other venous access site.

Right Heart Catheter:
Most SGC do not take a 0.035” wire. We use an Arrow Balloon Wedge Pressure catheter. Its specifications are:
- 6 Fr./110 cm.
- 2-lumen - one for balloon.
- Allows a 0.035” wire insertion through the central lumen.
- 1 cc stroke syringe.

TABLE 2- EQUIPMENT FOR PULMONARY ARTERY THROMBECTOMY	
Sheaths	• 6 Fr. /11 cm • 8 Fr. /11 cm
Wires	• 0.035” (260 cm) J-tip
Catheters	• Pigtail or Grollman • 6 Fr./110 cm right heart catheter (that will accomodate a 0.035” wire e.g. Arrow Balloon Wedge Pressure Catheter - see blue box)
Infusion	• EKOS Ekosonic catheter 12 cm/106 cm • 4 Fr./100 cm infusion catheter (e.g. Fountain catheter) with 5-10 cm infusion length
IVC Filter	• Retrievable filter such as Celect (Cook Medical, Bloomington, IN) or Optease (Cordis, Bridgewater NJ)
Drugs	• rt-PA • SK • IV UFH

IVC FILTER PLACEMENT

A optional (i.e. retrievable) IVC filter is deployed in a standard fashion.

RIGHT HEART CATHETERIZATION

- This is performed in the standard fashion.
- The catheter should be advanced under fluoroscopic guidance over a 0.035” wire through the filter. This reduces the likelihood of the filter being inadvertently tilted or dislodged by the catheter as it is advanced up the IVC.
- Pressures are measured in the RA, RV and PA.
- It is not advisable to obtain a pulmonary capillary wedge pressure as, firstly, it may be unreliable if the there is distal thrombus and, secondly, the act of pushing the catheter distally may lead to the thrombus being compacted in a sub-segmental pulmonary artery.

PULMONARY ANGIOGRAPHY

- The RHC is exchanged over the 0.035” 260 cm wire for a 5 Fr. pigtail catheter.
- The tip of the wire should be securely and safely parked in a first order pulmonary artery branch.
- Keep the pigtail in the main pulmonary, 1-2 cm above the pulmonic valve.

- Use a power injector for the pulmonary angiogram. We use 50% diluted contrast at 20 cc/sec for a total of 40 - 60 cc at 500 psi.
- In a cooperative patient use DSA with breath-hold for image optimization.
- Stay on the cine to see the dextro as well as the levo phase.

INSERTION OF INFUSION CATHETER

- After pulmonary angiography, the pigtail is exchanged out for the infusion catheter of choice over the 0.035" j-tip wire (figure 2A).
- If significant proximal (main pulmonary artery on either side/lower lobe branches) thrombus is visualized, an EKOS Ekosonic catheter (ultrasound assisted infusion catheter) placement is recommended instead of a generic infusion catheter. This high frequency, low-power ultrasound catheter allows better drug penetration within the thrombus thus reducing treatment duration and overall dosage.
- The Ekosonic catheter that is selected for this purpose is 6 Fr. /12 cm/106 cm working length (infusion length is 12 cm).
- This catheter has a radio-opaque marker at either end of the infusion length. Care is taken to keep both these markers above the pulmonic valve, ensuring drug and ultrasound delivery within the proximal thrombus.
- A bolus of 5 mg of r-tPA is given manually through the drug port of this catheter prior to starting the infusion.
- Depending upon the thrombus burden on the affected side and the overall risk of bleeding, the infusion is usually set at 1-3 mg/hour. The goal is to infuse 12-36 mg of r-tPA over 12 hours.
- The coolant (normal saline) is infused at 35 mL/hr through the 'Coolant' port of the catheter. If necessary, the coolant rate can be increased up to 150 ml/hour.
- IV UFH is infused through the sidearm of the venous sheath at 250-500 units/hour. While the r-tPA is being infused the goal PTT is 45-60 seconds.
- If there is bilateral significant thrombus, an EKOS catheter is placed on each side (figure 2B). The second access site is used to advance the SGC over the wire into the other pulmonary artery in the same manner as on the first side. The RHC is then exchanged out for the second EKOS catheter as in the above steps.
- If despite all the other abnormalities, there is no angiographically visible significant thrombus in the proximal PA bilaterally, r-tPA delivery via an infusion catheter (e.g. Fountain catheter, Merit Medical Systems, South Jordan, UT) is recommended. This is purely for cost-effectiveness purposes as infusion catheters are much cheaper than the EKOS catheter. 1-3 mg/hr of r-tPA is delivered through this catheter in the standard fashion.

MONITORING OF THERAPY

- The patient should be admitted to the ICU for close monitoring.
- Pulse oximetry and ECG should be monitored continuously.
- Non-invasive blood pressure should be monitored per ICU protocols.
- Frequent clinical assessment should be performed to promptly detect any complications of thrombolytic therapy such as intracranial or access site bleeding.
- Standard lab work for thrombolytic therapy is recommended (see chapter 13, table 5).
- A transthoracic echocardiogram should be performed within 24 hours.

TREATMENT END-POINTS

- Thrombolytic infusion is continued for 12-15 hours.
- In most patients, repeat pulmonary angiography or right heart catheterization is not necessary and the catheter(s) can be removed at the bedside.
- Parameters of clinical improvement include:
    - ▹ Improved oxygenation.

  - ▹ Improved RV function on follow-up echocardiogram.
  - ▹ Increase or normalization of systemic blood pressure.
- If these parameters do not show significant improvement, it is reasonable to repeat pulmonary angiography and assess the benefit of thrombolytic therapy.
- In the occasional case where we have encountered significant residual proximal thrombus, we have continued the therapy for an additional 6-12 hours and then removed the catheter at bedside. Continued thrombolytic therapy beyond this time-frame is not recommended due to diminishing returns and increasing risk of bleeding with a higher cumulative dose.

PERCUTANEOUS PULMONARY EMBOLECTOMY

- Percutaneous mechanical embolectomy is typically used in patients with submassive/massive PE that have contraindications to thrombolytic therapy and/or are poor surgical candidates for surgical embolectomy.
- While other available devices have been used for this purpose, one specific device stands out in current practice.
- This device is the AngioVac (AngioDynamics, Latham NY) (figure 3).
- Percutaneous pulmonary embolectomy is one application of this device.
- The AngioVac suction cannula is a 'venous' drainage system that requires a cardiopulmonary bypass circuit.
- The AngioVac suction cannula is a 22 Fr. coil-reinforced catheter with a balloon-actuated, expandable funnel-shaped distal tip used to remove thrombus in the pulmonary artery as well as the right atrium, right ventricle and inferior vena cava.
- The cardiopulmonary bypass circuit thus created removes the blood (and thrombus) through the AngioVac suction cannula(outflow), passing through a filter and re-infuses through another (17 Fr.) 'inflow' central venous cannula (femoral or rarely jugular).

OUTLINE OF PROCEDURE

- Confirmation of submassive/massive PE on echo/CTA.
- Right heart catheterization.
- Pulmonary angiography.
- Insertion of the AngioVac suction cannula.

EQUIPMENT

- A 4 Fr. micropuncture kit.
- Sheaths: 2 x 6 Fr./11 cm Cordis Brite-tip sheath.
- Wires: 0.035 " standard J-tip 260 cm wire and 0.035 " J-tip Amplatz Superstiff 260 cm wire.
- A 6 Fr./110 cm right heart catheter (RHC).
- A 5 Fr./100 cm non-angled pigtail catheter.
- A 26 Fr. Gore Dryseal® sheath.
- AngioVac suction cannula.
- AngioVac Extracorporeal circuit.

ACCESS

- Venous access is obtained in bilateral common femoral veins.
- A micropuncture kit is used under ultrasound guidance to obtain access to minimize bleeding complications.
- A 6 Fr./11 cm Cordis Brite-tip sheath is placed in bilateral common femoral veins (CFV).
- Right CFV offers a more direct path to the heart and hence the pulmonary arteries. It is, therefore, the preferred access site for PE cases. Alternatively, right IJ access may be used, if

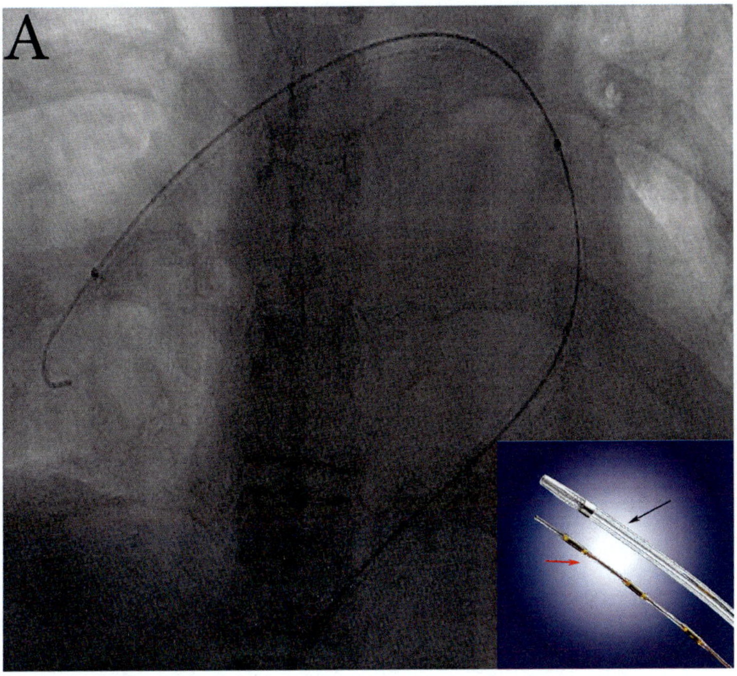

Figure 2A.

A 0.035" J-tip wire is present in the right main pulmonary artery. An EKOS catheter (black arrow in the inset) has been advanced over this wire into this vessel. The wire is then removed and the ultrasound catheter (MicroSonic® device; red arrow in the inset) is advanced through the wire lumen into this catheter until it cannot be advanced any further. This ultrasound catheter is then secured on the outside.

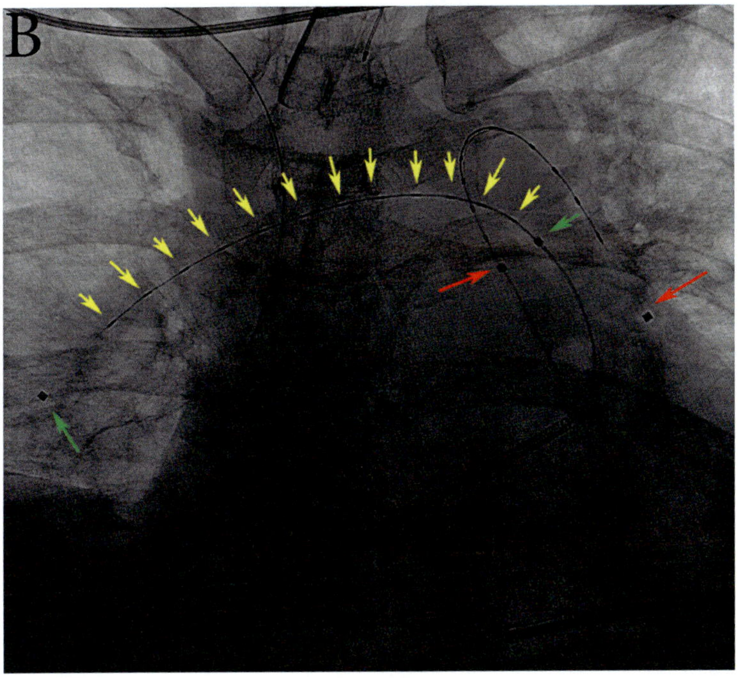

Figure 2B.

EKOS® catheters are noted in position in bilateral pulmonary arteries. The green arrows represent treatment zone markers of the catheter in the right pulmonary artery and the red arrows indicate the treatment zone markers of the catheter in the left pulmonary artery. The yellow markers indicate the ultrasound transducers on the inner catheter (MicroSonic® device).

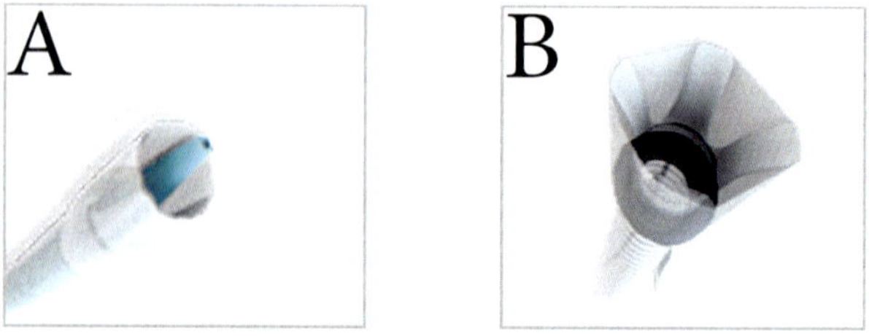

Figure 3.
The distal tip of the AngioVac cannula.

A. With the dilator in place and the balloon deflated.

B. With the dilator removed, the balloon (black ring) is inflated to 1.5 ATM, that converts the tip into a funnel shape.

femoral access fails to deliver the catheter to the PA.

- The right CFV wire is exchanged out for an Amplatz Superstiff wire. The skin track is serially dilated and the 26 Fr. Gore Dryseal sheath is advanced into the IVC under fluoroscopic guidance, over the stiff wire.
- The dilator and wire are removed and the sheath flushed.
- The left CFV sheath is similarly exchanged out for the inflow cannula.

ANTICOAGULATION

- After the Dryseal sheath and inflow cannula are in place, the patient is fully anticoagulated with IV UFH.
- Goal ACT is > 300 sec.
- This should be achieved prior to insertion of the AngioVac cannula.

RIGHT HEART CATHETERIZATION

- Standard technique is used to perform the right heart catheterization.
- RA, RV and PA pressures are measured.

PULMONARY ANGIOGRAPHY

- Over the 0.035 " 260 cm standard J-tip wire, the RHC is exchanged for a 5 Fr. non-angled pigtail catheter.
- Keep the pigtail catheter in the main pulmonary artery 1 to 2 cm above the pulmonic valve.
- Use a power injector for the angiogram with a 50:50 mixture of saline/contrast at 20 cc/sec for a total of 40 cc at 500 psi.
- Breath-hold and instructions are used with DSA.
- A long cine run should be performed to see the dextro as well as the levo phase.

WIRE EXCHANGE

- Exchange the non-angled pigtail catheter for a 6 Fr. RHC over the standard 0.035 " J-tip wire.
- Through the RHC, advance the 0.035" Amplatz Superstiff 260 cm wire into the pulmonary artery.
- Remove the RHC and leave the Amplatz Superstiff wire parked in the pulmonary artery.

ANGIOVAC EXTRACORPOREAL CIRCUIT

- A perfusionist team is needed to put the patient on cardiopulmonary bypass circuit.
- There are several steps in putting together the extracorporeal circuit.

INSERTION OF THE ANGIOVAC SUCTION CANNULA

- Over the 0.035" 260 cm Amplatz Superstiff wire, advance the AngioVac suction cannula through the Dryseal sheath and stop once you exit the sheath.

- Under fluoroscopic guidance, advance the AngioVac suction cannula over the Superstiff wire into the main pulmonary artery.
- Remove the dilator and wire.
- The AngioVac suction cannula is now in position for venous drainage to begin.
- The AngioVac suction cannula can be advanced to the bifurcation of the main pulmonary artery. However, the device is bulky and may prolapse back into the RV. If this happens, repeat the steps as mentioned above to place the suction cannula back into the main PA.
- Venous drainage during extracorporeal bypass is intended for up to 6 hours. The amount of time spent using the suction cannula depends upon the amount of thrombus retrieved.
- While the patient is on this extracorporeal bypass, the patient should have an ACT greater than 300 seconds.

COMPLICATIONS OF THE ANGIOVAC SUCTION CANNULA

- The main and most feared complication associated with the AngioVac cannula is perforation of the main pulmonary artery. In trying to advance the suction cannula with the dilator in place, there is a chance that the pulmonary artery is perforated. This may lead to PEA arrest secondary to cardiac tamponade which can be visualized under TEE. These procedures are usually performed in a hybrid operating room with a CT surgeon.

CASE 1

A 56-year old female was transferred from an outside hospital with severe respiratory distress. A few days prior, she had undergone hip replacement. On the second post-op day, she was diagnosed with submassive PE. She was treated with IV heparin initially which was switched to IV Lepirudin on the fourth day after a diagnosis of heparin-induced thrombocytopenia was made. She had also undergone an IVC filter placement at that hospital. Residual DVT was reported in the left popliteal vein. Her blood work showed some degree of liver dysfunction with elevated transaminases and an elevated INR of 5.0. She had an indwelling PICC line in the right upper extremity.

Markers of Submassive PE:

- TnI > 2.0 ng/ml
- BNP 1980 pg/ml

Echocardiographic Findings:

- Marked RV dilatation with severe hypokinesis except the apex (figure 4A & 4B).
- RV systolic pressure ~ 50 mm Hg.
- After transfer, she had to be emergently intubated and was taken to the cath lab. Her blood pressure was 95 mm Hg and IV dopamine was started.

Patient Preparation & Access:

- She was prepped and draped in a standard fashion.
- Her PICC line was exchanged over a 0.018" 300 cm SteelCore® wire for a 6 Fr./11 cm hydrophilic sheath.
- As the INR was high, ultrasound guidance was used to obtain right femoral vein access using a micropuncture needle. A 6 Fr. sheath was inserted in this vein.

Right Heart Catheterization:

- A right-heart catheter was advanced into the right heart circulation via the upper extremity venous sheath.
- Hemodynamic measurements were made and the catheter was advanced into the left pulmonary artery.
- The catheter was then exchanged out for a pigtail catheter which was placed in the main pulmonary artery.

Pulmonary Angiography:

- Pulmonary angiography was performed using a power injector at the following settings 20 cc/sec for a total volume of 40 cc @ 500 psi.
- A large embolus was seen in the right main pulmonary artery, extending into upper and lower pole branches (figure 5A). On the left, the thrombo-embolic burden was relatively less but multiple emboli were present in both upper and lower lobe branches.

EKOS Ekosonic Catheter Placement:

- Since an IVC filter was already in place, the next step was to insert the infusion catheters in each pulmonary artery.
- The pigtail was exchanged out and an Ekosonic catheter was inserted via the upper extremity access.
- Next, the right-heart catheter was introduced through the femoral access and advanced in a standard fashion into the right pulmonary artery. It was delivered across the IVC filter under fluoroscopy, over a wire and with the balloon deflated. These steps reduce the likelihood of

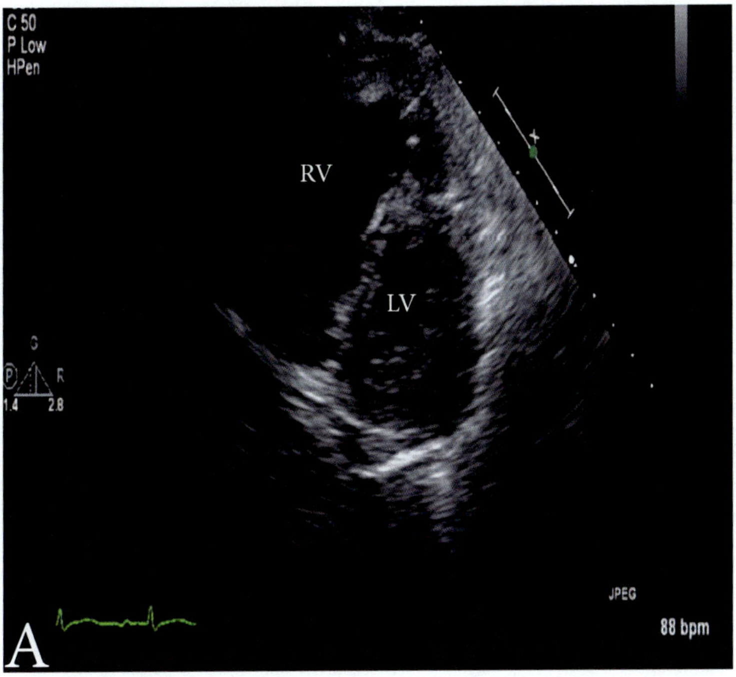

Figure 4A.

Transthoracic echo short axis view shows significant RV dilatation. The interventricular septum was flattened during systole suggesting pressure overload.

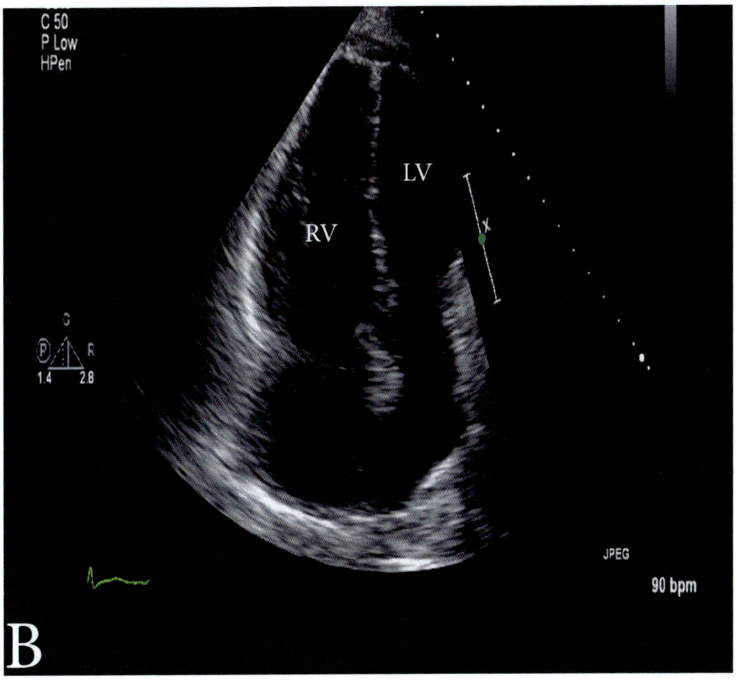

Figure 4B.

Apical four chamber view on transthoracic echocardiogram again shows significant dilatation of the RV. Classically, the RV becomes hypokinetic with sparing of the apex. This is often referred to as McConnell's sign.

Figure 5A.

Pulmonary angiography demonstrates a sizeable thrombus (red arrow) in the main right pulmonary artery, obstructing both the upper and lower right pulmonary artery branches. Very significant oligemia of the right lung is also noted.

This angiogram is peformed via a right femoral vein access. A right-heart catheter has been advanced via the right cephalic vein and is noted in the right atrium at this point.

An endotracheal tube is also present.

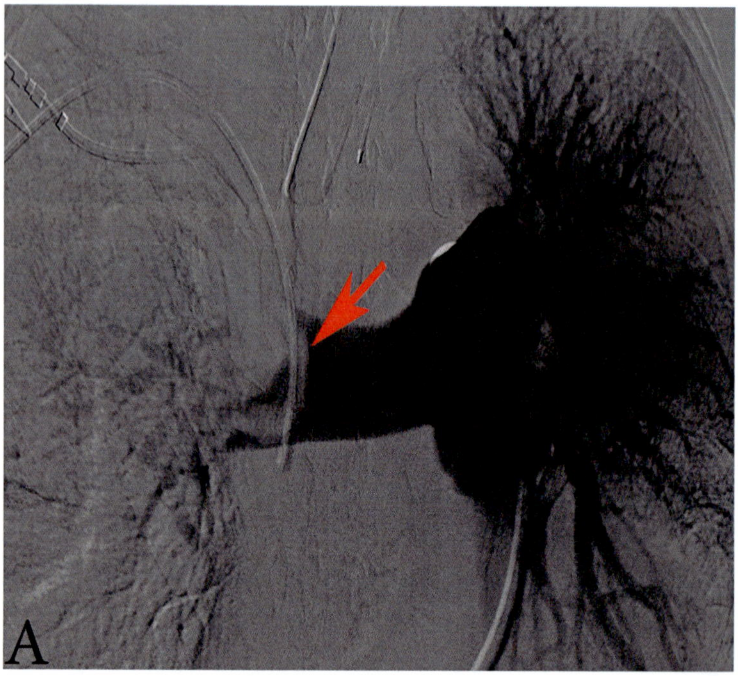

Figure 5B.

Post-thrombolytic therapy pulmonary angiography. This reveals complete resolution of the thrombus in the proximal right pulmonary artery. Additionally, no signficant thrombus is noted in the distal segment or subsegmental arteries.

An EKOS catheter is noted in the left pulmonary artery. This has been inserted via the right cephalic vein access. A right-heart catheter is now noted via the left internal jugular vein. This was being used for hemodynamic assessment during montoring of therapy in the intensive care unit.

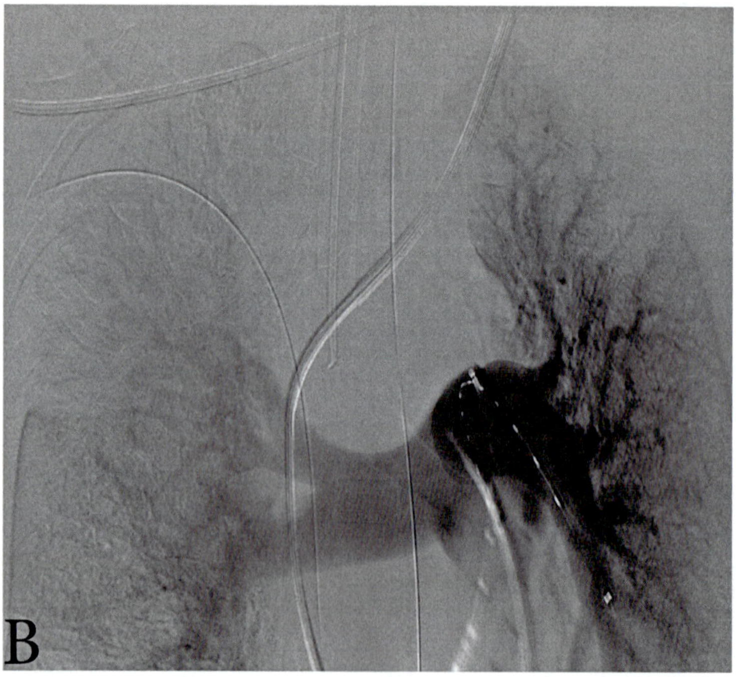

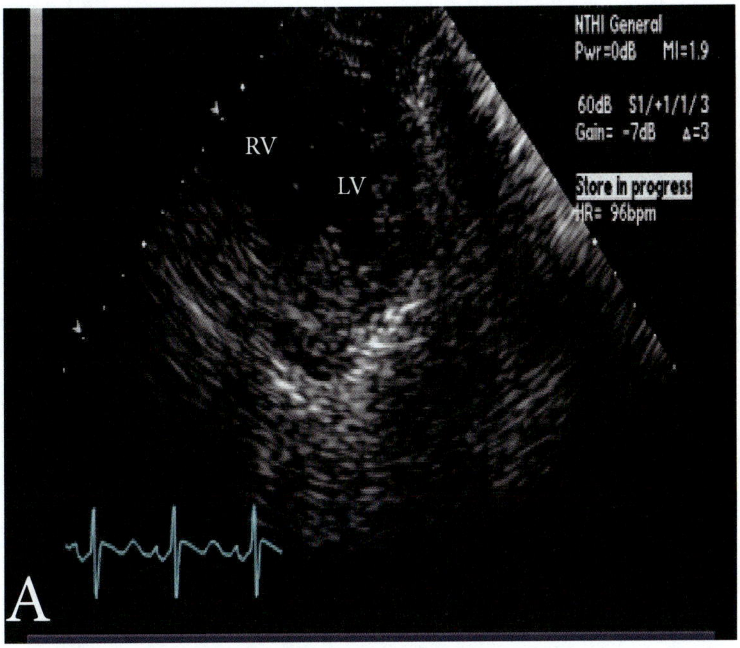

Figure 6A.

Follow-up echocardiogram performed after discontinuation of the catheter-directed thrombolytic therapy shows significant improvement in the size and function of the right ventricle.

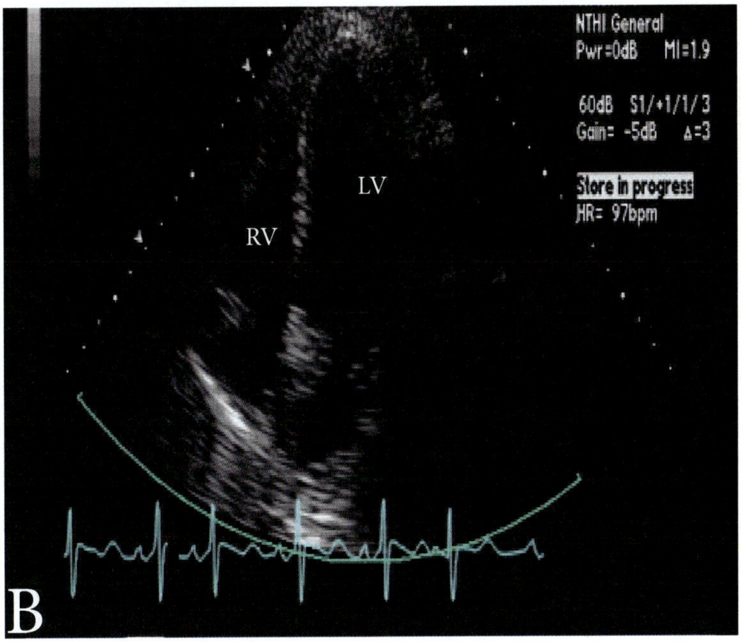

Figure 6B.

An apical four chamber view on transthoracic echocardiogram post-lytic therapy again shows complete normalization of RV strain.

dislodging the filter.
- This catheter was then exchanged out for the second Ekosonic catheter.
- A 5 mg bolus of r-tPA was given through the right pulmonary artery catheter and 3mg through the left pulmonary artery catheter.
- An r-tPA infusion was started using the Ekosonic catheter at 1 mg/hr. Coolant infusion was kept at 35 ml/hr and IV Lepirudin was continued.

ICU Course:

- Her oxygenation improved fairly quickly and O2 requirements were rapidly reduced.
- Systolic blood pressure improved and pressors were weaned off.
- Twelve hours later, the patient underwent repeat pulmonary angiography and right heart catheterization.
- Her CVP and PA pressures normalized. The thrombus in her right PA although still present was significantly smaller in size. On angiogram there was near total resolution of the thrombus burden within the left lung (figure 5B).
- Within 24 hours, the patient's hemodynamics improved considerably as well as normalization of the RV size and RV/LV end-diastolic diameter ratio on follow-up echocardiography (figure 6A & 6B).
- The patient was extubated shortly thereafter.
- She was continued on warfarin and IV Lepirudin until discharge.
- Her hypercoaguable work-up was unremarkable.
- At 1-month and 6-month follow-up, she was clinically asymptomatic and she continued to have a normal echocardiogram.
- Her warfarin was continued for a total duration of one year.

CASE 2

A 33-year old female was brought emergently to our cath lab by EMS for an acute anterior wall STEMI. This was diagnosed on a field EKG performed in the ambulance. Three weeks prior to her presentation this lady had driven from Florida to Connecticut over a period of 24 hours. A few days later, she developed bilateral pedal edema, right worse than left. This swelling resolved spontaneously. On the day of admission, she was feeling fine until the afternoon when she had a syncopal episode witnessed by her boyfriend. The patient regained consciousness spontaneously and experienced severe SOB and dizziness. En route to the hospital, she also developed severe chest discomfort.

Upon arrival to the cath lab, she was in extremis and severe respiratory distress. She was emergently intubated. Under the assumption of an anterior STEMI, coronary angiography was performed emergently.

Markers of Submassive PE:

- TnI > 12 ng/ml
- BNP 2500 pg/ml

Echocardiographic Findings:

- Due to the nature of her presentation this study was performed after coronary and pulmonary intervention.
- Marked RV dilatation with severe hypokinesis except the apex (figures 7A & 7B).
- RV systolic pressure 70 mm Hg.

Patient Preparation & Access:

- She was intubated emergently on the cath lab table.
- She was prepped and draped in a standard fashion.
- A 6 Fr. sheath was inserted in the right CFA.

Coronary Angiography & Left Ventriculography:

- Normal RCA and circumflex coronary arteries.
- Normal proximal ¾ of the LAD; abrupt cutoff was noted in the very distal/apical LAD suggestive of an embolic etiology.
- A left ventriculogram was performed and showed mild apical hypokinesis otherwise hyperdynamic LV (figure 8A).

Right Heart Catheterization:

- Suspecting a massive or submassive PE, emergent right heart catheterization was done via a 6 Fr right femoral vein access.
- This showed:
  - ▹ CVP : 35 mm Hg; RV: 70 / 35 mm Hg; PA: 70 / 40 mm Hg.
  - ▹ The right heart catheter was exchanged out for a pigtail catheter.

Pulmonary Angiography:

- Pulmonary angiography was performed using a power injector at the following settings 20 cc/sec for a total volume of 40 cc @ 500 psi.
- This showed a large embolus in the main right pulmonary artery with very little flow to the right lung (figure 8B).
- Partial thrombosis of the descending branch of the left pulmonary artery was also noted.

Figure 7A.

A pre-operative transthoracic echo showed severe RV dysfunction and enlargement.

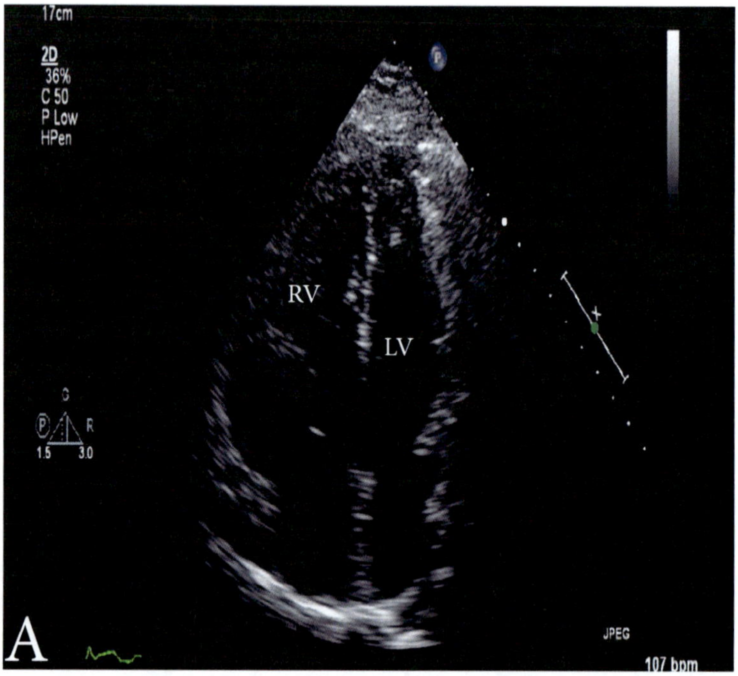

Figure 7B.

A short-axis view on pre-operative echo showed significant RV dilatation with flattening of the interventricular septum.

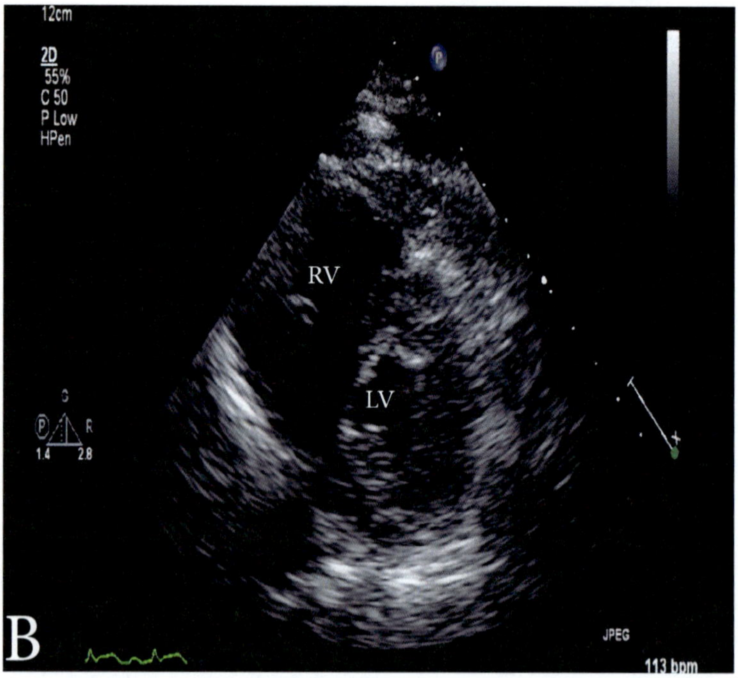

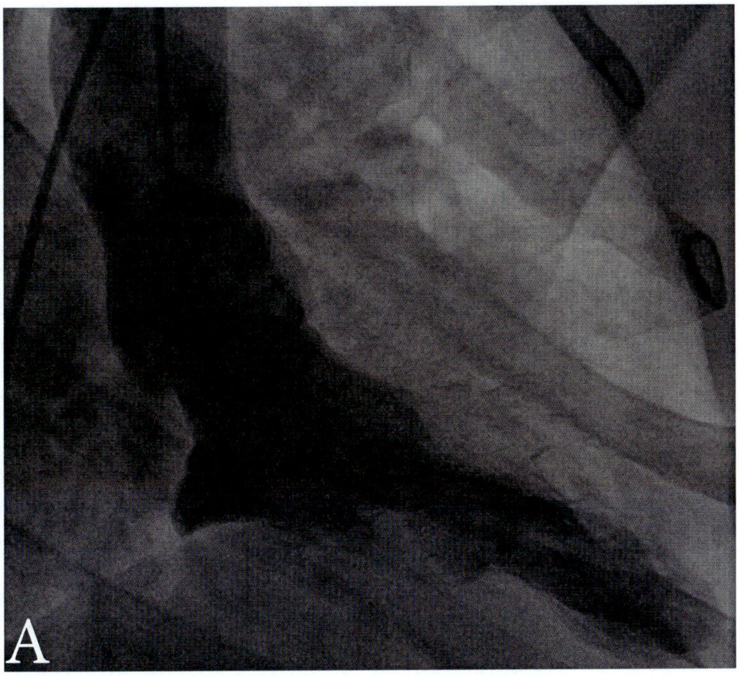

Figure 8A.

Left ventriculography revealed apical akinesis. Rest of the LV wall motion was noted to be within normal limits.

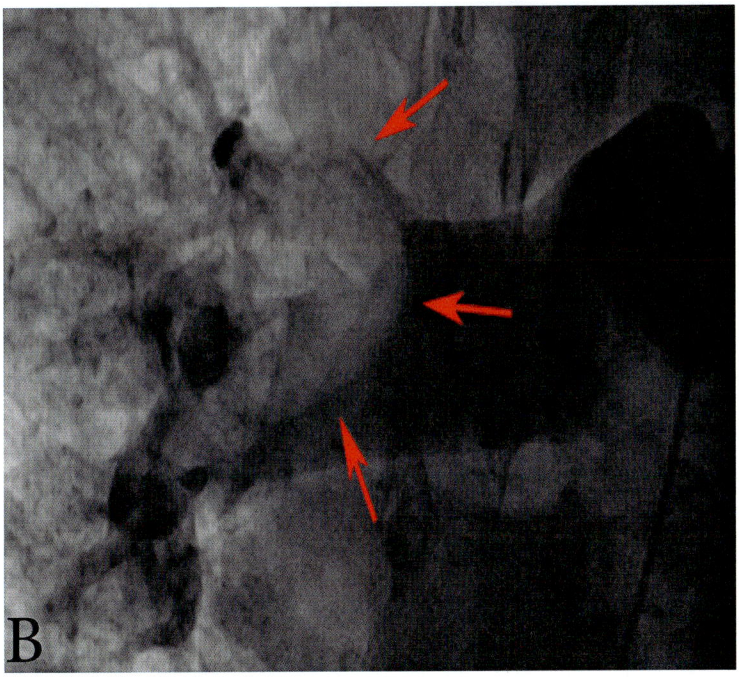

Figure 8B.

Pulmonary angiography (non-subtracted as DSA was not available in this room) revealed an impressive thrombus (red arrows) in the right pulmonary artery obstructing flow to both upper and lower divisions.

Figure 9A.

Post-thrombolytic trans-esophageal echocardiogram showed normalization of the right ventricular function. Not shown in this image, but this study showed the presence of a membranous VSD. This explained the right to left shunt and an embolic LAD apical infarct.

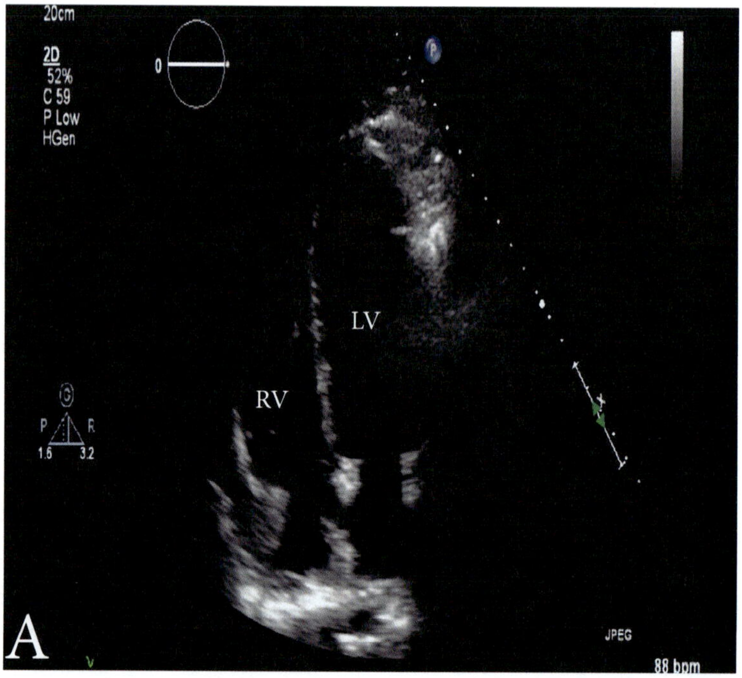

Figure 9B.

Short axis transgrastic view on TEE showed normalization of RV function post-thrombolytic therapy.

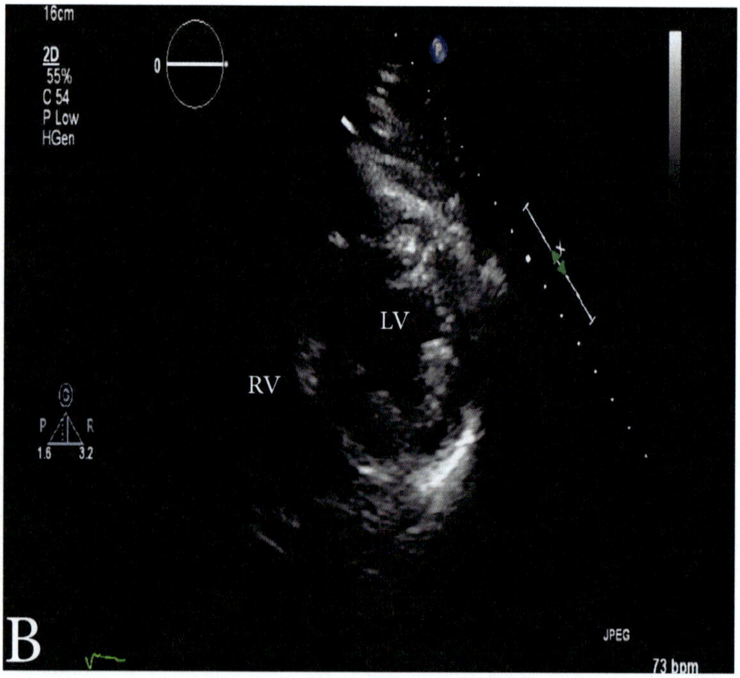

IVC Filter Placement:

- A retrievable IVC filter was deployed via the right femoral vein in a standard fashion after upsizing the sheath to 8 Fr.
- A 0.035" J-tip 260 cm wire was advanced through the filter under fluoroscopy. The right-heart catheter was advanced over it into the right pulmonary artery.

EKOS Ekosonic Catheter Placement:

- The pigtail was exchanged out and an Ekosonic catheter (6 Fr./12 cm/106 cm) which was parked in the right pulmonary artery.
- A 6 Fr./11 cm sheath was inserted in the left femoral vein.
- Over the 0.035" J-tip 260 cm wire, the right heart catheter was advanced over it into the left pulmonary artery.
- This catheter was then exchanged out for another Ekosonic catheter (6 Fr./12 cm/106 cm).
- A bolus of 5 mg of r-tPA was given through each catheter. Coolant was given at 35 ml/hr on each side and IV heparin was continued at 250 units/ hour on each side.

ICU Course:

- Her echocardiographic findings improved over the next 24 hours (figures 9A & 9B).
- Given her massive thrombus burden, it was decided to repeat pulmonary angiography.
- This showed near total resolution of the thromboembolic burden.
- A TEE was performed to determine the source of right to left shunting. This study revealed a small membranous ventricular septal defect.
- She was extubated the following day and transitioned slowly to warfarin.
- At 1-month and 6-month follow-up, she was clinically asymptomatic and she continued to have a normal echocardiogram.
- Her warfarin was continued for a total duration of one year.

CASE 3

Mrs. Z is a 55-year-old female with no significant past medical history was recently hospitalized at an outside institution for a parietal lobe intracranial hemorrhage with left-sided residual hemiparesis. She was sent to rehabilitation where she developed suddent onset chest discomfort and severe dyspnea. She was referred to the emergency department of the outside institution where she had a CTA of the chest that demonstrated a saddle pulmonary embolus with an RV/LV ratio > 1.0 (figure 10). At that time, the patient was hypoxic requiring 40% non-rebreather with a systolic blood pressure in the low 90s and tachycardic in the 110s. Given her recent history of an intracranial bleed, she was not a candidate for tPA. The neurologist at the outside hospital advised against the use of systemic heparin given her risk of further bleeding. CT surgery at that hospital was asked to evaluate for embolectomy but the patient was also deemed a poor candidate for surgical embolectomy. An IVC filter was placed and the patient was transferred to our facility for potential percutaneous mechanical embolectomy with the AngioVac suction cannula. Upon transfer to our institution, a stat echocardiogram was performed which showed severe RV dilatation and hypokinesis (figure 11).

Patient Preparation & Access:

- The procedure was performed in a hybrid operating room.
- The patient was prepped and draped using standard surgical technique.
- Right internal jugular vein access was obtained under ultrasound guidance using a 4 Fr. micropuncture kit. A 6 Fr./11 cm sheath was then inserted.
- Right common femoral vein access was obtained with a 4 Fr. micropuncture kit.
- Attempts were made to obtain access in the left common femoral vein. However, we were unable to pass the standard 0.035 " J-tip wire into the IVC. Therefore, that site was abandoned.

IVC Filter Retrieval:

- A 0.035" standard J-tip wire was advanced into the inferior vena cava through the right IJ access under fluoroscopic guidance.
- A 5 Fr. pigtail catheter was advanced into the IVC through the filter under fluoroscopic guidance.
- A venacavogram was performed to demonstrate the degree of thrombus burden within the IVC filter.
- Over the 0.035 " standard J-tip wire, the IJ sheath was up-sized to a 13 Fr. sheath. Through this sheath a filter retrieval sheath was advanced into the IVC and the filter removed in a standard fashion.
- The 13 Fr. sheath was maintained in the right IJ.

Right Heart Catheterization:

- A right heart catheter was advanced into the right main pulmonary artery via the right common femoral vein sheath.
- The catheter was exchanged out for a non-angled pigtail catheter and was placed into the main pulmonary artery.

Pulmonary Angiography:

- Pulmonary angiography was performed using a power injector at 20 cc/sec for a total of 40 cc at 500 psi.
- A large saddle embolus was seen at the bifurcation of the main pulmonary artery extending into the left and right main pulmonary arteries.

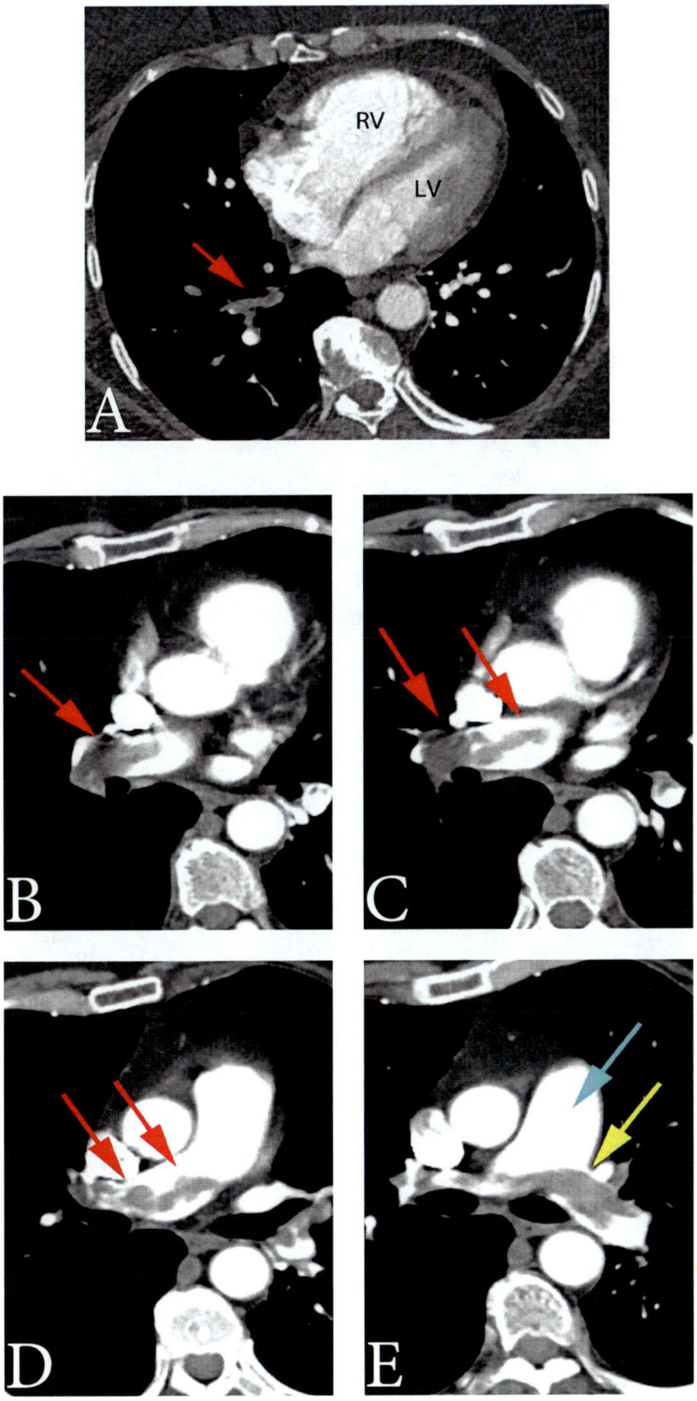

Figure 10.

Chest CTA (axial slices).

A. RV/LV ratio > 1.0. Thrombus is visualized in the right pulmonary artery distribution (red arrow).

B - D. Large thrombus burden is seen in the right main pulmonary artery (red arrows).

E. Extensive thrombus is seen in the left main pulmonary artery (yellow arrow) and extending into the right main pulmonary artery. The main pulmonary artery (blue arrow) is dilated significantly.

Figure 11.

Pre-op transthoracic echo:

A. An apical four chamber view shows significant RV dysfunction suggested by severe RV dilatation. An RV/LV ratio in this view clearly exceeds 1.0 and was suggestive of underlying massive pulmonary embolism.

B. A short-axis view of the RV and LV also shows significant dilatation of the RV compared to the LV. The interventricular septum in flattened.

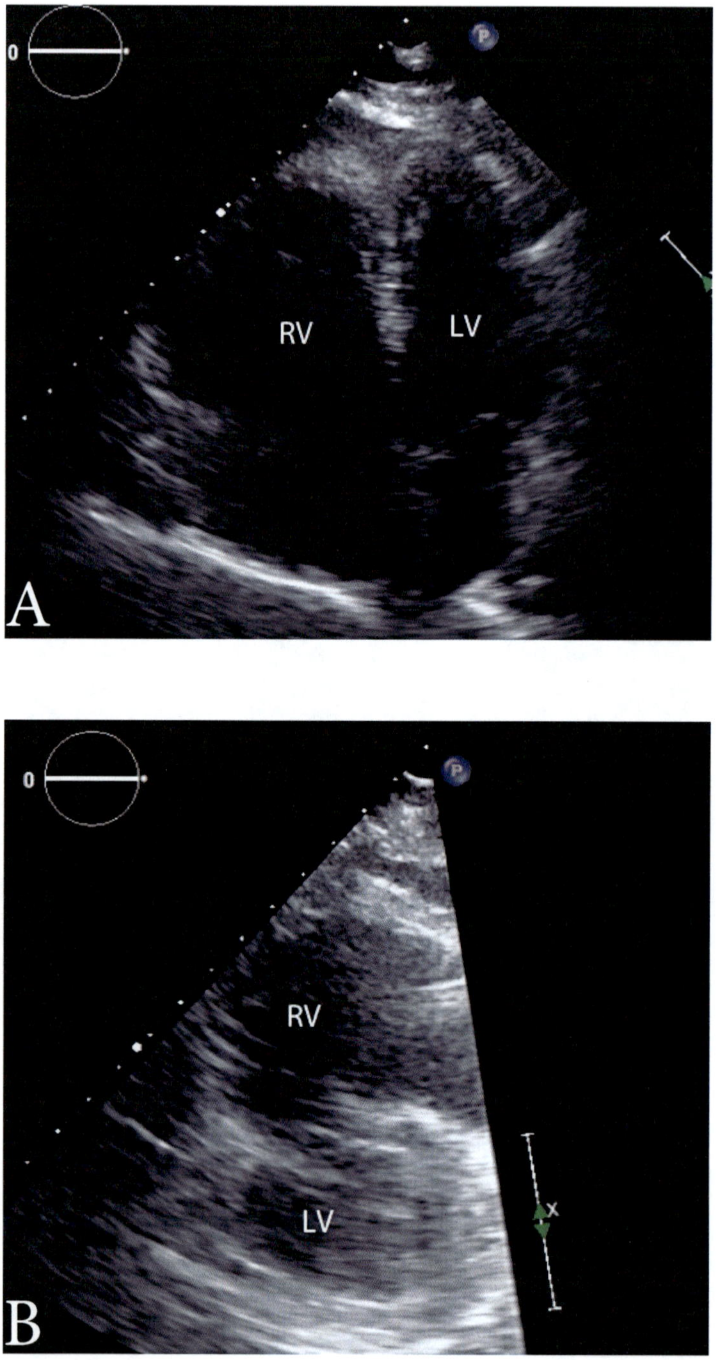

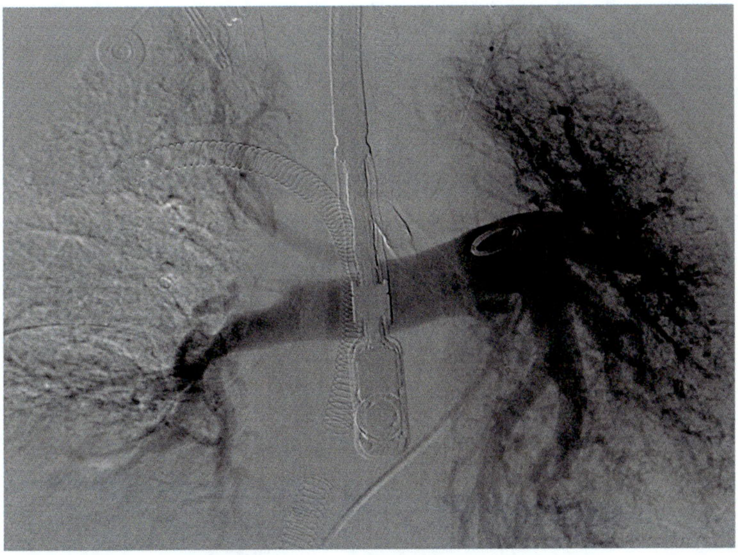

Figure 12.

Pulmonary angiography after pumonary artery embolectomy shows persistent severe oligemia of the right lung and the apical and basal segments of the left.

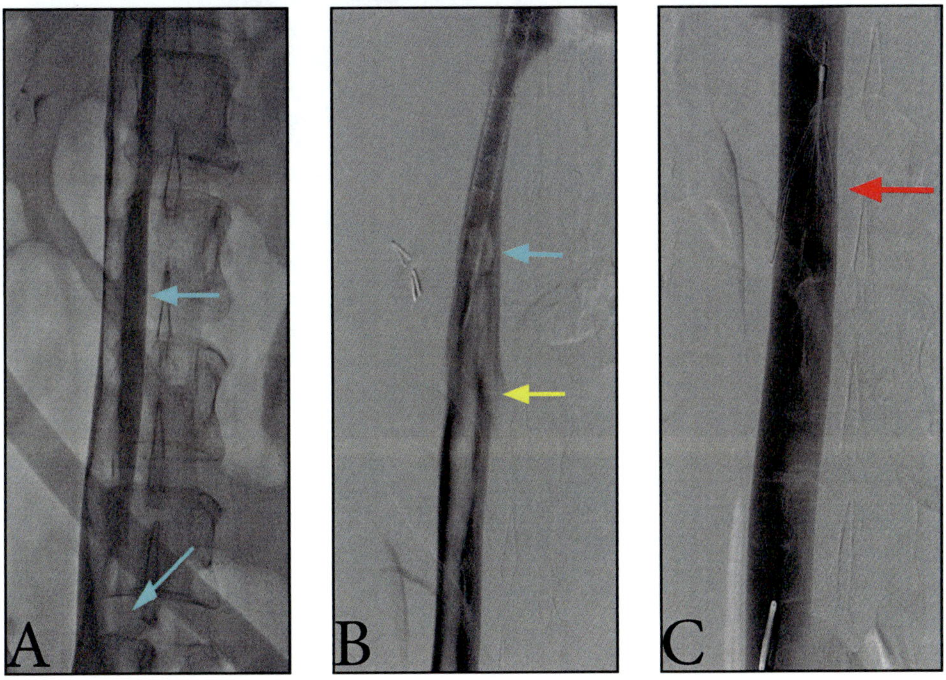

Figure 13.

A. IVC cavogram shows a large filling defect (blue arrows) extending from the left common iliac vein to the supra-renal IVC.
B. IVC cavogram on DSA shows the cranial extent of the thrombus, extending above the renal vein (yellow arrow) and into the hepatic IVC.
C. After removal of the thrombus with the Angiovac, a new IVC filter (red arrow) has been deployed.

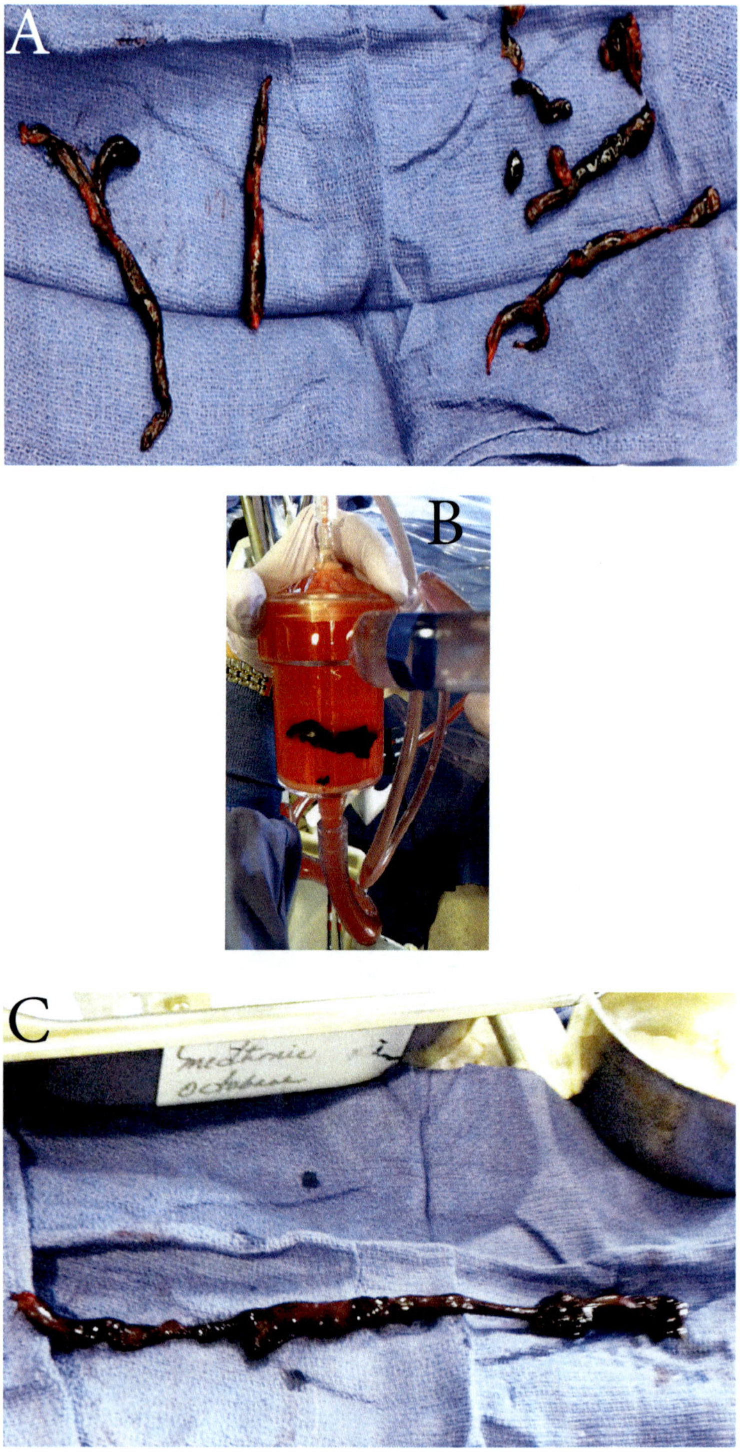

Figure 14.

Post-thrombectomy clot burden:

A. Thrombus removed from the pulmonary artery.

B. Thrombus in the filter of the AngioVac circuit.

C. Extensive thrombus removed en bloc from the IVC using the AngioVac catheter.

Insertion of the AngioVac Suction Cannula:

- The non-angled pigtail catheter was exchanged for a right heart catheter over the 260 cm 0.035" standard J-tip wire.
- Through the right-heart catheter, a 260 cm 0.035" J-tip Amplatz Superstiff wire was placed in the distal portion of the right main pulmonary artery.
- The 6 Fr./11 cm Cordis Brite tip sheath in the right common femoral vein was switched out for a 26 Fr. Gore Dryseal sheath, after successive pre-dilatations with increasing size dilators.
- The 13 Fr. sheath in the right internal jugular vein was switched out for the 17 Fr. Reinfusion cannula over a 0.035" standard J-tip wire.
- The extracorporeal circuit was being prepared by the perfusionist team.
- Over the Amplatz Superstiff wire, the AngioVac suction cannula with the dilator was advanced into the right ventricular outflow tract. At this point, the patient had a PEA arrest and required emergent surgical embolectomy.
- A significant amount of thrombus was extracted from the right and left main pulmonary arteries (figure 14A).
- Repeat pulmonary angiography demonstrated a significant decrease in thrombus burden with some residual clot in the middle lobe artery and superficial branch of the ascending branch of the right main pulmonary artery (figure 12).

IVC Filter Insertion:

- A venacavogram was performed from the right common femoral vein sheath.
- A large mobile thrombus was seen in the IVC extending from the left common iliac vein beyond the renal veins (figures 13A & 13B).
- The AngioVac suction cannula was used to remove this large thrombus within the IVC. While doing so, the IVC thrombus was dislodged from the cannula and embolized into the right atrium/ventricle.
- The AngioVac suction cannula was now advanced into the RA and removed a significant amount of thrombus under TEE guidance.
- A repeat venacavogram was performed and demonstrated complete resolution of thrombus within the IVC.
- Standard technique was used to deploy the IVC filter (figure 13C).

ICU Course:

- Post-op the patient was transferred to the ICU.
- She was maintained on IV UFH.
- Her oxygenation and blood pressure gradually improved.
- A repeat MRI of the brain showed a space-occupying lesion in the right fronto-parietal region.
- On the 7th post-op day, the patient was taken to the OR for a right fronto-parietal craniotomy.
- Microsurgical resection of a tumor was performed which was later confirmed to be a high-grade glioma.
- Post-op she underwent a repeat MRI of the brain. No additional bleeding was noted despite continued IV UFH.
- The patient was then switched to enoxaparin (100 mg subcutaneous twice daily and warfarin for a target INR of 2-3).
- The patient continued to do well and on the 14th post-op day she was discharged to a rehab facility. Her vitals were stable with O2 saturations of 99% on room air.
- Her only residual neurologic defect was 3/5 motor strength in the left upper and lower extremities. No sensory defecits were noted. Speech, memory and other intellectual functions

were intact.

- The enoxaparin was continued until an INR of 2-3 was achieved.
- Upon 4 week follow-up, she was doing well neurologically with improving strength.
- She is being evaluated for radiation therapy.
- Her thrombophilia work-up remained negative.

INDEX

A

B

C

D

INDEX

INDEX

Made in United States
North Haven, CT
04 November 2026

21730358R00265